STRATHCLYDE UNIVERSITY LI

30125 00088767 8

This book is to be returned on or before
the last date stamped below.

-5 APR 1988 9 MAR 2005

2 6 APR 1995

- 6 APR 1999

2 5 MAR 2003

2 2 APR 2004

DUE

2 7 MAR 2008

LIBREX —

ANDERSONIAN LIBRARY
★
WITHDRAWN
FROM
LIBRARY
STOCK
★
UNIVERSITY OF STRATHCLYDE

HYDROGEN ENERGY SYSTEM

Proceedings of the 2nd World Hydrogen Energy
Conference, held in Zurich, Switzerland,
21-24 August 1978

(In five volumes)

Volume 3

Advances in Hydrogen Energy I

Other Pergamon Titles of Interest

ANDRESEN & MAELAND	Hydrides for Energy Storage
BLAIR et al	Aspects of Energy Conversion
BOER	Sharing the Sun
HUNT	Fission, Fusion and the Energy Crisis
IAHE	Hydrogen in Metals
KARAM & MORGAN	Environmental Impact of Nuclear Power Plants
MCVEIGH	Sun Power
MURRAY	Nuclear Energy
SCHLEGEL & BARNEA	Microbial Energy Conversion
VEZIROGLU	First World Hydrogen Energy Conference Proceedings
VEZIROGLU	Remote Sensing Applied to Energy-Related Problems
VEZIROGLU	Energy Conversion - A National Forum
DE WINTER	Sun: Mankind's Future Source of Energy
ZALESKI	Nuclear Energy Maturity

Related Journals Published by Pergamon Press

International Journal of Hydrogen Energy
Annals of Nuclear Energy
Progress in Nuclear Energy
Solar Energy
Sun World
Progress in Energy and Combustion Science
Energy Conversion
Energy

HYDROGEN ENERGY SYSTEM

Proceedings of the 2nd World Hydrogen Energy Conference,
held in Zurich, Switzerland, 21-24 August 1978

Edited by

T. NEJAT VEZIROGLU
and
WALTER SEIFRITZ

PRESENTED BY:
International Association for Hydrogen Energy

HOSTED BY:
Swiss Federal Institute for Reactor Research
Wurenlingen, Switzerland

IN COOPERATION WITH:
Swiss Federal Institute of Technology
Zurich, Switzerland

Clean Energy Research Institute
University of Miami
Coral Gables, Florida, U.S.A.

Volume 3

PERGAMON PRESS
OXFORD · NEW YORK · TORONTO · SYDNEY · PARIS · FRANKFURT

U.K.	Pergamon Press Ltd., Headington Hill Hall, Oxford OX3 0BW, England
U.S.A.	Pergamon Press Inc., Maxwell House, Fairview Park, Elmsford, New York 10523, U.S.A.
CANADA	Pergamon of Canada, Suite 104, 150 Consumers Road, Willowdale, Ontario M2J 1P9, Canada
AUSTRALIA	Pergamon Press (Aust.) Pty. Ltd., P.O. Box 544, Potts Point, N.S.W. 2011, Australia
FRANCE	Pergamon Press SARL, 24 rue des Ecoles, 75240 Paris, Cedex 05, France
FEDERAL REPUBLIC OF GERMANY	Pergamon Press GmbH, 6242 Kronberg-Taunus, Pferdstrasse 1, Federal Republic of Germany

Copyright © 1979 International Association for Hydrogen Energy

All Rights Reserved. No part of this publication may be reproduced, stored in a retrieval system or transmitted in any form or by any means: electronic, electrostatic, magnetic tape, mechanical, photocopying, recording or otherwise, without permission in writing from the publishers.

First edition 1979

British Library Cataloguing in Publication Data

World Hydrogen Energy Conference,
2nd, Zurich, 1978
Hydrogen energy system. - (Advances in hydrogen energy; 1).
1. Hydrogen fuel - Congresses
I. Title II. Veziroglu, Turham Nejat
III. Seifritz, Walter IV. International
Association for Hydrogen Energy
V. Series
665'.81   TP359.H8   78-40507
ISBN 0-08-023224-8

In order to make this volume available as economically and as rapidly as possible the authors' typescripts have been reproduced in their original forms. This method unfortunately has its typographical limitations but it is hoped that they in no way distract the reader.

*Printed and bound in Great Britain by
William Clowes, Beccles and London*

D
665. 81
WoR

2ND WORLD HYDROGEN ENERGY CONFERENCE

Organizing Committee

H. L. Baetsle, Nuclear Energy Research Centre, Mol, Belgium

H. Barnert, Juelich Nuclear Research Centre, Federal Republic of Germany

G. Beghi, EURATOM J.N.R.C., Ispra, Italy

G. G. Carleson, AB Atomenergi, Studsvik, Sweden

P. Gelin, Saclay Nuclear Research Centre, France

P. Godin, Electricité de France, France

H. Marchandise, Commission of the European Communities, Brussels, Belgium

J. Pottier, Gaz de France, Paris, France

J. Rastoin, Saclay Nuclear Research Centre, France

E. Roth, Saclay Nuclear Research Centre, France

W. Seifritz (Chairman), Swiss Federal Institute for Reactor Research, Wurenlingen, Switzerland

T. Nejat Veziroglu (ex-officio), President, International Association for Hydrogen Energy

Local Organizing Committee from EIR

O. Antonsen, C. Bachmann, W. Bucher, K. H. Buob, E. Dubois, F. Etter, H. Hostettler, S. Huwyler, H. Kunz, B. Müller, T. Nordmann, A. Ray, R. Schmid, G. Schulzova, W. Seifritz, P. Sinha, R. von Schumacher, H. Weber, F. Widder, K. H. Wiedemann

Staff

Coordinators:	Carol Pascalis
	Deirdre Finn
	Janis Garren
Editorial Assistants:	Laxman Phadke
	John Sheffield

Paper Selection Committee

H. Barnert, Kernforschungsanlage Jülich GmbH, Jülich, F. R. Germany

K. D. Beccu, Battelle, Centre de Recherches, Genève-Carouge, Switzerland

R. E.Billings, Billings Energy Corporation, Provo, Utah, USA

J. K. Dawson, Atomic Energy Research Establishment, Harwell, UK

J. E. Funk, University of Kentucky, Lexington, Kentucky, USA

P. Godin, Electricité de France, Chatou, France

D. P. Gegory, Institute of Gas Technology, Chicago, Illinois, USA

K. C. Hoffman, Brookhaven National Laboratory, Upton, New York, USA

G. Kaske, Chemische Werke Hüls AG, Marl, F. R. Germany

K. F. Knoche, Rhein.-Westf. Technische Hochschule, Aachen, F.R. Germany
A. Lavi, Department of Energy, Washington, D.C., USA
C. Marchetti, Int. Institute for Applied Systems Analysis, Laxemburg, Austria
L. J. Nuttall, General Electric Company, Wilmington, Massachusetts, USA
T. Ohta, Yokohama National University, Yokohama, Japan
W. Peschka, Deutsche Forschungs- und Versuchsanstalt für Luft-und Raumfahrt, Stuttgart, F. R. Germany
F. J. Plenard, L'Air Liquide, Paris, France
J. L. Russell, General Atomic Company, San Diego, California, USA
G. Sandstede, Battelle Institut, Frankfurt, F. R. Germany
M. V. C. Sastri, Indian Institute of Technology, Madras, India
H. Säufferer, Daimler Benz AG, Stuttgart, F. R. Germany
J. H. Swisher, Department of energy, Washington, D.C., USA
J. A. Tatchell, Imperial Chemical Industries Ltd., Billingham, UK
H. Teggers, Rhein. Braunkohlenwerke AG, Köln, F. R. Germany
J. Weber, Gebrüder Sulzer AG, Winterthur, Switzerland

Sponsors

Department of Energy, Washington, D.C., USA
AGA AG, Gase, Schweiss- und Löttechnik, Pratteln
BBC AG Brown, Boveri & Cie., Baden
Bernsische Kraftwerke AG (BKW), Bern
Bonnard & Gardel Ingénieur-Conseils SA, Lausanne
Carbagas, Gase und Apparate, Bern-Liebefeld
Centralschweizerische Kraftwerke (CKW), Luzern
Elektrowatt Ingenieurunternehmung AG, Zürich
Gebrüder Sulzer AG, Winterthur
Jäggi AG Bern, Wärmetauscher, Kessel- und Behälterbau, Bern
Landis & Gyr AG, Zug
Motor-Columbus Ingenieurunternehmung AG, Baden
Oerlikon-Bührle Holding AG, Zürich
SA l'Energie de l'Ouest-Suisse (EOS), Lausanne
Sauerstoffwerk Lenzburg AG, Lenzburg
Schweizerische Aluminium AG, Zürich
Schweizerische Kreditanstalt, Zürich
Schweizerische Volksbank, Bern

FOREWORD

After the successful 1st World Hydrogen Energy Conference of March 1976, the 2nd World Hydrogen Energy Conference is now to be held. In the tradition established by this earlier meeting, it is going to provide an event worthy of an important Energy/Environment topic: Hydrogen Energy.

Hydrogen, produced by water-splitting reactions by means of harnessing the world's abundant non-fossil primary energy sources (solar, nuclear, ocean-thermal, wind and/or geo-thermal) can transcend the twilight decades of our present fossil fuel era in meeting the phenomenal energy-demand of tomorrow's world. The "Hydrogen Energy" concept is under serious study with all its ramifications around the world, backed up by experimental and demonstration activities in many instances. It provides an energy infrastructure to link the new primary energy sources to the energy consuming sectors.

The 2nd WHEC is providing a focus on this activity by way of its program of presentations and panel discussions. All facets of the hydrogen-energy system will be authoritatively covered. During the four day event there are to be an appropriate "mix" of topical sessions (sessions on primary energy sources, production, storage, transmission, utilization, environmental effects, economical problems, alternatives) in which invited lectures and formal papers are presented. It is expected that the 2nd World Hydrogen Energy Conference will be a landmark event in the world's accelerating search for new clean and abundant sources of energy. Appropriately, the theme of the meeting is "Progress Towards a Universal Hydrogen Energy System."

This proceedings presents by sessions the lectures and papers to be given at the 2nd World Hydrogen Energy Conference, with the exception of those which did not reach the editors on time. The latter will be considered for publication in the International Journal of Hydrogen Energy. It is expected that this proceedings will serve as reference volumes covering the latest developments in the field of Hydrogen Energy.

T. Nejat Veziroglu
President, IAHE

Walter Seifritz
Chairman, 2nd WHEC

ACKNOWLEDGMENTS

The Organizing Committee gratefully acknowledges the sponsorship of the Conference by the U.S. Department of Energy. We also thank the Swiss Federal Institute for Reactor Research, the Swiss Federal Institute of Technology and the Clean Energy Research Institute for their generous cooperation.

We wish to extend our appreciation to the Keynote Speaker, T. Nejat Veziroglu, Clean Energy Research Institute, University of Miami, Coral Gables, Florida, and to the Banquet Speaker, W. Haefele, International Institute for Applied Systems Analysis, Laxenburg, Austria.

Special thanks are due also to our authors and lecturers who provided the substance of the conference as published in the present proceedings

Our appreciation is owed to the Paper Selection Committee, the Session Chairmen and Co-Chairmen for organizing the technical sessions. In acknowledgment we list these session officials on the following pages.

SESSION OFFICIALS

TECHNICAL SESSION 1   —   PRIMARY ENERGY SOURCES FOR HYDROGEN
                            PRODUCTION

   Session Chairman: H. Barnert
                            Kernforschungsanlage Julich
                            Julich, Federal Republic of Germany

   Session Co-Chairman: J.L. Russell
                            General Atomic Company
                            San Diego, California, USA

TECHNICAL SESSION 2A —   ELECTROLYTIC HYDROGEN PRODUCTION I

   Session Chairman: P. Godin
                            Electricite de France
                            Chatou, France

   Session Co-Chairman: L.J. Nuttall
                            General Electric Company
                            Wilmington, Massachusetts, USA

TECHNICAL SESSION 2B —   ELECTROLYTIC HYDROGEN PRODUCTION II

   Session Chairman: P. Godin
                            Electricite de France
                            Chatou, France

   Session Co-Chairman: L.J. Nuttall
                            General Electric Company
                            Wilmington, Massachusetts, USA

TECHNICAL SESSION 3A —   THERMOCHEMICAL AND HYBRID
                            HYDROGEN PRODUCTION I

   Session Chairman: J.E. Funk
                            University of Kentucky
                            Lexington, Kentucky, USA

   Session Co-Chairman: K.F. Knoche
                            Rhein.-Westf. Technische Hochschule
                            Aachen, Federal Republic of Germany

TECHNICAL SESSION 3B -   THERMOCHEMICAL AND HYBRID
                          HYDROGEN PRODUCTION II

   Session Chairman   J.E. Funk
                          University of Kentucky,
                          Lexington, Kentucky, USA

   Session Co-Chairman: K.F. Knoche
                          Rhein.-Westf. Technische Hochschule
                          Aachen, Federal Republic of Germany

TECHNICAL SESSION 3C -   THERMOCHEMICAL AND HYBRID
                          HYDROGEN PRODUCTION III

   Session Chairman: J.E. Funk
                          University of Kentucky
                          Lexington, Kentucky, USA

   Session Co-Chairman: K.F. Knoche
                          Rhein.-Westf. Technische Hochschule
                          Aachen, Federal Republic of Germany

TECHNICAL SESSION 4 -   HYDROGEN PRODUCTION FROM FOSSIL FUELS

   Session Chairman: H. Teggers
                          Rhein. Braunkohlenwerke AG
                          Koln, Federal Republic of Germany

   Session Co-Chairman: J.A. Tatchell
                          Imperial Chemical Industries, Ltd.
                          Billingham, United Kingdom

TECHNICAL SESSION 5A -   HYDROGEN PRODUCTION ALTERNATIVES AND
                          OTHER INNOVATIVE PROCESSES I

   Session Chairman: T. Ohta
                          Yokohama National University
                          Yokohama, Japan

   Session Co-Chairman: A. Lavi
                          Department of Energy
                          Washington, D.C., USA

TECHNICAL SESSION 5B -   HYDROGEN PRODUCTION ALTERNATIVES AND
                          OTHER INNOVATIVE PROCESSES II

   Session Chairman: T. Ohta
                          Yokohama National University
                          Yokohama, Japan

   Session Co-Chairman: A. Lavi
                          Department of Energy
                          Washington, D.C., USA

TECHNICAL SESSION 6 - TRANSMISSION AND DISTRIBUTION

    Session Chairman: J. Kaske
                             Chemische Werke Huls AG
                             Marl, Federal Republic of Germany

    Session Co-Chairman: D.P. Gregory
                             Institute of Gas Technology
                             Chicago, Illinois, USA

TECHNICAL SESSION 7A - HYDROGEN STORAGE I

    Session Chairman: K.D. Beccu
                             Battelle Centre de Recherche
                             Geneve-Carouge, Switzerland

    Session Co-Chairman: W. Peschka
                               DFVLR
                             Stuttgart, Federal Republic of Germany

TECHNICAL SESSION 7B - HYDROGEN STORAGE II

    Session Chairman: K.D. Beccu
                               Battelle Centre de Recherche
                             Geneva-Carouge, Switzerland

    Session Co-Chairman: W. Peschka
                               DFVLR
                             Stuttgart, Federal Republic of Germany

TECHNICAL SESSION 8 - USE OF HYDROGEN IN TECHNICAL PROCESSES
                             AND ENERGY SECTOR

    Session Chairman: J.K. Dawson
                               Atomic Energy Research Establishment
                             Harwell, United Kingdom

    Session Co-Chairman: F.J. Plenard
                               L'Air Liquide
                             Paris, France

TECHNICAL SESSION 9 - HYDROGEN IN TRANSPORTATION

    Session Chiarman: R.E. Billings
                               Billings Energy Corporation
                             Provo, Utah, USA

    Session Co-Chairman: H. Saufferer
                               Daimler-Benz AG
                             Stuttgart, Federal Republic of Germany

TECHNICAL SESSION 10 -   SPECIAL APPLICATIONS

    Session Chairman: M.V.C. Sastri
                          Indian Institute of Technology
                          Madras, India

    Session Co-Chairman: G. Sandstede
                          Battelle Institute
                          Frankfurt, Federal Republic of Germany

TECHNICAL SESSION 11 -   MATERIALS ASPECTS IN PRODUCTION,
                          TRANSMISSION AND STORAGE

    Session Chairman: J.H. Swisher·
                          Department of Energy
                          Washington, D.C., USA

    Session Co-Chairman: J. Weber
                          Gebruder Sulzer AG
                          Winterthur, Switzerland

TECHNICAL SESSION 12 -   OVERALL SYSTEM ECONOMICS AND
                          ENVIRONMENTAL ASPECTS

    Session Chairman: C. Marchetti
                          International Institute for
                            Applied Systems Analysis
                          Laxenburg, Austria

    Session Co-Chairman: K.C. Hoffman
                          Brookhaven National Laboratory
                          Upton, New York, USA

ROUND TABLE DISCUSSION:

    Session Chairman: V.A. Legasov
                          I.V. Kurchatov Institute of
                            Atomic Energy
                          Moscow, U.S.S.R.

    Session Co-Chairmen: W.D. Van Vorst
                          University of California, Los Angeles
                          Los Angeles, California, USA

                          K.H. Weil
                          Stevens Institute of Technology
                          Hoboken, New Jersey, USA

CONTENTS

Volume 3

**TECHNICAL SESSION 5A   Hydrogen Production Alternatives
and Other Innovative Processes I**

TECHNICAL SESSION 5B **Hydrogen Production Alternatives and
Other Innovative Processes II**

TECHNICAL SESSION 6 Transmission and Distribution

TECHNICAL SESSION 7A Hydrogen Storage I

I. Jacob, D. Shaltiel
The Hebrew University of Jerusalem
Jerusalem, Israel

Volume 1

Review Papers

TECHNICAL SESSION 1   Primary Energy Sources for Hydrogen Production

TECHNICAL SESSION 2A Electrolytic Hydrogen Production I

TECHNICAL SESSION 2B Electrolytic Hydrogen Production II

Volume 2

TECHNICAL SESSION 3A Thermochemical and Hybrid Hydrogen Production I

TECHNICAL SESSION 3B Thermochemical and Hybrid Hydrogen Production II

TECHNICAL SESSION 3C Thermochemical and Hybrid Hydrogen Production III

TECHNICAL SESSION 4 Hydrogen Production from Fossil Fuels

Volume 4

TECHNICAL SESSION 8 Use of Hydrogen in Technical Processes and Energy Sector

TECHNICAL SESSION 9 Hydrogen in Transportation

TECHNICAL SESSION 10 Special Applications

TECHNICAL SESSION 11 Materials Aspects in Production, Transmission and Storage

TECHNICAL SESSION 12   Overall System Economics and Environmental Aspects

Volume 5

Opening and Banquet Addresses

TECHNICAL SESSION 1   Primary Energy Sources for Hydrogen Production

TECHNICAL SESSION 2A   Electrolytic Hydrogen Production I

TECHNICAL SESSION 2B   Electrolytic Hydrogen Production II

TECHNICAL SESSION 5B Hydrogen Production Alternatives and Other Innovative Processes II

TECHNICAL SESSION 7A Hydrogen Storage I

TECHNICAL SESSION 7B Hydrogen Storage II

TECHNICAL SESSION 8   Use of Hydrogen in Technical Processes and Energy Sector

TECHNICAL SESSION 9 Hydrogen in Transportation

TECHNICAL SESSION 10 Special Applications

TECHNICAL SESSION 12   Overall System Economics and Environmental Aspects

TECHNICAL SESSION 5A

HYDROGEN PRODUCTION ALTERNATIVES AND OTHER INNOVATIVE PROCESSES I

OPTIMIZATION OF A THERMOCHEMICAL WATER SPLITTING CYCLE : THERMODYNAMIC ANALYSIS - EXPERIMENTAL WORK WITH A SOLAR FURNACE

B. CHEYNET and C. BERNARD
Centre d'Information de Thermodynamique Chimique Minérale
Laboratoire de Thermodynamique et Physico-Chimie Métallurgiques
Associé au C.N.R.S., L.A. n° 29 E.N.S.E.E.G. BP 44
38401 Saint-Martin-D'hères, France.

M. DUCARROIR
Laboratoire des Ultra-Réfractaires, C.N.R.S.
BP 5, 66120 Font-Romeu, France

ABSTRACT

A computation method, based on the minimization of the total Gibbs energy, is presented for the study and selection of thermochemical cycles for hydrogen production. In the case of the "Fe-O-H-Br" system a three-step cycle is generated. The calculations show the characteristic behaviour of the hydrolysis reaction in relation with different parameters. The experimental results are in good agreement with the theoretical equilibrium and illustrate the reliability of the thermodynamic approach. This reaction is also experimentally studied in a fluidized-bed reactor heated by the concentrated radiation beam of an arc image furnace.

INTRODUCTION

A large number of laboratories are working on thermochemical hydrogen production processes, some of them having developed computing programs in order to generate cycles which could lead to the decomposition of water [1,2,3]. This method is important because it discards empirical combination of experimentally known reactions. Also, calculation of thermodynamic data at high temperatures is made possible, thus making for the lack of available experimental results at the considered temperatures.

It now appears unlikely that classical thermochemical production of hydrogen (maximum temperature available with a high temperature nuclear reactor) can compete in the future with electrolytic processes. Before reaching a final decision, a very detailed optimization of the process is needed. This is a complex work depending on numerous parameters (thermodynamic, kinetic, thermal, engeneering, technological and even

economic). Such a study implies a thorough knowledge of the
theory underlying the chemical reaction in relation with the
temperature, the pressure and the reactor composition inlet,
and this constitutes the thermodynamic optimization aspect of
the problem involved. From this analysis one obtains a set of
theoretical conditions leading to a maximum equilibrium conversion rate that can be used with additional data in order to
optimize the process. In short, an evaluation of a thermochemical hydrogen process seems to proceed following three
main stages : determination of the cycle, thermodynamical
optimization and process optimization, the latter taking
into account kinetic data obtained from experimentation.

The purpose of this paper is to present a computer aided method to solve the two first stages mentioned above and to
illustrate the interest of the method by means of a few experimental tests.
The authors take into account the complex chemical equilibria
in the "Fe-O-H-Br" system to determine and to optimize the
different chemical steps that build up a short and relatively
new cycle. They do not deal with energetic or engeneering
topics.
Furthermore the primary energy source to be used is the highly
concentrated solar energy, so that the classical constraint of
temperature does vanish. With an increase of the maximum temperature value at which heat is available, the probability to
find a short cycle (2 or 3 steps) is increasing.

THERMODYNAMIC ANALYSIS

Several methods of computer aided search for cycles have already been published. Generally from a master list of elements or
compounds the computer constructs all the possible reactions.
Then a subroutine is applied to combine them in such a way as
to form a sequence whose overall result is the dissociation of
water into its elements, hydrogen and oxygen. Then, the large
number of possible cycles is restricted by application of
several criteria. Often, the results still include a number of
reactions which are not suitable or realistic from a chemical
point of view, because the search :
     - is made with the constraint of stoechiometry of the
reaction with not more than a fixed number of components,
     - does not take into account the competing side reactions.
A search of this kind is prohibitively costly in computer time
because it requires numerous handlings of data files and tests.
Only a method which takes into account the complex multiphase
equilibria existing within an atomic system will allow to give
a more probabilistic idea of the reactions taking place [4].

PROCEDURE

Among the equilibrium computation methods [5] the best one is based on the minimization of the total Gibbs energy G.
For a system of Γ phases each of them including $N\gamma$ components, the expression for G is given by the following equation :

$$\frac{G}{RT} = \sum_{\gamma=1}^{\Gamma} \sum_{i=1}^{N\gamma} n_i^\gamma \left(\frac{\mu_{i\gamma}^\circ}{RT} + \phi_\gamma (\log P + \log \frac{n_i^\gamma}{n_\gamma}) + (1 - \phi_\gamma) \log a_i^\gamma\right) \quad (1)$$

where : n_i^γ : number of moles of the ith component in the γth phase.

$\mu_{i\gamma}^\circ$: standard chemical potential of the ith component in the γth phase

n_γ : total number of moles in the γth phase

ϕ_γ : 1 for a gaseous phase
0 for a condensed phase

a_i^γ : activity of the ith component in the γth condensed phase.

Assuming an ideal gas phase (this is the case because the total pressure in cycles is generally of an order of magnitude of one or several atmospheres) and pure condensed phases, the standard chemical potential is written as :

$$\mu^\circ = T \left[\frac{G^\circ - H^\circ 298}{T}\right] + H^\circ 298 = T (f.e.f)_T + H^\circ 298 \quad (2)$$

to minimize the G function for which the absolute value is unknown, one must calculate its variations in relation with the composition of the mixture. As the enthalpy of formation of the elements in their reference state is zero, the calculation works only with $\Delta\mu^\circ$ values of the following form :

$$\Delta\mu^\circ = T \left(\sum(fef_T)_p - \sum(fef_T)_r\right) + (\Delta H_F^\circ 298)_p - (\Delta H^\circ_F 298)_r \quad (3)$$

p : product
r : reactant

Thus the needed input data are the standard heat of formation at 298 ($\Delta H^\circ_F 298$) and the free energy function (f.e.f) in relation with the temperature for all the chemical species considered.
The program is written in PL/1 language and works according to the conversational mode CP/CMS of the "Centre Inter Universitaire de Calcul de Grenoble". Thermodynamic data are drawn automatically from a coupled bank of reviewed and selected values sponsored by the "Groupe Scientifique Thermodata Europe".

The temperature, the pressure, the input composition, the precision may be fixed by the user, from a communication terminal. Moroever the procedure provides a control for the iteration steps and the printing of the results. In general, all the considered cases include one to four condensed phases and numerous gaseous species ; an average of 30 iterations is required to obtain an accuracy of 10^{-8} moles, this implying a computation time of one second with an IBM 360/67.

Study of the Fe-0-H-Br system

This system was selected for two main reasons : iron is probably the best element able to fulfill some of the constraints of the closed thermochemical processes and a great variety of investigations using iron oxides and chlorides can be found in literature. Moreover, the difficult reduction $Fe^{3+} \rightarrow Fe^{2+}$ in the iron chlorine processes [6] does not exist in the iron bromine family due to the instability of $FeBr_3$.
In the computation of this system the input components are :

- in the gaseous state : H_2O, O_2, H_2, Fe, FeO, FeO_2H_2, Br, Br_2, HBr, $FeBr_2$, Fe_2Br_4.
- in the condensed state : Fe, FeO, Fe_2O_3, Fe_3O_4, FeO_2H_2, FeO_3H_3, $FeBr_2$.

In any cycle, water contributes necessarily as a reactive in a step of the type $A+nH_2O$ or $A+B+nH_2O$. We shall confine oneselves to the first type. All the components in the available list are exhaustively combined with water, one after the other. The results are presented in TABLE I.

Cases 1.2.3 : direct oxidation of iron requiring a strong reducing component to close the cycle leads to a sequence of five reactions.

Case 4 : leads to a two steps cycle previously reported by Nakamura, involving the decomposition of magnetite at high temperature.

Cases 5.6 : do not seem to be of any interest because of the difficulty to proceed towards the reverse oxide hydration reaction.

Case 8 : may initiate a cycle. However the transformation rate of bromine is very low : 2,5 % at 1000K, 16 % at 1500K, 28 % at 2000K.

Case 9 : leads to a four reactions cycles already proposed by
Bowman [7].

Case 10 : may be the basis of a cycle. This reaction goes
through a maximum yield with the temperature which is interesting to find experimentally.

- Hydrolysis of FeBr$_2$

The influence of temperature on the equilibrium composition
for an initial mixture of reactants (1 FeBr$_2$ + 6 H$_2$0) corresponding to a commercially available form of this bromide is
clearly shown on Fig. 1. The maximum yield (69 %), is obtained
at 970K about. At this temperature the yield increases as the
water amount is increased (Fig. 2). The influence of both
parameters is summarized on Fig. 3. The pressure is also an
important factor as it appears on Fig. 4 and Fig. 5, those
figures showing also the combined effect of temperature and
pressure.

The most favourable conditions for the hydrolysis reaction
can be derived from this study : large excess of water with
respect to stoechiometry, temperature around 1000K, low total
pressure.

The choice of this reaction as the first step of a closed
cycle implies that magnetite can be converted into bromide.

- Bromination of magnetite

The simplest way to obtain this transformation is the direct
reaction with bromine. Under atmospheric pressure and using
an initial mixture with bromine in excess (1 Fe$_3$0$_4$ + 10 Br$_2$),
Fig. 6 shows the variation of the equilibrium composition
versus temperature. Below 1700K there is a competition between
the formation of Fe$_2$0$_3$ and FeBr$_2$. At fixed temperature (1800K)
and pressure (1 atm) the ratio Br$_2$/Fe$_3$0$_4$ must be as high as
8 to allow total reaction Fig. 7.

- Closure step

Since bromhydric gas is produced in the first step and bromine
consumed in the second, the only way to close the cycle is

$$2 \ HBr + 1/2 \ O_2 \rightarrow H_2O + Br_2$$

This homogeneous gas phase reaction is well known. The calculation leads to a complete transformation at ambiant temperature, and furthermore the kinetics would be satisfactory at
500K [8].

- Conclusion

Following these results, the usefulness of the complex equilibrium method is clear. It has allowed the construction of a
three step cycle in the considered system. It has also determined the conditions leading to the maximum conversion rate for
each step. All the results are summarized on TABLE II. The
classical representation as a sequence of chemical stoechiometric reactions would be :

$$3\ FeBr_2 + 4\ H_2O \rightarrow Fe_3O_4 + 6\ HBr + H_2 \qquad 1000K$$

$$Fe_3O_4 + 3\ Br_2 \rightarrow 3\ FeBr_2 + 2O_2 \qquad 1800K$$

$$6\ HBr + 3/2\ O_2 \rightarrow 3\ H_2O + 3\ Br_2 \qquad 500K$$

When one compares this scheme with TABLE II, the lack of information consecutive to such a schematization is evident. One
must bear in mind that the founded theoritical conditions probably may be changed by an optimization of the process, but
the important fact is to know the behaviour of the reactions
in a large field of temperature pressure and composition in
view of further experimentations.

HYDROLYSIS REACTION EXPERIMENTS

Sealed tube

The previous equilibrium data are sharply dependent on temperature. A first experimental study has been initiated. An amount
of cristalline $FeBr_2,nH_2O$ is introduced in a silica tube in
order to obtain, by reference to the previous calculation, an
equilibrium pressure of two atmospheres. Then the tube is
sealed under vacuum and heated at the desired temperature
during 12 hours. After quenching in a water bath the products
($FeBr_2$ and Fe_3O_4) are spectrophotochemically analysed. The
results are presented on TABLE III and Fig. 8. The experimental behaviour of the system is very similar to the expected
equilibrium, except for the yields at the highest temperature
experimented.

Open tube

In order to demonstrate the possibility to use solar energy
as a primary energy source, experiments have been made in a
fluidized bed reactor heated by the concentrated beam coming
from an arc image furnace (Fig. 9.10). The furnace is a three
mirror paraboïdal system. The radiant energy source set at the
focus of the first mirror is a xenon arc lamp (6500 W) whose
emission spectrum is similar to the solar spectrum. Then the

parallel flux is received either directly on the second parabolic mirror of horizontal axis, or indirectly on the third
one of vertical axis, after reflexion on a retractable plane
mirror.

The reaction bed made of silica particules (15 g, 200 < Ø <
250 µ) impregnated with ferrous bromide (0,5 g of $FeBr_2$,
4 H_2O) is enclosed in a vertical quartz tube (Ø 26 mm).
Argon is then introduced, whose contribution is to fluidize
the bed and to complete, if necessary, the addition of the
required water. The latter is supplied by mean of a vaporizer
and metering pump, Fig. 11. The temperature is measured by a
thermocouple and adjusted by controlling the electrical power
applied to the lamp.
In all the experiments, a preheating time of 15 mn is needed
in order to reach the thermal equilibrium of the bed. During
this period, due to the water content of the hydrate, the
reaction is starting up without additional water in the gas
stream.
At 480°C, the conversion yield reaches 32,7 % at the end of the
preheating and afterwards increases as it can be seen on
Fig. 12 (curve 1), but is seems difficult to obtain in these
conditions a total transformation. On the contrary, when
water is added in the fluidizing stream a complete conversion
is rapidly reached, Fig. 12 (curve 2). The influence of the
temperature has been studied in the following conditions :
T = 500°C, H_2O/Ar = 1/25, 5 minutes duration after the preheating time. The reaction is favoured by an increase of temperature, Fig. 13, but it is important to remind that, as low
as 500°C, some $FeBr_2$ is removed from the bed by the argon
stream (15 % in 5 minutes). Therefore, a temperature increase
(yield increase) is associated with reactant losses and an
optimization process should be studied in this respect.

CONCLUSION

This study, consisting both of research and theoretical
optimization of the thermochemistry of water splitting cycles
is characterized by the fact that complex multiphase systems
in their entirety are taken into account. It avoids the use of
patterns, which tend to over-simplify and ignore competiting
side reactions.
The choice of solar energy as a primary source raises considerably the upper limit of temperature imposed by the use of
nuclear heat.
In spite of its low efficiency, which excludes possible industrial application, the Fe-O-H-Br system has been retained as
example because it has a hydrolysis step whose optimization
demonstrates the possibilities of the method.

The open tube experiments allowed a first approach to the
technological problems posed by high temperature solar
chemistry.

FOOTNOTE :-For figures 12 and 13 the percent conversion
of bromide into oxyde is calculated from the results of a
spectrophotochemical analysis of the species remaining
in the bed.
     - The solar fluidized bed reactor has been previously
studied by G. FLAMANT (Revue Internationale d'Héliotechnique
[COMPLES] 2ème Semestre 77, P. 39-43).

REFERENCES

1 RUSSEL J.C., PORTER Jr. and J.T., 1975, Hydrogen Energy,
(Plenum Press, New-York) part 1, 515/529.

2 YOSHIDA K. KAMEYAMA H., TOGUCHI K., 1975, U.S. Japan Joint
Seminar, 20-23 July (Ohta's Laboratory : YOKOHMA University)
45-54.

3 DONAT G., ESTEVE B. RONCATO J.P., 1976, Compte rendu de fin
d'étude - Contrat DGRST 75-7-0840.

4 BERNARD C., DENIEL Y., DUCARROIR M., JACQUOT A.,VAY P.,
1975, J. Less Common Metals, 40, 165/171.

5 F. VAN ZEGGEREN, S.H. STOREY, "The Computation of Chemical
Equilibria", 1970, CAMBRIDGE University Press.

6 HARDY-GRENA C., 1973, Progress Report n° EUR4958F
Joint Nuclear Research Centre, ISPRA.

7 BOWMAN P., 1976, A.I.M. International Congress on Hydrogen
and its Prospects, LIEGE, BELGIUM 15-18 Nov.

8 PASCAL P., 1960, Nouveau Traité de Chimie Minérale, t. XVI
(MASSON, PARIS), 378/411.

TABLE I - POSSIBLE REACTIONS WITH H_2O IN THE Fe-O-H-Br SYSTEM

Case N°	Reactants A	Reactants n H_2O	Equilibrium species	Range of studied temperature	Percent conversion of A
1	Fe	1	FeO, H_2	300-800K	100 %
2	Fe	4/3	Fe_3O_4, H_2	300-800K	"
3	Fe	>4/3	Fe_3O_4, H_2, H_2O	300-800K	"
4	FeO	10	Fe_3O_4, H_2, H_2O	300-800K	"
5	FeO_2H_2	10	Fe_3O_4, H_2, H_2O	300-800K	"
6	FeO_3H_3	10	Fe_2O_3, H_2O	300-800K	"
7	Fe_3O_4	1	Fe_3O_4, H_2O	300-1500K	0 %
8	Br_2	1	HBr, O_2, Br_2, Br, H_2O	1000-2000K	Weak
9	$FeBr_2$	1	FeO, HBr, $FeBr_2$, Fe_2Br_4, H_2O	600-1500K	Optimum versus T
10	$FeBr_2$	>4/3	Fe_3O_4, H_2, HBr, $FeBr_2$, Fe_2Br_4, H_2O	600-1500K	"

TABLE II - THERMODYNAMIC OPTIMIZATION OF THE Fe-O-H-Br SYSTEM

TK	INPUT COMPOSITION	EQUILIBRIUM COMPOSITION	PERCENT CONVERSION
1000	3,68 $FeBr_2$ 36,8 H_2O	1 Fe_3O_4 1 H_2 (G) 6 HBr (G) 0.569 $FeBr_2$ (G) 0.056 Fe_2Br_4 (G) 32.810 H_2O (G)	81.5 %
1800	1 Fe_3O_4 8 Br_2	3 $FeBr_2$ (G) 2 O_2 (G) 2.075 Br_2 (G) 5.849 Br (G)	100 %
500	6 HBr 1,5 O_2	3 H_2O (G) 3 Br_2 (G)	100 %

TABLE III - EXPERIMENTAL RESULTS OF HYDROLYSIS STEP

(SEALED TUBE)

T°C	Volume (cc) of the tube	M_{mg} FeBr$_2$,6H$_2$	P_{atm} Theoretical equilibrium pressure	Percent conversion	
				experimental	Theoretical (P=2)
600	59,3	75,8	1,8	13,5	16,5
700	56,3	71,1	1,99	36,4	36,5
800	57,7	71	2,14	51,3	51
900	62,3	76,9	2,35	27,6	46,8
900	53,5	67,08	1,96	26,24	46,8

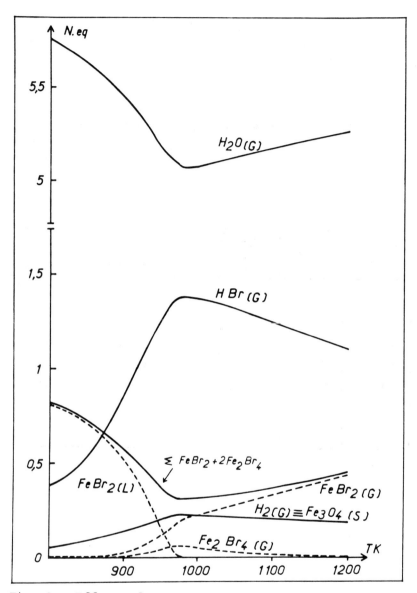

Fig. 1 : Effect of temperature on equilibrium characteristics of hydrolysis reaction - 1 $FeBr_2$ + $6H_2O$ -

P = 1 atm

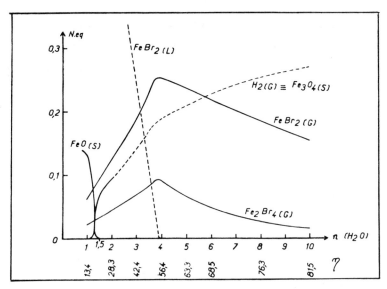

Fig. 2 : Effect of H_2O feed rate on equilibrium
characteristics of hydrolysis reaction -
1 $FeBr_2$ + nH_2O - T = 1000K, P = 1 atm.

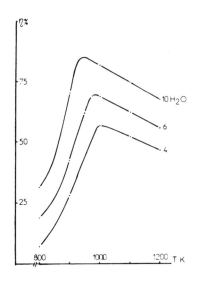

Fig. 3 : Percent conversion
of $FeBr_2$ versus T for different initial compositions

- 1 $FeBr_2$ + nH_2O - P = 1 atm. -

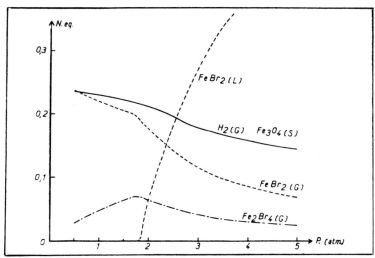

Fig. 4 : Effect of pressure on equilibrium characteristics of hydrolisis reaction - $1FeBr_2$+
$6H_2O$ - T = 1000 K

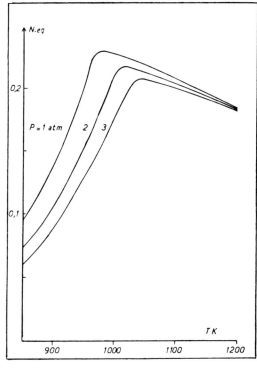

Fig. 5 : Effect of temperature
and pressure on magnetite
equilibrium quantity in
hydrolysis reaction.
-$1FeBr_2$ + $6H_2O$.

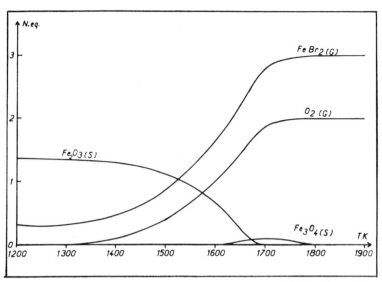

Fig. 6 : Effect of temperature on equilibrium characteristics
of bromination - $1Fe_3O_4 + 10Br_2$
P = 1 atm (Except Br_2, Br)

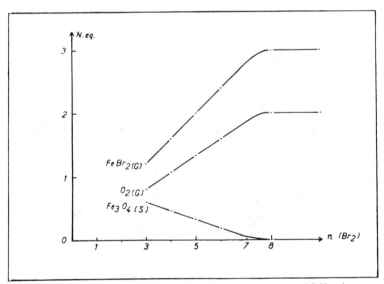

Fig. 7 : Effect of Br_2 feed rate on equilibrium
Characteristics of bromination - $1Fe_3 + nBr_2$ - T = 1800 K
P = 1 atm (Except Br_2, Br.)

1046

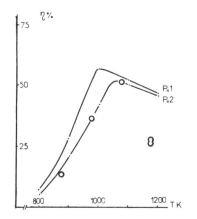

Fig. 8 : Theoretical (——·——)
and experimental (0) percent
conversion of $FeBr_2$ versus T
in hydrolysis reaction

$$1 \ FeBr_2 + 4 \ H_2O$$

Fig. 9 : Arc Image Furnace

1047

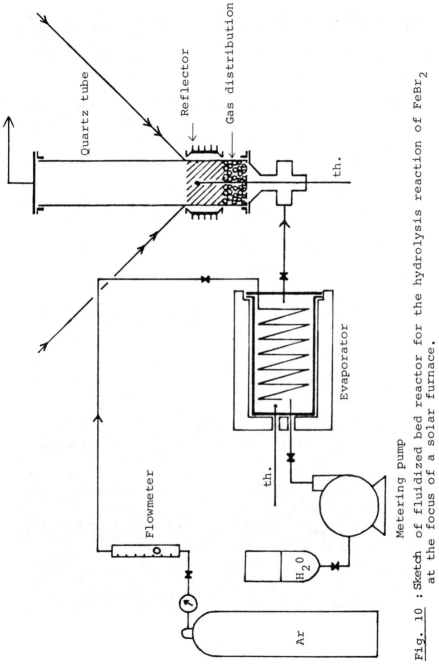

Fig. 10 : Sketch of fluidized bed reactor for the hydrolysis reaction of $FeBr_2$ at the focus of a solar furnace.

Fig. 11 : Experimental apparatus for hydrolysis reaction
at the focus of the arc image furnace.

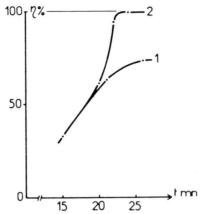

Fig. 12 : Percent conversion of bromide for $FeBr_2$, 4 H_2O hydrate.

1 - Without additional water

2 - With additional water

$$H_2O/_{Ar} = 1/25$$

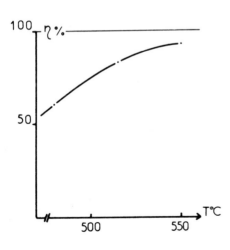

Fig. 13 : $FeBr_2$ Hydrolysis
Percent conversion of bromide versus T.

SOLAR-THERMOCHEMICAL PRODUCTION OF HYDROGEN FROM WATER

J. R. Schuster and J. L. Russell, Jr.
General Atomic Company
San Diego, California, U.S.A.

ABSTRACT

Hydrogen is a versatile element that is of current importance primarily as a chemical feedstock, but that has tremendous potential as a future replacement for fossil fuels. It is readily storable and transportable in many forms and upon reaction with oxygen produces only heat and water. This nonpolluting fuel could be of tremendous environmental benefit if it could be produced from recyclable water and solar energy. Of the several theoretical approaches to producing hydrogen from water, thermochemical water splitting using the sulfur-iodine cycle has the potential for being an efficient approach. This cycle utilizes high-temperature heat that could be supplied by a solar concentrator. The Fixed Mirror Solar Concentrator incorporates features that make it well suited to solar-thermochemical hydrogen production. It has adequate concentration and its modular design lends itself to supplying heat at different temperature levels to various portions of the process. The fixed, rugged module, constructed of concrete and solidly mounted glass mirrors, has the potential for being a reliable, low-cost performer compared with other solar concentrator concepts.

INTRODUCTION

Solar energy, as a primary energy form, can meet many of society's needs. Where the technology exists and the economics are attractive, solar energy systems are already being successfully implemented. An example of this is the solar heating and cooling of buildings. Considerable attention is also being given to the application of solar energy for producing electricity. Particularly in producing electricity, however, there is a major problem with solar energy: diurnal, interruptible sunlight makes energy storage a necessity. Due to the cost of energy storage and the cost of backup generating capability, central station solar electric plants are projected to have only limited application, and only in the intermediate load and peaking categories. For a solar energy plant to be base loaded it will have to have the capability for storing energy for periods in excess of 24 hr.

The potential for using solar energy could be greatly increased if methods were developed for converting it to other secondary energy forms that are storable and nonpolluting and that could be substituted for natural gas and petroleum-derived fuels. These developments would expand the role of

1051

solar energy in the residential, commercial, and industrial sectors as
well as provide clean solar-derived energy forms for use in transportation. Hydrogen is such a secondary energy form. Upon reaction with oxygen, the only products are heat and water. For electricity production it
can be used to fire boilers, run gas turbines, or power fuel cells. It
can also be economically transported by gas pipeline, and thus could be
available to pipeline gas customers as a substitute for natural gas or
for use as a chemical feedstock. It could also be used as fuel for
ground and air transportation. In fact, production of a versatile secondary form of energy such as hydrogen may be the only avenue for the
wide-spread implementation of solar energy.

HYDROGEN CONSUMPTION

The hydrogen consumption rate worldwide is rapidly increasing. In 1938
about 2.5 billion scf (71 million m^3) was produced, and by 1973 the estimated annual total had risen to over 9 trillion scf (250 billion m^3),
representing an order of magnitude increase every decade during the in-
tervening 25-year period [1]. In 1973, about one-third of the world's
total hydrogen was consumed in the U.S., requiring about 3% of the total
U.S. energy consumption for its production. Consumption of hydrogen in
the U.S. has grown by a factor of 40 since 1945, and has tripled in the
last decade.

Recent Users

The distribution of hydrogen among the major consumers in the U.S. in
1972 is shown in Fig. 1 [2]. This is based on a total consumption of
3 trillion scf (85 billion m^3). Petroleum refining (hydrotreating and
hydrodesulfurization, and hydrocracking) accounted for 42%, ammonia synthesis for fertilizer production accounted for 37%, chemicals (primarily
for the manufacture of cyclohexane) accounted for 9%, methanol production
accounted for 7%, and miscellaneous uses accounted for 5%.

The miscellaneous uses included the production of foodstuffs and soap,
refining and annealing of some metals, welding, semiconductor manufacture, uranium extraction and processing, corrosion control in steam power
plants and nuclear reactors, nylon and polyurethane manufacture, use as a
rocket fuel, and cooling of large electric generators in utility power
stations.

Future Uses

Many studies have examined the potential future uses of hydrogen. These
uses are summarized in TABLE I. In addition to historic uses, an expanded scenario might include using hydrogen as a gaseous fuel for both
home and industry, as a fuel to provide peaking power for electric utilities by burning it directly in gas turbines or reacting it in fuel cells,
as a reducing agent in iron ore reduction and steelmaking, as a liquid
fuel for high-speed aircraft, as a liquid, gaseous, or hydrided fuel for

ground transportation, in the synthesis of hydrocarbon fuels using a Fischer-Tropsch type process, and as a coolant to achieve superconductivity.

A recent study [1] projected future hydrogen consumption in the U.S. based on two scenarios:

1. A reference scenario that assumes continued historic uses with some new uses in coal gasification and liquefaction. This scenario was derived from estimates of future national energy requirements made by the Ford Foundation Energy Policy Project [3].

2. An expanded scenario that includes the major uses shown in TABLE I. This scenario was derived from estimates of future national energy requirements made by the Westinghouse Corporation [4].

Figure 2 shows the hydrogen growth history and the projections plotted from 1950 to 2000 on a logarithmic scale. The reference projection, with less than the historical growth rate, shows an increase by 1985 to 1.7 times current usage, and by the year 2000 to 5.5 times. Growth factors for the expanded projection, with approximately the historic growth rate, are 3.4 by 1985 and more than 20 by the year 2000.

To satisfy the large future demand for hydrogen, it is important that the hydrogen be produced economically. This implies large production plants to obtain economies of scale. For each plant multiple customers will be necessary, and some of these customers will be industries that now use only captive hydrogen, i.e., hydrogen consumed where it is produced or in nearby chemical plants as a feedstock or fuel. Thus, a shift in the ratio of captive to merchant hydrogen users can be expected to occur (merchant hydrogen is produced for sale by industrial gas companies). Based on the analyses of Ref. 1, by the year 2000 merchant hydrogen could supply as much as 75% of the total hydrogen demand.

HYDROGEN PRODUCTION

Recent Sources

The distribution of energy sources for hydrogen production in the U.S. during 1972 is shown in Fig. 3 [2]. Petroleum refining operations, particularly catalytic cracking and catalytic reforming, supplied 45%, reforming of natural gas supplied another 45%, and the remaining 10% came from other sources such as by-products of coal processing, coke manufacture, fuel oil refining, water electrolysis, and by-product gases from ethylene manufacture or other chemical processes.

Most of this hydrogen was captive hydrogen, while only about 7% was merchant hydrogen. Most merchant hydrogen is presently shipped in cylinders; however, a small portion is transported by pipeline. There is currently a

50-mile (80.5 km) hydrogen pipeline in the Houston area serving ten industrial customers. Hydrogen is also being supplied by pipeline to customers in the Geismar, Louisiana, area. The longest hydrogen pipeline in the world is in the Ruhr area of Germany. It is about 127 miles (204 km) long and has been in operation since the late 1930's. For many applications the use of merchant hydrogen may be the best approach, as pipeline transmission of hydrogen remains economical for large distances [5].

Future Production Methods

TABLE II lists possible future production methods for hydrogen. Those based on fossil feedstocks include methods that have been used historically as well as coal gasification with steam-methane reforming and hydrogen production directly or indirectly from cultured biomass. The future will require a nonfossil, renewable source of hydrogen such as can be derived from water.

TABLE II includes five water-splitting methods for producing hydrogen: electrolysis, thermal decomposition, thermochemical decomposition, hybrid (combined electrolysis and thermochemical steps), and photolysis. These methods, with the exception of photolysis, will require a high-temperature heat source to achieve reasonable energy conversion efficiency. The success of photolysis hinges on the discovery of a chemical cycle with broadband adsorption and high quantum yields. Direct thermal decomposition of water requires temperatures on the order of 5000°F (3033 K). Although it is being studied, there are substantial separation and materials obstacles and it currently appears impractical from an engineering standpoint.

Whereas direct thermal decomposition may be impractical, thermochemical decomposition (water splitting) reacts water chemically with other materials to yield hydrogen or intermediate hydrogen-bearing compounds. These compounds can in turn be thermally decomposed at relatively low temperature to obtain hydrogen. A closed chemical cycle can be formed wherein the intermediate compounds are regenerated at various temperature levels, and the only feedstocks required for the process are water and an external supply of high-temperature heat.

In addition to thermochemical water splitting being performed at a much lower temperature [Ref. 6 indicates a practical upper limit of 1600°F (1144 K)] than direct thermal decomposition, it potentially is a more efficient process than electrolysis. As illustrated in Fig. 4, electrolysis first requires a power cycle to convert thermal energy obtained from a high-temperature heat source to electricity. If the electricity is alternating current, it must be conditioned for use in the electrolysis cell where hydrogen is produced. Electrolysis net efficiency is therefore lowered due to the thermodynamic inefficiency of the power cycle, the inefficiency of the power conditioning equipment, and the inefficiency of the electrolysis cell.

THE SULFUR-IODINE THERMOCHEMICAL WATER-SPLITTING CYCLE

Under gas and electric industry sponsorship, a program to investigate thermochemical water splitting has been under way at General Atomic Company since October 1972. Progress is described in Ref. 7.

General Atomic has discovered a promising thermochemical water-splitting cycle and is developing the thermochemical water-splitting process to where it can be demonstrated as a commercially feasible source of hydrogen.

In contrast to many other thermochemical cycles that have been discovered, this cycle requires no solids handling, and its temperature requirements [maximum process temperature of 1500 to 1600°F (1089 to 1144 K) occurs during production of SO_2] can probably be met using metallic construction materials. Every reaction step necessary for the completion of the cycle has been demonstrated, and it is now fairly certain that a hydrogen production process can be devised that is within the range of current and near-term technology. This includes chemistry, chemical engineering, materials of construction, and high-temperature heat source technology. The net overall thermal efficiency for the process, based on the current version of the flow sheet, is 41.4%. Portions of the flow sheet are being reworked to incorporate new chemical data and alternative, more energy-efficient, process methods, and the eventual efficiency is expected to be in the range of 45% to 50%.

Cycle Description

In 1973, a computer program was developed to search thermodynamic and thermochemical data for combinations of chemical reactions that could form closed hydrogen-producing cycles. Reference 8 describes the methods upon which the computer program is based. Thousands of cycles were generated, and the computerized search was supplemented by manual evaluations and feasibility demonstrations in the laboratory. From the screening during 1973 and 1974, four cycles were found that seemed to warrant detailed study. The cycle regarded as the most promising is characterized by three reactions:

$$2 H_2O + SO_2 + I_2 \rightarrow H_2SO_4 + 2HI \tag{1}$$

$$H_2SO_4 \rightarrow H_2O + SO_2 + 1/2 O_2 \tag{2}$$

$$2HI \rightarrow I_2 + H_2 \tag{3}$$

This sulfur-iodine cycle has been considered by other investigators who dismissed it because of problems involved in separation of the two acids formed in the initial reaction.

What has made the cycle practical is the discovery in GA laboratories that if excess iodine is added to the first reaction, a stratification of the liquid phases will occur. The reactions are characterized by

$$2H_2O + SO_2 + xI_2 \rightarrow H_2SO_4 + 2HI_x \qquad \text{Aqueous} \qquad (4)$$

$$2HI_x \rightarrow xI_2 + H_2 \qquad \lesssim 571°F \ (573 \ K) \qquad (5)$$

$$H_2SO_4 \rightarrow H_2O + SO_2 + 1/2 \ O_2 \qquad \lesssim 1600°F \ (1144 \ K) \qquad (6)$$

The HI_x in Reaction (4) represents the average of several polyiodides formed in the reaction. The HI is a good solvent for iodine, and the resulting HI_x solution is not miscible in H_2SO_4. As the reaction proceeds, H_2SO_4 rains upward, leaving behind a lower phase rich in HI_x. The acids can be readily separated and thermally decomposed to obtain the hydrogen and oxygen products and the iodine and SO_2 for recycle.

Possible Processing Method

In reality, the mode of accomplishing this cycle is more complex than the three equations shown earlier. One of the modes (probably the simplest for actually completing this cycle) is presented in the following set of transformations and in the block diagram of Fig. 5:

Step

1    $2H_2O(\ell) + SO_2(g) + xI_2(s) \rightarrow H_2SO_4(sol) + 2HI_x(sol)$

2    $H_2SO_4(sol) \rightarrow H_2SO_4(\ell)$

3    $H_2SO_4(\ell) \rightarrow H_2SO_4(g)$

4    $H_2SO_4(g) \rightarrow H_2O(g) + SO_3(g)$

5    $SO_3(g) \rightarrow SO_2(g) + 1/2 \ O_2(g)$

6    $2HI_x(sol) \rightarrow 2HI(g) + (x-1) \ I_2(\ell) + H_2O(\ell)$

7    $2HI(g) \rightarrow H_2 + I_2(g)$

8    $I_2(g) \rightarrow I_2(\ell)$

9    $xI_2(\ell) \rightarrow xI_2(s)$

Recycle is required in steps 2 through 5 because of the incomplete decomposition of SO_3 in step 5, incomplete decomposition of HI in step 7, and imperfect removal of HI from the HI_x solution in step 6. Removal of impurities (i.e., HI, I_2, and SO_2) from the H_2SO_4 solution and of H_2SO_4 and

SO_2 from the HI_x solution is another complication of this cycle. Purifying the H_2SO_4 appears simple, but removing it and SO_2 from HI_x is complicated by the reduction of H_2SO_4 and SO_2 to lower valence states such as free sulfur and H_2S. However, if step 1 is carried out under the right conditions, sulfur species contamination of the HI_x solution is not a problem. Other impurity problems will be recovery of I_2 and HI vapors from the product H_2 and of SO_2 vapors from the product O_2.

SOLAR POWER CONCEPTS

Large-scale production of hydrogen from water using solar energy will require systems that highly concentrate solar energy. Both central receiver and distributed receiver concepts are undergoing rigorous development, principally through Department of Energy (DOE) funded programs.

Central Receiver

Central receiver concepts employ a single heat receiver upon which sunlight from a field of mirrored heliostats is reflected. These heliostats require two-axis steering to maintain proper orientation to the constantly moving sun. According to the number of heliostats used, concentrations in excess of 1000 suns can be obtained at the heat receiver, and this heat can raise high-temperature steam. The major emphasis for central station solar electric power development has been on central receiver systems. These same systems could be used to run an electrolysis plant, or the high-temperature heat could be used to drive the chemical reactions for thermochemical water splitting.

Distributed Receiver

Distributed receiver systems fall into two categories: line focusing and point focusing types. They generally employ a concentrating mirror module that has its own heat receiver. Either the mirror module or the heat receiver is steered to keep the heat receiver in focus as the sun moves. Single-axis tracking is usually employed, although two-axis tracking is possible and is required for point focusing types.

Distributed receivers are currently being developed with DOE funding, primarily for total energy applications in which they supply heat to an intermediate-temperature [around 600°F (589 K)] power cycle fluid that drives a turbogenerator, and the cycle waste heat then heats and air-conditions buildings.

There are a few distributed receiver concepts capable of very high concentration; however, most concepts produce moderate concentration in the range of 20 to 100 suns. They employ a pipeline network to transport collected energy from individual receivers to a central site. In so doing, they suffer thermal line losses. Nevertheless, the collector construction and tracking control of most distributed receivers are simpler than for central receivers.

THE FIXED MIRROR SOLAR CONCENTRATOR

General Atomic is developing a line focusing distributed receiver device
called the Fixed Mirror Solar Concentrator (FMSC).

Design Principles

The FMSC concentrates direct sunlight even though the sun's position with
respect to the mirror varies daily and seasonally. The FMSC, which uses
a particular fixed shape to concentrate the sun's light, is an array of
long, narrow, flat reflecting slats arranged on the segment of a cylinder.
The array of fixed reflecting slats produces a narrow focal line that
follows a circular path as the sun moves, and the focal line is tracked
by the moving heat receiver pipe. The minimum image width is equal to the
slat width plus an increment due to the subtended angle of the sun. To
achieve this sharp focal line, each slat must be installed with its normal
at an angle with respect to the reference or tangent slat normal that is
one-fourth of the angle between that slat and the reference slat. These
principles are illustrated in Fig. 6 and described in Ref. 9.

General Atomic is developing the FMSC for both solar electric power production and solar total energy applications. It is capable of a theoretical concentration of about 206 suns and is able to efficiently collect
solar energy at temperature levels sufficient to drive the sulfur-iodine
thermochemical water-splitting cycle.

Construction

The solar mirror strips can be mounted in many ways. Figure 7 shows an
80-ft long by 6.6-ft (24.4 m by 2.0 m) aperture FMSC with mirror strips
mounted on a metal frame. This unit is under test at the Georgia Institute of Technology.

The construction method involving mounting the mirror strips on a metal
frame results in modules that are easy to transport. If an application
called for remote location of a small number of modules, they could be
made at a factory and transported to the site. For larger installations,
it is less expensive to make them of precast concrete on-site.

Figure 8 shows a 25-ft long by 7.0-ft wide (7.6 m by 2.1 m) aperture,
precast concrete mounted mirror module built by GA and on test at Sandia
Laboratories. The combination of a concrete substrate and solidly supported glass mirrors results in a rugged structure that is easy to clean
and that has excellent weathering characteristics.

The heat receiver, through which a fluid is circulated to absorb the concentrated solar energy, can be seen mounted above the mirror in Fig. 8.
Greater detail of the heat receiver is illustrated in Fig. 9. It uses a
secondary concentrator of the compound parabolic type and contains a

single absorber tube of rectangular cross section. The tube is exposed to air beneath the glass cover and is insulated at the back and sides.

The potential advantages of the FMSC are (1) the inherent economy of fabricating fixed mirrors in contrast to the moving mirrors of other systems, (2) the low-cost header system resulting from the long in-line layout of the cylindrical mirror modules, and (3) flexibility of plant size deriving from the modular character of the distributed receiver approach.

THERMOCHEMICAL HYDROGEN PRODUCTION WITH SOLAR HEAT

The basic problem in coupling a solar heat source to water splitting is that of devising systems that can use the time-varying solar heat input to permit a uniform level of hydrogen production over a 24-hr day. A thermochemical approach may readily lend itself to solar heating because many intermediate compounds are storable. Figure 10 illustrates a possible system arrangement for the sulfur-iodine cycle. The mirror field is divided into high- and intermediate-temperature portions. The low-temperature solution reaction is conducted 24 hr/day, producing H_2SO_4 and HI. The H_2SO_4 is stored and during daylight is pumped from storage, preheated, and then decomposed in the heat receiver pipes of the high-temperature mirror field. The acid decomposition products are then heat exchanged with the incoming acid, the oxygen and water are removed, and the liquid SO_2 is stored for use in the low-temperature solution reaction.

The intermediate-temperature mirror field heats a heat transport fluid, such as a eutectic salt. During daylight the salt charges a heat storage reservoir, generates steam to drive compressors and pumps, and provides heat for concentration and cracking of the HI to yield the hydrogen product and iodine for recycle to the main reaction. At night the heat transport fluid continues to perform these functions, but bypasses the mirrors and is heated by the heat storage reservoir.

REFERENCES

1. Kelley, J. H., and E. A. Laumann, "Hydrogen Tomorrow, Demands & Technology Requirements," Report of the NASA Hydrogen Energy Systems Technology Study, Jet Propulsion Laboratory Report JPL 5040-1, December 1975.

2. Peterman, D. D., et al., "Studies of the Use of High-Temperature Nuclear Heat from an HTGR for Hydrogen Production," National Aeronautics and Space Administration Report NASA CR-134919, General Atomic Company, September 30, 1975.

3. A Time to Choose: America's Energy Future, Ford Foundation Energy Policy Project, Ballinger Publishing Co., Cambridge, 1974.

4. Ross, P. N., "The Nuclear Electric Economy — 1972-2000 AD," Westinghouse Electric Corp., Power Systems Planning, East Pittsburgh, Pa., 1973.

5. Kakac, S., and T. N. Veziroglu, "Economics of Nuclear-Electrolytic Hydrogen," First World Hydrogen Energy Conference Proceedings, Miami Beach, March 1-3, 1976.

6. Russell, J. L., Jr., et al., "Water Splitting — A Progress Report," First World Hydrogen Energy Conference Proceedings, Miami Beach, March 1-3, 1976.

7. Schuster, J. R., et al., "Status of Thermochemical Water Splitting Development at General Atomic," Ninth Synthetic Pipeline Gas Symposium Proceedings, Chicago, October 31 - November 2, 1977.

8. Russell, J. L., Jr., and J. T. Porter, "A Search for Thermochemical Water-Splitting Cycles," Hydrogen Economy Miami Energy Conference Proceedings, Miami Beach, March 18-20, 1974.

9. Russell, J. L., Jr., E. P. DePlomb, and R. K. Bansal, "Principles of the Fixed Mirror Solar Concentrator," General Atomic Company Report GA-A12902 (Rev.), February 1, 1977.

1061

TABLE I
POSSIBLE FUTURE USES FOR HYDROGEN

● Historic uses

● Gaseous fuel in home and industry

● Peaking power for electricity

● Steelmaking

● Liquid fuel for high-speed aircraft

● Fuel for ground transportation

● Production of synthetic hydrocarbon fuels

● Cooling fluid for superconductivity

TABLE 2
POSSIBLE FUTURE HYDROGEN PRODUCTION METHODS

● Historic methods

● Coal gasification/steam-methane reforming

● Biomass

● Water-splitting methods

   − Electrolysis

   − Thermal decomposition

   − Thermochemical decomposition

   − Hybrid

   − Photolysis

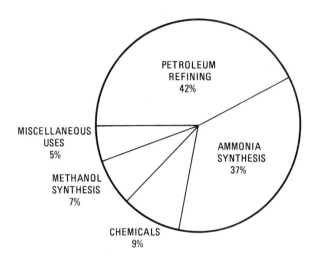

Fig. 1. Hydrogen Consumption in the U.S. During 1972

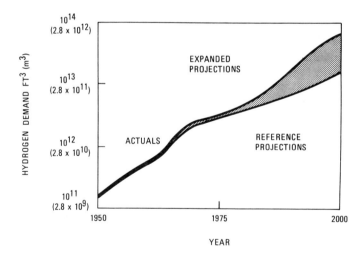

Fig. 2. Growth in Hydrogen Demand in the U.S.

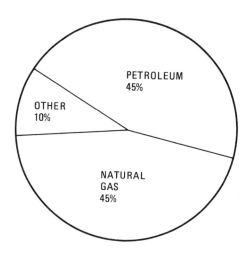

Fig. 3. Hydrogen Production Feedstocks
in the U.S. During 1972

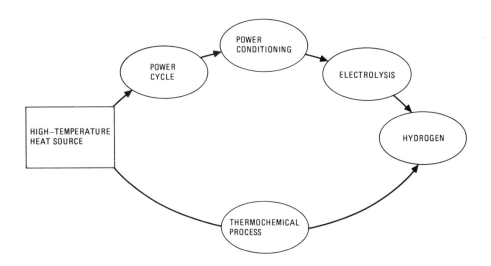

Fig. 4. Two Paths for Producing Hydrogen from Water

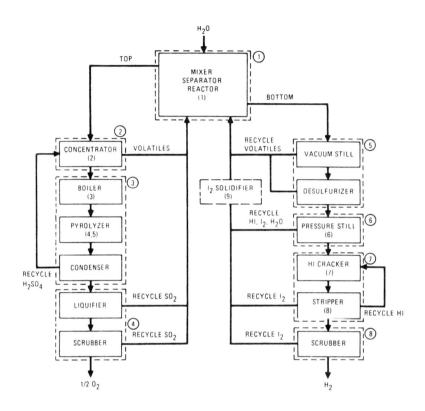

Fig. 5. Sulfur-Iodine Thermochemical Water-Splitting
Cycle Process Steps

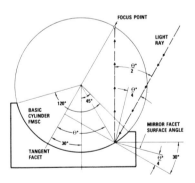

Fig. 6. FMSC Geometry

Fig. 7. Metal-Frame-Mounted FMSC Installed at Georgia
Institute of Technology

Fig. 8. Precast Concrete Mirror Module
at General Atomic

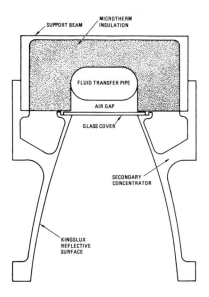

Fig. 9. Heat Receiver Assembly
Cross Section

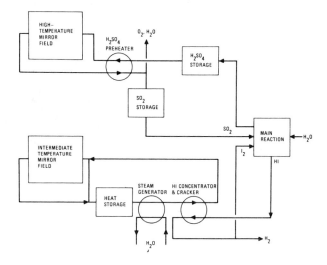

Fig. 10. Possible Solar/Thermochemical
Water-Splitting Configuration

SOLAR BEAM-ASSISTED ELECTROLYSER APPLIED TO YOKOHAMA MARK 5 & 6

T. Ohta, N. Kamiya, T. Otagawa and M. Suzuki
Yokohama National University
Hodogaya-ku, Yokohama, Japan
&
S. Kurita and A. Suzuki
IHI Technical Laboratories
Toyosu, Koto-ku, Tokyo, Japan

ABSTRACT

An electrochemical concept was used in calculation of the energy storage through the photochemical reaction and the conditions that give the maximum storage were examined. The potential difference of the two systems, i.e., Fe^{3+}/Fe^{2+} and I_3^-/I^-, was considered as the stored energy. Taking into consideration of the electrolysis that follows the photochemical reaction, not only a large magnitude of the energy storage but a large amount of Fe^{3+} formation are required to give a high overall efficiency. The concentration of iodine should be raised up in order to keep the high concentration of Fe^{3+} so that the extra overvoltage can be avoided in electrolysis. When the more concentrated I_3^- solution is used, however, the photochemical cell should be thinner to complete the reaction. On the other hand the more positive the equilibrium electrode potential before photolysis is kept, the higher the total concentration of Fe^{3+} becomes. When the ratio of the concentrations of Fe^{2+} to I_3^- is large the photochemical reaction proceeds fast, but the electrode potential shifts rather small from the equilibrium potential. Considering all of these, it was concluded that the photochemical cell should be thin and the concentration of I_3^- should be high with appropriate ratio of Fe^{2+}/I_3^-.

INTRODUCTION

The authors have developed the hydrogen producing systems called Yokohama Mark 5 and 6 which include photochemical and electrochemical, and photochemical, electrochemical and thermochemical hybrid processes [1,2,3]. The system consists of the photochemical reaction between ferroussulfate and iodine and the succesive electrolysis of the products. The first photochemical reaction can be written substantially as

$$2Fe^{2+} + I_3^- \xrightleftharpoons{h\nu} 2Fe^{3+} + 3I^-. \tag{1}$$

As relatively large amount of KI is added in the solution, most of I_2 molecules are converted to I_3^- and the absorption of light is mainly caused by I_3^- which absorbs over a wide range of the wavelength up to 600nm.

The products, Fe^{3+} and I^- are introduced to the cathode compartment of the electrolyser where only ferric ion is reduced and iodide is separated as hydrogen iodide through the intermediate compartment. The electrolyser has another compartment, where oxygen evolves.

$$2Fe^{3+} + 3I^- \xrightarrow{\text{cathode}} 2Fe^{2+} \quad \text{cathode compartment}$$
$$\longrightarrow 2HI + I^- \quad \begin{array}{l}\text{intermediate}\\ \text{compartment}\end{array} \tag{2}$$
$$H_2O \xrightarrow[\text{anode}]{} 1/2O_2 \quad \text{anode compartment}$$

Electrochemical technique was applied [4] as for the concept of the energy conversion in the photochemical process, i.e., the difference of the electrochemical potential of ferric/ferrous and iodine/iodide systems is considered as the amount of the stored energy. There is of course no difference of the potentials before irradiation. The electrode potential of both systems are represented as

$$E_1 = E_1^\circ + \frac{RT}{F} \ln \frac{a_{Fe^{3+}}}{a_{Fe^{2+}}} \tag{3}$$

$$E_1^\circ = 0.771 \text{ V vs.NHE at } 25°$$

$$E_2 = E_2^\circ + \frac{RT}{2F} \ln \frac{a_{I_3^-}}{(a_{I^-})^3} \tag{4}$$

$$E_2^\circ = 0.536 \text{ V vs.NHE at } 25°,$$

where E_1° and E_2° are the standard potentials of both systems, and a is the activity of each component. Although the activity of tri-iodide, iodide ions does not differ much from the concentration in the presence of the other components, and the activity can be replaced by the concentration, ferric and ferrous ions are affected much with anions or other complex forming ligands. So in the present case the concentrations should be multiplied by the activity coefficients.

$$E_1 = E_1^\circ + \frac{RT}{F} \, ln \, \frac{\alpha C_{Fe^{3+}}}{\beta C_{Fe^{2+}}} \tag{5}$$

$$= E_1^{\circ\prime} + \frac{RT}{F} \, ln \, \frac{C_{Fe^{3+}}}{C_{Fe^{2+}}} \tag{6}$$

$$E_1^{\circ\prime} = E_1^\circ + \frac{RT}{F} \, ln \, \frac{\alpha}{\beta} \tag{7}$$

Now we get the Eqs. (5), (6) and (7), where α and β are the activity coefficients, $E_1^{\circ\prime}$ is the formal standard potential and C is the concentration.

Of the phosphate buffer solution, the ratio of the activity coefficients α/β is very small and the most of Fe^{3+} ions form complexes. The formal potential (7) is therefore closer to E_2° so that the photochemical reaction (1) takes place rather easily, while in the case of Cl^- buffer, there exist much bare ferric ions which cause the fast back reaction and retard apparently the forward photochemical reaction. In the former case the amount of the energy stored is small inversely to the fast photochemical reaction. On the other hand in the latter case, even though the photochemical reaction does not take place completely, the energy will be stored even more in the system because the potential difference, $E_1 - E_2$ has the $E_1^{\circ\prime} - E_2$ term which is larger in the latter case. In this way the amount of the energy stored is expected to increase with incresing of α/β value and it will reach the maximum at an appropriate value of α/β. The sulfate containing solution of pH 1 which has the α/β value of 0.0180 gave the maximum ΔE [4] and so it was used in all experiments in this report.

The purpose of this investigation is to get the optimum condition of the photochemical and electrochemical hybrid system.

EXPERIMENTAL

A 1 Kw Xenon lamp was used as a light source for the photolysis. The schematic layout was shown in Fig.1. The mixed solution of iodine, potasium iodide and ferric and ferrous sulfate was pumped into the photochemical cell of 15cm (width) x 25cm (height) x 0.1 or 0.2cm (thickness), and the irradiated solution was drained off after passing through the spectrophotometer which detected the concentration of the iodine. The photochemical cell was equipped with a platinum and a reference electrode (Ag/AgCl) which monitored the photochemical reaction. The electrode potential obtained in this case shows the mixed potential of the iodine/iodide and ferric/ferrous systems.

The iodine/iodide and the ferric/ferrous solutions the concentrations of which are the same as the irradiated solution
were made up separately and examined by a rotating disk
electrode of The Hokuto Denko's. The mixed potential was
examined by overlapping the voltammograms of two systems. As
for the electrode materials, commercially available platinum
and titanium foils were used. The surface of the titanium
foil was metallic but brown. From the X-ray spectra it was
found consisted of TiN and Ti_2N which are conductive enough
for electrolysis. As for the chemicals, reagent grade materials were used without further purification except iodine.
Iodine was purified by sublimation.

Quantitative determination of each component by the spectro-
photometric technique

Iodine molecules dissolve in the KI solution forming such as
I_{2aq}, I_3^- or I_5^-. When the concentration of iodine is low,
I_{2aq} and I_3^- are the main species. However as the solution
becomes more concentrated, the havier species like I_5^- and
I_7^- appear in the solution. All the ratio of such components
obey the eqilibrium eqation, and the equilibrium constants
are as [5,6],

$$K_3 = [I^-][I_2]/[I_3^-] = 660.7 \qquad (8)$$

$$K_5 = [I^-][I_2]^2/[I_5^-] = 1.85 \times 10^5 \quad at\ 25°. \quad (9)$$

The [I_2] in the above equations denotes the aqueous iodine
and the maximum value is 1.32×10^{-3} mol/1 at 25°. The aqueous
iodine has its maximum absoption at 460nm. The molecular
extinction coefficient was determined to be 582 [$l \cdot mol^{-1} \cdot cm^{-1}$].
On the other hand the maximum wavelength of I_3^- locates around
350nm and its molecular extinction coefficient is about 2.5×10^4.
Varing the total concentration of I_2 from 10^{-4} to 4×10^{-3} mol/1,
keeping the KI to I_2 ratio 20, the concentrations of [I_2],
[I_3^-] and [I_5^-] were obtained. The [I_3^-]/Σ[I_2] value (
Σ[I_2] = [I_2] + [I_3^-]) calculated by using only K_3, Eq.(8)
and the {[I_3^-] + [I_5^-]}/Σ[I_2] value (Σ[I_2] = [I_2] + [I_3^-]
+ [I_5^-]) calculated by applying both K_3 and K_5, Eqs.(8) and
(9), are almost the same and the difference is only 0.1~0.5%
so that the total mumber of multicomplexed iodine ions can be
calculated simply by using Eq.(8). It is obvious that the concentration of I_7^- or I_9^- is negligibly small in our experiments.
The absorbancy obtained by Eq.(8)was plotted in Fig.2 against
the concentration of I_3^-. The absorbancy of the solution containing iodine does not obey the Lambert-Beer's law if plotted
against the total number of iodine. Fortunately a linear
relation was obtained at 400 ~ 500 nm, using 1 mm thickness
flow cell in this case. Ferric and ferrous ions don't have

a large number of molecular extinction coefficient in this region as to interfere the determination of I_3^- except the case of very small quantity of I_3^- and very large amount of Fe^{3+} ion. In this way all the species contained in the solution can be determined.

RESULTS AND DISCUSSIONS

Effects of the concentration and its ratio of each component upon the photochemical reaction

An example of the photolysis was given in Fig.3, showing the decay curve of the absorbancy of tri-iodide at 400nm. The absorbancy decreased rapidly by irradiation and reached the minimum point and again increased gradually according to the back reaction in the dark. The data obtained in this manner were summarized in Figs.4 and 5. Fig.4 shows the electrode potential and the rate of Fe^{3+} formation as a function of [Fe^{2+}]/[I_3^-]. Here the initial ratio of the concentrations [KI]/[I_3^-] was kept 25 and so the equilibrium potential decreased as the concentration of I_3^- increased. The case, on the other hand, holding the initial equilibrium potential at 0.600 V was shown in Fig.5, where the [KI]/[I_3^-] was varied.

Electrode potential

Before irradiation the Fe^{3+}/Fe^{2+} and I_3^-/I^- couples are in the equilibrium state and their potential E_1, E_2 is equal to the equilibrium potential E_{eq}.

$$E_1 = E_2 = E_{eq} \tag{10}$$

The ratio of the redox components of both systems is related to the Eqs.(4) and (6), and is given as,

$$\left(\frac{C_{Fe^{3+}}}{C_{Fe^{2+}}}\right)^2 \frac{(C_{I^-})^3}{C_{I_3^-}} = \exp\left\{\frac{2F}{RT}(E_2^\circ - E_1^{\circ'})\right\} \tag{11}$$

From the Eqs.(10) and (11) the ratio of the redox components was given as a function of E_{eq} in Table 1.
The larger the E_{eq} becomes, the more oxidative the I_3^-/I^- system is and the concentration of I^- is negligiblly small compared to I_3^- at the higher E_{eq} than 0.600 V. On the other hand at 0.540 V (E_2° = 0.538 V) the ratio, is as small as 1.37 which shows that the concentrations of I_3^- and I^- are almost the same. As for the Fe^{3+}/Fe^{2+} system, the ratio of the both species is inclined extremely to reductive at each point in this range and so even by a small change in the concentration of Fe^{3+} ion, E_2 will shift much to positive. If E_{eq} is kept

at a certain point, the potential shift E_1 - E_{eq} becomes
smaller with increasing of the ratio of the concentration of
Fe^{2+} to I_3^-. These speculations were proved by the experiments
as shown in Figs.4 and 5. Although enhancing the energy storage at the photochemical process is indeed valuable in improving the overall efficiency, both the rate of formation of Fe^{3+}
and the total Fe^{3+} concentration are also important in order
to avoid the extra overvoltage in the electrolysis process.

The rate of Fe^{3+} formation and the electrolysis condition

The dependence of [Fe^{2+}]/[I_3^-] on i-E curves were shown in
Fig.6 with the case of (c) of Fig.5. C_1 and C_3 are related to
the point of [Fe^{2+}]/[I_3^-] of 2.5 and 10, respectively. When
the ratio, [Fe^{2+}]/[I_3^-] is small, the shift of E_1 is large
so that the electrolysis voltage decreases, the curve, however,
is rather gentle. On the contrary it is steap at the high
ratio, because of the higher concentrations of the total Fe^{3+}.

Thickness of the photochemical cell

The average penetrating depth of photons is inversly proportional to the concentration of I_3^- and 2.7×10^{-3} mol/1 and 5.3
$\times 10^{-3}$ mol/1 were obtained as the appropriate concentraion for
the thickness of the cell of 2 mm and 1 mm [4]. The results
show in Fig.4 that the photochemcial reaction takes place
fairly rapidly even with the higher concentration and the rate
of Fe^{3+} formation increases with increasing of the concentration of I_3^-. The reaction proceed at the same rate with both
1 and 2 mm cell up to 2×10^{-3} mol/1 but beyond this point that
is close to the appropriate concentration for 2 mm cell, the
reaction takes place more slowly with 2 mm cell than with 1
mm's.

The electrode materials

The photochemical reaction of I_3^--Fe^{2+} system was monitored
by Pt and Ti-N electrodes. As mentioned above, each of I_3^-/
I^- and Fe^{3+}/Fe^{2+} system will have the individual potential
under irradiated condition and we can only observe the mixed
potential E_m, when the i_c of the i-E curve of Fe^{3+}/Fe^{2+}
system is just equal to the i_a of I_3^-/I^- system so that the
potential is largely affected by the electrochemical reacti-
vity on the electrode.

In Fig.7 was shown an example of the potential shift of I_3^--
Fe^{2+} system under the irradiation condition using Pt and Ti-
N electrodes. Both electrodes showed the same potential
before irradiation, the potential shifted negative with the
Pt electrode and positive with the Ti-N electrode. By turning off the light, both potentials gradually returned to

the previous point. As for the Ti-N, it was recognized by the other experiments that the electorde does not excite with the photon irradiation under these conditions.

The i-E curves of I_3^-/I^- and Fe^{3+}/Fe^{2+} systems, the concentration of which is the same as at the photostationary state were taken separately by means of the rotating disk electrode, as shown in Fig.8. With Pt electrode, an extremely steap curve was obtained for I_3^-/I^-, being in marked contrast to the gentle curve for Fe^{3+}/Fe^{2+} system. Overlapping these two i-E curves gave a little more negative value of E_m than the observed one. On the other hand almost the same shape of i-E curves were given for both systems with Ti-N electrode. This is due not to the high reaction rate for Fe^{3+}/Fe^{2+} system but the retarded phenomena of the reaction of I^-, and Fe^{3+} can be reduced relatively faster than I^- is oxidized. The factor which controls the oxidation of I^- on the Ti-N is not clear but it may be possible to make up an electrode that reduces Fe^{3+} in extremely high selectivity.

Designing the electrolyser

The scheme of the solar beam-assisted electrolyser was shown in Fig.9. It is necessary to electrolyse the Fe^{3+} as fast as possible in order to avoid the back reaction and loss of the stored energy. Among the products of the photochemical reaction, only Fe^{3+} should be reduced. As shown in the previous section, however, Pt electrode is more sensitive to the I_3^-/I^- system rather than Fe^{3+}/Fe^{2+} system and the extra overvoltage will be required in the electrolysis of the Fe^{3+}, I^- mixture. Even though the Ti-N electrode showed the positive shift of the electrode potential under the irradiation, it is still difficult to reduce only Fe^{3+}. Here we assume the ideal electrode which reacts only Fe^{3+}.

If $\Delta C_{Fe^{3+}}$[mol/cm^3] of Fe^{3+} ion are produced in the photochemical cell in the average time of t [sec], the electricity of It coulombs per V [cm^3] of the product solution is required to the electrolysis,

$$(\Delta C_{Fe^{3+}})FV = It , \qquad (12)$$

where I is the current [A], t is the time required by the electrolysis and F is the Faraday's constant. Assuming that t is the time passing through the cell and during the time, $\Delta C_{Fe^{3+}}$ of Fe^{3+} ion are produced by the photochemcial reaction, the following eqations are derived,

$$I = \frac{v(\Delta C_{Fe^{3+}})FV}{S\delta} \qquad (13)$$

and

$$vt = S\delta , \tag{14}$$

where v is the flow rate of the solution at the inlet or outlet of the cell and S and δ are the surface area and the thickness of the cell, respectively. Further V is controlled to be equal to vt, then

$$I = (\Delta C_{Fe}3+) \cdot Fv . \tag{15}$$

Now let's consider the electrolytic reaction on the electrode surface. The current density i [A/cm^2] is related to the concentration of Fe^{3+} in the solution ($C_{Fe}^S 3+$) and at the surface of the electrode ($C_{Fe}^e 3+$) as

$$i = \frac{FD (C_{Fe}^S 3+ - C_{Fe}^e 3+)}{\delta'} \tag{16}$$

where D is the diffusion constant [cm^2/sec] and δ' is the thickness [cm] of the diffusion layer. As the current i becomes large, $C_{Fe}^e 3+$ decreases gradually down to 0, when the current reaches the limiting current i_{l-},

$$i_{l-} = \frac{FD \cdot C_{Fe}^S 3+}{\delta'} \tag{17}$$

For designing of the electrolyser, the condition of the experiment (Fig.4,E) was taken as an example. From the data D~10^{-5}, δ'=10^{-3} and $_I C_{Fe}^S 3+$=6.74x10^{-6}, $_I i_{l-}$ of 6.50x10^{-3} [A/cm^2] was obtained for the inlet condition (Fig.10). As for the outlet $_o i_{l-}$ of 4.74x10^{-4} was obtained.

On the other hand the limiting current density of the oxidation of Fe^{2+}, i_{l+} is obtained analogous to the i_{l-}, as

$$i_{l+} = \frac{FD \cdot C_{Fe}^S 2+}{\delta'} \tag{18}$$

The i_{l+} for the inlet and the i_{l+} for the outlet were 0.0616 [A/cm^2] and 0.0676, respectively. Substituting these data and k_s of 0.05 (this is the maximum one reported hitherto [7]), the exchange current densities at the inlet and the outlet were derived.

$$_I i_o = Fk_s (_I C_{Fe}^S 3+)^{0.5} (_I C_{Fe}^S 2+)^{0.5} \tag{19}$$

$$_o i_o = F k_s (_o C^S_{Fe3+})^{0.5} (_o C^S_{Fe2+})^{0.5} \qquad (20)$$

$_I i_o = 0.100$ [A/cm^2], $_o i_o = 0.0283$.
If the electrolysis is carried out with the small overpotential ($\eta < 75$ mV), the relation of the η to the current density i is given by

$$\eta = \frac{RT}{F} (\frac{1}{i_o} + \frac{1}{i_l+} + \frac{1}{i_l-}) \, i \qquad (21)$$

from which the reaction resistance (η/i) of 4.62 [V/A] for the inlet and 55.4 for the outlet were obtained.

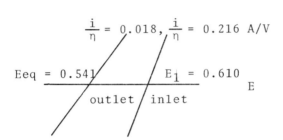

$$\frac{i}{\eta} = 0.018, \frac{i}{\eta} = 0.216 \ A/V$$

Eeq = 0.54

$E_1 = 0.610$

E

outlet / inlet

As the curve is steeper at the inlet than at the outlet, smaller value of η is required in the former case. In order to electrolyze Fe^{3+} of $(\Delta C_{Fe}3+)v$ [mol/sec], the surface area of S' is required, if the electrolysis is carried out with the current density of i_c,

$$i_c S' = (\Delta C_{Fe}3+) \, vF \qquad (22)$$

However, as the η value is different in each point of the electrolyser, the sum of the small electrodes, the electrolysis voltage is varing gradually, might give the best performance of the electrolysis.

Now, if the η value is taken as $(E_1-E_{eq})/2$, the current density of 7.45×10^{-3} [A/cm^2] is obtained from the Eq.(21) and also the surface area of 22.5 cm^2 is given by replacing the appropriate values in Eq.(22). δ', k_s and i_o were taken of the best values and in practical case, much efforts must be made to get such values.

CONCLUSION

In order to get the large amount of the energy storage by the photochemcial reaction, [Fe^{2+}]/[I_3^-] and the total concentration of I_3^- and Fe^{2+} should be kept low. However the better performance of electrolysis was shown in the case of large rate of Fe^{3+} formation and in the presence of the large concentration of the total Fe^{3+}. When the cells with two kinds of the thickness, i.e., 1 and 2 mm were used, a distinct difference appeared in the rate of photochemical reaction with the concentration of I_3^- of more than 3×10^{-3} mol/l. With 1 mm cell, the solution of as high as 7×10^{-3} mol/l of I_3^- could be treated, when the electrolysis current density of 7.45×10^{-3} A/cm^2 was expected.

ACKNOWLEDGMENTS

The authors are indebted to the Ministry of International Trade and Industry for the financial support.

REFERENCES

[1]. T.OHTA & N.KAMIYA, *Proc.10th IECEC*, p.772, 1975.
[2]. T.OHTA & N.KAMIYA, ACS Centenial Symposium on Foreign Technologies, Apr. 4-9, 1976, N.Y.
[3]. T.OHTA *et al*, *Int. J. Hydrogen Energy*, 1, 255 (1976).
[4]. T.OHTA *et al*, submitted to *Int. J. Hydrogen Energy*.
[5]. J.S.CARTER, *J. Chem. Soc.*, 1928, 2228.
[6]. G.N.PEARCE & W.G.EVERSOLE, *J. Phys. Chem.*, 28, 245 (1924).
[7]. N.TANAKA & R.TAMAMUSHI, *Electrochim. Acta*, 9, 963 (1964).

TABLE 1. RELATIONSHIP BETWEEN E_{eq} AND THE RATIO OF THE REDOX COMPONENTS

E_{eq} V vs.N.H.E.	$C_{I_3}-/(C_{I}-)^3$	$C_{Fe^{3+}}/C_{Fe^{2+}}$
0.620	696	0.154
0.600	147	0.0707
0.580	30.8	0.0324
0.560	6.49	0.0149
0.540	1.37	6.83×10^{-3}
0.520	0.287	3.13×10^{-3}

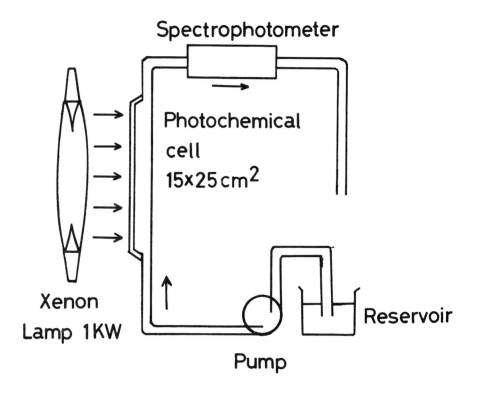

Fig.1   Schematic layout of photolysis

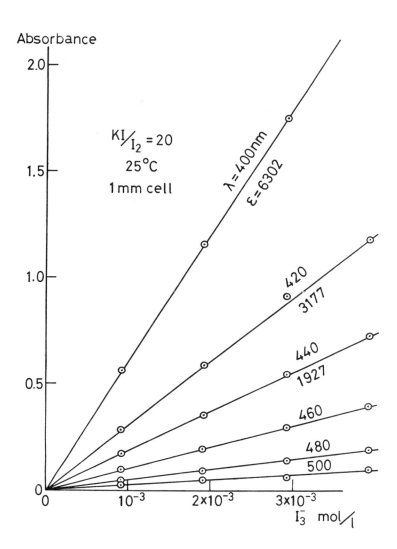

Fig.2 Absorbancy and concentration of I_3^-

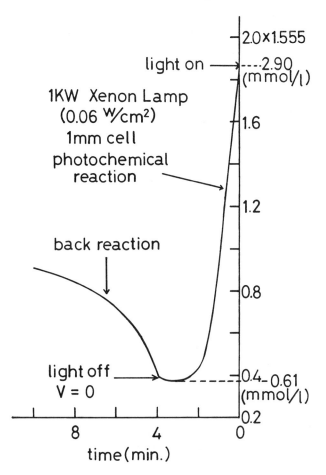

Fig.3   Absorbancy decay curve of I_3^- - Fe^{2+}
system under irradiation

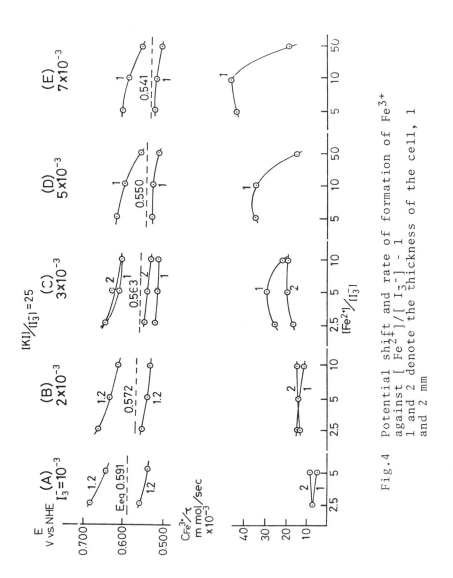

Fig.4 Potential shift and rate of formation of Fe^{3+} against $[Fe^{2+}]/[[I_3^-]-1$ 1 and 2 denote the thickness of the cell, 1 and 2 mm

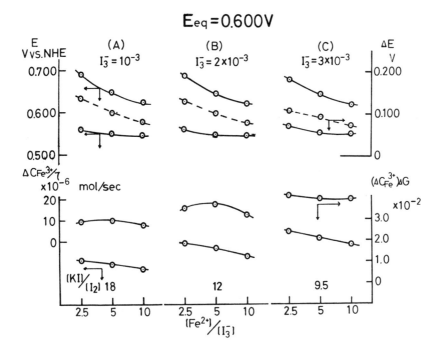

Fig.5 Potential shift and rate of formation of Fe^{3+}
against $[Fe^{2+}]/[I_3^-] - 2$

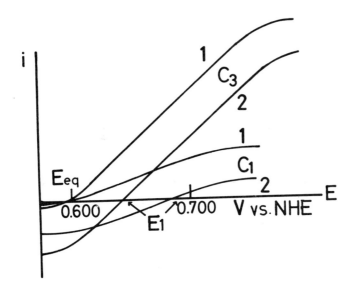

Fig.6 i - E curves of Fe^{3+}/Fe^{2+} system
the case of Fig.5 (c)

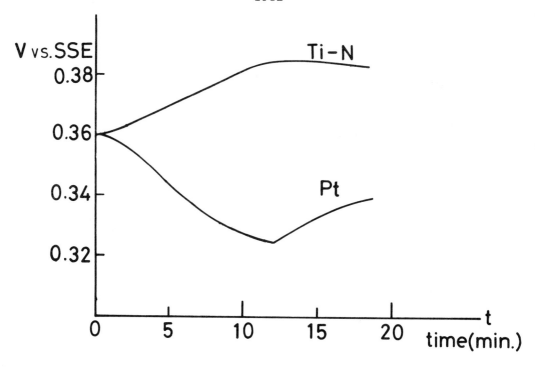

Fig.7　Potential shift with Pt and Ti-N electrodes

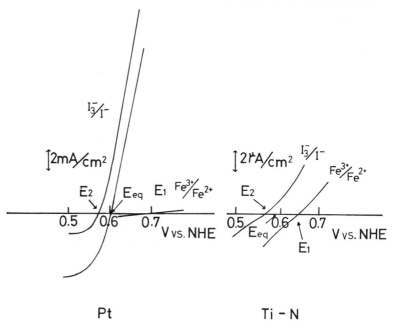

Fig.8　i - E curves of I_3^-/I^- and Fe^{3+}/Fe^{2+} systems

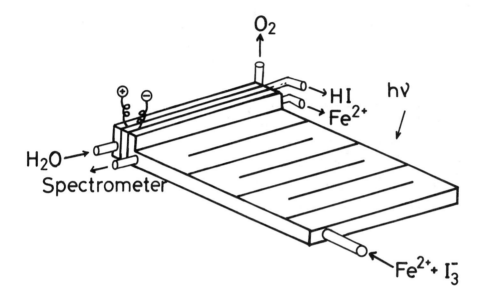

Fig.9   Solar beam-assisted electrolyser

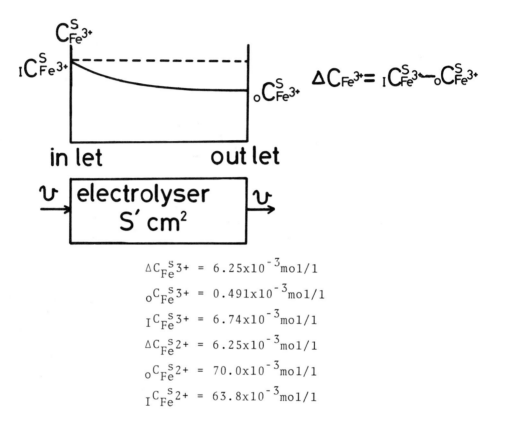

$\Delta C_{Fe}^S 3+ = 6.25\times10^{-3} mol/l$

$_o C_{Fe}^S 3+ = 0.491\times10^{-3} mol/l$

$_I C_{Fe}^S 3+ = 6.74\times10^{-3} mol/l$

$\Delta C_{Fe}^S 2+ = 6.25\times10^{-3} mol/l$

$_o C_{Fe}^S 2+ = 70.0\times10^{-3} mol/l$

$_I C_{Fe}^S 2+ = 63.8\times10^{-3} mol/l$

Fig.10 Concentration distribution in the electrolyser

SEPARATION OF HYDROGEN FROM THE MIXTURE OF HYDROGEN IODIDE, HYDROGEN AND IODINE IN THERMO-GRAVITATIONAL COLUMN

S. Tanisho, N. Wakao and T. Ohta
Yokohama National University
Hodogayaku, Yokohama, Japan

ABSTRACT

Thermogravitational column is expected to be successfully employed for the separation of hydrogen from hydrogen iodide-
hydrogen-iodine gas mixture. The advantages of the thermogravitational column are that the decomposition of hydrogen
iodide into hydrogen and iodine proceeds in the bulk space and
the iodine deposited on the cold wall is easily recovered.
Separation of hydrogen and argon from the mixture is also
reported in this paper. From the measurements it is found that
the recovery efficiency largely depends upon the feed location,
width of the annular spacing(between hot and cold walls),
temperature difference between the walls and the temperature
distribution in the axial direction.

INTRODUCTION

If there exists a temperature gradient in a homogeneous gas
mixture, a concentration gradient is also established. This
phenomenon is called the thermal diffusion. If a binary gas
mixture is at rest, at an equilibrium state a lighter constituent is enriched in the hot place and a heavier in the cold.
The equilibrium state is maintained by a coupling effect of
the two fluxes, thermal diffusion flux and ordinary diffusion
flux.

In 1938, Clusius and Dickel[1] showed that convective currents
have a cascading effect in a separation column and the thermal
diffusion effect is so magnified that the thermal diffusion
becomes a powerful separation method. In this column, lighter
constituent moved to the hot wall is carried away by natural
convection currents to the top of the column, while heavier
constituent moved toward the cold wall is enriched at the
bottom.

Since then the thermogravitational column(Clusius-Dickel
column) has widely been employed for the thermal diffusion
separation [2,3,4,5,6,7]. Thesedays, thermal diffusion separation is being paid much attention in the separation of radio-

1085

active inert gas mixtures produced in nuclear reactor[8].

As far as the authors know, however, the thermal diffusion separation has not been applied when the system is accompanied by a chemical reaction and/or phase change. Let us consider a thermal diffusion separation for the system of hydrogen iodide, hydrogen and iodine.

Suppose the cold wall is kept at a temperature at which iodine becomes solid state. Iodine will desublimate from the cold wall, and the concentration gradient will result in the same direction as the temperature gradient. The thermal diffusion flux and the ordinary diffusion flux are superimposed and the separation efficiency is greatly enhanced. On the hot wall, hydrogen iodide will decompose into hydrogen and iodine, because of the lower concentration of gaseous iodine. The thermal diffusion separation will, therefore, provide a good decomposition-separation method for the gaseous mixture of hydrogen iodide, hydrogen and iodine.

Prior to the experiments with hydrogen iodide gas, the measurements with the hydrogen-argon mixture were made.

THEORETICAL TREATMENT OF THERMOGRAVITATIONAL COLUMN

When reaction or phase change does not occur in the mixture, the flux of species 1 in binary gas mixture is expressed as

$$\mathbf{J}_1 = \rho x_1 \mathbf{v} + \rho D [-\nabla x_1 + \alpha x_1 x_2 \nabla \ell n T] \tag{1}$$

where ρ is the density of the gas mixture, x_1 is the mole fraction of the lighter constituent, $\mathbf{v}$ is the convection velocity, D is the coefficient of molecular diffusion, α is the thermal diffusion constant, T is the absolute temperature. The corresponding equation for species 2 is

$$\mathbf{J}_2 = \rho x_2 \mathbf{v} + \rho D [-\nabla x_2 + \alpha x_1 x_2 \nabla \ell n T] \tag{2}$$

As a matter of convenience, we assume that
(1) Convective flow produced by the density gradient is laminar and there are no end effects,
(2) Radial temperature distribution is established by thermal conduction and there is no temperature gradient in the axial direction,
(3) Temperature-dependent quantities of the gases are evaluated at the average temperature, and
(4) Steady state is maintained in the column.

With the assumptions, the radial temperature distribution is given by

$$\frac{d^2T}{dr^2} + \frac{1}{r}\frac{dT}{dr} = 0 \tag{3}$$

Then,

$$T(r) = -\frac{T_H - T_L}{\ln R_L - \ln R_H}\ln r + \frac{T_H - T_L}{\ln R_L - \ln R_H}\ln R_H + T_H \tag{4}$$

$$= A\ln r + B \tag{5}$$

where subscripts H and L mean the hot and cold temperature walls, respectively.

The velocity distribution is determined by the equations of motion and energy:

$$\frac{d^2v}{dr^2} + \frac{1}{r}\frac{dv}{dr} = -\frac{\bar{\rho}\bar{\beta}g}{\bar{\mu}}(T - \bar{T}) \tag{6}$$

Then,

$$v(r) = \frac{a_1 A}{4}r^2\ln r - \frac{a_1}{4}(A-B+T)r^2 + a_2\ln r + a_3 \tag{7}$$

where $\bar{\beta}$ is the coefficient of volume expansion at $\bar{T}$, which is the temperature at the point where the sign of v changes, g is the acceleration of gravity, $\bar{\mu}$ is the viscosity of the mixture at $\bar{T}$ and a_1, a_2, a_3 are the constants. $\bar{T}$ and a_1, a_2, a_3 are determined by the boundary conditions:

$$2\pi\int_{R_H}^{R_L} \rho v r\, dr = \sigma \tag{8}$$

$$v(R_H) = v(R_L) = 0 \tag{9}$$

Each component of flux J_1 is expressed as

$$J_{Dr} = -\rho D\frac{\partial x_1}{\partial r} \tag{10}$$

$$J_{Tr} = \rho D\alpha x_1 x_2\frac{1}{T}\frac{dT}{dr} \tag{11}$$

$$J_{Cz} = \rho x_1 v \tag{12}$$

$$J_{Dz} = -\rho D\frac{\partial x_1}{\partial r} \tag{13}$$

and

$$J_{1r} = J_{Dr} + J_{Tr} \tag{14}$$

$$J_{1z} = J_{Dz} + J_{Cz} \tag{15}$$

Since div $\mathbf{J}_1$ is zero, Equation(1) reduces to

$$-\frac{d}{dr}(rJ_{1z}) = r\frac{dJ_{1z}}{dz} \tag{16}$$

The net flow due to both thermal and ordinary diffusion is zero at either wall surface, a solution to Equation(16) is subject to the boundary conditions.

$$J_{Dr} + J_{Tr} = 0 \qquad \text{at} \quad r = R_H \quad \text{and} \quad r = R_L \tag{17}$$

Integration of Equation(16) becomes

$$-r(J_{Dr} + J_{Tr}) = \int_{R_H}^{r} r\frac{dJ_{1z}}{dz} \, dr \tag{18}$$

Substituting Equations(10),(11) into Equation(18), one obtains

$$\frac{\partial x_1}{\partial r} = \frac{\alpha x_1 x_2}{T}\frac{dT}{dr} + \frac{1}{\rho Dr}\int_{R_H}^{r} r\frac{dJ_{1z}}{dz} \, dr \tag{19}$$

The net vertical transport of species 1 is

$$\tau_{1z} = 2\pi \int_{R_L}^{R_H} rJ_{1z} \, dr \tag{20}$$

$$= x_1 (R_L) - 2\pi \int_{R_H}^{R_L} \frac{\alpha x_1 x_2}{T}\frac{dT}{dr}\int_{R_H}^{r} \rho vr \, drdr$$

$$- 2\pi \int_{R_H}^{R_L} \frac{1}{\rho Dr}\int_{R_H}^{r} r\frac{dJ_{1z}}{dz}\int_{R_H}^{r} \rho vr \, drdrdr$$

$$- 2\pi \int_{R_H}^{R_L} \rho Dr \frac{\partial x_1}{\partial z} \, dr \tag{21}$$

For the convenience, Equation(21) is integrated under the following assumptions:
(1) constant concentration gradient in z-direction
(2) $\partial x_1/\partial z$ is independent of r
(3) $x_1 x_2$ is independent of r
(4) ρv is independent of z

With these assumptions, we have an equation identical with that derived by Furry, Jones and Onsager in 1939[9].

$$\tau_{1z} = x_1 (R_L)\sigma + Hx_1 x_2 - (K_C + K_D)\frac{\partial x_1}{\partial z} \tag{22}$$

where

$$H = -2\pi \int_{R_H}^{R_L} \frac{\alpha}{T} \frac{dT}{dr} \int_{R_H}^{r} \rho vr \; drdr \tag{23}$$

$$K_C = 2\pi \int_{R_H}^{R_L} \frac{1}{\rho Dr} \left(\int_{R_H}^{r} \rho vr \; dr \right)^2 dr \tag{24}$$

$$K_D = 2\pi \int_{R_H}^{R_L} \rho Dr \; dr \tag{25}$$

Since the radial concentration variance is small compared with the concentration differences between the top and bottom ends, we may approximate the concentration x as a mean concentration $\overline{x}$. Equation(22) can, therefore, be written as

$$\tau_z = \overline{x}\sigma + H\overline{x}(1-\overline{x}) - (K_C + K_D)\frac{d\overline{x}}{dz} \tag{26}$$

At the equilibrium state in batch system, i.e., $\tau_z = 0$, $\sigma = 0$, the solution of Equation(26) is

$$Q_e \equiv \frac{x_e}{1-x_e} \cdot \frac{1-x_s}{x_s} = exp\left[\frac{HL}{K_C+K_D}\right] \tag{27}$$

where Q_e is the equilibrium separation factor, x_e, x_s are the concentrations, respectively, at the end of enriching and stripping section, and L is the column length.

In a continuous separation system, thermogravitational column is considered to provide a coupled effect of two different kinds of separation process: enriching and stripping separation processes. The part above the feed location is the enriching section for lighter constituent. The material balance at the top end of the column is given from Equation(26) as

$$\sigma_e x_e = \sigma_e \overline{x} + H\overline{x}(1-\overline{x}) - (K_C+K_D)\frac{d\overline{x}}{dz} \tag{28}$$

The material balance at the bottom end of the column is

$$-\sigma_s x_s = -\sigma_s \overline{x} + H\overline{x}(1-\overline{x}) - (K_C+K_D)\frac{d\overline{x}}{dz} \tag{29}$$

Equations(28),(29) can be written as

$$-(K_C+K_D)\frac{d\overline{x}}{dz} = \overline{x}^2 - (1+\gamma)\overline{x} + \gamma x_p \tag{30}$$

where $\gamma=\sigma_p/H$, $\sigma_p=\sigma_e$ in the enriching section and $\sigma_p=-\sigma_s$ in the stripping section. Writing the two roots of the Equation(31) α and β,

$$\xi^2 - (1+\gamma)\xi + \gamma x_p = 0 \tag{31}$$

$$\alpha,\beta = \frac{1}{2}\{(1+\gamma) \pm \sqrt{((1+\gamma)^2 - 4\gamma x_p)}\} \tag{32}$$

Equation(30) is, then, integrated as

$$\frac{1}{\alpha-\beta}\ell n\frac{(x_p-\alpha)(x_f'-\beta)}{(x_f'-\beta)(x_p-\alpha)} = -\frac{HL'}{K_C + K_D} \tag{33}$$

where x_f' is the concentration at the feed location and L' is the column length of enriching(or stripping) section.

EXPERIMENTAL EQUIPMENT

Schematic diagram of the equipment is shown in Fig.1. The details of the top and bottom sections are illustrated in Fig.2. The thermal diffusion column is a concentric tube of which dimensions are listed in TABLE 1. The main section of the column consists of stainless steel tubes and inner heating glass tube having a nichrome wire in it. The surface tempera-ture of the heater is measured at three locations(L/4,L/2 and 3L/4) with Chromel-Alumel thermocouples. The temperature at the half length, L/2, is kept constant by PID type temperature regulator. The cold wall is cooled by circulating the water at 25°C. The temperature increase at the outlet is kept less than 3°C. The compositions of the feed gas, top and bottom products are analyzed in gas chromatograph. Constant flow rates of the feed, top and bottom products are maintained by needle valves. The pressure in the column is about five to ten mmH_2O above the atmosphere.

EXPERIMENTAL RESULTS

There are several formulae for evaluating the separation effi-ciency of the thermogravitational column. Equation(34) shows the equilibrium separation factor derived by Furry et al., Equation(35) is the concentration difference between the top and bottom products and Equation(36) is the Newton efficiency:

$$Q = \frac{x_e}{1-x_e}\frac{1-x_s}{x_s} \tag{34}$$

$$\Delta = x_e - x_s \tag{35}$$

$$\eta = \frac{\sigma_e x_e}{\sigma_f x_f} - \frac{\sigma_e}{\sigma_f}\frac{1-x_e}{1-x_f} \tag{36}$$

The equilibrium separation factor Q indicates the separation degree and the Newton efficiency of separation means a recovery efficiency. In the continuous flow system, the maximum separation is obtained from the material balance in terms of the partition ratio σ_e/σ_f:

$$\Delta_{max} = x_{emax} = x_f \frac{\sigma_f}{\sigma_e} \qquad (\text{ at } \frac{\sigma_e}{\sigma_f} \geqq x_f) \qquad (37)$$

$$\eta_{max} = -\frac{1}{1-x_f} \frac{\sigma_e}{\sigma_f} + \frac{1}{1-x_f} \qquad (\text{ at } \frac{\sigma_e}{\sigma_f} \geqq x_f) \qquad (38)$$

$$= \frac{1}{x_f} \frac{\sigma_e}{\sigma_f} \qquad (\text{ at } \frac{\sigma_e}{\sigma_f} < x_f) \qquad (39)$$

Feed Location

Figure 3 shows the relation between the equilibrium separation factor and the partition ratio at various feed locations. The graph shows the importance of the feed location. If the feed is at the bottom of the column, the equilibrium separation factor increases considerably. On the other hand, if the feed location is at the top, the equilibrium separation factor increases in the region where the partition ratio is larger than the mole fraction of feed gas. When the feed location is at the half length of the column, the equilibrium separation factor seems to have a maximum value at certain partition ratio. Figure 4 shows the concentration of the products as a function of partition ratio. From this graph we can realize the reason why the separation factor behaves as illustrated in Fig.3. In the case of bottom feed, the lighter constituent is easily enriched in the top product at the certain partition ratio smaller than the mole fraction of the feed gas. The separation factor, therefore, increases even to infinity. However, in the case of top feed, the product is not as much enriched as to pure hydrogen. In the case of top feed, the product is considerably stripped to heavier constituent in the region where partition ratio is larger than the mole fraction of feed gas. In the case of bottom feed, the product is not highly stripped, but almost constant concentration is maintained over the wide range of partition ratio.

Therefore, if we use the process as a highly enriching process for a lighter constituent, the feed should be located at the bottom of the column and the partition ratio should be less than the mole fraction of the feed gas. On the other hand, if we use the process for a highly stripping process, we had better locate the feed at the top of the column and the partition ratio should be higher than the mole fraction of the feed gas.

As far as the recovery efficiency is concerned, however, some-

what different conclusions may be derived from Fig.5. This
graph shows the Newton efficiency of separation as a function
of the partition ratio. The graph indicates that the maximum
Newton efficiency of separation is attained when the feed is
located at the half length of the column.

Distance between Walls

According to the theory, the separation depends strongly upon
the wall distance as $H \propto (\Delta R)^3$ and $K \propto (\Delta R)^7$. In the case that the
diameter of the cold tube is constant, the equilibrium separa-
tion factor in batch system is theoretically calculated as a
function of the diameter of the hot tube. The result is illustrated in Fig.6. When cold tube diameter is 43mm, the optimum
hot tube diameter is found to be about 30mm, where the distance
between the walls is about 6.5mm. Figures 3 and 7 show the cases
of D_i=30mm and D_i=20mm, respectively. The experimental results
show good agreement with the theoretical lines except large
wall distance column E.2. During the experiment with large
wall distance column E.2, we observed violent oscillation of
flow meters. In the case of small wall distance column E.1,
the oscillation was not observed. The reason for the oscillation seems due to the wall distance. Onsager and Watson[10]
point out that the free convective flow between the concentric
vertical cylinders changes from laminar to turbulent at
Reynolds number about 150. According to their equation the
Reynolds numbers of our runs are 166 for run E.1 and 858 for
run E.2. The column with large wall distance E.2 must have
been operated in the turbulent region. Because of the turbulence, rapid density fluctuation would have resulted locally
in the column and this would have probably caused the oscillation of the flow meters. The discrepancies will be due to the
turbulence and density oscillation, which are contrary to the
assumption we made before.

Temperature Distribution between Cold and Hot Walls

The effect of temperature difference between the cold and hot
walls was investigated with small column E.3. The results
are shown in Fig.8. In the case of bottom feed, the increase
in the equilibrium separation factor was rather small, although
the temperature difference increased. On the other hand, in
the case of top feed, the effect of temperature difference was
large in the region of large partition ratio. In this case,
the concentration in the bottom product was too low to be
measured in gas-chromatograph. This will be due to the temperature distribution resulted from concentration gradient in
the vertical direction.

So far theoretical research in thermal diffusion column has
been made by many investigators under the assumption of no

temperature gradient being in the vertical direction. When the thermal conductivities of pure gas differ considerably from each other, temperature gradient is established in the vertical direction under the condition of constant heat flux.

Thermal conductivity of hydrogen is 0.180kcal/m.hr°C at 100°C and that of argon 0.0183kcal/m.hr°C. The ratio is about ten times. The large difference will cause temperature gradient in the vertical direction. For an illustration for experimental results, Fig.9 shows the surface temperatures at three locations on the heater wall(L/4,L/2 and 3L/4) with several partition ratios in the case of top feed. Large temperature gradient, as much as 130°C at certain partition ratio, is seen in the graph. The temperature increase at lower part enhances the effect in the bottom product and large effect of temperature difference in the separation factor is observed.

Hydrogen Iodide-Hydrogen-Iodine System

Several experiments were made with hydrogen iodide system. The results are shown in Figs.10 and 11. The separation characteristics for hydrogen iodide-hydrogen-iodine gaseous mixture was similar to that of hydrogen-argon binary gas mixture. In the case of small distance column E.1, the equilibrium separation factor was large compared to the hydrogen-argon system. On the other hand, in the case of large wall distance column E.2, the equilibrium separation factor was not high.

With regard to the decomposition of hydrogen iodide, the concentration of hydrogen in the top product was 0.074 in mole fraction at $\sigma_e/\sigma_f = 0.25$ in the case of small distance column, the mean concentration of products is then 0.02 in mole fraction. As the feed gas was pure hydrogen iodide and the feed gas rate was 0.4ℓ/hr, this indicates that hydrogen was produced at the rate of 8cc/hr in the column. The column volume was 1.2ℓ, so that the mean residence time for hydrogen iodide was about three hours. In three hours, at the mean temperature 59°C, the volume of hydrogen decomposed is not small. If the heater surface is coated with catalyst, more decomposition will be expected.

CONCLUSIONS

From the present measurements we come to the conclusions:
(1) Thermogravitational column provides an effective separation method for gas mixture.
(2) Sufficient separation effect is attained at small temperature difference.
(3) Separation performance is sensitive to feed location.

(4) Optimum recovery efficiency is obtained when the feed
    location is at the half length of the column.
(5) There is an potimum wall distance.
(6) Large temperature gradient exists in a vertical direction.
(7) Considerable decomposition is observed in the case of
    hydrogen iodide system.

References

[1] Clusius, K. and Dickel, G., Naturwiss.,26,546(1938)
[2] Vasaru, G.,Müller, G., Reinhold, G., Fodor, F., " The The-
    rmal Diffusion Column",VEB Deutsch. Verlag Wissenschaft.
    (Berlin),1969
[3] Jones, R. C. and Furry, W. H., Rev. Mod. Phys.,18,151(1946)
[4] Corbett, J. W. and Watoson, W. W., Phys. Rev.,101,519(1956)
[5] Shimizu, M. and Takashima, Y., J. Nucl. Sci. Technol.,7,
    574(1970)
[6] Tanisho, S., Wakao, N. and Ohta, T., Preprints of the Annu-
    al Meeting of the Soc. of Chem. Engrs. Japan,E313(1977)
[7[Tanisho, S., Wakao, N. and Ohta, T., Preprints of the
    Autumnal Meeting of the Soc. of Chem. Engrs. Japan,D207
    (1977)
[8] Kitamoto, A., Takashima, Y. and Shimizu, M., J. Nucl. Sci.
    Technol.,13,574(1976)
[9] Furry, W. H., Jones, R. C. and Onsager, L., Phys. Rev.,55,
    1083(1939)
[10] Onsager, L. and Watson, W. W., Phys. Rev.,56,474(1939)

TABLE 1. SIZES OF COLUMNS AND
EXPERIMENTAL CONDITIONS OF HYDROGEN-ARGON SYSTEM.

Column	Outer Diameter [mm]	Inner Diameter [mm]	Column Length [mm]	Hot Wall Temp. [°C]	Cold Wall Temp. [°C]	Feed Rate [ℓ/hr]	Feed Mole Fraction x_f [-]
E.1	43	30	1600	100	25^{+3}_{-0}		
E.2	43	20	1600	100		1.08	0.145
E.3	20	6	1000	(100 / 200)		~ 1.12	~ 0.156

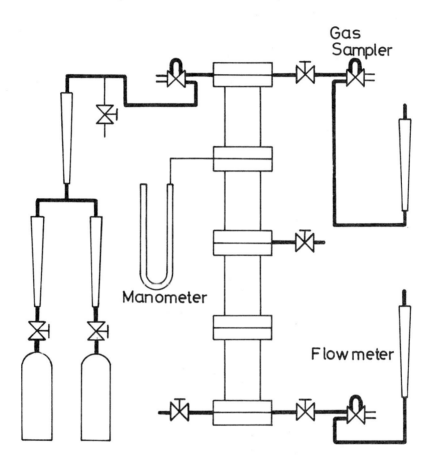

Fig. 1 Schematic diagram of the equipment

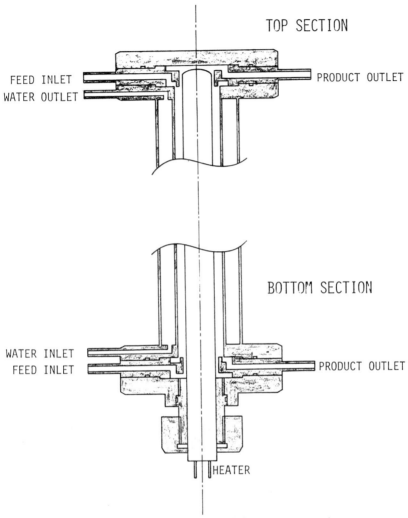

Fig. 2   Detailes of the top and bottom sections

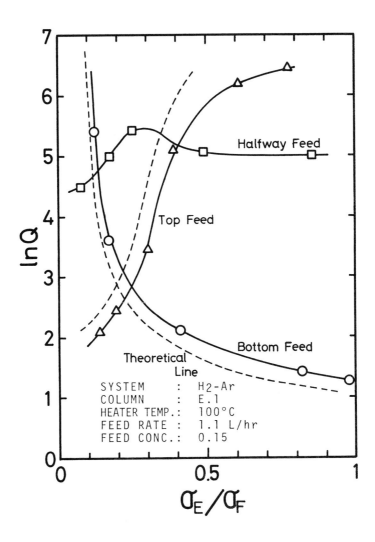

Fig. 3 Separation factor in small wall-
distance column E.1

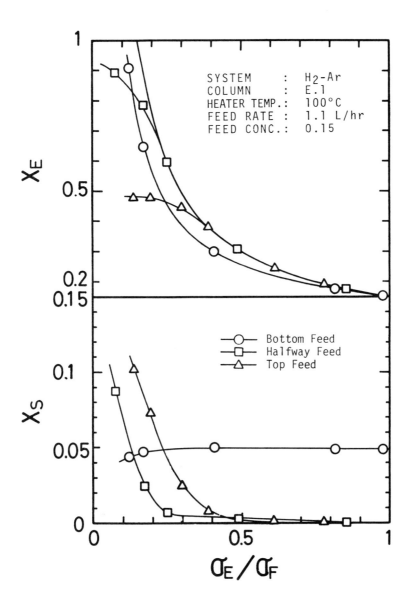

Fig. 4 Concentrations of top and sottom
products in small wall-distance column E.1

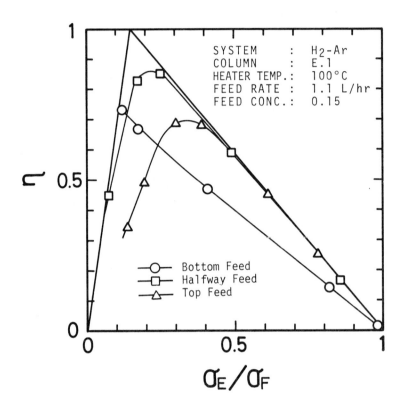

Fig. 5   Newton separation efficiency

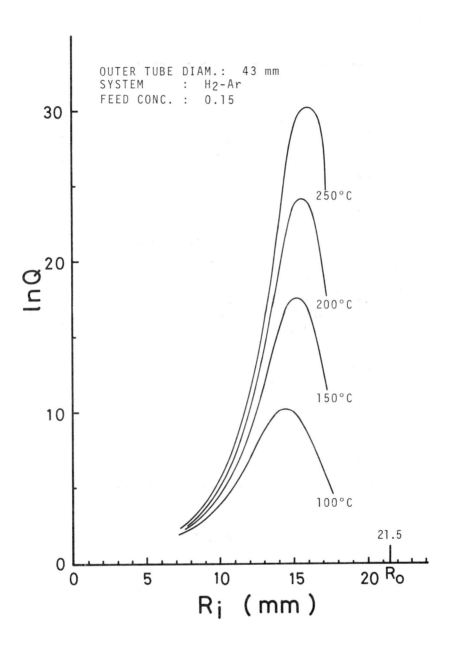

OUTER TUBE DIAM.: 43 mm
SYSTEM   : H_2-Ar
FEED CONC. : 0.15

Fig. 6   Effect of wall distance on heater diameter

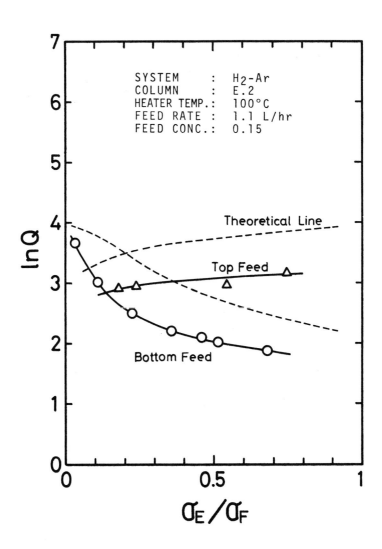

Fig. 7 Separation factor in large wall-
distance column E.2

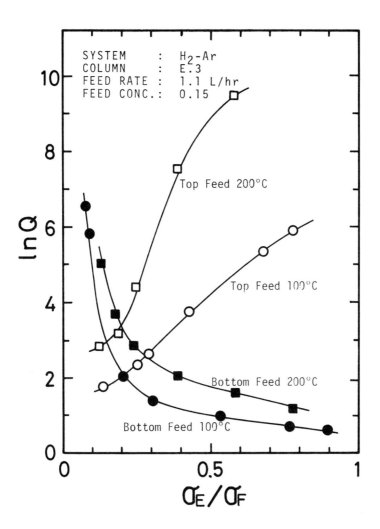

Fig. 8 Effect of temperature level on
separation factor

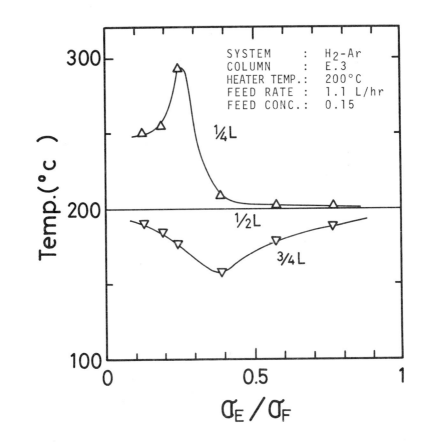

Fig. 9   Temperature distribution in the axial
direction of the column

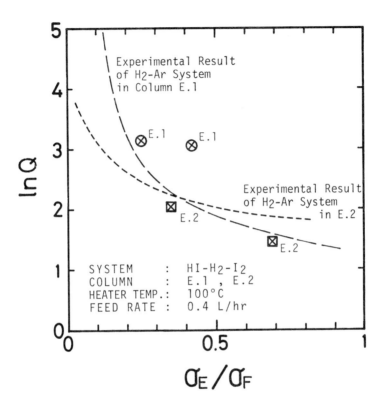

Fig. 10 Separation factor for hydrogen
iodide system

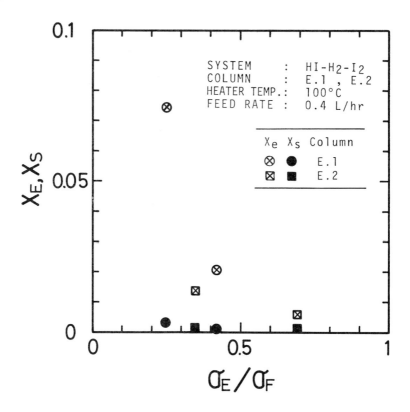

Fig. 11 Hydrogen concentration in products
of hydrogen iodide system

HYDROGEN PRODUCTION AT INTERMEDIATE TEMPERATURES BETWEEN 220 °C AND 320 °C,
BY ELECTROLYTICAL DECOMPOSITION OF INJECTED WATER AT ATMOSPHERIC PRESSURE
IN A MIXTURE OF MOLTEN HYDROGENOSULPHATES AND SULPHATES

B. Para
Ecole Centrale des Arts et Manufactures
92290 Châtenay-Malabry, France
J. Galland
Université Paris XI
P. Azou
Ecole Centrale des Arts et Manufactures
92290 Châtenay-Malabry, France

ABSTRACT

The proposed technique of hydrogen production is an electrolytical decomposition of injected water at atmospheric pressure in a mixture of molten
hydrogenosulphates and sulphates, at intermediate temperatures (220°C –
320 °C).
We have studied the influence of the temperature and hydratation rate, the
influence of continuous cycles of polarization and potential sweeping rate;
we have also looked for the ohmic drop problem.
At last, behaviour of different metals and alloys has been studied.
With respect to the free corrosion, only tantalum seems up to now to be
suitable to the reactor protection.
With respect to the cathodic electrode behaviour, polarization curves have
been achieved at different sweeping rates. All the studied metals and alloys
show a fast cathodic current decay occuring for different potential and
current conditions for every studied metal. This great disadvantage for hydrogen production could be overcome by appropriate current chopping method.
Moreover, if the formation of an hydrogenated compound would be confirmed,
a new opportunity for this technique of electrolysis could be envisaged,
which would allow an important and reversible hydrogen storage in solid
phase on massive materials.

I. PRINCIPLE

It is known that during the electrolysis of various acid molten electrolytes, hydrogen is formed on a metallic cathode.
In the case of potassium hydrogenosulphate, Tajima and al. [1] have observed that during the electrolysis of this salt, hydrogen and oxygen are
formed on the electrodes in the ratio of 2 to 1 ; this electrolytical production is achieved with a reversible deshydratation of the hydrogenosulphate;
so a continuous hydratation could allow an electrolytical production of
hydrogen with interesting energetical conditions.

1107

II. MAIN ADVANTAGES

Electrolytical production of hydrogen in molten salts baths at atmospheric
pressure and intermediate temperature offers two main advantages compared
with the classical electrolytical productions in aqueous solutions :

Firstly, the work range of temperatures is included between 220 °C and
500 °C.
- According to the thermodynamic rules, an increase in temperature allows
to decrease thermodynamical voltage of water decomposition ; from 1.23 V
at 25 °C to 1.05 V at 227 °C (at p= 1 atm).
- According to kinetic rules, an increase in temperature considerably decreases the electrodes overvoltages related to the irreversibilities of
this system.

Secondly, high current densities are achieved ; molten salts are always
highly ionised media ; their ionic resistivity often falls under 1 Ω-cm.

III. ELECTROLYSIS CHARACTERISTICS

III.1. Nature and composition of the bath

The choice of the nature and composition of the bath (NaHSO4 39%, KHSO4
44,3%, Na2SO4 16,7%, in weight) has been determined from the following
considerations :

The sodium and potassium hydrogenosulphates have good potential qualities
of hydrogenation ; actually their partial dissociation can be written :

$$HSO_4^- \rightleftharpoons H^+ + SO_4^{--}$$

The transformation in disulphate with water elimination must be added :

$$2\ HSO_4^- \rightleftharpoons S_2O_7^{--} + H_2O\uparrow$$

The experimental achievement of the stability of the molten salts media is
easy to obtain with stationnary conditions of hydratation.

The melting points of the alcaline hydrogenosulphates are low : the sodium
hydrogenosulphate melts at 186 °C, the potassium hydrogenosulphate at 207°C,
but their eutectic melts at 125 °C [2] , so we could hope to work in a
large range of temperatures since the sulphates decomposition in SO3 only
happens at 500 °C [2] , according to the reaction :

$$S_2O_7^{--} \rightleftharpoons SO_4^{--} + SO_3\uparrow$$

However, the corrosion problems arising from the high acidity of the medium
have led us to work at 300 °C in conditions close to the
acid eutectic saturation by the neutral sodium sulphate [4].
So, the useful range of temperatures decreases to 220 - 500 °C ; this disad-

vantage is counterweighted by the increase of the corrosion resistance of
the materials in the bath.
We have chosen the neutral sodium sulphate the conductivity of which is
much higher than that of the potassium sulphate, as the values here under
mentioned show [2] :

$$K_{KHSO_4, \; 300 \; °C} = 0,141 \; \Omega^{-1} \; cm^{-1}$$

$$K_{NaHSO_4, \; 310 \; °C} = 0,256 \; \Omega^{-1} \; cm^{-1}$$

$$K_{K_2SO_4, \; 1200°C} = 2,304 \; \Omega^{-1} \; cm^{-1}$$

$$K_{Na_2SO_4, \; 1200°C} = 1,494 \; \Omega^{-1} \; cm^{-1}$$

III.2. Electrochemical aspects

We have considered the hydrogen production by electrolysis in the medium
$KHSO_4$, $NaHSO_4$, Na_2SO_4, between two metallic inalterable electrodes of Pt.
Na^+ or K^+ reductions are not observed in the studied potential range (above
cathodic polarization of -3 volts refered to Ag/Ag_2SO_4 reference electrode).
Water is a weak acide in this medium, so the gazeous release of hydrogen
at the cathode can only be assigned to the HSO_4^- reduction [5] according
to :

$$2 \; HSO_4^- \; + \; 2e \longrightarrow H_2\uparrow + \; 2 \; SO_4^{--}$$

On the anode, the oxydation reaction leads to an oxydation release (sulphur
at oxydation level VI can't be oxydized); a disulphate ion is formed during
oxydation of the oxygen bound in a SO_4^{--} ion :

$$2 \; SO_4^{--} \longrightarrow S_2O_7^{--} \; + \; 1/2 \; O_2\uparrow + \; 2 \; e$$

As soon as the salts melt, the bath deshydratation occurs as :'

$$2 \; HSO_4^- \rightleftharpoons S_2O_7^{--} \; + \; H_2O\uparrow$$

The combined action of this deshydratation and that of the electrolysis,
give rise to fast decomposition of the bath ; water injection allows to
stabilize the composition by displacing the previous equilibrium ; actually, with each imposed partial pressure of water vapour, are associated
fixed proportions of disulphate and sulphate.
We can write :

for the cathodic reaction :

$$2 \; HSO_4^- \; + \; 2 \; e \longrightarrow H_2\uparrow + \; 2 \; SO_4^{--}$$

for the anodic reaction :

$$2 \ SO_4^{--} \longrightarrow S_2O_7^{--} + 1/2 \ O_2\nearrow + 2 \ e$$

a water injection regenerates the bath :

$$S_2O_7^{--} + H_2O \longrightarrow 2 \ HSO_4^-$$

(1) + (2) + (3) an finally equivalent to :

$$H_2O \longrightarrow H_2\nearrow + 1/2 \ O_2\nearrow$$

III.3. Experimental technique

Electrolysis cell

A glass cell holding about 1.8 litre of the mixed salts is put in a furnace allowing to work up to 500 °C ± 5 °C. The temperature is measured in the bath.
The injected water brings about stirring ensuring an uniform temperature in the cell.
The flowmeter regulates the water flow between 0.2 ml/mn and 38 ml/mn ; the effective pressure of injected water is about 3.10^3 Pa i.e. 3% of the atmospheric pressure.

Reference electrode

We have used a second kind electrode made of a silver wire in equilibrium with a sodium and potassium hydrogenosulphate solution saturated with sodium sulphate and silver sulphate [5], so the potential electrode indicates the SO_4^{--} activity ; the electrolytical function is realised by a micro-capillary.
To prevent the solvent deshydratation, the electrode is filled so as to form a barrier of solid salt at its upper part.
At last, because of the temperature gradient which induces a thermopile in the absence of stirring, the silver wire is protected by an isolating ceramic.

IV. STUDY OF THE MAIN PARAMETERS OF THE HYDROGEN PRODUCTION

IV.1. Temperature

The polarization curves set up on a platinum wire (1 cm^2 of side surface) show that, when the temperature rises, the thermodynamical voltage of water decomposition decreases as well as the overvoltages related to the reversibilities of the system (see fig.1).
We must notice that for the lower temperature (220 °C), neutral sulphate precipitates and the chemical composition of the molten salt is no longer the previously given composition in weight.

IV.2. Hydratation rate

The hydratation influence is shown on figure 2.
Measurements of electrolyte resistivity realized in an hydrated and non
hydrated medium are not accurate enough to show any difference. Probably,
the expected decrease of the resistance electrolyte with increasing hydratation is counterweighted by the bubble effect of the injected water.

IV.3. Continuous polarization cycles

A frequent problem, during experiments on solid metallic electrodes in
molten salts media, is the large increase of hydrogen absorption by metal
under cathodic polarization; as we will see it later this absorption may
highly alter the electrode behaviour.
So, we have tried to get regularly the same initial state of the metallic
surface, carrying out one anodic polarization.
Figure 3 shows that, repeated cathodic and anodic cycles, for high sweeping rate, are likely to alter the surface properties.
In the anodic range, curves dispersion is of little account; but in the
cathodic range, we observe good reproductibility of characteristic values
of potential as the upper current increases.
We can remark that for other study conditions, (sweeping rate or electrode
nature), the evolution of potential-current curves may be more important.

IV.4. Sweeping rate of the potential

Under cathodic polarization, we observe on the figures 4, 5 and 6 that the
upper current density is so much lower as the sweeping rate is smaller.
Otherwise, whatever the sweeping rate is, (600 mV/mn, 60 mV/mn, steady
stade curve) we observe a large decay of current density ; we will come
back on this fact during the comparative study of different metals
(§V.2), but we can already now notice that the tantalum behaviour differs
from that of platinum and iron, since the upper current achieved on the
steady state curve (reached by steps of 250 mV each twenty minutes) is much
higher.
Under anodic polarization, for the curves of the liberated oxygen get on
the platinum (see fig.4) the current densities are much less altered by the
sweeping rate variation.
All these results can be understood as follows : for potentiokinetic curves,
the electrochinecial kinetics is imposed by the slower process ; so for the
different sweeping rates, deviations appear on the obtained curves since
the sweeping rate changes the relative magnitude of the different limiting
states to be considered, such as the mass transport near the electrode. In
the present state of our study the nature of these various states is not
yet exactly known.

IV.5. Ohmic drop correction

We have get polarization curves on platinum with electrodes whose shape are
cylinders or sheets ; we can notice appreciable differences about the

obtained current densities (see fig.7).
Measurements of ohmic drops realized by classical appropriate transient
method using a current interruptor [6] have been achieved to specify the
part of the electrode shape and especially of its radius, in term of electrolyte resistance Re ; results show an increase of Re when we use a platinum electrode in shape of wire of 1 cm^2 (Re $\simeq$ 0.8 Ω) or a platinum
electrode in sheet shape, of the same surface (Re $\simeq$ 1.2 Ω).
In the same way, electrolyte resistance is higher for the cylinder than
for the wire, since the measured values are of the same range as the cylinder surface is twice the wire surface.
We have experimentally noticed that whatever are the relative positions of
electrodes, Re value is not changed ; that is consistent to the low nume-
rical values of Re.
At last, we can notice that the energetical losses in the cell by JOULE
effect is small (about 0.2 watt for steady state).

V. BEHAVIOUR OF SOME METALS AND ALLOYS

V.1. Behaviour under free corrosion

Few tests have been made to precise what materials could compose the cell
reactor ; results are given in table 1.
Only tantalum, whose chemical corrosion resistance is well known following
oxyde formation, seems up to now to be suitable for electrolysis cell
achievement.
We have not taken into account platinum because of its high cost.

V.2. Relative behaviour of different metals and alloys for cathodic
overpotentials

To compare the behaviour of different metals and alloys, we have plotted
polarization curves at 600 mV/mn for an hydratation rate of 200 ml/h ;
in these conditions it is unlikely that a too high deshydratation of the
baths occurs which will modify the kinetics of the processes involved.
Moreover, as seen in § IV.3, the anodic polarization insures comparable
studying conditions.
Figures 8 and 9 give the obtained curves for :
    Pt sheet (99,95 %)
    Fe sheet (Armco)
    Ni sheet (99,9 %)
    Ti sheet (pure industrial)
    Hastelloy B (C 0,05% ; Mn 1% ; Si 1% ; Fe 5% ; Mb 28%
            Cr 1% ; Co 2,5% ; Ni 61,45%)
    Hastelloy C (C 0,08% ; Mn 1% ; Si 1% ; Fe 5% ; W 4% ;
            Mb 16% ; Cr 15% ; Co 2,5% ; Ni 56,42%)
Figure 10 gives the obtained curves for :
    Pt cylinder (99,99%)
    Fe cylinder (Armco)
    Ta cylinder (99,99%)

These experimental curves are reproductible after a few cathodic and anodic cyles.
In the anodic range, for iron, nickel and Hastelloy alloys, we obtained a small anodic dissolution current (below 10 mA cm^2) , for the highly passivable metals, titanium and tantalum, we notice a residual current smaller than one mA/cm^2, probably resulting from an oxyde formation; at last, in the case of platinum the anodic curve is that of the oxygen release.
In the cathodic range, the general shape of the curves is the same :

The potentials of the begining of cathodic reaction, are not much changed for a metal or another, consistent to our knowledge of the minimal over-voltages of the hydrogen production on these metals[7].

The different slopes observed on the curves cannot be connected to a characteristic property of the studied materials.

We have systematically observed a current decay above a determined cathodic overpotential ; the values of potential and current for which the phenome-nom appears, highly change from a metal to another.

We could connect this breakdown of the electrolyse to the phenomenom called "cathodic effect" which is observed in somes electrolytes molten salts [8]. Various hypothesis are done about this fact ; for example, it seems, the occurence of electrodes effects is due to electrolyte heating by JOULE effect, until a water vapour coverage is formed. Then, an insulation of the electrode from the bath is created.
However, that hypothesis is not sufficient to explain the difference of behaviour between two electrodes of same shape and surface for our studied metals ; for example, if we compare the platinum and tantalum behaviour : the measurements of the electrolyte resistance done for these two electrodes give very close values when the conditions of the cathodic current decay are not at all comparable (see fig.10).
For platinum, the observed decay is very fast, only lasts for a few seconds, while for tantalum we notice a gradual decay for several hours ; the small difference in thermal conductivity between these two metals (for Pt 0.17 c.g.s, for Ta 0.13 c.g.s.) cannot explain this different behaviour.
Finally, the figure 9 for iron, shows the obtained curves if the values of the decay potential is not reached ; the reverse curve is practically super-imposed to the direct one ; we may suppose that a hydrogenated compound (may be an hydride) is formed, which would play a similar part to that of an oxyde in the case of passivation mechanism. However, the existence of such compound is very difficult to prove since when we return to the room temperature and atmospheric pressure conditions,thermodynamic instability occurs.
Yet, the fact that blocking keeps on along the reverse curves allows to make the study of such hydrogenated compound feasible.

CONCLUSION

This new approach of the behaviour of different metals and alloys, able to compose the electrodes or the cell, for hydrogen production, by the electrolysis of injected water in molten sulphates allows to draw the following conclusions :

All the studied metals and alloys show a fast current cathodic decay occuring for potential and current conditions different from a metal to another; if it could be confirmed that this decay is bound to the formation of an unsteaday hydrogenated compound, it could be necessary to study the influence of a chopping current method. Somes tests have shown that a short current interruption or getting the electrode out of the bath for a few seconds were sometimes enough to allow to reach again much higher current densities.

This technique of hydrogen production seems to meet with a great obstacle due to the cathode blocking during electrolysis.

However, if effectively this blocking arises from a great absorption of hydrogen, complementary measurements of the absorbed hydrogen could corroborate it. Then, a new opportunity for this technique of electrolysis could be considered, which would allow an important and reversible storage of hydrogen in solid phase of massive materials.

This technique would offer some advantages compared with the powders one generally used in the case of gazeous hydrogenation.

At last, with respect to the technology for the intended working temperatures (250 - 320°C) tantalum could compose an interesting protection material for the electrolysis cell.

REFERENCES

1. S. Tajima, M. Soda, T. Mori and N. Baba. Electrochim. Acta 1, 205 (1959)
2. G.J. Janz, Molten oalto Handbook. Academic Press, New York, 1966.
3. Ben Hadid, Thèse 3° cycle, Chimie analytique, Université Paris VI (1975)
4. Elkholy, thèse 3° cycle, Université Paris VI (1977)
5. J.P. Vilaverde, thèse 3° cycle, Chimie analytique, Université Paris VI (1973)
6. P. Morel, thèse de Doctorat d'Etat, Sciences Physique, Université Paris (1968)
7. J.O.M. Bockris, S. Srinivasan. Fuel cells Their Electrochemistry - Mc Graw Hill, New-York (1970)
8. C. Guilpin, J. Garbaz Olivier, Thermochimica Acta, 13, 467-470 (1975)

TABLE 1

METAL	LOSS WEIGHT mg/dm^2/day
Pt	0
Ta	+ 16
Ti	- 900
HC	- 6000
Mo	-11400

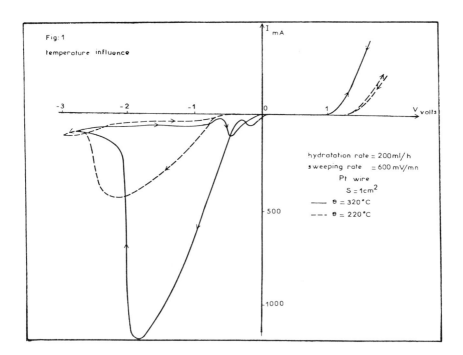

Fig: 1

temperature influence

hydratation rate = 200ml/h
sweeping rate = 600 mV/mn
Pt wire
S = 1cm²
——— θ = 320°C
--- θ = 220°C

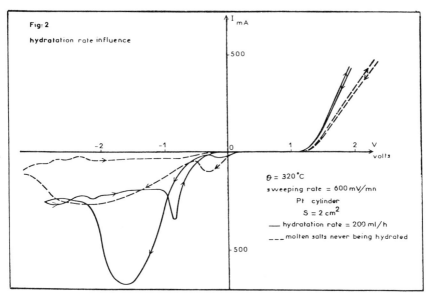

Fig: 2

hydratation rate influence

θ = 320°C
sweeping rate = 600 mV/mn
Pt cylinder
S = 2 cm²
——— hydratation rate = 200 ml/h
--- molten salts never being hydrated

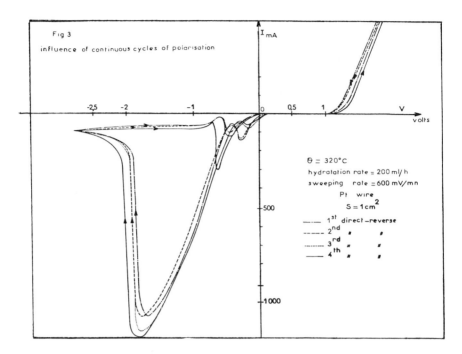

Fig 3

influence of continuous cycles of polarisation

$\theta = 320°C$

hydratation rate = 200 ml/h

sweeping rate = 600 mV/mn

Pt wire

$S = 1 cm^2$

—·— 1st direct –reverse

––––– 2nd " "

·········· 3rd " "

——— 4th " "

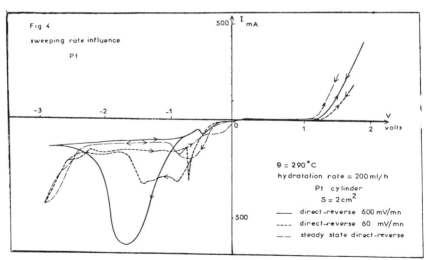

Fig 4

sweeping rate influence

Pt

$\theta = 290°C$

hydratation rate = 200 ml/h

Pt cylinder

$S = 2 cm^2$

——— direct –reverse 600 mV/mn

––––– direct –reverse 60 mV/mn

—— steady state direct –reverse

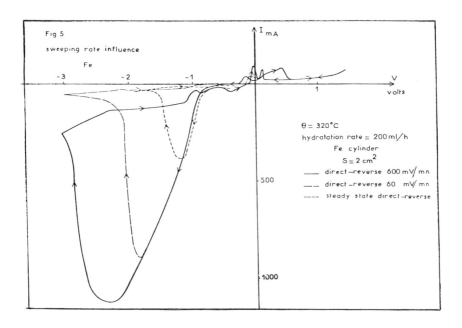

Fig 5

sweeping rate influence

Fe

$\theta = 320°C$

hydration rate $= 200\,ml/h$

Fe cylinder

$S = 2\,cm^2$

—— direct—reverse 600 mV/mn

— — direct—reverse 60 mV/mn

----- steady state direct—reverse

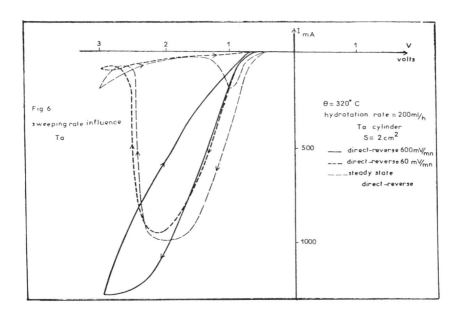

Fig 6

sweeping rate influence

Ta

$\theta = 320°\,C$

hydration rate $= 200\,ml/h$

Ta cylinder

$S = 2\,cm^2$

—— direct-reverse 600 mV/mn

— — direct—reverse 60 mV/mn

— — steady state
direct—reverse

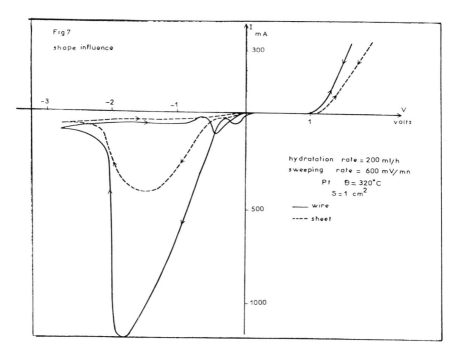

Fig 7

shape influence

hydratation rate = 200 ml/h
sweeping rate = 600 mV/mn
Pt   θ = 320°C
S = 1 cm²
—— wire
---- sheet

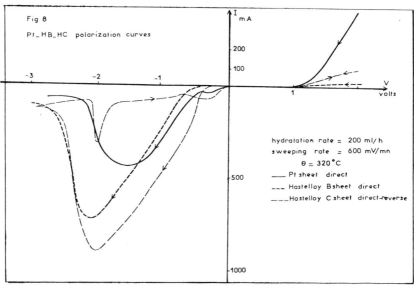

Fig 8

Pt_HB_HC polarization curves

hydratation rate = 200 ml/h
sweeping rate = 600 mV/mn
θ = 320°C
___ Pt sheet direct
___ Hastelloy B sheet direct
___ Hastelloy C sheet direct-reverse

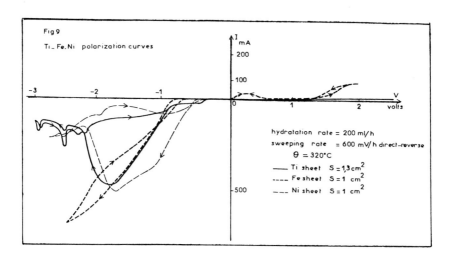

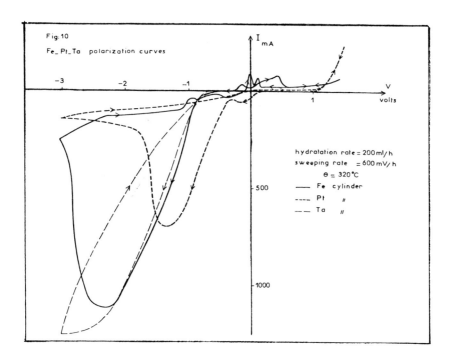

ON THE STUDY OF HYDROGEN PRODUCTION FROM
WATER USING SOLAR THERMAL ENERGY

S. Ihara
Electrotechnical Laboratory
Tanashi, Tokyo, Japan

ABSTRACT

Theoretical thermal efficiency of hydrogen production by one-step water splitting utilizing solar heat at high temperatures is calculated. Carnot efficiency is assumed for the conversion of effective work input, and the solar collection efficiency is considered for the total energy input. The overall efficiency shows its maximum in the range of temperature between 1500 and 2700 K depending upon the solar concentration ratio and the method of product separation. The technical feasibility of direct splitting method is discussed on the basis of those calculated results.

INTRODUCTION

Various methods to produce hydrogen from water utilizing solar energy have been proposed and assessed in recent years. Direct thermal splitting of water, a one-step high temperature process, has been mentioned among the rest [1-5]. But the efficiency of the process has not been well defined, partly because of the lack of a consistent discussion on its thermodynamics.

In this paper, the thermodynamic state of dissociated water vapour system and the thermal efficiency of one-step process to produce hydrogen are described. Also the feasibility of such process when solar heat is used as the input energy is discussed.

THERMODYNAMICS OF THE PROCESS

It is known that the dissociation of water proceeds in two steps:

$$H_2O \rightarrow HO + H \quad ; \quad HO \rightarrow H + O$$

In the dissociated system in a certain range of temperature, atomic particles are recombined to produce molecular particles:

$$2H \rightarrow H_2 \quad ; \quad 2O \rightarrow O_2$$

1121

The equilibrium constants of these four reactions, mol fractions of the six components in the dissociated water, and the thermodynamic properties of each component can be calculated once the partition functions of all components are known. The calculation methods are described in elsewhere [2,6].

According to the elementary thermodynamics, enthalpy, entropy, and free energy of the dissociated water vapour are calculated by

$$H_r = \sum_1^6 x_i H_i \tag{1}$$

$$S_r = \sum_1^6 \{x_i S_i - R x_i \cdot \ln(x_i)\} \tag{2}$$

$$G_r = H_r - TS_r \tag{3}$$

where R is the universal gas constant, x the mol fraction, and the subscript i designates the component as follows:

$$1 \rightarrow H_2O; \quad 2 \rightarrow HO; \quad 3 \rightarrow H; \quad 4 \rightarrow O; \quad 5 \rightarrow H_2; \quad 6 \rightarrow O_2 \tag{4}$$

Molecular weight of the dissociated water is

$$M = 18.016 x_1 + 17.0074 x_2 + 1.008 x_3 + 16 x_4 +$$
$$2.016 x_5 + 32 x_6 \tag{5}$$

Then, the moles of the dissociated water vapour system, N, and the moles of each component in the system, ni, are

$$N = 18.016/M, \qquad n_i = x_i N \tag{6}$$

The thermodynamic state of the products resulting from water splitting, i.e. hydrogen and oxygen which are also dissociated in a certain temperature range, can be calculated by the similar way. The sum of the enthalpy of the products is

$$H_p = \sum_3^6 n_{pi} H_i \tag{7}$$

in which npi designates the moles of the product component i. The entropy and the free energy of the product system are

$$S_p = \sum_3^6 n_{pi} S_i + S_m \tag{8}$$

and

$$G_p = \sum_3^6 n_{pi} G_i - S_m \tag{9}$$

where Sm is the entropy of mixing. When a stoichiometric mixture of the products is assumed it is

$$S_m = -R\sum_{3}^{6} n_{pi} \ln(n_{pi}/n_p), \qquad n_p = \sum_{3}^{6} n_{pi} \qquad (10)$$

and when hydrogen atoms and molecules are separated from oxygen atoms and molecules, it is

$$S_m = -R[n_{p3}\ln\{n_{p3}/(n_{p3} + n_{p5})\} + n_{p5}\ln\{n_{p5}/(n_{p3} + n_{p5})\}$$

$$+ n_{p4}\ln\{n_{p4}/(n_{p4} + n_{p6})\} + n_{p6}\ln\{n_{p6}/(n_{p4} + n_{p6})\}] \qquad (11)$$

The computed results for the pressure of 1 atm. is illustrated in Fig.1 to show the energy requirement of an isobaric hydrogen production system. The energy is measured from -68.3 Kcal, which is the enthalpy of liquid water at 298 K, in the figure. The lower and the upper solid curves are respectively the enthalpy of water vapour Hr and that of products Hp. ΔHr shows the required thermal energy for heating water up to a specified temperature, and the difference ΔHpr (= Hp - Hr) shows the required energy input to produce hydrogen and oxygen from the water vapour. This change in enthalpy can be brought about by the combination of work and heat [7] since

$$\Delta H_{pr} = (G_p - G_r) + T(S_p - S_r) = \Delta G_{pr} + T\Delta S_{pr} \qquad (12)$$

The dashed lines between the two solid curves show the calculated quota of work and heat. The vertical distance designated as TΔSm is obtained by multiplying the temperature to the difference of equation (10) and (11), which means the minimum theoretical work of separation. The uppermost dashed line is drawn in order to exemplify the amount of the work of compression: the vertical distance designated as Wc shows the input power required for reciprocating machines to compress 1 mol of hydrogen and 0.5 mol of oxygen from 1 atm. to 30 atm. at 300 K. The power required is

$$W_c = \{nRT\gamma/(\gamma - 1)\}\{(P_2/P_1)^{(\gamma-1)/\gamma} - 1\} \qquad (13)$$

in which T = 300 K, P2 = 30 atm., P1 = 1 atm., γ = 1.405 for hydrogen and 1.395 for oxygen, and n = 1 mol for hydrogen and 0.5 mol for oxygen have been assumed in the calculation.

The key to the research on one-step water splitting process is to find out the method how to bring about the change in free energy, ΔGpr. We cannot specify at present a technique to decompose water and to separate the products, but it is suggested that electrical and/or mechanical energy input can cause the change in free energy. Probably a technique, so to speak a water vapour electrolysis at very high temperature would be a comprehensive example. In such case, the efficiency of electric power generation must be taken into account to estimate the total input energy. Another example to use pumping work through a selective membrane suggests itself. In this case, the input is a mechanical power, and the net amount of the required energy can be estimated by using the well known expression for isothermal work of pumping, $nRT \cdot \ln(P2/P1)$. Replacing P2 and P1 respectively with the partial pressure of the product component, $x_{pi}P$, and with that of the reactant component, $x_i P$, we obtain for an ideal condition

$$W_i = \sum_3^6 n_{pi} RT \cdot \ln(x_{pi}/x_i) \tag{14}$$

It should be noted that the work of pumping, W_i, is identical with the change in free energy, ΔG_{pr}, within the accuracy of calculation. This implies that, even from partly dissociated water vapour, stoichiometric amount of hydrogen and oxygen can be produced by an isothermal and isobaric continuous extraction of the products. But, an actual pumping process is influenced by volumetric efficiency, pressure drop, and friction loss, all of these will make the required work of pumping far larger than W_i especially at the state of low degree of dissociation.

THERMAL EFFICIENCY

The definition of thermal efficiency of an one-step water splitting process used here is

$$\eta = \Delta H_0/(Q + W/\varepsilon_1\varepsilon_2 + W_c/\varepsilon_1) \tag{15}$$

where $\Delta H0$ is the heat of formation of water at 298 K and 1 atm. (=68.3 Kcal/g mole H2), Q the thermal energy input, W the effective work input, Wc the work of compression (Eq.(13)), ε_1 the efficiency of converting thermal energy to work, and ε_2 the efficiency of work input. The first and second law of thermodynamics may be written to obtain [8]

$$Q + W = \Delta H_0 + q_r \tag{16}$$

$$(q_r + T_0 \Delta S_0)/T_0 \geq (Q + q_u)/T \tag{17}$$

where qr designates the heat rejected from the process at the temperature of T0, $\Delta S0$ the entropy change for hydrogen and oxygen recombination at 1 atm. and T0, and qu the heat recovered and reused in the process. As are discussed in the preceding section and summerized in Fig.(1) it can be shown that

$$W = \Delta G_{pr} \tag{18}$$

$$Q + q_u = \Delta H_r + T\Delta S_{pr} = H_r + 68.3 + T\Delta S_{pr} \tag{19}$$

Combining equations (15) to (19) yields

$$\eta_p = \Delta H_0 / \{\Delta H_0 + q_r + \Delta G_{pr}(1 - \varepsilon_1 \varepsilon_2)/\varepsilon_1 \varepsilon_2 + \bar{W}_c/\varepsilon_1\} \qquad (20)$$

$$q_r \geq (T_0/T)(H_r + 68.3 + T\Delta S_{pr}) - T_0 \Delta S_0 \qquad (21)$$

Equation (20) gives the efficiency of upper limit which corresponds to the reject heat of lower limit given by equation
(21). Assuming $T0 = 298$ K, $\Delta S0 = 0.039$ Kcal/g mole K, and
$\varepsilon_1 = 0.7(T - T0)/T$, computation has been made and the results
are shown in Fig.(2) to Fig.(5). The coefficient 0.7 of the
Carnot efficiency was taken so as to approximate it to a more
realistic value.

Fig.(2) shows the results when the efficiency of work input,
ε_2, is assumed to be unity. It also shows the effect of system pressure. It is seen that work of compression, Wc, becomes larger when the system pressure is decreased and, thus,
the efficiency is lowered.

Fig.(3) shows the results when the effective work input, W,
is supplied by an isothermal pumping of products. The efficiency of the pumping process may be written

$$\varepsilon_i = n_{pi} RT \cdot \ln(x_{pi}/x_i)/\{n_{pi} RT \cdot \ln(x_{pi}/x_i) + W_1(n_{pi}/n_i)\} \qquad (22)$$

where W1 designates the associated losses. Combining equations
(14) and (22), the efficiency of work input, ε_2, may be written

$$\varepsilon_2 = \Delta G_{pr}/[\sum_3^6 \{1 + W_1/n_i RT \cdot \ln(x_{pi}/x_i)\} n_{pi} RT \cdot \ln(x_{pi}/x_i)] \qquad (23)$$

Assumption has been made for the calculated results shown in
Fig.(3) that W1 is the order of 1% of the isothermal pumping
work for 1 mole of product. Notwithstanding such an optimistic
assumption, mechanical work input seems to be inefficient at
the temperature where the degree of dissociation of water is
too small.

The calculated results show that high temperature energy of
above 1000 K is required to obtain high thermal efficiency at
the one-step water splitting process. Only a solar furnace
can provide such energy at present. Therefore, solar energy
utilization efficiency by the process should be evaluated here.
The efficiency of a solar concentrator is defined as [9]

$$\eta_s = 1 - \varepsilon \sigma T^4/\alpha \rho IC \qquad (24)$$

where ε is the emissivity, σ the Stefan-Boltzmann constant of
5.67×10^{-8} W/m^{2}K^4, α the black body absorption coefficient of
a water heating surface, ρ the mirror efficiency, and I the
intensity of the solar energy redirected by the mirror. C is
the concentration ratio defined by the total area of the concentrator mirror divided by the solar image size at the water

heating surface. If all of the required energy input is supplied by the solar concentrator, the solar utilization efficiency, η, of the hydrogen production system can be estimated
by the product of equation (20) and (24):

$$\eta = \eta_p \cdot \eta_s \qquad (25)$$

Fig.(4) shows the calculated solar utilization efficiency for
the system pressure of 1 atm. for various value of solar concentration ratio, C. Assumed constants are $\varepsilon = \alpha = 0.9$, $\rho =$
0.75, and $I = 600$ W/$\bar{m}^2$. The upper curves are the results when
$\varepsilon_2 = 1$ is assumed, and the lower ones are that when ε_2 is calculated by equation (23). The efficiency of one-step process
itself increases by a temperature increase as is shown in Fig.
(2) and (3), whereas according to equation (24) radiation loss
detoriorates the system efficiency at a very high temperature.
Therefore, it is seen that there is an optimum temperature
where the efficiency of a specified system reaches maximum.

Fig.(5) shows the efficiency calculated with the same procedure for the system pressure of 0.1, 1, and 10 atm., in which
concentration ratio $C = 13000$ is fixed. Since the pumping
work is largely affected by the degree of dissociation of
water, the efficiencies of lower curves decrease in the order
of increasing system pressure, while those of upper ones
increase in the order of increasing system pressure.

DISCUSSION

The thermal efficiency of one-step water splitting process
could be made high if an efficient way to bring about the
change in free energy in the process. We cannot specify a concrete scheme to do so at present, but the efficiency can be
estimated by the equilibrium thermodynamics of water vapour.
The calculated results shown in Fig.(2) are for the assumed ef-
fciency of work input of $\varepsilon_2 = 1$, for which a concept of electrolysis of water vapour at high temperature might be suggested in principle. In such case the reactant is not necessarily
be a dissociated system, and high efficiency is calculated
even at the temperature around 1000 K. The process which works
at such temperature range would be desirable from the viewpoint of solar energy utilization, since the solar collector
can provide the heat input at high efficiency with a moderate
concentration ratio as is shown in Fig.(4).

When a pumping power is used to provide the effective work
input for the process, the temperature must be made much higher
as is shown in Fig.(3), and it can attain high efficiency above
2000 K. But, above this temperature the efficiency of solar
collector decreases with higher temperature until it reaches

zero at

$$T = (\alpha\rho IC/\epsilon\sigma)^{0.25} \tag{26}$$

Therefore, as is seen in Fig.(4), the concentration ratio should be made higher than 4000 in order to maintain high overall efficiency. The solar collector must have the concentration ratio higher than 1000 to 4000 since the required temperature is as high as a few thousands degree Kelvin. At the same time, the required land area for the collector must be the order of 10^5 m^2 since the hydrogen production rate of a commercial plant should be the order of tens of megawatts of hydrogen product. A system of large number of field mirrors with central-receiver tower may achieve such high concentration ratio at the high power capability. For example, the concentration ratio of order 1000 to 2000 will be attained by the superposition of images from the heliostat mirrors. An additional concentration ratio will be attained by the paraboloid placed at the tower, or by the curvature of each heliostat itself.

Also, Fresnel lenses would be available as the solar energy collector. High temperature solar image is produced at each focal point of the lens, and optical guides should be used to integrate these dispersed images to use their heat at a single reactor for hydrogen production. It has been estimated by calculation that fused-silica-core optical fibers transmit solar radiation over distances of about 40 m with the efficiency higher than 80% [10]. The power carrying capability of the present high quality optical fiber is said to be the order of 10 Kw/cm^2, though the price is expensive. The most important aspect of the application of optical guides is that it offers the possibility of eliminating the central-receiver tower from a solar collector system, and thus the problems associated with the accuracy of the construction and control for tracking the sun and focusing its images can be alleviated. The device for terminating optical guides, as well as the cost reduction, would be the major subject of its future development.

As for the separation of products at high temperatures, the possibility of Knudsen flow diffusional separation has been analyzed [5], and the use of absorption and diffusion in non-porous metals and ceramics has been suggested [2]. Both are now being tested by the author using a Xenon-arc high temperature image furnace to simulate the solar collector. The materials tested so far are Nb, Ta, W, Zr_2O_3-CaO, and Al_2O_3, and a small amount of hydrogen has been separated from the dissociated water vapour of about 1800 K. However, the feasibility of the application of these materials to the membrane is still uncertain since the metals are chemically unstable in the water vapour of above 1000 K, and the ceramics are easily broken by a thermal stress.

REFERENCES

[1] N. C. Ford and J. W. Kane: "Solar Power", Bulletin of the Atomic Scientists, 27, Oct. 1971, p.p.27-31.

[2] S. Ihara: "Feasibility of Hydrogen Production by Direct Water Splitting at High Temperature", Paper presented at the 1st World Hydrogen Energy Conference, 1976.(Conference Proceedings p.p.5B55-5B70). Also, International Journal of Hydrogen Energy, 3, 1978(in Press).

[3] E. Bilgen, M. Foex, F. Sibieude and F. Trombe: "Use of Solar Energy for Direct and Two-Step Water Decomposition Cycles", International Journal of Hydrogen Energy, 2, No.3, 1977, p.p.251-257.

[4] T. Nakamura: "Hydrogen Production from Water Utilizing Solar Heat at High Temperatures", Solar Energy, 19, No.5, 1977, p.p.467-475.

[5] E. A. Fletcher and R. L. Moen: "Hydrogen and Oxygen from Water", Science, 197, No.9, 1977, p.p.1050-1057.

[6] S. Ihara: "Approximations for the Thermodynamic Properties of High-Temperature Dissociated Water Vapour", Bulletin of the Electrotechnical Laboratory, 41, No.4, 1977, p.p.259-280.(in Japanese).

[7] R. E. Chao: "Thermochemical Water Decomposition Processes", Ind. Eng. Chem., Prod. Res. Develop., 13, No.2, 1974, p.p. 94-81.

[8] J. E. Funk: "Thermochemical and Electrolytic Production of Hydrogen from Water", Paper presented at the United States-Japan Joint Seminar on Key Technologies for the Hydrogen Energy System, 1975(Conference Proceedings p.p.1-34).

[9] A. F. Hildebrandt and L. L. Vant-Hull: "A Tower Top Focus Solar Energy Collector", Paper presented at the Winter Annual Meeting of the American Society of Mechanical Engineers, 1973(ASME Publication 73-WA/Sol-7).

[10] D. Kato and T. Nakamura: "Application of Optical Fibers to the Transmission of Solar Radiation", J. Appl. Phys., 47, No.10, 1976, p.p.4528-4531.

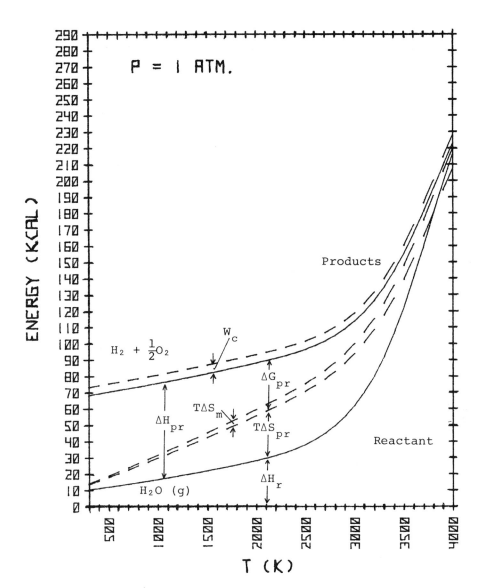

Fig. 1. Required energy for one-step
hydrogen production from water vapour
of 18 grams.

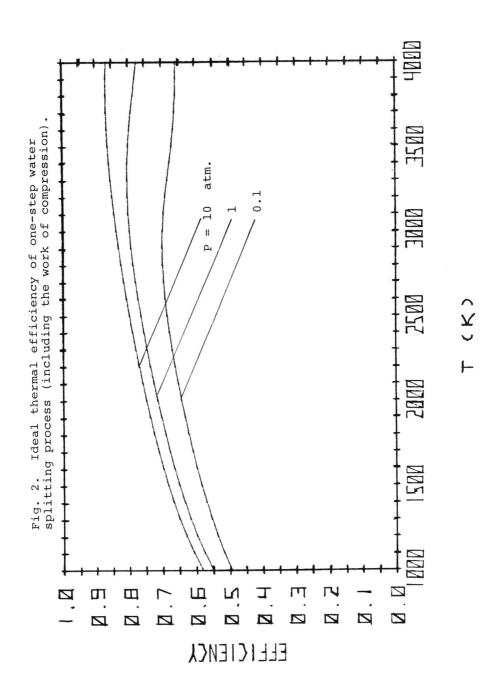

Fig. 2. Ideal thermal efficiency of one-step water splitting process (including the work of compression).

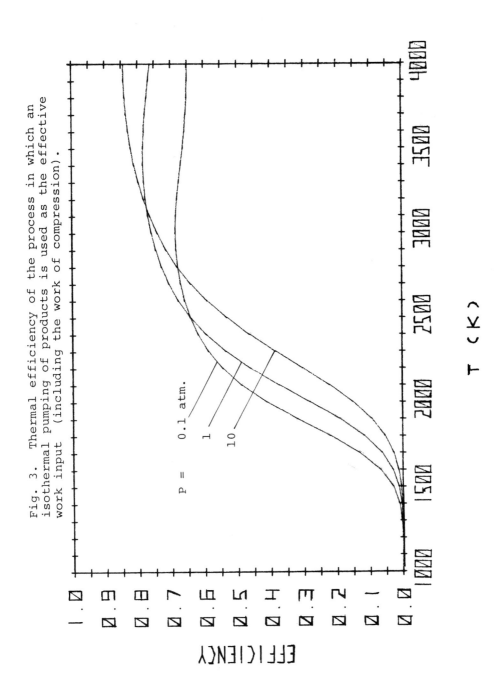

Fig. 3. Thermal efficiency of the process in which an isothermal pumping of products is used as the effective work input (including the work of compression).

1132

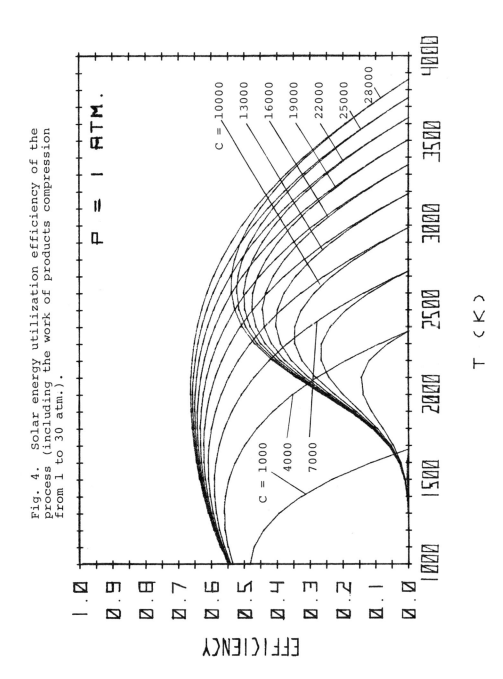

Fig. 4. Solar energy utilization efficiency of the
process (including the work of products compression
from 1 to 30 atm.).

1133

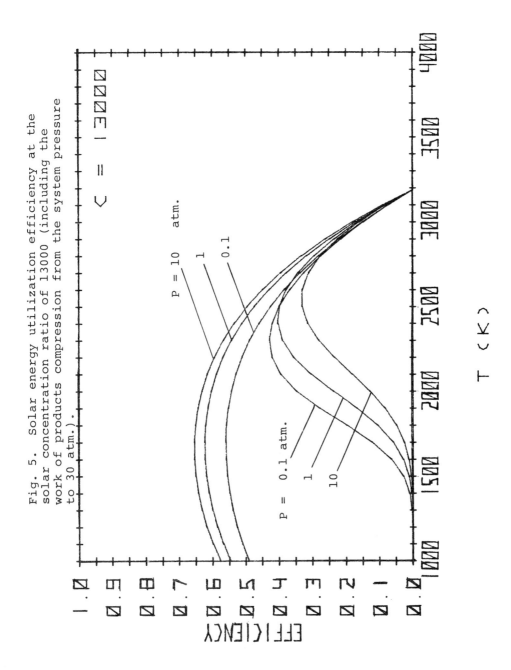

Fig. 5. Solar energy utilization efficiency at the solar concentration ratio of 13000 (including the work of products compression from the system pressure to 30 atm.).

HYDROGEN PRODUCTION FROM FUSION REACTORS COUPLED
WITH HIGH TEMPERATURE ELECTROLYSIS*

J. A. Fillo, J. R. Powell, M. Steinberg, F. Salzano,
R. Benenati (Consultant, Polytechnic Institute of NY),
V. Dang, S. Fogelson (Consultant, Burns and Roe), H.
Isaacs, H. Kouts, M. Kushner (Consultant, Burns and Roe),
O. Lazareth, S. Majeski, H. Makowitz, T. V. Sheehan
(Consultant, Brookhaven National Laboratory)
Brookhaven National Laboratory
Upton, New York 11973 USA

ABSTRACT

The decreasing availability of fossil fuels emphasizes the need to develop
systems which will produce synthetic fuel to substitute and compliment the
natural supply. An important first step in the synthesis of liquid and
gaseous fuels is the production of hydrogen. Thermonuclear fusion offers
an inexhaustible source of energy for the production of hydrogen from water.
Processes which maybe considered for this purpose include electrolysis,
thermochemical decomposition or thermochemical-electrochemical hybrid cycles.
Preliminary studies at Brookhaven indicate that high temperature electrolysis
has the highest potential efficiency for production of hydrogen from fusion.
Depending on design electric generation efficiencies of ∿40–60% and hydrogen
production efficiences of ∿50–70% are projected for fusion reactors using
high temperature blankets.

INTRODUCTION – FUSION AS A SYNTHETIC FUEL PRODUCER

Need for Synthetic Fuels Using Some Type of Inexhaustible Resource

World reserves of oil and gas are being rapidly depleted and the general
consensus is that they will be essentially gone in a short period of time.
Many of the industrialized nations, the United States for example, are now
importing a large fraction of their petroleum needs, with consequent worries
about sources of supply to maintain continued growth patterns, future prices,
and effects on the balance of payments.

It is generally believed that future energy demands which have in the past
relied on oil and gas will in the future have to be supplied by alternate
sources. This will mean an increased reliance on established energy sources
such as coal and nuclear (LWR) to meet the projected energy deficit.

Coal can supply both electricity and synthetic portable fuels, but there
appear to be concerns as the degree to which it can meet future demands, in
terms of production rates, total available resources, and possible harmful
environmental effects. For example, the potential long-range climatological
effects of large additions of CO_2 to the atmosphere are being studied.

---

* Work performed under the auspices of The Department of Energy.

Nuclear (LWR) sources, though, will supply energy primarily to generate electric power. This will help to abate some of the demands for oil and gas. However, for a number of demand sectors, practical technology has not yet been developed (and may never be) for direct electrical use. Also, nuclear (LWR) may be found wanting as an energy source in a few decades due to heavy pressure on uranium reserves. The long-range prospects for nuclear energy deployment also depend upon demonstration of terminal waste disposal technology and resolution of broader energy policy questions.

These conditions have led to an increased interest to identify a way to produce substitute fuels from the inexhaustible energy sources, that is, solar, geothermal, fission (breeder), and fusion. Thermonuclear fusion as a future inexhaustible energy source appears to have the advantage of relatively high power density as compared to alternate inexhaustible energy sources such as solar or geothermal, with the potential of favorable economics. Further, it has no apparent geographic or climatological constraints. The supply of deuterium and lithium fuels for fusion reactors appears to offer no significant resource concern. If the synthetic fuel derived from fusion energy is hydrogen, it can serve multiple functions; as a chemical feedstock for much more efficient coal liquification and gasification; ammonia production and metal-ore reduction; as an efficient source of electricity by using fuel cells; and ultimately, as a primary transportable fuel. The supply of hydrogen would be virtually unlimited since it is obtained from water.

Inexhaustible fission technologies, such as the LMFBR or GCFR can also be considered as a source for the production of synthetic fuels based on hydrogen. Here, however, the primary product is expected to be electricity, and production of synthetic fuels will probably have to follow the route of conventional electrolysis, at a relatively low efficiency compared to the potential for fusion reactors.

In contrast, fusion reactors have the potential to supply unique energy forms (e.g., radiation and very high temperature heat) that may lead to significantly increased efficiency for the production of synthetic fuels. An evaluative study of the application of fusion energy to synthetic fuel production has been recently carried out [1] which identifies a number of promising approaches.

Basic Options of Hydrogen Production from Fusion

Three basic candidate methods of hydrogen production from neutron energy have been identified: a) direct decomposition including high temperature electrolysis; b) thermochemical cycles including partial radiolysis; and c) direct radiolysis. While comparision of these methods are difficult to make, engineering judgment can be exercised so that expected relative efficiencies of the processes and relative engineering feasibility, i.e., complexity, materials compatibility and temperature limitations, mechanical design requirements, etc., can be made. Based on these subjective comparisons, preferential ranking of the above methods is possible and is

discussed in some detail in the panel study [1] on fusion synfuels
production. One of the most promising is high temperature electrolysis
which is expected to have the highest efficiency [∿50–55% based on a
conventional power cycle (40% efficient) and as high as 70% with an advanced
power cycle (60% efficient)]. Its technology is also more near at hand
compared, for example, with thermochemical or radiolysis.

The preliminary evaluations developed in Ref. [1] were the basis for choosing
high temperature electrolysis for the fusion synfuels conceptual design [2]
reported here.

Synthetic Fuels Using Fusion

The high energy neutrons from DT fusion reactions can penetrate very deeply
into materials before their kinetic energy is transformed to heat. This
unique feature of fusion energy, and the fact that ∿80% of the energy released per DT fusion reaction is carried by 14 MeV neutrons, can dramatically
increase the efficiency of electric power generation, as well as produce H_2
and H_2 based synthetic fuels at high efficiency.

This deep penetration of the primary neutrons makes two temperature region
blankets feasible. In this concept, a relatively low temperature metallic
structure is the vacuum/coolant pressure boundary, while the interior of the
blanket, which is a simple packed bed of non-structural material, operates
at very high temperatures. Separate coolant circuits are required for the
two temperature regions, as well as a thermal insulator between them.

Materials for the hot interior are capable of much higher temperatures than
HTGR type conditions (800°C). Further, the coolant for the hot interior
need not be helium, but can be a process fluid like steam or CO_2. This
direct heating feature eliminates the transfer of high temperature heat
across a metallic primary heat exchanger, which could severely limit the
maximum temperature and choice of coolant.

Table I shows the melting point of some candidate high temperature refractory
materials for the hot interior. All appear compatible with helium or argon
coolant. Only the oxide refractories, and perhaps some carbides (e.g., SiC),
would be compatible with steam or CO_2 coolant. The hot interior probably
would be a packed bed of small diameter (1-2 cm) rods or balls. The low
peak power densities (∿10 MW/m^3) and the large surface area in the blanket
should result in relatively low temperature differences (on the order of
100°C) between the coolant and the packed bed.

In high temperature blankets for electric production tritium is bred in a
solid lithium compound (e.g., LiAlO$_2$) in the high temperature interior. The
characteristic time for release into the inert gas coolant is only a few
minutes, from which it is recovered. With steam coolant, however, this
mode of tritium breeding is not feasible, since the tritium cannot be
readily extracted from the steam circuit. Instead, a solid lithium compound
can be placed on the outer surfaces of the module. The bred tritium will

then diffuse to the vacuum chamber and be recovered from the plasma exhaust. Breeding ratios of ∿0.4 to ∿0.6 can be achieved with MgO or Al_2O_3 interiors, but breeding ratios of ∿1.0 require BeO interiors. This implies that two types of blanket modules may be required to satisfy tritium breeding require- ments - the steam cooled type just described, as well as blanket modules which have a high tritium breeding ratio. These could be inert gas cooled, high temperature solid breeding blankets, capable of achieving breeding ratios of ∿1.3 to 1.8.

Direct heating of steam by neutron energy to high temperatures in refractory oxide fusion blankets appears practical, and is the mode of generating process heat for the high temperature electrolysis process. Circulating high-temperature steam through the blanket poses some engineering design problems, but these are not deemed insurmountable. A high temperature electrolysis process should generate H_2 from water at an overall efficiency of ∿50-70 percent depending on design. Since hydrogen production cost by electrolysis is primarily determined by electric power cost, the fact that high temperature process blanket steam is used for directly decomposing water to H_2 and that high overall efficiency (50 percent compared with 30-35 per- cent by conventional electrolysis) can be achieved, implies significant potential economic advantages of a fusion-HTE system.

The previously noted unique ability of fusion neutrons to directly heat the interior of a blanket to very high temperatures, offers great potential for the high efficiency power cycles using fusion heat. The FAST (Fusion Augmented Steam Turbine) power cycle (discussed in Section 4) can admit steam, superheated directly in the blanket, to the turbine inlet at temper- atures of ∿2000°F. Power cycle efficiencies of ∿60 percent appear possible under these conditions with a power cycle efficiency of 60 percent and HTE units operating at 1800°C. The potential efficiency for generation of H_2 from fusion energy is very high, on the order of 70 percent.

FUSION-HTE SYSTEM

The hydrogen production process couples fusion, the primary inexhaustible energy source, with high temperature electrolysis, i.e., the direct decom- position of water at temperatures whereby a significant fraction of the decomposition energy is provided by thermal energy. Steam is transported from the blanket system and distributed to the electrolyzers, the balance of energy being made up by electrical work supplied to the electrolytic cells. The latter source of work is a result of converting a fraction of the fusion energy to thermal energy which is then converted to electricity in a standard thermal cycle. A simplified process flow sheet which couples the three basic elements of the system with hydrogen and oxygen as products is shown in Fig. 1.

A conceptual design depicting the system is shown in Fig. 2. The elec- trolyzers are contained in pressure vessels which surround the Tokamak fusion reactor and are housed outside the reactor assembly in separate com- partments. They are fed steam directly from the high temperature blankets.

The last compartment contains the heat exchangers for H_2 separation as well as the hydrogen storage vessels. To maintain a high steam/H_2 mixture temperature, the mixture is sent back to the blankets for reheat. The number of electrolyzers is a function of temperature drop across the electrolyzers and hydrogen production requirements. Electrical requirements for the electrolysis process are served by the power conversion unit. The electrical generating plant houses the turbines and steam generators as well as auxiliary systems. Power conversion is based on a helium-steam loop. Not shown are the tritium facilities which must also be accounted for plasma fueling.

Blanket

With reference to Fig. 1 the high and low temperature regions refer to the fusion blanket, the high temperature region being further subdivided into two regions. The blanket must satisfy three basic functions: a) to convert neutron energy to process heat (steam) for electrolysis; b) to convert neutron energy to thermal energy for electrical needs; and c) to breed tritium. The latter two functions are carried out in a helium cooled blanket, the tritium diffusing into the helium stream to be trapped out by absorption or conversion to gas and then trapped out by absorption. It may also be possible to breed tritium on the surface of the process heat (steam) module.

The low temperature region is common to both types of blankets while the subdivision of the high temperature region refers to either the process heat or electric production modules.

In either blanket design the structural shell is stainless steel. The process heat (steam) modules contain a high temperature oxide such as MgO, ZrO_2, Al_2O_3, or SiC. The shell is thermally insulated from the high-temperature refractory material and is independently cooled by a bank of coolant tubes. The insulator may be a fibrous material, compatible with the high temperature oxide.

The shell structure, ~ 0.75 cm thick, is maintained at ~ 350 to $400°C$. Typically, 30 percent of the fusion energy is deposited in this region assuming no thermal leakage from the high temperature zone. The coolant tubes are compatible with the shell structural material, and in the stainless steel design, the tubes are typically ~ 0.5 cm ID, 0.2 cm wall thickness. Low temperature steam and/or water mixture is the coolant. Operating conditions are ~ 2000 psi. The high temperature refractory zone is either a packed ball bed or rods through which steam, ~ 1500 to $2000°C$, is passed. If tritium breeding is required in the high temperature blanket, $LiAlO_2$ may be placed between the coolant tubes and shell with tritium diffusing into the plasma chamber. Preliminary calculations indicate that breeding ratios of ~ 0.5 to 0.7 may be achieved. The deficit in tritium is provided by high breeding blankets designed for electric power production. Actually, calculations indicate that the high breeding blankets would provide sufficient tritium so that additional make-up would not be required.

A promising fabrication approach to the design of the process heat modules
is shown in Fig. 3. Several low temperature shells, typically 30 cm in
width and several meters in length, are placed side by side on a strong
structural backing plate. Modules are inserted and removed through small
access ports in the plasma chamber. Fig. 4 shows a typical access port
arrangement for a Tokamak reactor.

Structurally the electric production and high breeding blankets are essentially the same as the high temperature blankets, the outside shell being
cooled by a bank of coolant tubes which operate at the same process conditions as noted before. A neutron multiplier, Be or PbO, and a solid
lithium compound, $LiAlO_2$ and neutron moderator replace the high temperature
refractory. These materials are thermally insulated from the structural
shell by a fibrous graphite cloth. The multiplier-breeder-moderator zone
is cooled by helium at 700 to 800°C and 30 atm.

In these designs tritium breeding occurs in the hot interior of the module
with the bred tritium rapidly diffusing out of the solid lithium compound
(e.g., $LiAlO_2$) into the He coolant from which it is trapped out by absorption
in a metal hydride or by conversion to T_2O and absorption in molecular sieve
material. The neutron multiplier (e.g., PbO, Be) is used in the portion of
the module near the first wall in order to maximize the tritium breeding
ratio. The breeding ratios are relatively high with a PbO neutron multiplier
(1.3) and very high with the Be multiplier (1.8).

The fabrication approach to the design of the electric production modules
(Fig. 5) is essentially the same as the process heat modules. Insertion
and/or removal is through small access ports in the plasma chamber.

No special operations and/or maintenance problems are foreseen for the HTE
and electric generation modules. The steam circuit will probably be strongly
activated by Na^{24} (15 hours half life) released from the hot interior of the
modules, but this will decay to negligible levels in a few days. Other
relatively long-lived activations from blanket impurities and/or crud from
low temperature piping systems will be present but should present less
problems than now faced in LWR's.

The key issues for the blanket appear to be primarily related to materials.
The module structure has to maintain vacuum integrity in the radiation/
thermal cycling environment for several years. This problem is common to
all fusion reactor blankets and no new class of problems appears to be
generated by fusion reactors using the HTE process. The materials effort
now underway in the fusion program should lead to the development of satisfactory structural materials for these applications. More specialized
material problems related to the HTE applications appear, however. These
are conceptual with the stability of oxides such as MgO and Al_2O_3 in the
high temperature steam under radiation and thermal cycling conditions. Such
materials will be used both in the form of solid rods or balls, as well as
a low density solid block or fibrous thermal insulation. The principal
requirement is that these materials not crumble or inject excessive amounts

of fines to the coolant system and that thermal insulation capability be maintained during the life of the module. Because of the more specialized aspects of these materials, it may be advisable to ensure that they will also be considered in the fusion materials development program.

The general design concepts involved for such high temperature blankets appear feasible. Given satisfactory materials, there appear to be no bars to the development of high temperature blankets for fusion reactors using the HTE process to produce synfuels.

Design of High Temperature Electrolyzers

High temperature steam leaving the high temperature refractory zone of the blanket is distributed to the electrolyzers in ceramic-lined ducts. The high temperature solid oxide electrolyzer consists of a cooled steel symmetrical pressure vessel, internally insulated, which operates at 10 atmospheres. Steam or mixture of steam and hydrogen is fed to the electrolyzer where water is reduced to hydrogen on one side of the electrolyte and oxygen is liberated on the other. The existing H_2/H_2O mixtures are recycled back to the blanket system for reheating so that steam can be electrolyzed at a high temperature. When the hydrogen concentration reaches the design value at the highest temperature, the gas mixture passes into a series of lower temperature electrolyzers. The endothermic electrolysis reaction cools the gases to a temperature where conventional heat exchangers can be used.

The oxygen generated in the first series of high temperature electrolyzers passes directly to the low temperature electrolyzers without reheating. The low temperature electrolyzers, have oxygen inlet and outlet ports, while the high temperature vessels require only an oxygen exit port for the oxygen produced in the electrolyzer.

A schematic diagram of the HTE is shown in Fig. 6. The diameter of the vessel is 3.5 m and the length is 6.8 m. The central plenum receives the high temperature steam entering the HTE. Two adjacent plenums collect the gas after electrolysis in the tubes. These are next to the large regions in which the oxygen is produced. The two end sections of the HTE unit are at low temperature, and house electrial connections at the end of the electrolyte tubes.

A major factor in the design of the electrolyzers is the minimization of thermal stresses due to heating and cooling and temperature cycling during operation. The outer cooled region of the containment vessel is held at virtually constant temperature at all times and will not experience any significant problem. Internal components, on the other hand, will experience temperature changes of over 1400°C and large dimensional changes when the HTE unit starts up from or shuts down to room temperature. The tubes, for example, are designed so that one end is fixed, with the other free to move, to accommodate the dimensional changes.

The gas entering the HTE is H_2O in the first electrolyzer and a mixture of H_2O and H_2 in subsequent electrolyzers. This gas is distributed from the central plenum through feed tubes. The feed tubes are 2.4 m long and 5 mm in diameter, and can be joined tubes that are sealed together. The feed tube has spacers to center it in the electrolyte tube and can support the outer tube at high temperatures if required. The feed tubes are centered within the electrolyte support tubes as shown in Fig. 7. The gas passes through the feed tube to the base of the electrolyte support tube. It then passes between the tube and feed tubes to the second plenum. During passage the gas diffuses across the porous support tube to the cathode where it is electrolyzed to hydrogen and cooled by the endothermic reaction.

The two tube sheets which hold the feed tubes are both constructed as interconnecting sections which are sealed together with a ceramic braze. Each section is 0.5 m square with 46 tube rows of 45 each. The shapes of the sections are designed to conform with the ceramic insulation and support around the walls.

The electrolyte support tubes are constructed as shown by the schematic in Fig. 7. The length of each cell depends on the resistivities of the electrodes and interconnection materials but is approximately 1 cm on the 1 cm diameter tubes. The tubes are closed at one end. Each end has a conduction oxide layer. At the closed end the conducting surface oxide is in electrical contact with an oxide cap which passes through the insulation. The end of the cap is at a temperature where metal can be used in the oxidizing atmosphere. The metal continues the electrical path to a conducting plate at each end of the vessel. The plate in turn is connected to an insulated terminal through the pressure vessel. These components are shown schematically in Fig. 7. The open end of the electrolyte support tube with its conducting oxide is sealed into the tube sheet section with a conducting seal. The seal in turn makes contact with a conducting oxide facing the tube sheet sections which are also sealed together with a conducting ceramic seal. These details of the electrolyte support tube are also shown in Fig. 7.

The high temperature solid oxide electrolyzer is supported on a porous zirconia tube as shown in Fig. 8. The thickness of the electrolyte can be reduced to about 10 μm with the supported electrolyte constructed in this manner. The electrolyte must be as thin as practical in order to reduce I^2R losses or overvoltages. At temperatures close to 1650°K a thickness of $\sim$ 1.0 mm is acceptable whereas at 1000°K a 10 μ electrolyte layer would be required. The materials of construction are, for example, $ZrO_2 - Y_2O_3$ electrolyte, doped In_2O_3 anode, a perovskite for high temperatures and nickel for low temperature cathodes, doped $LaCrO_3$ for the interconnection materials and conducting sides.

There are a large number of variables which must be considered in the optimization of solid oxide electrolytes. These include the diameter and length of the electrolyte support tubes, the current density, the flow rate of the steam, etc.

There are losses due to excess voltage that may occur during the flow of current. These include electrochemical polarization losses and resistance losses. The resistance losses arise because of the resistivity of the electrodes, electrolyte, and interconnection materials. These can be reduced by proper cell design and geometry. The polarization losses are related to the electrochemical reaction kinetics. They may be due to the intrinsic activation energy of the reaction which at high temperature may be negligible but which will have to be checked experimentally. The second polarization loss arises due to concentration gradients or slow migration of reactants to the interface. The latter have been considered and discussed by Tedmon, et. al. [3] and are not expected to be a problem at the temperatures of interest.

While the R&D requirements are yet to be fully defined, the issues are centered about:

1) strength of materials as it relates to the porous support tube; temperature limits of the oxides and metals; pore size, sintering and mass transport effects;

2) electrochemical properties as they relate to high temperature electronic conduction in electrolytes; electrochemical kinetics; and

3) high temperature electrodes.

Process Design

The thermal efficiency of hydrogen produced increases as the electrolysis reaction temperature increases. In order to maintain a high reaction temperature, the outlet stream of steam/hydrogen from the HTE is sent back to the blanket to absorb heat for temperature reheat. The mixture is then sent to the next electrolyzer unit, and the reheat process repeated. When the hydrogen concentration has built up to the required level and for a given temperature drop along the electrolyzer tube, the gas mixture then passes into lower temperature electrolyzers where the endothermic reaction cools the gases during the last stages of electrolysis to a temperature at which conventional heat exchangers are used for water-hydrogen separation. For a maximum steam temperature of 1377°C, there are nine electrolyzers--in series--operating at the maximum reaction temperature. These are followed by three electrolyzers operating at lower temperatures, decreasing by 150°C per electrolyzer to an outlet temperature of 727°C. For the maximum steam temperature equal to 1827°C, the number of electrolyzers are six and six, respectively. The oxygen generated in the high temperature electrolyzers is passed directly to the low temperature electrolyzers without reheating. Maximum temperatures will be fixed by material limitations which in turn will set limits on optimum values. Electrical input to the electrolyzers is supplied from the power conversion cycle.

The outlet oxygen and steam/hydrogen mixture from the last set of HTE's is sent to heat exchangers where heat is recovered by the inlet make-up water stream. The make-up water stream from the outlet of the heat exchanger will take up heat from the helium loop of the superheater before returning to the breeding blanket of the fusion reactor. Hydrogen is separated out from steam in the heat exchanger. The separated water will combine with the make-up water for recycle. Oxygen produced can be used as an oxidizing agent.

There are alternative methods to recover heat from the oxygen and steam/hydrogen streams. For example, one can take saturated steam from the turbine as make-up steam and send it to the O_2 heat exchanger and hydrogen/steam heat exchanger to recover the heat. The remaining heat of oxygen and steam/hydrogen streams would be recovered from water in a hydrogen separation unit where hydrogen is separated out from steam by condensation. The coolant water from the hydrogen separation unit would be returned to the boiler feed water of the power cycle loop.

The electrical power generation plant is similar to a conventional power plant. Low temperature steam from the low temperature blanket region is pumped through a superheater where the temperature of steam is raised by heat absorption from the helium loop. Power is generated through high or intermediate pressure turbines. Steam from the turbine is condensed in a low pressure condenser before returning to the blanket. The cycle efficiency is shown to be $\sim$ 38 percent. The unique ability of fusion neutrons to directly heat the interior of a blanket to very high temperatures offers great potential for high efficiency power cycles using fusion heat. Typical values may be $\sim$ 60 percent with a FAST (Fusion Augmented Steam Turbine) cycle (discussed in the next section). This implies that the overall hydrogen process efficiencies would be high, $\sim$ 65 to 70 percent, for high temperature electrolysis as well as low temperature electrolysis, $\sim$ 60 percent.

FAST(FUSION AUGMENTED STEAM TURBINE) CYCLE

For the FAST cycle steam is superheated directly in the hot interior of the blanket module. Open cycle gas turbines presently operate at much higher inlet temperatures than the conventional steam turbine. The latter has been held at an inlet temperature of $\sim$ 1100°F for many years because of material temperature limitations in the steam generator/superheater. The present inlet temperature for gas turbines is $\sim$ 2000°F (1090°C) with a projection of $\sim$ 2400°F (1320°C) by the early 1980's [4]. With direct superheating of steam in fusion blankets, overall cycle efficiency can be raised from the $\sim$ 38% level achieved in fossil fuel steam plants to a level of $\sim$ 60% assuming that the turbine inlet temperature is 2000°F (1090°C).

High temperature turbine cycles based on the combustion of hydrogen and oxygen have been extensively investigated [5,6] with projected turbine inlet temperatures up to 3000°F. Operation at these temperatures will require development of water cooled or ceramic blades. This would increase the efficiency of the FAST cycle to $\sim$ 70%; however, the increase may not warrant the major turbine development program that would be required.

Figure 9 shows a flow sheet for the FAST cycle. Approximately 30-50% of the steam in the turbine circuit flows through the blanket, emerging at a high temperature, typically 1500-1800°C. It then mixes with the main steam flow; the resultant mixed temperature is controlled to the desired turbine inlet temperature by the relative flow proportions. Bypassing most of the steam flow around the blanket reduces blanket pressure drop, flow velocity, piping dimensions, and the carry over of blanket fines. Fig. 10 shows the efficiency of the FAST cycle as a function of turbine inlet temperature from 1600°F (870°C) to 2400°F (1320°C), for the case of three reheats and the limiting case with continuous reheat, turbine inlet pressure of 2000 psia, and exhaust pressure of 1 psia. The FAST cycle efficiency of ∿ 60% which could be achieved with essentially developed turbine technology, can be very important for fusion. It will greatly reduce unit $/KW(e) capital costs for a fusion power plant, as well as reduce the thermal pollution levels per KWH by a factor of 3, as compared to LWR.

An additional feature of the FAST cycle is the efficient use of the hot/cool energy splits in the blanket. The fusion energy from the cool structure produces high pressure saturated steam, and the high temperature heat from the interior superheats it. Alternate advanced power cycles that only use the high temperature need very high efficiency to match the FAST cycle, since the low temperature heat from the structure only produces ∿ 33% efficiency in a separate conventional steam cycle. For a 70/30 hot/cool split, an alternate cycle would require an efficiency of 70% to achieve an overall average efficiency of 60% for the reactor.

RESULTS OF DESIGN STUDY

Efficiency Considerations

The basis for the energetics of a high temperature electrolysis process for hydrogen production is the thermodynamics of the decomposition of water.

$$H_2O_{(\ell)} = H_2O_{(g)} \tag{1}$$

$$H_2O_{(g)} = H_{2(g)} + 1/2\ O_{2(g)} \tag{2}$$

ℓ refers to liquid state

g refers to gaseous state

The degree of decomposition of water as a function of temperature is expressed by the Gibbs free energy

$$\Delta G = \Delta H - T\Delta S \tag{3}$$

where

ΔG = free energy change for reaction (2), Kcal/mol

ΔH = enthalpy change for reaction (2), Kcal/mol

T = absolute temperature, °K

ΔS = entropy change for reaction (2), Kcal/mol °K

The electrolytic decomposition of water is controlled by the relationship

$$\Delta G = nfE \tag{4}$$

n = number of gram equivalents per atom, $n = 2$

f = Faraday's constant 96,500 Coulomb/gm equiv.

E = emf or voltage potential of cell, volts

In an electrolyzer, the electrical energy supplied to the cell is related to ΔG, the non-work energy needed for the decomposition is expressed by the $T\Delta S$ term, and the total energy is related to the total enthalpy change for the system, ΔH.

The thermodynamic values for the water decomposition reaction, (2), as a function of temperature, is given in Fig. 11. This information is well known and readily available from many sources; the best thermodynamic compilation dating back to the NBS Circular 500 (1952). It is noted that for this system, ΔH remains almost constant with temperature, increasing only very slightly up to temperatures as high as 4200°K. The $T\Delta S$ term steadily rises and as a result the ΔG term steadily decreases until it reaches zero at a temperature on the order of 4200°K. The quantity ΔG relates to the amount of electricity fed to the cell and $T\Delta S$ the thermal energy. What is important is that as the temperature of the cell is raised, the $T\Delta S$ term increases. The fraction of the thermal energy input required increases and the electrical energy needed decreases. Fig. 11 shows the fraction of thermal energy input and indicates that at 2160°K (1886.8°C) 50 percent of the energy input for water decomposition would be thermal energy and 50 percent electrical energy. Theoretically, at 4200°K all the energy can be put in as thermal energy for water decomposition and no electrical energy would be required. This would be a true thermal splitting of water. However, this temperature cannot be reached because other dissociative chemical reactions take place, forming the free radicals, H, 0, and OH which are energy absorbing and limit the upper temperature.

The constraint placed on the system is that it must be energetically balanced. This means that the energy generated by the fusion reactor is only used to produce hydrogen gas. Depending on the temperature that can be achieved in the high temperature region of the fusion reactor blanket, the ratio of thermal energy to electrical energy is fixed by the thermodynamic conditions established in Fig. 11. This means that not all the energy in the high temperature region can be used in the HTE so that the remainder must be used in the conventional power cycle to generate electrical power for the HTE. This provides the rationale for subdividing the high temperature region into two regions; one marked HTRE for the electrolyzer circuit and the other HTRP for the power cycle, the latter incorporating the helium cooled tritium breeding blanket.

To gain some estimate of the effect of temperature, a simplified energy analysis is considered to determine the efficiency of the conversion of the total fusion energy to hydrogen fuel gas energy. The analysis assumes that:

1) 100 units of thermal energy total is generated by the fusion reactor,

    70 units of thermal energy generated in the HTR,
    30 units of thermal energy generated in the LTR (x-rays, etc.)

    let x = fraction of the 70 units in HTR which can be used in HTE; this is the fraction absorbed in HTE,

    then $1 - x$ = fraction of the 70 units in HTR which is used in the HTRP part of the conventional power cycle.

2) The conventional power cycle operates at a maximum of 40 percent efficiency (conversion of thermal energy to DC electrical energy).

3) The high temperature electrolyzer operates at close to 100 percent current efficiency which is a fairly reasonable assumption at elevated temperatures.

For a balanced system:

$$\frac{\text{Thermal energy}}{\text{Electrical energy}} = \frac{T\Delta S}{\Delta G} = \frac{70x}{0.4\,[30 + (1-x)\,70]} \tag{5}$$

Assuming an average temperature for the HTE, the ratio of $T\Delta S/\Delta G$ can be obtained from Fig. 11. From Eq. (5) x can be computed. The cycle efficiency of the entire system defined as conversion of total fusion energy to hydrogen fuel gas energy is then:

$$\% \text{ Cycle efficiency (LHV)} = \frac{0.4\,[30 + (1-x)\,70] + 70x}{100} \times 100 \tag{6}$$

Cycle efficiency is based on the lower heating value (LHV) of hydrogen since the thermodynamics are based on Eq. (2) for water in the gaseous state. The convention for the sale of fuel is based on the higher heating value (HHV).

The HHV efficiency is given in Table II along with the LHV and a plot of both LHV and HHV efficiency for both conventional (40%) and advanced (60%) power cycles are shown in Fig. 12. There is approximately 7 to 8 percentage points improvement in HHV over LHV. It is also noted that based on the assumptions made, ideally the highest efficiency that can be obtained is at an HTE temperature of 3600°K when x = 1 and all the HTR heat (70%) goes to the HTR and the LTR heat (30%) goes to providing the electricity. The efficiencies as calculated are idealized in that pumping and heat losses from the various process units will reduce the efficiency. In addition at very high temperatures (>2300°K) material temperature limitations may impose practical operating limits. Below a 1000°K the over-voltage problem may impose a lower operating limit. Furthermore 40 percent efficiency may be too high a value for the conventional power cycle. All these additional inefficiencies are taken into account in the actual process design of the reference system.

The value of the HTE system can be compared to conventional fusion power with low temperature or conventional electrolyzers. The well-known Lurgi electrolytic cells operate at 30 atm and 80°C at an efficiency of 80 percent. The advanced GE-solid polymer electrolyzers (SPE) operate at 125°C and are reported to yield efficiencies of 90 percent. When combining these efficiencies with the 40 percent conventional power cycle, the range of LHV efficiency values of 32 to 36 percent are obtained and the HHV efficiencies are 42.2 to 45.4 percent. At 1700°K (1426.8°C) which is a reasonably high temperature for HTE cells, a HHV efficiency of 56.9 percent can be obtained. Thus, the HTE cycle yields from 14.5 to 17.7 percentage points higher than a conventional electrolyzer cycle and thus yields an improvement of from 31.2 to 42.0 percent in efficiency over conventional systems on a comparable basis. This improved efficiency should be translated to lower operating and capital cost for a synthetic fuel process and would seem to justify a development program for HTE.

In addition to the H_2 efficiency gains as a consequence of high temperature, a substantially higher H_2 efficiency can be achieved with the inclusion of an advanced power cycle (60% efficiency) in the system. For example, at 1700°K a HHV efficiency of 72 percent can be realized (compared with 57% with a lower power cycle efficiency). This improved efficiency should like-wise be translated into lower operating and capital cost for a synthetic fuel process and would seem to justify a development program for advanced power cycles based on high temperature blankets.

Results of the conceptual design study are summarized in Table III for two maximum steam temperatures, 1377°C and 1827°C. The power cycle efficiency is found to be ∿38% (the net efficiency accounting for pumps, etc.). It is assumed that the reactor operates in an ignited state, with long plasma burn and minimal extra recirculating power for special portions of the fusion reactor, i.e., beams, magnets, tritium recycle, etc., that would not be included in the recirculating power requirements associated with the power conversion. This accounts for 2% of electrical requirements. The hydrogen thermal process efficiencies are 49% and 51% for the maximum steam temperatures, 1377°C and 1827°C. While the efficiency increases with temperature, the increase is not that great between ∿1400°C and 1800°C as born out by the theoretical calculations as well. Overall conversion from steam to hydrogen is 89.4% and 94.3%.

ECONOMICS

While complete economic studies of the system were not attempted, estimates of the capital investment costs as well as fuel production cost evaluations were made. This phase of the study relies on cost estimate assumptions for individual components, such as the fusion reactor, coal-synthetic fuel plant, etc. These results are summarized in Tables IV through VI.

Before any costing can be done, some idea as to the fuel production capacity for a given fusion reactor size is necessary. Table IV includes a summary of the hydrogen fuel production capacity of the reference design HTE system based on a 2000 MW(th) fusion reactor with a conventional power cycle (CP) efficiency of $\sim$40 percent. These results are compared with a system operating with an advanced high efficiency power cycle (AP) operating at $\sim$60 percent efficiency. The maximum HTE temperature was fixed at 1600°K in both cases.

For a fixed reactor thermal rating, the hydrogen produced, i.e., standard cubic feet/day (scf/d), increases in direct proportion with the system efficiency. In terms of equivalent gasoline production in barrels/day, a 2000 MW(th) fusion reactor-HTE system is a relatively small fuels plant. Such a plant, operating at 70 percent efficiency for H_2 production would produce the energy equivalent of 20,000 bbl/day, which would fuel $\sim$500,000 autos with average driving patterns. A factor of three reductions in coal feed (tons/day) is achieved in syngas (methane) production if fusion produced hydrogen is used, as compared to a conventional syngas plant fed by coal. This large savings in coal usage realized with the fusion produced hydrogen would greatly extend coal resources and reduce environmental effects.

Table V summarizes estimates on capital investment costs. Note that the syngas production rate is three times that reported in Table IV and is based on a 6000 MW(th) fusion reactor. The assumed costs of the fusion reactor plus electrolyzers are taken to be in the range of $400 to 800 KW(th) [$1000 - 2000/KW(e) equivalent] based on reference designs for fusion reactors producing electricity at conventional efficiency (30 - 40 percent). A conventional syngas fuel plant costs $\sim$ one billion dollars and an additional one billion dollars is needed for coal feed operating costs which can be considered as a tradeoff for the additional capital investment for the fusion reactor process.

Results show that the fusion-HTE system based on the lower fusion costs is slightly more than the total cost of a syngas system at the lower efficiency (50 percent) and slightly less at the higher efficiency (70 percent). Doubling the fusion plus electrolyzer costs increase the total costs accordingly.

Table VI is an evaluation of fuel production costs. Assuming fixed charges to be 15 percent in a fusion-synfuels plant, the fuel costs based on the lower fusion costs are competitive with those based on a conventional coal-synfuels plant. Fuel costs resulting from the fusion-synfuels plant would be approximately one-half that of a comparative fission electrochemical system. Looked at from another perspective, the hydrogen produced from a fusion-synfuels plant is equivalent to an energy cost corresponding to $\sim$ 45¢ to 60¢/gallon of gasoline. Since all of these comparisons are based on assumed costs, no definitive conclusions can be drawn except that if the cost per unit of thermal output of a fusion HTE plant is comparable to that projected for fusion electric plants, fusion produced hydrogen should be economically competitive.

SUMMARY

Based on results obtained from the study as well as comparisons with other methods of hydrogen production, the following tentative conclusions reached are:

1) HTE has the highest potential efficiency for production of synfuels from fusion; a fusion to hydrogen energy efficiency of $\sim$ 70 percent appears possible with 1800°C HTE units and 60 percent power cycle efficiency; an efficiency of $\sim$ 50 percent appears possible when 1400°C HTE units and 40 percent power cycle efficiency;

2) relative to thermochemical or direct decomposition methods HTE technology is in a more advanced state of development, e.g., single cell units have been built and tested at 1000°C;

3) based on efficiency results HTE methods would appear to have potentially lower unit process or capital costs compared with thermochemical or direct decomposition methods;

4) while design efforts are required HTE units offer the potential to be quickly run in reverse as fuel cells to produce electricity for restart of Tokamaks and possible spinning reserve for a grid system.

REFERENCES

1. "Fusion Energy Applied to Synthetic Fuel Production", DOE Report, CONF-770593, October, 1977.

2. "Fusion Reactors-High Temperature Electrolysis (HTE)", DOE Report, HCP/T0016-01, January, 1978.

3. Spacil, H. S., and Tedmon, Jr., C. S., Electrochemical Dissociation of Water Vapor in Solid Oxide Electrolyte Cells", J. Electrochem. Soc., 116, 1618, 1627 (1969).

4. Armstrong, C. H., "Effect of Recent Advancements in Gas Turbine Tech-- nology on Combined Cycle Efficiency", ASME Paper 74-PWR-8, Joint Power Generation Conference, Miami Beach, California (Sept. 15-19, 1974).

5. Kelley, J., "Hydrogen Tomorrow - Demands and Technology Requirements", Jet Propulsion Laboratory, Pasadena, Calofirnia (Dec., 1975).

6. Hausz, W., "Hydrogen Systems for Electric Energy", 72-TMP-15, General Electric Company - TEMPO Center for Advanced Studies, Santa Barbara (1972).

TABLE I

MELTING POINTS OF SOME HIGH TEMPERATURE REFRACTORIES

Carbides	M.P. (°K)	Oxides	M.P. (°K)	Nitrides	M.P. (°K)
HfC	4161	ThO_2	3573	TaN	3361
TaC	4148	MgO	3098	BN	3273
NbC	3773	HfO_2	3085	TiN	3205
ZrC	3533	ZrO_2	2973	ZrN	3203
TiC	3523	CaO	2843		
SiC	3100	BeO	2725		
VC	3083	Al_2O_3	2323		

TABLE II

FUSION ENERGY-HIGH TEMPERATURE ELECTROLYSIS
WITH CONVENTIONAL POWER CYCLE FOR
PRODUCTION OF HYDROGEN FUEL GAS

100% HTE efficiency
 40% Conventional power cycle efficiency
 70% of fusion energy generated in high temperature region
 30% of fusion energy generated on low temperature region

Temperature °K	°C	Hi-Temp. Ht. to total energy ratio $T\Delta S/\Delta H$	Hi-Temp. Ht. to elec. energy ratio, $T\Delta S/\Delta G$	Fraction of HTR for HTE, x	Overall Cycle Efficiency LHV, %	HHV, %
1000	726.8	0.298	0.295	0.1508	47.0	54.8
1460	1186.8	0.333	0.500	0.238	50.0	57.3
1700	1426.8	0.390	0.639	0.292	53.0	59.9
2000	1726.8	0.464	0.866	0.355	55.5	62.0
2160	1886.8	0.500	1.000	0.408	57.2	63.5
2600	2326.8	0.605	1.530	0.542	62.7	69.0
3000	2726.8	0.712	2.470	0.710	69.8	74.6
3600	3326.8	0.854	5.830	1.000	82.0	84.6

Conv. cells

| 353 | 80 Lurgi-80% cell eff. | | | | 32.0 | 45.2 |
| 398.2 | 125 GE/SPE-90% cell eff. | | | | 36.0 | 45.4 |

LHV=% efficiency based on lower heating values of hydrogen

HHV=% efficiency based onhigh heating value of hydrogen

TABLE III

SYNTHETIC FUELS FROM FUSION REACTORS
TABLE OF PROCESS FLOW PARAMETERS

Electrolytic Cells	Max. Inlet Temp. to Electrolyzers	
	1650°K	2100°K
ΔT/Cell	150°C	150°C
Electrical Input to Cells	598.4 MW(e)	573.7 MW(e)
Thermal Energy to Cells	392.9 MW(t)	459.8 MW(t)
Output Temp. of Steam + H_2 (last HTE)	727°C	727°C
Overall Steam Conversion to H_2	94.3%	89.4%
H_2 Separation		
H_2 Quality – at 9 atm and 50°C (water impurity)	98.65%	98.6%
Power Conversion		
High Temp. Region – T&P of Helium	800°C, 30 atm	800°C, 30 atm
Low Temp. Region – T&P of Steam (sat.)	350°C, 2400 psia	350°C, 2400 psia
Energy Split (high temp. region)	1007.1 MW(t)	940.2 MW(t)
(low temp. region)	600 MW(t)	600 MW(t)
Power Conversion Eff.	38%	38%
Total Plant Efficiency (HHV H_2)	49.3%	51.2%
Hydrogen Production	23.5 MT/HR	25.9

TABLE IV

HIGH TEMPERATURE ELECTROLYSIS WITH CONVENTIONAL AND ADVANCED POWER CYCLE
FUEL PRODUCTION CAPACITY FOR 2000 MW(t) FUSION REACTOR

	HTE-CP	HTE-AP (FAST)
High Temperature Region - °K	1600°K	1600°K
Cycle Efficiency - %	52.5%	71.5%
Hydrogen Production - SCF/D	267×10^6	364×10^6
Hydrogen Production - MT/D	638	868
Equiv. SNG. (Methane) - SCF/D	86×10^6	117×10^6
Coal Hydrogenation with HTE H_2 - T/D	1,914	2,600
SNG. Production with HTE H_2 - SCF/D	134×10^6	182×10^6
Conventional SNG Plant Equiv. - Coal Feed T/D	5,960	8,100
HHV Equiv. Gasoline - BBL/D	14,800	20,130

TABLE V

CAPITAL INVESTMENT COST FOR A FUSION–SYNTHETIC FUEL PRODUCTION PLANT

[Basis: Hydrogen Production Rate has Fuel Equivalent Value to a 250 x
10^6 SCF/D SNG Plant]

Efficiency: Output H_2 Fuel Energy/Input Fusion Energy	Capital Cost of Plant
50%	$2.4 x 10^9
70%	$1.7 x 10^9

Cost Assumption: Fusion Reactor + Electrolyzers = $400/KW(th) [$1000/KW(e) Equivalent]

Efficiency: Output H_2 Fuel Energy/Input Fusion Energy	Capital Cost of Plant
50%	$4.8 x 10^9
70%	$3.3 x 10^9

Cost Assumption: Fusion Reactor = Electrolyzers = $800/KW(th) [$2000/KW(e) Equivalent]

Coal Synthetic Fuel – SNG Plant	Capital Cost of Plant
Conventional SNG	$1.0 x 10^9 *

* Coal Feed Cost Equivalent to Additional $1.0 x 10^9 Investment at $25/Ton

TABLE VI

PRODUCTION COST EVALUATION

Fuel Cost
($/10^6 BTU)

Fusion-Synthetic Fuel Production Plant
Assumption: 15% Fixed Charge - Fusion Reactor
             + Electrolyzers = $400/KW(th)
             [$1000/KW(e) Equivalent]

Efficiency: Output H$_2$ Fuel Energy/Input Fusion Energy

50%	4.70
70%	3.40

Assumption: 15% Fixed Charge - Fusion Reactor
             + Electrolyzers = $800/KW(th)
             [$2000/KW(e) Equivalent]

Efficiency: Output H$_2$ Fuel Energy/Input Fusion Energy

50%	9.40
70%	6.80

Coal Synthetic Fuel Plant (based on $25/ton coal)

Syncrude ($24/BBL)	4.00
SNG	3.00-4.00

Fission - Electrolytic Synfuel Plant

Assumption: Reactor + Electrolyzers - $1000/KW(e)

30% Efficient System	7.85

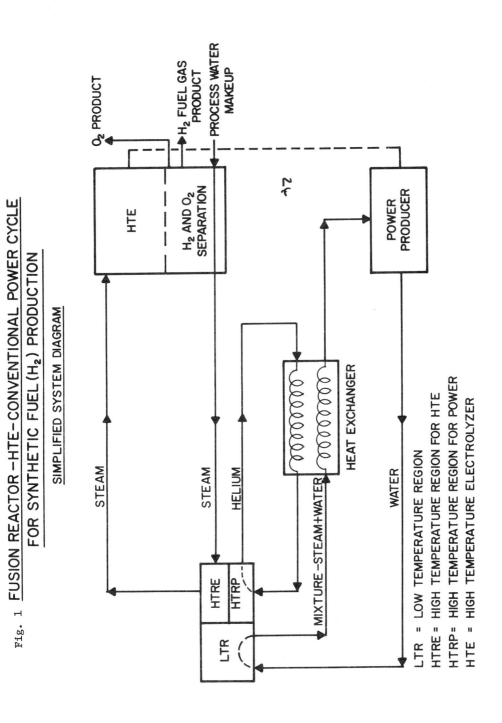

Fig. 1 **FUSION REACTOR – HTE – CONVENTIONAL POWER CYCLE FOR SYNTHETIC FUEL (H₂) PRODUCTION**

SIMPLIFIED SYSTEM DIAGRAM

LTR = LOW TEMPERATURE REGION
HTRE = HIGH TEMPERATURE REGION FOR HTE
HTRP = HIGH TEMPERATURE REGION FOR POWER
HTE = HIGH TEMPERATURE ELECTROLYZER

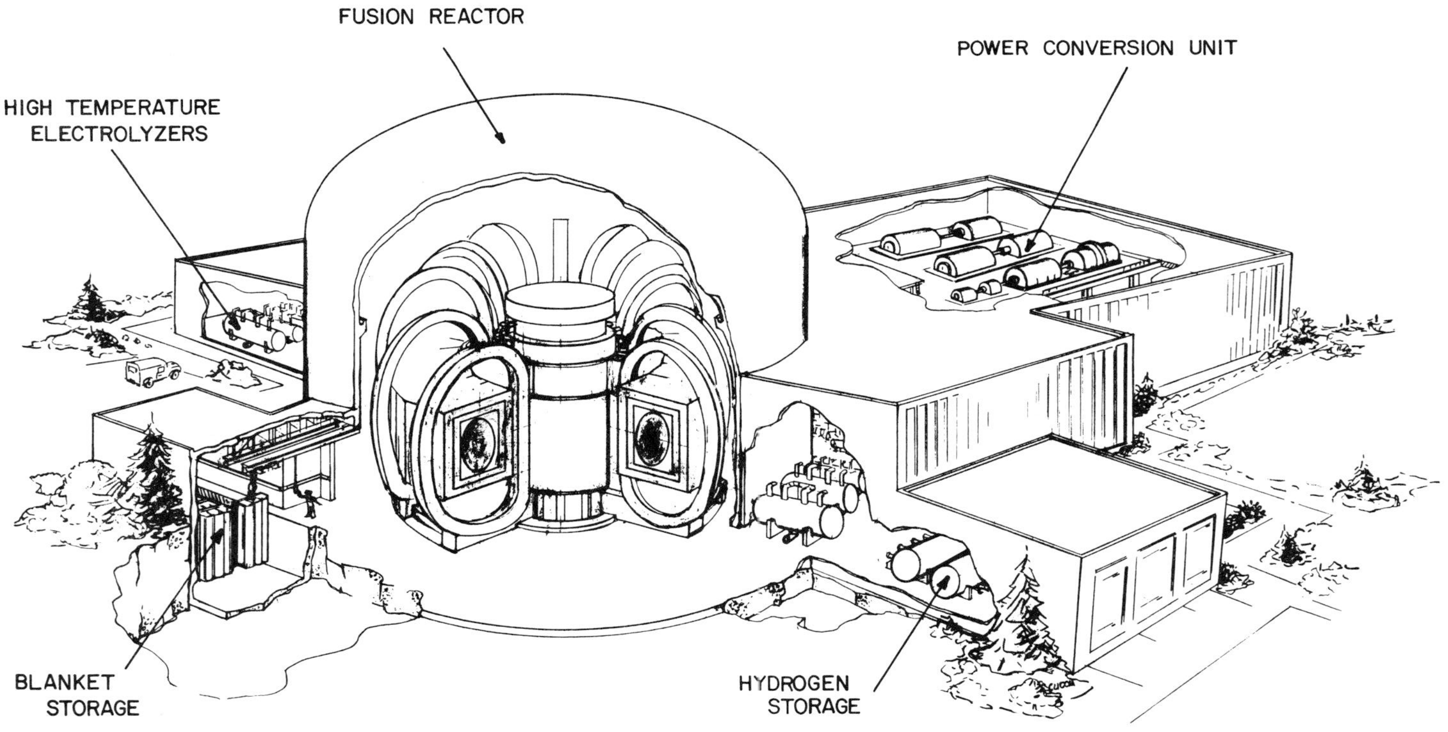

Fig. 2 FUSION REACTOR-HIGH TEMPERATURE ELECTROLYSIS SYSTEM (FR-HTES)

1159

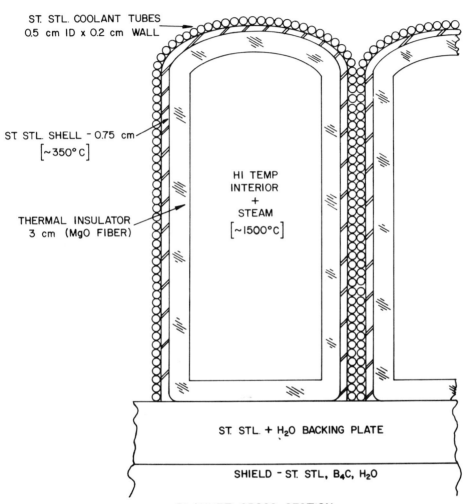

BLANKET CROSS SECTION

Fig. 3   HTE PROCESS HEAT MODULE

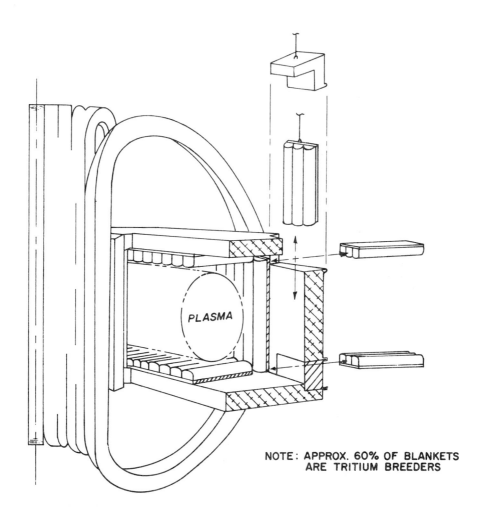

NOTE: APPROX. 60% OF BLANKETS
ARE TRITIUM BREEDERS

Fig. 4   SCHEMATIC SHOWING MODULE REMOVAL

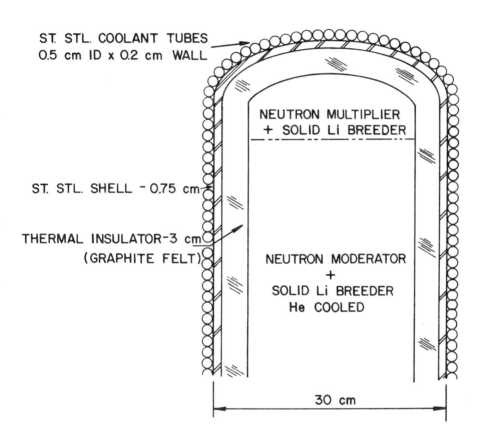

ST. STL. COOLANT TUBES
0.5 cm ID x 0.2 cm WALL

NEUTRON MULTIPLIER
+ SOLID Li BREEDER

ST. STL. SHELL - 0.75 cm

THERMAL INSULATOR-3 cm
(GRAPHITE FELT)

NEUTRON MODERATOR
+
SOLID Li BREEDER
He COOLED

30 cm

Fig. 5   ELECTRIC PRODUCTION MODULE - T BREEDING

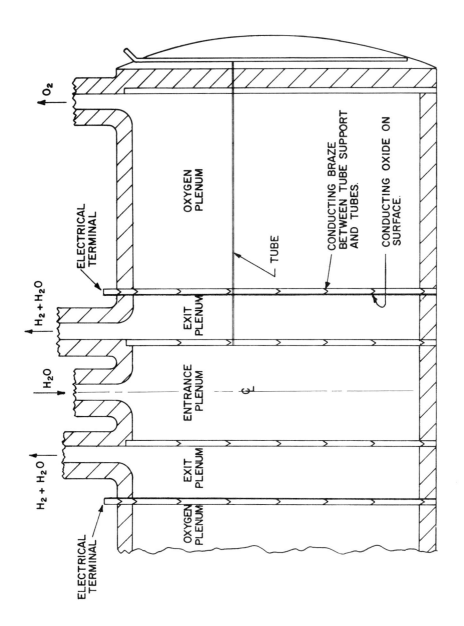

HTE VESSEL

Fig. 6

1163

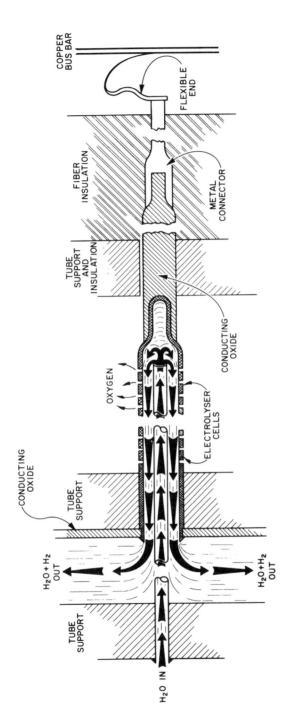

COPPER BUS BAR

FLEXIBLE END

FIBER INSULATION

METAL CONNECTOR

TUBE SUPPORT AND INSULATION

CONDUCTING OXIDE

OXYGEN

ELECTROLYSER CELLS

CONDUCTING OXIDE

TUBE SUPPORT

H₂O + H₂ OUT

H₂O + H₂ OUT

TUBE SUPPORT

H₂O IN

DETAILED CROSS-SECTION OF HTE CELL

Fig. 7

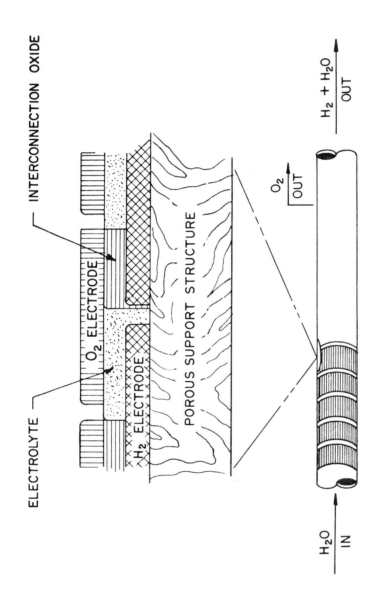

HTE CELL DESIGN
(WESTINGHOUSE FUEL CELL)

Fig. 8

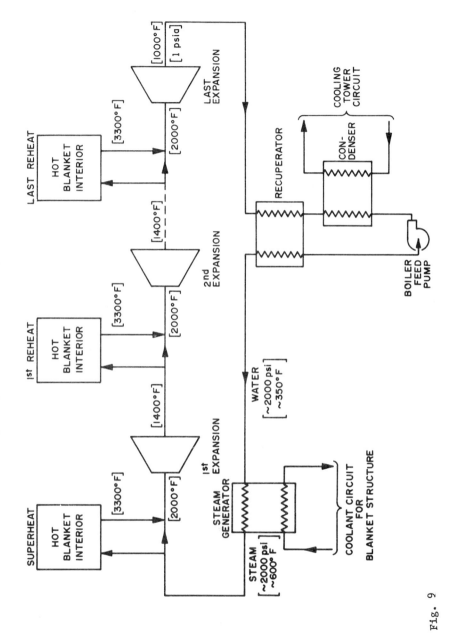

FAST POWER CYCLE
(FUSION HEAT SOURCE)

Fig. 9

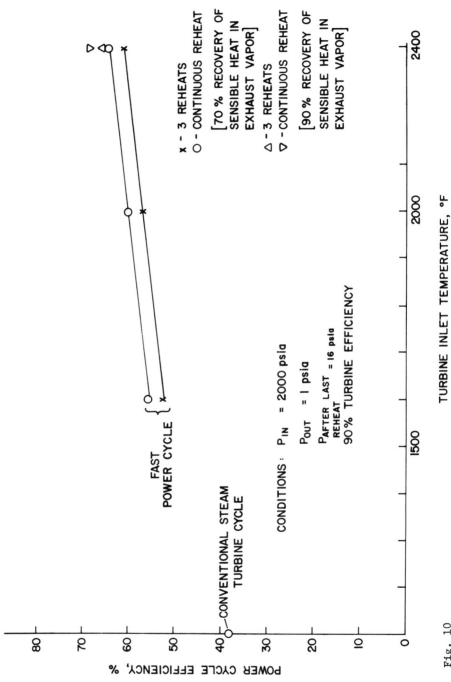

Fig. 10

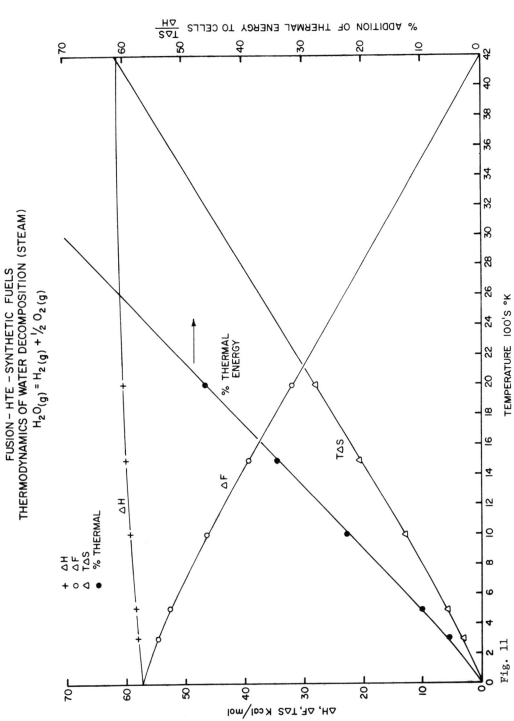

FUSION – HTE – SYNTHETIC FUELS
THERMODYNAMICS OF WATER DECOMPOSITION (STEAM)
$H_2O_{(g)} = H_2(g) + \frac{1}{2} O_2(g)$

Fig. 11

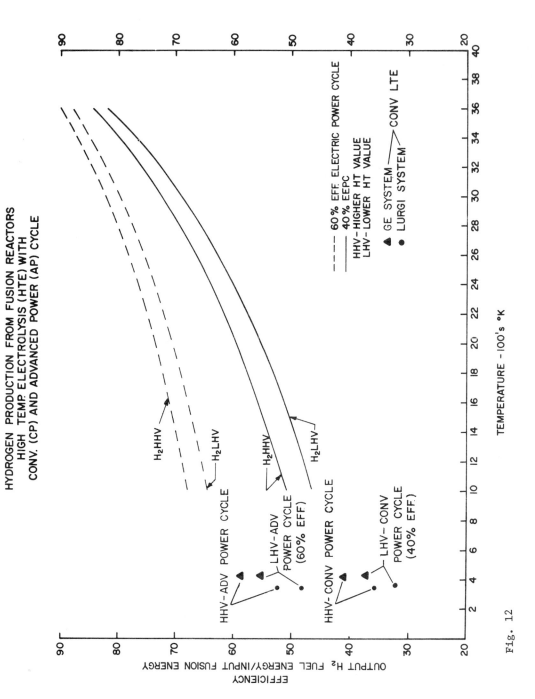

Fig. 12

PLASMOCHEMICAL CYCLE OF HYDROGEN PRODUCTION FROM THE WATER

I.G.Belousov, V.A.Legasov, V.D.Rusanov
I.V.Kurchatov Institute of Atomic Energy,Moscow,USSR

ABSTRACT

The paper is devoted to the two-stage plasmochemical cycle of hydrogen generation from the water, based on decomposition of the carbon dioxide in an nonequilibrium gas-discharge plasma with the subsequent hydrogen production from the carbon oxide by steam conversion.

Comparison is given of the plasmochemical cycle with the electrolysis and thermochemical ones. Some general question of choice of the cycle for the industrial technology are briefly considered.

INTRODUCTION

The problem of development of a highly efficient method for conversion of nuclear energy into chemical energy of "synthetic" fuel seems to be very urgent at present. Difficulties of its realization lie in variety of possibilities suggested by chemical, electrolysis, radiolysis, and plasmochemical technologies. These difficulties are essential since a large-scale production can be based only on very few solutions ensuring economy and long-term competitiveness.

In the present work we shall confine ourselves with the closed cycles of hydrogen production from the water, realized by means of an external (e.g. nuclear) energy source. Therefore, a general approach in estimation of possible closed water decomposition cycles would be of interest, which would allow to make a choice of versions for the large technology. In this

case any particular arguments in favor of one or another cycle give no opportunity to narrow the search. For this purpose we use the following general considerations:

- the maximum efficiency of the decomposition cycles depends,
  as is known, on their stage structure as well as on the
  extent of perfection of individual stages.

For example, efficiency of the n stage cycle[I] of decomposition of substance AB

1. $AB + C_1 \longrightarrow AC_1 + B \qquad T_1$
2. $AC_1 + C_2 \longrightarrow AC_2 + C_1 \qquad T_2$
. . . .

n. $AC_{n-1} + C_n \longrightarrow AC_n + C_{n-1} \qquad T_n$ $\qquad\qquad$ (1)

$\quad n + 1.\ AC_n \longrightarrow A + C_n \qquad T_{n+1} = T_{AC_n}$

$\qquad\qquad \sum AB \longrightarrow A + B$

with the canonic order of temperatures of performing reactions, T_i, and temperatures T_{AC} of decomposition of intermediate substances

$$T_1 < T_2 < \dots < T_n < T_{AC_n} < T_{AC_{n-1}} \dots < T_{AC_1} < T_{AB} \quad (2)$$

does not exceed

$$\eta = \frac{\left(1 - \frac{T_1}{T_{AC_1}}\right) \dots \left(1 - \frac{T_n}{T_{AC_n}}\right)}{\left(1 - \frac{T_1}{T_{AB}}\right) \dots \left(1 - \frac{T_n}{T_{AC_{n-1}}}\right)} < \frac{1 - \frac{T_1}{T_{AC_n}}}{1 - \frac{T_1}{T_{AB}}} \quad (3)$$

which is notoriously lower than the maximum efficiency corresponding to the temperature of decomposition of substance
AB, T_{AB}, maximum temperature of the heat source, $T_{AC} = T_{n+1}$,
and minimum temperature of the heat discharge reservoir, T_1.
This statment is true for any thermochemical cycle where
the number of stages is more than two, if one or several stages realize within a temperature range limited with extreme

and only for the two-stage cycles the maximum efficiency coincides with the theoretical limit of efficiency of decomposition of the substance into components.

The maximum efficiency of the two-stage cycles is equal to the efficiency of an ideal electrolysis process of substance decomposition provided by the current from the "Carnot machine" if the limiting maximum temperatures of the two-stage cycle and "Carnot machine" are equal.

A practical consequence of this particular case of the equivalence theorem is the possibility to make a preliminary estimation of perspectiveness of the hydrogen production technology in complete lack of experience of industrial realization.

We know beforehand that the thermochemical technology has no advantages over the electrolysis technology as far as the closed thermodynamic cycles of substance decomposition is concerned. Hence, decisive become the questions of practical realization of the processes with high efficiency and high specific capability. It is believed that the efficiency of the heat-electricity conversion cycles will not be, generally speaking, lower than that of the thermochemical cycles of heat-chemical energy conversion as the extreme temperatures of the cycles differ little from each other. Moreover, due to a relative freedom in the choice of the coolant for the heat-work conversion machines, it can be hoped to obtain a wider temperature range in this case compared to the thermolysis case. As for the capability, it depends significantly on intensity of heat transfer from the external source through constructive shells to the working bodies and in the general case can be considered equal for heat machines and thermolysers. An additional factor limiting intensities of the thermolyser processes is the kinetic features of the reactions proceeding. Thus, the energy processes of the heat machines are potentially more productive and efficient and, if there were no limitations in the electrolysis part, the electrical trend in the hydrogen industry energetics would be beyond competition. However, since the electrolysis of water is involved in these processes, it is here that difficulties in organization of highly productive and highly efficient processes on the electrodes are encountered, which have not been overcome yet as much as it would be desirable.

The electrolysis process, however, can be replaced by the plasmochemical process which is significantly more efficient and, thus, the problem of comparison of the water splitting cycles will be solved in favor of the electrical cycle.

Plasmochemical Cycle.

The plasmochemical processes occuring in a plasma are of a volume character and have a very high intensity. Many of them, however, have a comparatively low efficiency and in some cases cannot be recommended for practical application. This particularly concerns the equilibrium processes.

The non-equilibrium process of plasmochemical destruction of the carbon dioxide into the carbon oxide and oxygen

$$CO_2 \longrightarrow CO_2^* \longrightarrow CO + \frac{1}{2} O_2 \qquad (4)$$

has both high productivity and high efficiency under appropriate organization of the discharge. Below the mechanism of reaction (4) will be analyzed in detail and the nature of favorable combination of useful factors clarified Prior to this we shall consider a simple solution of the problem of electrical decomposition of the water into hydrogen and oxygen using the short-term and efficient reaction (4). Indeed, adding the "shift" reaction

$$CO + H_2O \longrightarrow CO_2 + H_2 \qquad (5)$$

to reaction (4) and ensuring separation of the gaseous components with carbon oxide-dioxide recirculation we shall obtain
plasmochemical cycle (2) whose final result is a highly efficient and economic process of water vapor decomposition into
hydrogen and oxygen by means the electrical energy from an
external power source.

Thus, the plasmochemical cycle includes three stages: generation of electricity, use of the electricity in the stage of
destruction of the intermediate substance and chemical stage
of shift.

The stages of generation and use of the electricity are equivalent to two stages of a certain thermolysis cycle. The shift
stage is purely chemical. Hence, the efficiency of the plasmochemical cycle can be compared with that of a certain three-
stage thermolysis cycle. In other words, the theoretical limit of the efficiency of the plasmochemical cycle under consideration lies lower than that of the electrolysis cycle due
to one extra stage, though practically the difference does not
seem to be significant.

The productivity and economy of the plasmochemical cycle, generally speaking, exceed those of the electrolysis cycle in
the large-scale production, due to the volume character and
high rates of the processes.

As for the comparison of the low power systems a more detailed
analysis is required.

The "shift" reaction proceeds very intensively at 250-400°C
with participation of extensively used catalysts based on
iron or other oxides and presents no problem for industrial
application. The problem of industrial application of the plasmochemical process (4) has not been solved so far, but it seems that having been successfully tested in a powerful laboratory device, the plasmochemical cycle of water decomposition

can be used as a basis of the large-scale technology of hydrogen generation. A schematic diagram of the technological equipment is given in Fig.1. (Fig.1. The schematic diagram of the technological chain. 1 - 2 - nuclear energetic blocks 3 - plasma reactor; 4 - heat exchanger; 5 - separator; 6 - aggregate; 7 - heat exchanger; 8 - separator.)

As is shown by the estimations, the efficiency of the two-stage cycle of water destruction given in Fig.1 differs from the theoretical one by a factor of about 0.65-0.7. The cycle may be improved by utilization of the heat from the plasmochemical and chemical reactors and by reduction of heat spent for separation of gas mixture components. In prospect the efficiency of about 0.85-0.9 can be considered as attainable one. Equally, increase in the NPP electrical efficiency results automatically in rise of the total efficiency of the technological power plant so that, on the whole, the nuclear heat-chemical energy conversion efficiency of about 35-40% can be expected. It should be noted in this connection that high efficiencies of the thermochemical cycles mentioned in the literature [1,2,3] are most likely obtained not on large experimental or industrial thermochemical plants and should be taken with care.

Let us consider the plasmochemical cycle and molecular mechanism of its realization in more detail. The main reason, why the two-stage plasmochemical cycle of hydrogen production from the water should be chosen, is a high power efficiency of the first stage (4), which can exceed 80%. Attempts to organize a direct plasmochemical cycle of water decomposition meet difficulties associated with a high reaction power of the H_2-O_2 system and high rate of the vibrational relaxation of the H_2O molecules [4].

The high efficiency of the plasmochemical stage (4) is reached due to a non-equilibrium nature of the CO_2 molecule destruction which proceeds efficiently on a condition that $T_e/T_o \gg 1$, where T_o is the "temperature" of neutral gas. For the typical experiment conditions with non-equilibrium SHF and HF discharges [5,7] the ratio T_e/T_o is 10^{-2}-10^{-1} ($T_e \sim 1 \div 3$ ev, $T_o \sim 5 \cdot 10^{-2} \div 10^{-1}$ ev). The concentration of charged particles, n_e , is usually low ($n_e \sim (1 \div 5) \cdot 10^{12} cm^{-3}$) and the degree of the scale ionization $n_e/n_o \sim 10^{-6}$ and 10^{-4}-10^{-3} for the SHF and HF induction discharges, respectively.

If a pressure of not less than several torrs is maintained in the core, the main channel of power losses for the electron gas will be excitation of the CO_2 molecule oscillations in the elementary act:

$$CO_2 + e \longrightarrow CO_2^* + e \qquad (6)$$

which is characterized by a high constant speed:

$$K_{ev} \geq 10^{-18} cm^3/sec$$

Molecules vibrationally excited in the process of VV relaxation exchange quanta effectively, which results in establishment of the equilibrium distribution over the oscillation levels due to the "diffusion flow" from small to large values of V. This process is characterized by $k_{vv} \approx 10^{-12}$ cm^3/sec, the VV relaxation constant, which exceeds the vibrational relaxation constant k_{vT} by several orders of magnitude in our case.

In the simplest case the Boltzman function $f_v \sim e^{-\frac{\hbar \omega v}{T_v}}$ with the vibrational temperature T_v can be chosen as equilibrium distribution in V .

For highly excited states (V≫1) the following elementary reactions are energetically resolved

$$CO_2^* + CO_2^* \longrightarrow O + CO + CO_2 \quad \Delta Q_1 = 5.5 \frac{ev}{mol} \qquad (7)$$

$$CO_2^* + O \longrightarrow CO + O_2 \qquad \Delta Q_2 \simeq 0.5 \frac{eV}{mol} \qquad (8)$$

as well as the reactions with close energy parameters through the intermediate radical CO_3

$$CO_2^* + CO_2^* \longrightarrow CO_3 + CO \qquad (9)$$

$$CO_3 + CO_2^* \longrightarrow CO + CO_2 + O_2 \qquad (10)$$

From the point of view of energetics both versions account for the summary reaction of CO_2 decomposition with energy consumption of about $3\frac{eV}{mol}$.

The speed constant for this stage can be written as [5] :

$$k_R(T_v, T_o) = k_o(T_o) \left(\frac{\mathcal{D}}{T_v}\right)^{S-1} e^{-\frac{\mathcal{D} - \Delta(T_o)}{T_v}} \qquad (11)$$

where S=4 is the number of oscillation degrees of freedom of the CO_2 molecule, D is the activation barrier, $\Delta(T_o)$ is the small correction connected with allowance for progressive and rotational degrees of freedom.

The oscillation energy balance in the stationary case is written as

$$k_{ev} n_e n_o \hbar\omega_1 - 2k_R(T_v, T_o) n_o^2 \Delta Q - k_{vT} n_o n_v \hbar\omega_2 = 0 \qquad (12)$$

from which, in particular, reduction for the minimum value of the concentration of charged particles follows

$$\frac{n_e}{n_o} \gg \frac{k_{vT}(T_o)}{k_{ev}} \cdot \frac{\hbar\omega_1}{\hbar\omega_2} \qquad (13)$$

Using (12) and (13) the rates of chemical acts (7) and (8) can be estimated: $V_R \simeq k_{ev} n_e n_o \frac{\hbar\omega_1}{\Delta Q}$

where n_0 is the CO_2 molecule concentration in the discharge.

Substituting the typical parameters of the SHF non-equilibrium discharge $n_e \gtrsim 10^{12} cm^{-3}$, $n_0 \simeq 10^{19} cm^{-3}$, we have $V_R \gtrsim 3 \cdot 10^{22}$ cm^{-3}/sec, which exceeds the rate of dissociation through the electron-excited state [6] by more than an order of magnitude.

The efficiency of the destruction of the CO_2 molecules becomes high beginning from a certain value of the oscillation temperature T_V . According to (12) this value may be determined as:

$$T_{vmin} \simeq \left[\mathcal{D} - \Delta(T_0) \right] \ell n^{-1} \left[\frac{k_0}{k_{VT}} \cdot \frac{2 \Delta Q}{\hbar \omega_2} \left(\frac{\mathcal{D}}{T_{vmin}} \right)^{S-1} \right] \quad (14)$$

If $T_V < T_{V \, min}$ then the larger part of the oscillation energy of molecules transforms into heat due to the VT relaxation escaping the chemical act.

In accordance with (12) dependence of the decomposition efficiency on the specific energy contribution W is given by [7] :

$$\eta = \frac{1}{W} \left[W + \frac{\alpha X}{2} \left(T_v - W \right) \theta \left(W - T_v \right) + \right. \quad (15)$$
$$\left. + \frac{x \hbar}{2} \left(\omega_{CO_2}^{\Sigma} - \omega_{CO} - \frac{1}{2} \omega_{O_2} \right) + \frac{2 \hbar \omega_2}{e^{\hbar \omega_2 / T_0} - 1} - \frac{2 \hbar \omega_2}{e^{\frac{\hbar \omega_2}{T_{vmin}} - 1}} \right]$$

Here $\alpha \gtrsim 1$ is the ratio of the speed constants of the oscillation excitation, CO and CO_2, ω_{CO} and ω_{O_2} are the oscillation constants of the corresponding molecules, $\omega_{CO_2}^{\Sigma}$ is the sum of the frequencies of all normal oscillations of CO_2. Similarly for the conversion coefficient X we have:

$$X = \frac{1}{\Delta Q} \left[W + \frac{\alpha X}{2} \left(T_v - W \right) \theta \left(W - T_v \right) + \right.$$
$$\left. + \frac{x \hbar}{2} \left(\omega_{CO_2}^{\Sigma} - \omega_{CO} - \frac{1}{2} \omega_{O_2} \right) + \frac{2 \hbar \omega_2}{e^{\frac{\hbar \omega_2}{T_0}} - 1} - \frac{2 \hbar \omega_2}{e^{\frac{\hbar \omega_2}{T_{vmin}} - 1}} \right] \quad (16)$$

Dependences η , X obtained from the calculation are shown in Fig.2.(Fig.2 Dependence of the efficiency (1) on conversion (2) in the CO_2 decomposition in a non-equilibrium plasma

a) - calculation results;

b) - data from the experiments with the SHF discharge;

c) -   data from the experiments with the HF discharge in
the magnetic field.) The experimental values of these
quantities corresponding to SHF and HF non-equilibrium discharges are also plotted in this figure.

High values of the efficiency of the process occuring in the
SHF discharge (up to 80%) are due to the notorions difference of the vibrational temperature from the translational
one, as is shown experimentally. The direct measurements of
T_0 show that this value lies within the range 600-800°K
(while under the optimal conditions $T_V \sim$ 2000-2500°K), simultaneously the ratio T_e/T_0 becomes $\sim$ 20.

It should be also noted that the speed of the reverse process in the $CO-O_2$ system is low since within the range 600-
800°K the mixture does not react by the radical chain mech-
nism, and the process of "slow" oxidation of CO has a noticeable speed only in the presence of H_2 or H_2O.

The kinetic and energy features of the plasmochemical reaction (4), confirmed experimentally, are the main physical factor of its practical applicability for the technology of the industrial generation of hydrogen from the water.
The latter circumstance, naturally, will become more concrete after gaining the experience in operation of the
pilot technological unit.

REFERENCES

1. Сборник "Вопросы атомной науки и техники", серия "Атомно-
   водородная энергетика", вып.I, ИАЭ им.И.В.Курчатова,М.,(1976)

2. I-st World Hydrogen Energy Conference, 1-3 March, 1976,
   Miami Beach, Florida (USA), Proceeding of the Hydrogen
   Economy Miami Energy Conference (THEME).

3. Pangborn J.B., Gregory D.P. Nuclear Energy Requirements
   for Hydrogen Production from Water. 749014 (1974).

4. Бочин В.П., Легасов В.А., Русанов В.Д., Фрицман А.А.,
   Шолин Г.В. Сборник "Вопросы атомной науки и техники",
   серия "Атомно-водородная энергетика", вып.I(2),2(3),М.,1977г.

5. Легасов В.А.,Русанов В.Д., Фрицман А.А., Шолин Г.В.,
   Ш Международный симпозиум по плазмохимии, Лимож, Франция,
   , 5.18 (1977г.)

6. Теоретическая и прикладная плазмохимия", под редакцией
   Полака Л.С., М., "Наука", (1957).

7. Легасов В.А., и др. ДАН СССР, 238, № I, 66 (1978г.)

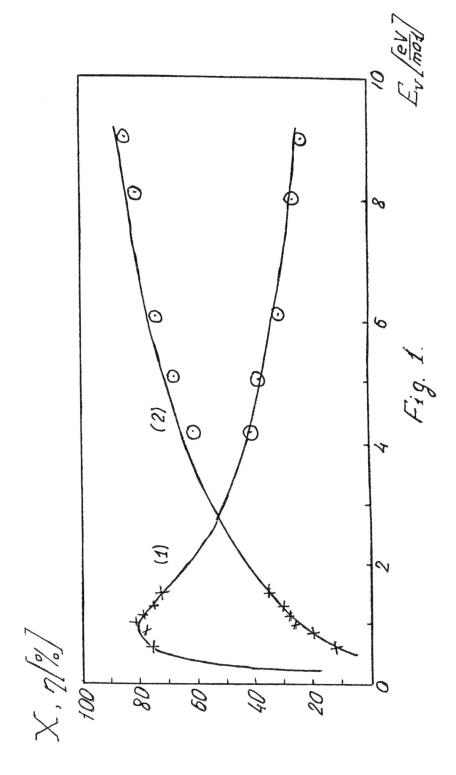

Fig. 1.

1181

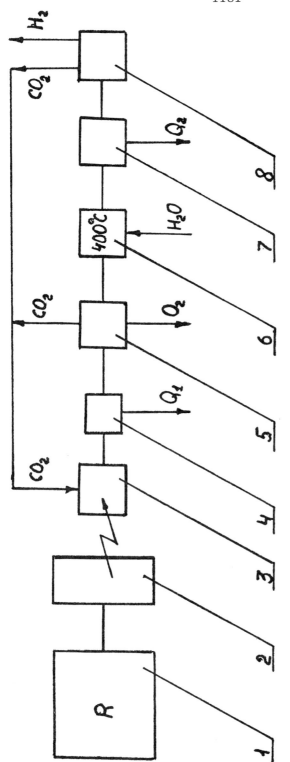

Fig 2

HYDROGEN PRODUCTION IN NON-EQUILIBRIUM
PLASMOCHEMICAL SYSTEMS

V.P.Bochin, A.A.Fridman, V.A.Legasov,
V.D.Rusanov, G.V.Sholin
I.V.Kurchatov Institute of Atomic Energy,
Moscow, USSR

ABSTRACT

One stage processes of water decomposition and hydrogen production in a non-equilibrium plasma have been considered. It has
been shown that the method described in the paper has an energy efficiency close to the characteristics of the traditional
electrolytic and thermochemical methods and, besides, is characterized by a higher productivity, comparatively low metal-
capacity and simplicity of technological realization.

1. Plasmochemical Method of Hydrogen Production.

The method of H_2O vapor decomposition and hydrogen production
in a non-equilibrium plasma attract considerable interest since it can allow to attain high productivity and efficiency
with simple technological equipment[1-3]. An important advantage of the plasmochemical systems is a volume nature of the
processes proceeding there. High volume flow speeds and high
speeds of homogeneous reactions in the gaseous phase permit to
attain high productivity on the systems which currently are simply realized. This feature of the plasmochemical methods for
hydrogen production makes it more attractive than other electrical methods, in particular, electrolysis (both with the liquid and (high-temperature) solid electrolytes. Another important feature of the non-equilibrium plasmochemical processes is
the fact that the energy contribution from the discharge is concentrated only in strictly determined degrees of freedom of

the molecules and given reaction channels. For this reason,
heat up of the gas, on the whole, is usually insignificant,
small are also heat losses associated with the inverse reaction. Possibility of attainment of high energy efficiency
(60-70%) at a comparatively low temperature of the gas is
an important positive feature of the plasmochemical method [4], which makes it different from both thermochemical
methods and high temperature electrolysis. A disadvantage
common for all the plasmochemical method for hydrogen production is the problem of separation of the products and isolation of the desired component from the final mixture. In
principle, this problem can be solved directly within the
plasmochemistry framework using systems with rotating plasma.
This entails, however, some technological difficalties connected with difference in pressures in the optimal operating
conditions of the plasmotrons and systems with rotating plasma.

The analysis and comparison of various plasmochemical schemes
show that, from the point of view of energy consumption and
productivity, optimal are the non-equilibrium processes of hy
drogen production, accomplished in a slightly ionized plasma
through vibrationally excited states of the reagents and dissociative attachment [2] . Realization of such optimal conditi
ons requires knowledge of the electron temperature $T_e \gg T_o$
(T_e=1-3 eV) and fulfillment of the conditions imposed on the
ionization degree. In particular, for the efficient direct
decomposition of the water vapor such a condition proves to
be rather stringent. For this reason, in addition to the pure
water vapor decompositions, we shall discuss here the question
on hydrogen production in a non-equilibrium plasma in the mixtures containing carbon oxides for which the limits on the ionization degree are found to be less rigorous.

2. Water Vapor Decomposition in a Non-equilibrium Plasma

The process of decomposition of water vapor in the plasma can
be accomplished by various mechanisms. The highest decomposition efficiency should be expected in the systems where the
electron temperature is not sufficient for intense excitation
of the electron states [4] and the larger portion of the energy contributed to the discharge is expended for excitation
of the vibration and for dissociative attachment. The relative contribution from these processes is compared at an electron temperature

$$T_{eo} = \mathcal{E}_a \, \ell n^{-1} \left[\frac{k_{amax}}{k_{ev}} \cdot \frac{\mathcal{E}_a}{\hbar \omega} \right] \qquad (1)$$

Here $\mathcal{E}_a$ =6 ev is the energy corresponding to the maximum
cross section of dissociative attachment K_a^{max} =10^{-9} $\frac{m}{sec}$ is
the speed constant of the process, corresponding to this ene-
rgy; K_{ev} is the speed constant of the vibrational excitation
of the water molecules by the electron impact, which at
T_e=1-3 ev can be estimated as K_{ev}=10^{-9} $\frac{cm^3}{sec}$ [5] ; $\hbar \omega$ =0.2ev
is the characteristic vibrational quantum of H_2O. At the abo-
ve parameters $T_{eo} \approx 1.7$ ev.

a) H_2O decomposition via vibrationally excited states of the molecules.

Consider the mechanism, kinetics and energetics of the H_2O decomposition in this case ($T_e < T_{eo}$). Dissociation of H_2O and
hydrogen production is then accomplished by the successive
stages of H_2O oscillative excitation, population of highly excited states during the V-V relaxation and, finally, by the
reactions with participation of H_2O^*. The reaction is initiated in the bimolecular act:

$$H_2O^* + H_2O^* \longrightarrow H + OH + H_2O \qquad (2)$$

$$V_Q = k_Q [H_2O]^2 \exp\left[- \frac{D(H_2O)}{T_v}\right] \qquad (3)$$

Here $K_o = 3 \cdot 10^{-10} \frac{cm^3}{sec}$ is the collision constant; $D(H_2O) \approx 5$ ev is the dissociation energy (2), T_v is the oscillation tempe-rature of the water molecules. Note that the alternative channel of decomposition initiation is the dissociative at-tachment:

$$e + H_2O \longrightarrow H^- + OH \qquad (4)$$

whose speed, however, is lower than that of process (2) on a condition

$$\frac{k_a^{max}}{k_o} \cdot \frac{n_e}{[H_2O]} \exp\left[- \frac{E_a}{T_e} + \frac{D(H_2O)}{T_v}\right] \ll 1 \qquad (5)$$

which usually is fulfilled at the T_e value under considera-tion.

The obtained radicals H and OH (2) initiate the chain reac-tion with participation of the oscillatively excited H_2O molecules

$$H + H_2O^* \longrightarrow H_2 + OH \;;\; \Delta H_1 = 15 \frac{kkat}{mot} \;;\; E_{a1} = 21 \frac{kkal}{mol} \qquad (6)$$

$$OH + H_2O^* \longrightarrow H + H_2O_2 \;;\; \Delta H_2 = 61 \frac{kkal}{mol} \;;\; E_{a2} = 70 \frac{kkal}{mol} \qquad (7)$$

The termination of the chain results mostly from the three-particle recombination (in the systems with a sufficiently high concentration of H_2O [6])

$$H + OH + H_2O \longrightarrow H_2O + H_2O \;;\; k_a = 3 \cdot 10^{-31} \frac{cm^6}{sec} \qquad (8)$$

The chain decomposition mechanism considered is characteri-zed by a chain length

$$\gamma = \sqrt{\frac{k_Q}{k_a[H_2O]}} \exp\left[\frac{D(H_2O) - E_{a1} - E_{a2}}{T_v}\right] \qquad (9)$$

Numerically at $T_v = 0.5$ ev and $[H_2O] = 3 \cdot 10^{18}$, $y = 10^2$. On the basis of (3) and (9) the speed of the unbranched chain process, decomposition is found as

$$V_R = V_0 y = k_0 [H_2O]^2 \sqrt{\frac{k_0}{k_a [H_2O]}} \exp\left[-\frac{\not{D}(H_2O) + E_{a1} + E_{a2}}{2T_v}\right] \quad (10)$$

It should be noted that the parallel channel of the chain extension:

$$OH + H_2O \longrightarrow H_2 + HO_2 \; ; \; \Delta H = 51 \frac{kkal}{mol} \; ; \; E_a = 75 \frac{kkal}{mol} \quad (11)$$

$$HO_2 + H_2O \longrightarrow H_2O_2 + OH \; ; \; \Delta H = 27 \frac{kkal}{mol} \; ; \; E_a = 30 \frac{kkal}{mol} \quad (12)$$

has a poorer kinetics than reactions (6) and (7) because of a higher activation barrier of the limiting process [6].

The total efficiency η of the process depends on the energy losses in the discharge η_{ex} efficiency η_{chem} associated with the negative contribution from the break reaction and heat losses of reactions (6) and (7) and, finally on the efficiency η_{vT} of the reaction relative to the oscillative relaxation.

The energy efficiency of the vibrational excitation by the electron impact is determined as

$$\eta_{ex} = k_{ev} \hbar\omega \Big/ \left[k_{ev}\hbar\omega + k_{et}\frac{2m}{M} T_e + k_{e2}\varepsilon_z + k_a^{max}\varepsilon_a e^{-\frac{\varepsilon_a}{T_e}}\right] \quad (13)$$

The terms in the denominator represent the energy contributions to oscillations, translational motion, H_2O rotation, and dissociative attachment, respectively. Under conditions considered $\eta_{ex} \simeq 90\%$.

The energy efficiency of the chain process proposed depends on the chain length:

$$\eta_{chem} = \frac{\Delta H_1 + \Delta H_2}{\dfrac{\not{D}(H_2O) - E_{a2}}{y} + E_{a1} + E_{a2}} \quad (14)$$

It is seen that when $\gamma = 10^2$ (see (9)) $\eta_{chem} = 85\%$, while in the absence of the chain process η_{chem} ($V=1$) $\simeq 50\%$.

In the vibrational relaxation the energy is lost both in the active and passive ($n_e = 0$) zone of the discharge. This energy loss channel may prove to be the most noticeable when the discharge parameters are chosen in accurately because of an anomalously high constant K_{VT} of the relaxation of the H_2O molecule within a wide temperature range ($K_{VT} \simeq 10^{-12} \frac{cm^2}{sec}$[7]) The vibrational relaxation in the active zone of the discharge can be neglected when a sufficiently tough criterion for the ionization degree is fulfilled:

$$\frac{n_e}{[H_2O]} > \frac{k_{vT}}{k_{ev}} \tag{15}$$

Only if condition (15) is fulfilled, the oscillation temperature T_V can be risen over the critical value $T_{V\ min}$ when the speed of reaction (10) becomes equal to that of the vibrational relaxation [4]. It follows from (10) that the minimum vibrational temperature which is high enough for an efficient decomposition of H_2O can be estimated as:

$$T_{vmin} = \frac{\mathcal{D}(H_2O) + E_{a1} + E_{a2}}{2\ell n \left[\frac{k_o}{k_{vT}} \frac{(E_{a1}+E_{a2})}{\hbar\omega} \sqrt{\frac{k_o}{k_a[H_2O]}} \right]} \tag{16}$$

At the above parameters $T_{V\ min} \simeq 0.3$ ev. It should be noted that in the absence of the chain process $T_{V\ min}$ ($\gamma = 1$) $\simeq 0.5$ev.

The relaxation losses in the passive zone of the discharge ($n_e = 0$) are due to the "residual" amound of the vibrationally excited molecules [3] :

$$\eta_{VT} \simeq \frac{E_V - T_{vmin}}{E_V} \tag{17}$$

where E_V is the total energy contribution per one water molecule. It is seen that η_{VT} rises with increasing E_V. However, increase in the energy contribution is limited by heatup of the mixture up to the ignition temperature $T_{0\ max}$

$(E_{V\ max} = \dfrac{4\ T_{0\ max}}{1-\eta} \simeq 1.2$ ev). It follows from foregoind that the maximum value of η_{VT} is reached when $E_V \simeq 1.2$ and is $\eta_{VT} \simeq 75\%$. Note that in the absence of the chain mechanism $\eta_{VT} = 30\%$.

From the calculations made using (14), (16), (17), the total maximum efficiency $\eta = \eta_{ex}\eta_{chem}\eta_{VT}$ is found to be $\sim 57\%$. For comparison, in the absence of the chain mechanism, the total effieicncy is $\eta_{max}(V=1) \simeq 14\%$. Dependence $\eta (E_V)$, in the presence or absence of the chain mechanism, is shown in Fig.1. Under the optimal conditions considered the relative hydrogen yield is about 20%.

b) H_2O decomposition via dissociative attachment.

Put $T_e \gtrsim T_{eo}$ in relaxationship (1). Within this energy range the energy contribution of the discharge can be concentrated mainly in the dissociation attachment of electrons to the water molecules. In each act of the dissociative attachment the electron disappears and a negative ion is produced. In the discharge the energy $W(H_2O) \simeq 30$ ev (with allowance for the electron excitation of H_2O) is spent up for electron production. Hence, the plasmochemical reaction proceeding via the dissociative attachment become energetically effective only when each electron produced in the plasma can repeatedly participate in the dissociative attachment. Multiple use of the electron becomes possible due to a high rate $K_d = 10^{-6}\ \dfrac{cm^3}{sec}$ of negative ion collapse by the electron impact [8,9]. The chain process initializes:

$$e + H_2O \longrightarrow H^- + OH \tag{4}$$

$$H^- + e \longrightarrow H + e + e \tag{18}$$

The chain termination occurs because of fast processes of the ion-ion recombination

$$H^- + H_2O^+ \longrightarrow H_2 + OH \tag{19}$$

$$H^- + H_2O + M \rightarrow H + H_2O + M \tag{20}$$

(When $[H_2O] = 3 \cdot 10^{18}$ the ion-ion recombination constant reaches $k_z^{ii} \simeq 10^{-7} \frac{cm^3}{sec}$) and due to the ion-molecular reaction

$$H^- + H_2O \longrightarrow H_2 + OH^- \tag{21}$$

having the rate $K_{io} \simeq 10^{-9} \frac{cm^3}{sec}$. Taking into account (19)-(21) the dissociative sticking chain length is [3] :

$$\nu_a \simeq \frac{k_d + k_z^{ii} + k_{io} \frac{[H_2O]}{n_e}}{k_z^{ii} + k_{io} \frac{[H_2O]}{n_e}} \tag{22}$$

Relation (22) imposes a restriction on the ionization degree such as

$$\frac{n_e}{[H_2O]} > \frac{k_{io}}{k_d} \tag{23}$$

Putting condition (23) fulfilled, we shall obtain the chain length $\nu_a = 10$ at the above parameters. Under the conditions considered the efficiency of the process is calculated from [3] :

$$\eta \simeq \frac{2W(H_2)}{\mathcal{E}_a + \frac{1}{\nu_a}W(H_2O)} \cdot \frac{k_a^{max} \mathcal{E}_a \exp(-\mathcal{E}_a/T_e)}{k_a^{max}\mathcal{E}_a \exp(-\frac{\mathcal{E}_a}{T_e}) + k_{ev}\hbar\omega + k_{et}\frac{2m}{M}T_e + k_{ez}\mathcal{E}_z} \tag{24}$$

Here $W(H_2)$ is the energy required for H_2 production from H_2O. When $T_e = 3$ ev and $\nu_a = 10$, $\eta \simeq 50\%$. For comparison, in the absence of the chain process η ($V_a = 1$) $\simeq 15\%$. Dependence η (V_a) is plotted in Fig.2. At the energy contribution $1\frac{ev}{mol}$ the above efficiency gives a relative hydrogen yield of about 20%.

c) Main restrictions on the process parameters

As has been shown above, decomposition of H_2O can occur in a non-equilibrium plasma with an efficiency of 50-70%. However, to attain such a high energy efficiency rather tough requirements to the discharge must be imposed. The main restriction in the decomposition, is the requirement of a high ionization degree. For different physical reasons, conditions (15) and (23) lead to the same restriction of the electron concentration, $n_e/[H_2O] > 3 \cdot 10^{-4}$. Requirement for such a high ionization degree for $T_e = 1-3\,ev$ is a rigorous condition. Fulfillment of this condition will require to use non-self-maintained discharges sustained by the high-current relativistic electron beam [10]. Mitigation of the stringent requirements imposed may be reached by adding an inert diluent to the water vapor, which, from the one hand, can rise the ionization degree and, from the other hand, to reduce the relative rate of relaxation of the excited water molecules. In this case it is necessary that in a sufficient dilution

$$\frac{n_e}{n_o} > \frac{K_{VT}(H_2O/Xe)}{K_{ev}^{H_2O}} \qquad (25)$$

where $K_{VT}(H_2O/Xe)$ is the constant of the H_2O molecule relaxation with an inert diluent (e.g. xenon). Putting $K_{VT}(H_2O/Xe) \simeq 3 \cdot 10^{-14} \frac{cm^3}{sec}$ we obtain for Xe: $n_e/n_o > 10^{-5}$ (for other inert diluents Ne- $n_e/n_o > 10^{-4}$; Ar: $n_e/n_o > 10^{-4}$; Kr: $n_e/n_o > 10^{-5}$).

In the water decomposition via the vibrational excitation of the reagents restrictions arise due to effects of the reactions with participation of free radicals involved in the chain process (6), (7). Because of a rather high threshold of reaction (7) the OH radical concentration can prove to be relatively high. In the approximation of stationary concentrations of the intermediate reagents it is

$$\frac{[OH]^2}{[H_2O]^2} = \frac{k_o}{k_a n_o} \exp\left[\frac{E_{a2} - E_{a1} - \mathcal{D}(H_2O)}{T_v} \right] \qquad (26)$$

A relatively high concentration of the OH radicals results in the intense-radical reactions and establishment of the equilibrium $OH+OH \rightleftharpoons H_2O+O$ leading to reduction in the energy efficiency. This process can be neglected only if the following condition is fulfilled:

$$\frac{K_{OH}}{k_a n_o} \exp \left[\frac{3E_{a2} - E_{a1} - \mathfrak{D}(H_2O)}{T_v} \right] \qquad (27)$$

where $K_{OH} \simeq 3 \ 10^{-12} \ \frac{cm^3}{sec}$ is the constant of the speed of the above reaction. Condition (27) proves to be rather stringent and means that for $n_o = 3 \ 10^{+19} cm^{-3}$, $T_v > 0.7$ ev which can result in explosion of the mixture $H_2-O_2-H_2O$. Condition (27) can be mitigated increasing the gas pressure.

The above mentioned high concentration of the OH radicals may also ruin the hydrogen produced in the reaction

$$OH + H_2 \longrightarrow H + H_2O \qquad (28)$$

with the activation barrier $E_{(-)} \simeq 0.5$ ev. This reaction can be only neglected if the following condition on the relative yield of hydrogen is fulfilled:

$$X = \frac{[H_2]}{[H_2O]} \ll \exp \left[\frac{E_{(-)} T_v - E_{a2} T_o}{T_v T_o} \right] \qquad (29)$$

In particular, for $T_v = 0.3$ ev this condition means $X < 0.01$ ($T_o \sim 10^3 K$). Requirement (29) is the main condition of stability of the hydrogen produced regarding the back reactions. If it is expressed as a limitation on T_o, then it will lead to a limitation similar to that shown in Fig.6. The negative contribution of reaction (28) proves to be stronger than the explosion hazard of the $H_2-O_2-H_2O$ mixture which at higher pressure of the mixture is inhibited by the H_2O vapor serving as a diluent (see (42)-(44)). It should be noted that the restrictions due to the OH radical ((27),(29)) relates, first of all, to the reactions of hydrogen production through the vibrational excitation of the reagents, while the H_2O decomposition via the dissociative attachment is mainly limited by the requirements on the ionization degree.

3. Hydrogen Production in Non-equilibrium Plasmachemical
   Systems H_2O-CO_2

As has been noted in the previous Section the direct decomposition of H_2O vapor in a non-equilibrium plasma meets restrictions connected both with the OH radical effect and requirements imposed on the ionization degree. Both these restrictions are essentially mitigated when adding CO to the
CO_2 system. Indeed, presence of the carbon oxide in the system leads to reduction of the OH free radical concentration
due to the reaction $OH+CO \rightarrow H + CO_2$ which practically has
no threshold. On the other hand, CO and CO_2 have the cross
sections of the vibrational excitation by the electron impact larger than that of H_2O by two orders of magnitude,
which allows to reduce significantly the requirements on the
ionization degree. As a matter of fact, CO_2 can serve as the
catalyst in direct production of hydrogen from the water in
the plasma. Indeed, the carbon oxide is not spent up entirely, but only permits to overcome difficulties encountered
in the decomposition of the pure water vapor. Such a process
can be accomplished both in one and two stages. In the first
stage the two-stage cycle CO_2 is decomposed to CO in the
plasma and then the water is reduced to hydrogen in the thermocatalytic shift reaction $(CO+H_2O \rightarrow CO_2+H_2)$ [1] . In the
present work the one-stage process of hydrogen production in
the CO_2-H_2O mixture as well as hydrogen production in the
plasmochemical system $CO-O_2-H_2O$ are discussed.

a) Non-equilibrium plasmochemistry of the CO_2-H_2O mixture.

The plasmochemical synthesis in the CO_2-H_2O mixture represents a complicated physical-chemical process which depending on the discharge parameters and mixture composition,
can produce various products such as hydrogen, carbon oxide, saturated hydrocarbons, olefines, aldehydes, organic
acids, alcohols etc [11] . Under the energetically optimal

conditions when the main energy contribution is concentrated in the vibrational degrees of freedom these determining parameters are the vibrational temperature T_V and ratio of concentrations $[CO_2]:[H_2O]$. We shall analyze the parameter range where the molecular hydrogen in the predominant product of the synthesis.

Let us consider first restrictions of the ionization degree. Addition of the carbon dioxide mitigates these restrictions since the CO_2 molecules are excited by the electron impact much more intensively and relaxate more weakly. Fig.3 shows the limiting degree of ionization in the CO_2–H_2O mixture, the dashed line indicates the restriction of the ionization degree in pure water vapor.

When the above condition on the electron concentration is fulfilled, the hydrogen is generated in the CO_2–H_2O mixture by the following mechanism. The carbon oxide is produced in the reactions

$$CO_2^* + CO_2^* \longrightarrow CO + O + CO_2 \qquad (30)$$

$$O + CO_2^* \longrightarrow CO + O_2 \ ; \ E_x \simeq 0,5 \, eV/mol \qquad (31)$$

at a rate which, in accordance with $[12]$, can be estimated as

$$\left(\frac{d[CO_2]}{dt}\right)_+ = 2k_0 [CO_2]^2 exp\left[-\frac{D(CO_2)}{T_V}\right] \qquad (32)$$

where $D(CO_2) \simeq 4$ ev is the effective activation barrier of the CO_2 dissociation which differs from the binding energy in this molecule both in the contribution of entire set of freedom degrees and in essentially nonadiabatic nature of the CO_2 dissociation. It should be pointed out that production of CO in the reactions via the intermediate radical CO_3 can be described by the similar kinetics:

$$CO_2^* + CO_2^* \longrightarrow CO_3 + CO \qquad (33)$$

$$CO_3 + CO_2^* \longrightarrow CO + O_2 + CO_2 \tag{34}$$

The main portion of O atoms (or, similarly, CO_3 radicals) react with CO_2^*, however, a part of them react with the water vapor producing the OH radicals in the process:

$$O + H_2O^* \longrightarrow OH + OH \quad \left(E_y \approx 1 \frac{eV}{mol} \right) \tag{35}$$

The OH radical produced initiates the chain process of reduction of the molecular hydrogen from the water with participation of the carbon oxide:

$$OH + CO \longrightarrow H + CO_2 \tag{36}$$

$$H + H_2O^* \longrightarrow H_2 + OH \tag{37}$$

In the calculation of the kinetics of the above non-branched chain reaction the chain length is determined as

$$V = \frac{[CO_2]}{[H_2O]} \exp\left[\frac{E_y - E_x}{T_v} \right] \tag{38}$$

It is clear that when $V \gg 1$ the hydrogen is effectively produced in the system. However, as has been noted, the vibrational excitation of the CO_2-H_2O mixture can result in formation of a number of different products. The region of the parameters at which the hydrogen is generated is restricted first of all by the reactions

$$OH + OH \longrightarrow H_2O + O \,, \ H + CO_2 \longrightarrow OH + CO \,, \ H + O_2 \longrightarrow OH + O$$

Fig.4 lists the permissible values of $[CO_2]:[H_2O]$ and T_v limiting the above hydrogen production mechanism from extraneous reactions.

The total efficiency of the hydrogen production process in the mixture under consideration mostly depends on the relaxation losses and losses in the above chemical reactions. The calculation of the efficiency gives the results presented in Fig.5. Thus, under the optimal conditions which correspond to

$1\frac{ev}{mol}$ CO_2 the energy consumption for the synthesis amount to about $5\frac{ev}{mol}$ H_2. Under these conditions the degree of water decomposition can reach about 60%.

The molecular hydrogen produced in the system is kept on the condition of relative smallness of the speeds of the back reactions:

$$OH + H_2 \longrightarrow H + H_2O \qquad (39)$$

$$O + H_2 \longrightarrow OH + H \qquad (40)$$

Both these reactions are characterized by the activation barrier $E_{(-)} \simeq 0.5$ ev and their speeds are found to be comparatively low due to the low translational temperature of the mixture ($T_0 \ll T_V$).

The condition for stability of the hydrogen produced relative to the above reactions in the requirement

$$T_0 < E_{(-)} \Big/ \left[\frac{D(CO_2)}{T_V} - 2\ell n \left(\frac{[CO_2]}{[H_2O]} \cdot \frac{1}{\sqrt{X\, f(OH)}} \right) \right] \qquad (41)$$

where X is the relative hydrogen yield from the H_2O, $f(OH) = [OH]/[H_2O]$. The restrictions of T_0, required for avoiding the chain blow described by the reaction competition, prove to be less stringent:

$$H + O_2 \longrightarrow OH + O \; ; \; E_0 \simeq 0.7 \frac{eV}{mol} \qquad (42)$$

$$H + O_2 + M \longrightarrow HO_2 + M \; ; \; k_3 \simeq 3 \cdot 10^{-31} \frac{cm^6}{sec} \qquad (43)$$

The condition on the temperature is then written as

$$T_0 < E_0\, \ell n^{-1} \left[\frac{k_0}{k_3 n_0} \right] \qquad (44)$$

The restrictions on the translational temperature, which are required to be fulfilled for stability of the hydrogen produced relative to the back reactions shown in Fig.6 ($[CO_2]$:
: $[H_2O] \simeq 10$, $X = 0.1$, $n_0 = 10^{19} cm^{-3}$).

b) Nonequilibrium plasmochemistry of the $CO-O_2-H_2O$ mixture.

Let us consider only the features which differentiate the process of hydrogen production in the $CO+O_2+H_2O$ mixture, from that in the CO_2-H_2O mixture. Hydrogen production in this mixture is the second stage of the cyclic process where, first, the reaction $CO_2 \rightarrow CO + \frac{1}{2}O_2$ is accomplished and then the hydrogen is reduced by the carbon oxide from the water. Unlike the thermocatalytic shift reaction $CO+H_2O \rightarrow CO_2+H_2$, the separation $CO+O_2$ is not required here before the second stage, which essentially simplifies the technology of the hydrogen production process on the whole. Describing the reactions proceeding in the $CO+O_2+H_2O$ mixture, we can distinguish three qualitatively different conditions. The first condition corresponds to the branched chain explosion $CO-O_2$ [13] , the second one-to slow oxidation of CO to O_2 with heterogeno-catalytic effect of H_2O [14] , and finally, the third condition specific for the nonequilibrium plasma, where hydrogen production can proceed effectively. The molecular hydrogen is also produced in the catalytic oxidation of CO, however, it rapidly vanishes on the surface ($H_2+O_2 \xrightarrow{S} H_2O_2$). H_2 can be effectively produced only because in the nonequilibrium plasma the speed of hydrogen generation in the volume will appreciably exceed the speed of its consumption for the catalytic oxidation. Production of H_2 in the nonequilibrium plasma follows a rapid process of atomic hydrogen production in the reactions:

$$CO^* + O_2 \longrightarrow CO_2 + O \qquad (45)$$

$$O + H_2O \longrightarrow OH + OH \qquad (46)$$

$$OH + CO \longrightarrow H + CO_2 \qquad (47)$$

The atomic hydrogen produced initiates the nonbranched chain reaction of H_2 production:

$$H + H_2O \longrightarrow H_2 + OH \tag{48}$$

$$OH + CO \longrightarrow H + CO_2 \tag{49}$$

A detailed calculation of the kinetics and energetics of the process considered allows to estimate the energy consumption for hydrogen production in the above system. Dependence of this energy consumption on the specific power contributed (energy per one CO molecule) is shown in Fig.7. The optimal energy consumption is then $\sim 1\frac{ev}{mol}$ H_2 and the relative hydrogen yield is about 5%. Thus, the above process of hydrogen production in the $CO-O_2-H_2O$ mixture is an interesting example of the plasmochemical reaction where the consumption of the electrical energy prove to be less than consumption of "chemical energy". Indeed, in this reaction formation of one H_2 molecule requires about 1 ev of electrical energy, while the hydrogen production enthalpy is about $3\frac{ev}{mol}$. The nonequilibrium plasma is used here not to contribute the energy but only to direct the reaction by the appropriate channel (in the equilibrium heatup of the $CO-O_2-H_2O$ mixture the predominant process is oxidation of CO). The electrical efficiency of the process reaches 300%. The reaction described above is, in this sense "a chemical explosion controlled by the nonequilibrium plasma", which is similar to some extent to the situation occuring in the chemical lasers. Naturally, taking into account the energy consumption for CO_2 decomposition in the plasma we shall obtain the total efficiency of about 70% (which, naturally, does not exceed unity). Thus, it has been shown that addition of CO_2 as a catalyst to the water vapor allows to reduce the requirements on the plasma ionization degree and decrease the contribution from the back reactions connected with the effect of the OH radical.

4. CONCLUSION

It has been shown that the above mentioned plasmochemical systems allow to produce hydrogen from the water with close va-
lues of energy efficiency, 60-70% (4,3-5 $\frac{ev}{mol}$, 5-6 $\frac{kw \times hr}{M^3}$)
The direct decomposition of water vapor is the simplest one
physically and allows to attain an energy efficiency up to
80% (4 $\frac{kw \times hr}{M^3}$). However, it imposes stringent restrictions on
the degree of plasma ionization. Additions of the catalysts
(in particular, CO_2) permit to mitigate requirements to the
electron concentration but, on the other hand, it narrows the
region of the permissible composition of the mixture and im-
pedes removal of the useful products from the mixture. The
capabilities of the plasmochemical systems described is very
high because of high volume rates of the gas flow. For example, when using the generators with the stationary power of
10MW, the capability of one block of the nonequilibrium plas-
motron reaches 2000 $\frac{M^3}{hr}$ H_2. Because of a high energy efficiency heatup of the gas is insignificant and the mixture does
not seem to be explosion hazardous (see Fig.6). In the case
of pure water vapor decomposition isolation of the hydrogen
from the products is reduced to separation of the hydrogen
from the oxygen (the water vapor is condensed). This problem
is not bery complicated because of a significant difference
between the H_2 and O_2 molecule masses. Such a separation can
be accomplish comparatively easyly both in the common and in
plasma cetrifuge.

In conclusion we shall compare the described plasmochemical
method for hydrogen production with the electrolysis. As has
been shown above, in the case considered 5 kw $\times$ hr of electrical energy can be expended for production of $1m^3 H_2$; in the
low-temperature electrolysis of water up to 6kw $\times$ hr of electrical energy is spent up for production of 1 $m^3 H_2$ because
of polarization of the oxygen electrode [15] . In addition,
the capability of the method described was estimated as

as 2000 $\frac{m^3}{hr}$, while the common electrolysers produce 500$\frac{m^3}{hr}$ on one block.

Increase in the energy efficiency of the electrolysis can be reached at the expense of a sharp reduction of the polarization effects with increasing the temperature.

Such high-temperature electrolysis systems allow to expect a high energy efficiency of 4-5 kwXhr per 1 m^3 of hydrogen, with the water supplied in the form of vapor at 850°K [16] . However, such devices have a limited capability from the unit surface (the capability of the elementary block of volume 1 cm^3 does not exceed $5\frac{1}{hr}$ [16]) because of their heterogeneous character. The required capability of the process can be attained then only by provision of many blocks, which might make the operation of the whole system very unstable. Besides the high-temperature electrolyses require application of a specific solid electrolyt, e.g. ZrO_2 + Y_2O. Such an electrolyt is expensive, as a rule, because the necessary conductivity is attained by use of scarce materials and complicated technology.

Thus, the result obtained permit to assert that the main electrical methods for hydrogen production, electrolysis and plasmochemical ones, are characterized by close values of the energy efficiency. Relative advantages of the electrical methods can be decided from the comparison of their capabilities and simplicity of technological realization of the corresponding processes. This comparison shows that the non-equilibrium plasmochemical systems are promising for decomposition of the water and hydrogen production.

REFERENCES

1. И.Г.Белоусов, В.А.Легасов, В.Д.Русанов "Вопросы атомной науки и техники" серия "Атомно-водородная энергетика", в.2 (3), стр. 158, 1977г.

2. В.П.Бочин, В.А.Легасов, В.Д.Русанов, А.А.Фрицман, Г.В.Шолин "Вопросы атомной науки и техники". Серия "Атомно-водородная энергетика", в.2(3), стр.93, 1977г.

3. В.П.Бочин, В.А.Легасов, В.Д.Русанов, А.А.Фрицман, Г.В.Шолин "Вопросы атомной науки и техники". Серия "Атомно-водородная энергетика" в.1(2), стр.55, 1977г.

4. В.А.Легасов, В.Д.Русанов, А.А.Фрицман, Г.В.Шолин. Ш Международный симпозиум по плазмохимии, Лимож, стр.5.18,1977г.

5. H.S.W.Massey "Electronic and ionic impact phenomena" Vol II Oxford at the Clarendon Press, 1969

6. В.Н.Кондратьев "Константы скорости газофазных реакций". М., "Наука", 1970г.

7. Progress in Reaction Kinetics, 5, 1965.

8. Б.М.Смирнов, М.И.Чибисов, ЖЭТФ, 49, 841, 1965г.

9. G.C.Tisone, L.M.Branscome Phys.Rev., 170, 1969, 169.

10. А.К.Вакар, В.П.Денисенко, Г.П.Максимов, В.Д.Русанов Ш Международный симпозиум по плазмохимии, стр.1.10,1977г.

11. Д.Н.Андреев "Органический синтез в электрических разрядах" АН СССР, М.-Л., 1953г.

12. В.А.Легасов, В.К.Животов, Е.Г.Крашенинников и др. ДАН СССР, 238, 66, 1978г.

13. С.Бенсон "Основы химической кинетики". М., "Мир", 1964г.

14. Льюис, Эльбе "Горение, пламя и взрывы в газах", М.,"Мир", 1968г.

15. А.П.Егоров и др. "Общая химическая технология неорганических веществ", М., "Химия", 1964г.

16. J.O.'M.Bocris I.Int.Conf.Hydrogen Energy, p.371, 1974.

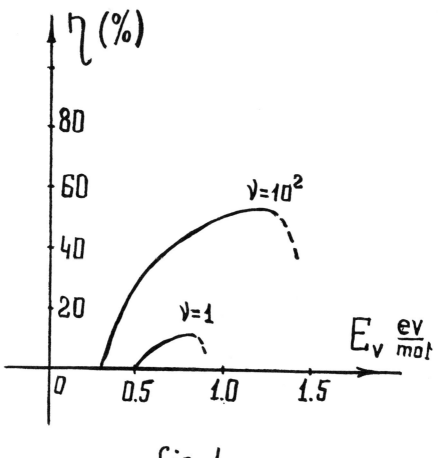

fig. 1

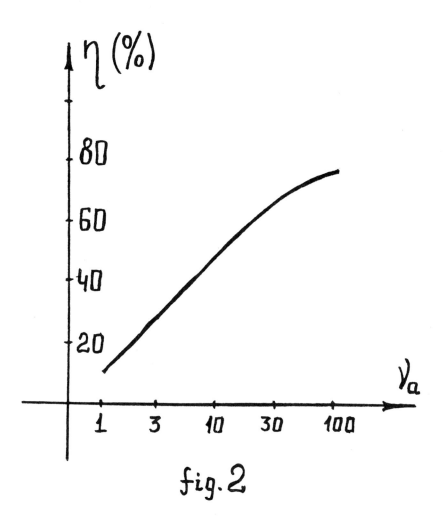

fig. 2

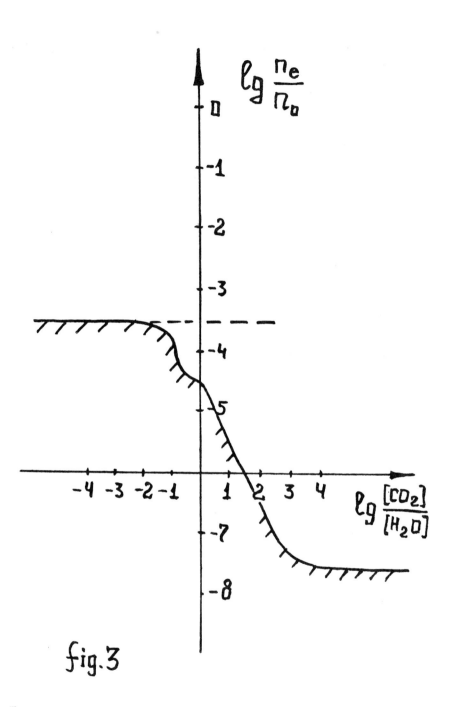

fig.3

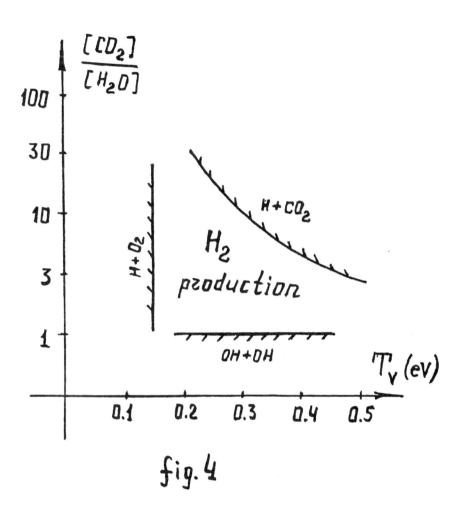

fig. 4

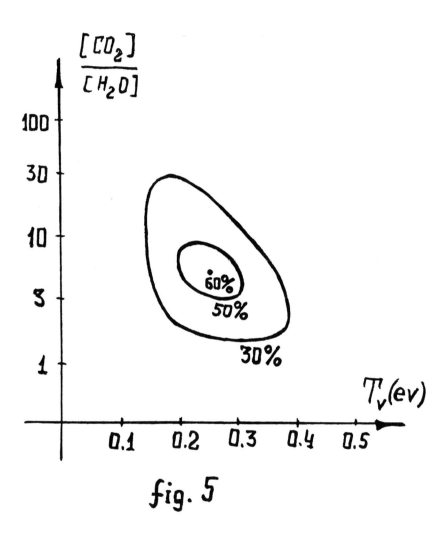

fig. 5

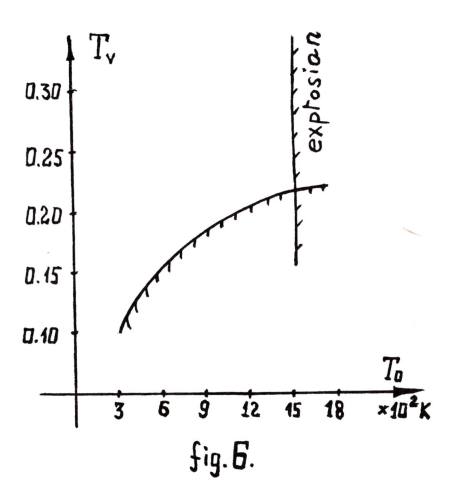

fig. 6.

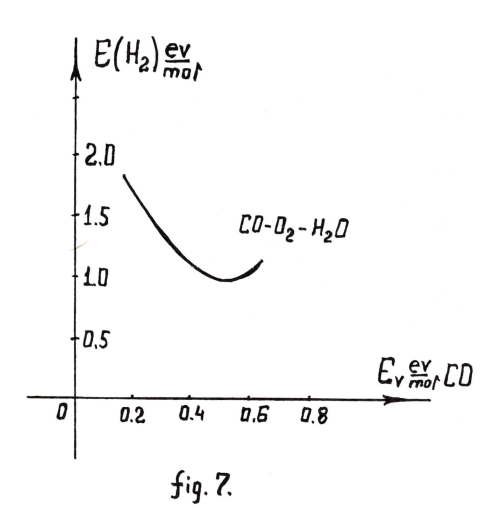

fig. 7.

SPIN-ALIGNED HYDROGEN

W. C. Stwalley
University of Iowa
Iowa City, Iowa, U.S.A.

ABSTRACT

Spin-aligned hydrogen is a thusfar hypothetical form of hydrogen which consists of hydrogen atoms maintained exclusively in their lowest electronic
energy magnetic sublevels under low temperature, high magnetic field conditions. Because the "spin-flip" (or Zeeman) energy is large compared to
the thermal energy under these conditions, highly exothermic recombination
of hydrogen atoms to form hydrogen molecules should not occur. This paper
briefly reviews the exotic properties of this unique substance and the current prognosis for its experimental preparation.

INTRODUCTION

Most atomic species have a nonzero magnetic moment, i.e. because not all
their electrons are paired, they behave like tiny magnets. Just as bar
magnets align themselves with a magnetic field to obtain a minimum energy,
so atomic species tend to align themselves with a magnetic field in a state
of minimum energy. Under even high magnetic field conditions (say 100 kG =
10T) at ordinary temperatures, however, the thermal energy of atomic motion
disaligns a great many of the atoms. What is the result of this disalign-
ment?

Let us first recall the interactions between bar magnets (Table I). If
two bar magnets are both aligned in the same direction by a separate magnetic field and are then pushed together with their alignment maintained,
one will feel a strong repulsive force between the two magnets. In a similar fashion, the forces between two spin-aligned hydrogen atoms (symbolized
H↑) are predominantly repulsive. However, if magnets with opposite align-
ments are brought together, they will strongly attract one another and,
upon making contact, will strongly adhere to each other. Likewise, if a
spin-aligned hydrogen atom encounters a disaligned hydrogen atom, they will
strongly attract each other and form a chemically-bonded hydrogen molecule
H_2 (assuming there is a means of dissapating the molecular bond energy released by this highly exothermic atomic recombination reaction).

Thus hydrogen ordinarily occurs in its molecular and not its atomic form
because no one has yet succeeded in totally preventing atomic spin disa-
lignment. However, if one could, one would have a metastable material with

a uniquely high energy content, particularly on a per gram basis, as illustrated by Table II (alternatively, the usable energy content per gram of the solid should be 3.4×10^6 joules/liter or about 100 times that of gunpowder!). This possibility was indeed discussed some twenty years ago [1,2], but recently there has been a revival of theoretical and experimental interest in spin-aligned hydrogen, described in the next sections.

THEORETICAL PROPERTIES OF SPIN-ALIGNED HYDROGEN

The repulsive forces between spin-aligned hydrogen atoms are known accurately from elaborate first principles calculations. Using this information, one can show conclusively that H↑ will remain a gas no matter how low the temperature, i.e. even at absolute zero [3,4]. This is unlike the behavior of any currently known substance, and indeed helium is the only substance which is even liquid at a temperature of absolute zero.

Moreover, H↑ should show extremely interesting "quantum fluid" behavior; in particular, it should be superfluid under certain conditions. Helium is the only similar "quantum fluid", but its behavior is far less "quantal" than is expected for spin-aligned hydrogen.

It appears also that spin-aligned hydrogen will not form a liquid under any circumstances, but it should be possible to form a solid phase by applying a pressure of ~50 atmospheres [3].

GENERAL STABILITY CONSIDERATIONS

As noted above, a necessary condition for the preparation of H↑ is that the thermal energy of atomic motion cause no disalignments, which lead to molecule formation and rapid heating and destruction of the system. In quantitative terms, one expects that a magnetic field to absolute temperature ratio of about one million gauss per degree kelvin is needed to completely prevent disalignments in a macroscopic sample over long periods of time (say years) [5]. Such field to temperature ratios can indeed now be attained with commercial instrumentation (e.g. a superconducting magnet yielding >10 kg magnetic field and a helium dilution refrigerator yielding <0.1 °K). We hope to have a system capable of such conditions in operation at Iowa for spin-aligned hydrogen experiments this year.

A second problem associated with preparation of H↑ is the containing wall [5], which could catalyze the recombinations of atoms in several ways. However, it does appear that certain walls (e.g. helium-coated) will not absorb H↑ or will absorb it only very weakly, so that the wall problem can be eliminated. The impurity problem [5] can also be fairly readily eliminated.

Somewhat more stringent field and temperature conditions have been suggested for solid or high-density spin-aligned hydrogen [6].

EXPERIMENTAL APPROACHES

At the present time, no experimental preparation of pure spin-aligned hydrogen has been carried out. Several extremely interesting experiments
[7,8] involving a small fraction of H atoms in a H_2 matrix have suggested
significant spin-alignment.

Recently, a practical method of H↑ preparation has been proposed [9]. It
is based on sending a relatively cold (say 1.5 °K) atomic beam of initially
unaligned H atoms (produced by e.g. a microwave discharge) into the high
magnetic field/low temperature region mentioned above. Only aligned H
atoms (i.e. H↑) will enter the region of high magnetic field; moreover,
they will readily be cooled to ~0.1 °K by a single wall collision (based on
similar behavior for helium atoms). It should be possible to prepare macroscopic samples (perhaps a millimole of H↑) in this way.

CONCLUSIONS

It appears likely to this author that macroscopic amounts of spin-aligned
hydrogen will be prepared within the next five years. The cost of doing so
will not be small, since the required low temperature/high magnetic field
technology is relatively expensive at the present time. The long-range
economics of these technologies will be strongly influenced by uncertain
factors such as the degree of helium conservation and whether or not superconducting transmission lines become practical. Nevertheless, the uniquely
high energy content (Table II) of this still unproven form of hydrogen does
suggest that long-term applications in energy storage are indeed possible.

ACKNOWLEDGMENT

Support received from the Petroleum Research Fund, administered by the
American Chemical Society, is gratefully acknowledged.

REFERENCES

1. J. T. Jones, M. H. Johnson, H. L. Mayer, S. Katz, and R. S. Wright,
   Characterizations of Hydrogen Atom Systems, Aeroneutronics Systems,
   Inc., Publication U-216 (1958).

2. M. W. Windsor, A. M. Bass and H. P. Broida, editors, Formation and
   Trapping of Free Radicals (Academic Press, N. Y., 1960), p. 400f.

3. W. C. Stwalley and L. H. Nosanow, Phys. Rev. Lett. 36, 910 (1976);
   M. D. Miller, L. H. Nosanow, and L. J. Parish, Phys. Rev. B 15, 214 (1977);
   L. H. Nosanow, in S. B. Trickey, E. D. Adams and J. W. Dufty, editors,
   Quantum Fluids and Solids (Plenum, N. Y., 1977), p. 279.

4. J. V. Dugan and R. D. Etters, J. Chem. Phys. 59, 6171 (1973);
   R. D. Etters, J. V. Dugan and R. W. Palmer, J. Chem. Phys. 62, 313
   (1975);
   R. L. Danilowicz, J. V. Dugan and R. D. Etters, J. Chem. Phys. 65,
   499 (1976).

5. W. C. Stwalley, Phys. Rev. Lett, 37, 1628 (1976);
   W. C. Stwalley, in S. B. Trickey, E. D. Adams and J. W. Dufty, editors, Quantum Fluids and Solids (Plenum, N. Y., 1977), p. 293.

6. A. J. Berlinsky, R. D. Etters, V. V. Goldman and I. F. Silvera, Phys.
   Rev. Lett. 39, 356 (1977);
   A. J. Berlinsky, Phys. Rev. Lett. 39, 359 (1977).

7. R. Hess, Adv. Cryog. Eng. 18, 427 (1973);
   R. Hess, doctoral dissertation, Universtiy of Stuttgart, 1971 (unpub-
   lished);
   R. Hess, Deutsch Luft- and Raumfahrt, Forschungsbericht 73-74: Atomarer
   Wasserstoff (Institut fur Energie Wandlung und Elektrische Antriebe,
   Stuttgart/Braunschweig, 1973);
   W. Peschka, Bull. Am. Phys. Soc. 23, 85 (1978).

8. R. W. H. Webeler, J. Chem. Phys. 64, 2253 (1976);
   G. Rosen, J. Chem. Phys. 65, 1795 (1976);
   R. W. H. Webeler and G. Rosen, Bull. Am. Phys. Soc. 23, 85 (1978);
   R. Danilowicz and R. D. Etters, Bull. Am. Phys. Soc. 23, 85 (1978);
   K. Sugawara and J. A. Woollam, Bull. Am. Phys. Soc. 23, 86 (1978).

9. W. C. Stwalley, Bull. Am. Phys. Soc. 23, 85 (1978).

TABLE I.   INTERACTIONS BETWEEN BAR MAGNETS AND BETWEEN HYDROGEN ATOMS

N/S + N/S	repulsive H↑ + H↑
N/S + S/N	attractive H↑ + H↓

TABLE II.   SPIN-ALIGNED HYDROGEN REACTIONS PRODUCE
FAR MORE HEAT PER GRAM THAN COMMON COMBUSTION REACTIONS

Reactions	$-\Delta H$ (kcal/gram fuel)	
	O_2 from air	O_2 as fuel
H↑ + H↓ → H_2	52	52
4H↑ + O_2 → $2H_2O$	86	9.6
$2H_2$ + O_2 → $2H_2O$	34	1.9
CH_4 + $2O_2$ → $2H_2O$ + CO_2	13	2.6

HYDROGEN PRODUCTION ALTERNATIVES AND OTHER INNOVATIVE PROCESSES II

HYDROGEN GENERATION VIA PHOTOELECTROLYSIS OF WATER-RECENT ADVANCES

A. J. Nozik
Materials Research Center
Allied Chemical Corporation
Morristown, New Jersey   07960
USA

ABSTRACT

In the past few years, a burgeoning international research effort has developed to produce hydrogen by the photoelectrolysis of water using sunlight. The status of this research is reviewed. Advances in the theoretical understanding of the energetics and mechanisms of photoelectrolysis is described, as well as progress in the development of new semiconductor electrodes materials. The major problems associated with the achievement of an economically viable photoelectrolysis system are also discussed.

INTRODUCTION

The work of Fujishima and Honda [1] on the photoelectrochemistry of TiO_2 single crystal anodes, published between 1970 and 1972, subsequently precipitated a major international research effort in the field of solar energy conversion using photoelectrochemical systems. Thus, as documented in a recent review of the field [2], about 150 papers have been published since 1975 which deal with photoelectrochemistry and applications to solar energy conversion; prior to this period a total of about 50 papers had been published in the general field of of photoelectrochemistry.

Two modes of solar energy conversion with photoelectrochemical systems are possible. In one mode, optical energy is converted into chemical energy, while in the second mode, optical energy is converted into electrical energy. We will be concerned here only with the former mode, and in particular with the photodecomposition of water into hydrogen and oxygen - a process called photoelectrolysis. The progress made in this field since the last report [3] at the First World Hydrogen Energy Conference will be reviewed and discussed.

1217

THEORETICAL

Review of Energetics of Semiconductor-Electrolyte Interfaces

All phenomena associated with photoelectrochemical systems are
based on the formation of a semiconductor-electrolyte junction
when an appropriate semiconductor is immersed in an appropriate electrolyte. The junction is characterized by the presence of a space charge layer in the semiconductor adjacent to
the interface with the electrolyte. A space charge layer generally develops in a semiconductor upon contact and equilibration with a second phase whenever the initial chemical potential of electrons is different for the two phases. For
semiconductors, the chemical potential of electrons is given
by the Fermi level in the semiconductor. For liquid electrolytes, the chemical potential of electrons is determined by
the redox potential of the redox couples present in the electrolyte; these redox potentials are also identified with the
Fermi level of the electrolyte.

If the initial Fermi level in an n-type semiconductor is above
the initial Fermi level in the electrolyte (or any second
phase), then equilibration of the two Fermi levels (or chemical potentials) occurs by transfer of electrons from the semiconductor to the electrolyte. This produces a positive space
charge layer in the semiconductor (also called a depletion
layer since the region is depleted of majority charge
carriers). As a result, the conduction and valence band edges
are bent such that a potential barrier is established against
further electron transfer into the electrolyte (see Figure 1).

An inverse but analogous situation occurs with p-type semiconductors having an initial Fermi level below that of the electrolyte.

A charged layer also exists in the electrolyte adjacent to the
interface with the solid electrode - the well known Helmholtz
layer. This layer consists of charged ions from the electrolyte adsorbed on the solid electrode surface; these ions are
of opposite sign to the charge induced in the solid electrode.
The width of the Helmholtz layer is generally of the order of
a few Å units. The potential drop across the Helmholtz layer
depends upon the specific ionic equilibrium obtaining at the
surface.

A very important consequence of the presence of the Helmholtz
layer for semiconductor electrodes is that it markedly affects the band bending that develops in the semiconductor when
it equilibrates with the electrolyte. Without the Helmholtz
layer, the band bending would simply be expected to equal the
difference in initial Fermi levels between the two phases

(i.e., the difference between their respective work functions). However, the potential drop across the Helmholtz layer modifies the net band bending as shown in Figure 1. This effect is similar to the well known situation in Schottky junctions formed between semiconductors and metals, where the potential barrier and band bending are usually strongly influenced by the presence of semiconductor surface states.

In Figure 1, the energy scales commonly used in solid state physics and in electrochemistry are shown for comparison. In the former, the zero reference point is vacuum, while in the latter it is the standard redox potential of the hydrogen ion-hydrogen (H^+/H_2) redox couple. It has been shown [4,5] that these scales are related in that the effective work function or Fermi level for the standard (H^+/H_2) redox couple at equilibrium is -4.5 eV with respect to vacuum. Hence, using this scheme, the energy levels corresponding to any given redox couple can be related to the energy levels of the valence and conduction bands of the semiconductor electrode.

To make the connection between the energy levels of the electrolyte and the semiconductor it is necessary to introduce the flat-band potential, U_{fb}, as a critical parameter characterizing the semiconductor electrode. The flat-band potential is the electrode potential at which the semiconductor bands are flat (zero space charge in the semiconductor); it is measured with respect to a reference electrode, usually either the standard normal H^+/H_2 redox potential (NHE) or the standard calomel electrode (SCE). Hence, the band bending is given by

$$V_B = U - U_{fb} \qquad (1)$$

where U is the electrode potential (Fermi level) of the semiconductor. At equilibrium in the dark, U is identical with the potential of the redox couple in the electrolyte.

The effect of the Helmholtz layer on the semiconductor band bending is contained within the flat-band potential. This important parameter is a property both of the semiconductor bulk and the electrolyte, as seen from the following relationship:

$$U_{fb} \text{ (NHE)} = (\chi + \Delta E_F + V_H) - 4.5 = (\phi_{SC} + V_H) - 4.5 \qquad (2)$$

where χ is the electron affinity of the semiconductor, ϕ_{SC} is the work function of the semiconductor, ΔE_F is the difference between the Fermi level and majority carrier band edge of the semiconductor, V_H is the potential drop across the Helmholtz layer, and 4.5 is the scale factor relating the H^+/H_2 redox level to vacuum.

Flat-Band Potentials

As will be described later, the flat-band potential of the semiconductor electrode establishes the external bias requirements for photoelectrolysis cells. Therefore, methods for predicting and measuring flat-band potentials are very important.

A significant advance in the prospects for predicting flat-band potentials has been made by Butler & Ginley [6]. Their method is based on Equation 2. The electron affinity, χ, is predicted from the atomic electronegativities of the constituent atoms of the semiconductor. V_H is accounted for by first establishing the point of zero zeta potential (PZZP); this is the condition where the number of adsorbed positive and negative ions is equal, so that $V_H=0$. For many semiconductors in aqueous electrolyte, H^+ and OH^- are the relevant adsorbed ions, and the PZZP is thus the pH at which $V_H=0$. In these cases, the change of V_H with pH follows a simple relationship (2), so that

$$V_H \text{ (volts)} = 0.059 \ (pH_{PZZP}-pH) \qquad (3)$$

It is possible to calculate pH_{PZZP} if the standard state electrochemical potentials of the adsorbed H^+ and OH^- ions are known [6]. If this is not possible, then pH_{PZZP} can be experimentally determined by a simple potentiometric titration of an aqueous suspension of a powder of the semiconductor in a solution of known ionic strength. Finally, the ΔE_F term in Equation (2) can be calculated knowing the carrier density of the semiconductor and its density of states effective mass [7].

Butler & Gingley [6] tested their method with many oxide semiconductors and found good agreement with their theoretical model. They also found that the pH_{PZZP} correlates with the bulk electronegativity of the semiconductor, so that it may be possible to determine pH_{PZZP} empirically rather than experimentally or theoretically. The general applicability of this method to non-oxide semiconductors and adsorbing ions other than H^+ and OH^- has not yet been established.

The most accurate compilation of experimental flat-band potentials for many semiconductors, measured for both n and p-type materials at several pH values and accounting for frequency dependent effects, has been presented by Gomes & Cardon [8]; the U_{fb} values are reported to be accurate within 0.05 to 0.1V. From the flat-band potential, and a knowledge of the carrier density, the effective mass, and the band gap of the semiconductor, the conduction and valence band edges can be determined with respect to the standard redox scale. These data are presented in Figure 2 for several semiconductors at pH = 1.0.

The effect of light intensity on the U_{fb} of TiO_2 has been investigated by Nozik [9]. No dependence on light intensity was
found, indicating that the equilibrium for ionic adsorption is
unaffected by light. This is an important characteristic which
simplifies the analysis of photoelectrochemical effects at
semiconductor-electrolyte interfaces.

Energetics of Photoelectrolysis Cells

Two types of photoelectrolysis cells can be distinguished. In
the first type, one electrode is a semiconductor and the second
electrode is a metal; this cell has been labeled [10] a
Schottky-type photoelectrolysis cell. In the second type, one
electrode is an n-type semiconductor and the second electrode
is a p-type semiconductor; this has been labeled [10] a p-n
type photoelectrolysis cell. These labels originate from the
idea that the cells can be visualized as Schottky junctions and
p-n junctions which have been split and electrolyte interposed
between the halves. However, it must be understood that the
critical junctions of all types of photoelectrolysis cells are
based on semiconductor-electrolyte junctions, which are always
analogous to Schottky junctions. For this reason it would be
better to designate the second type of cell described above as
p/n type photoelectrolysis cells to avoid confusion with certain properties specific to solid state p-n junctions.

1. Schottky-type cells

The energetics of the Schottky-type cell is represented in
Figure 3 for the photoelectrolysis of water into H_2 and O_2
using n-type electrodes (analogous analyses apply to p-type
electrodes). Since there are two redox couples in the electrolyte, the initial Fermi level in the electrolyte can be any-
where between them depending upon the initial relative concentrations of H_2 and O_2 in the cell. In Figure 3(a), the initial
Fermi level in the electrolyte (i.e., the metal electrode potential) is arbitrarily drawn just above the O_2/H_2O redox level.
Upon equilibration in the dark (Figure 3(b)), the Fermi level in
the semiconductor equilibrates with the electrolyte Fermi level,
producing a band bending, V_B, in accordance with Equation 1.

Under illumination (Figure 3(c)), the Fermi level in the semiconductor rises toward U_{fb}, producing a photovoltage, V_{ph}. V_{ph}
can be measured between the two electrodes with an external po-
tentiometer. However, the value of V_{ph} depends upon the
initial metal electrode potential, which depends upon the initial relative concentrations of H_2 and O_2; V_{ph} can vary from
zero volts to some finite value. Except under very special
circumstances (initial metal electrode potential and the valence band edge of the semiconductor electrode both at the
O_2/H_2O redox potential), V_{ph} is not the potential energy available for photoelectrolysis.

With the two electrodes shorted together, the maximum Fermi
level possible in the cell is the flat-band potential [11-13].
In Figure 3, U_{fb} is below the H^+/H_2 redox potential. Hence,
even with illumination intensity sufficient to completely
flatten the semiconductor bands, H_2 could not be evolved at
the counter-electrode because the Fermi level is below the
H^+/H_2 potential (Figure 3(c)). In order to raise the Fermi
level in the metal counter-electrode above the H^+/H_2 potential,
an external anodic bias, E_B, must be applied, as shown in
Figure 3(d). This bias also provides the overvoltage at the
metal cathode, η_c, required to sustain the current flow, and
it increases the band bending in the semiconductor to maintain
the required charge separation rate.

The situation depicted in Figure 3 is the one that describes
most of the n-type semiconductors studied to date. For these
semiconductors, an external bias is required to generate H_2 and
O_2 in a Schottky-type photoelectrolysis cell; the further U_{fb}
lies below H^+/H_2, the greater the bias. The bias can be applied
either by an external voltage source, [14-16] or by immersing
the anode in base and the cathode in acid [14-16] (the two compartments being separated by a membrane). In the earliest work
of Fujishima and Honda the presence and need for a bias for the
photoelectrolysis of water using a TiO_2/Pt cell was not explic-
itly stated.

Several semiconductors such as $SrTiO_3$ [12,17,18], $BaTiO_3$ [19].
$KTaO_3$ [20], Nb_2O_5 [21], and ZrO_2 [21], have U_{fb} above the H^+/H_2
potential; therefore, no external bias is required to generate
H_2 and O_2 in a Schottky-type cell. This has been confirmed for
$SrTiO_3$ [12,17,18], $BaTiO_3$ [19], and $KaTaO_3$ [20]. Unfortunately
these oxides all have large band gaps (3.3 to 3.5eV), which result in essentially zero solar absorbtivity and, hence, they
are ineffective in systems for solar energy conversion.

For purposes of discussion of Figure 3, the difference between
the O_2/H_2O redox potential and the valence band edge at the interface is defined as the intrinsic overpotential of the semiconductor anode, η_a (9, 22). This overpotential is not the
usual overpotential or overvoltage of conventional electrochemistry since it is current independent and is determined only
by the band gap, the flat-band potential, and the redox potential of the electrolyte donor state.

A simple identity can be derived [3,9,10,14] from the energy
diagram of Figure 3(d), and it is given at the bottom of
Figure 3. This identity can be considered to represent an
energy balance describing the distribution of the energy produced by the absorption of a photon by the semiconductor. In
the absence of a bias, the input energy of an absorbed photon
is equal to the semiconductor band gap (E_g has been shown to be

equivalent to the free energy of formation of an electron-hole pair [23]). Energy loss terms in the semiconductor resulting from the movement of electrons and holes from their point of creation to the respective electrolyte interfaces are the band bending (V_B), and the difference between the conduction band edge in the bulk and the Fermi level (ΔE_F). The net electron-hole pair potential available at the electrode interfaces is thus $E_g - V_B - \Delta E_F$. In the electrolyte, a portion of this potential energy is recovered as the free energy $(\Delta G/nF)$ of the net endoergic reaction in the electrolyte. The rest of the potential energy is lost through the irreversible, entropy-producing terms η_a, η_c, and iR (ohmic heating). One loss term which had been included in previously published energy balance equations [3,10,14] was the potential drop across the Helmholtz layer due to adsorbed dipolar molecules of the electrolyte. However, as first pointed out by Gerischer [24], this term is already contained within other terms of the energy balance, and hence is eliminated from the present energy balance expressions.

Some degree of controversy has arisen [25] over the physical significance of the energy balance equation described above, and the nature of the energetics of photoelectrolysis, in general [26]. The essence of the controversy relates to whether the steady state distribution of photogenerated electrons and holes should be described in terms of statistical thermodynamics, or not. In the former case, the energetics would be treated using normal thermodynamic arguments based on the collective behavior (statistical average) of all the electrons and holes of the system.

Gerischer [26-28] has used a statistical approach and invoked the quasi-Fermi level concept for electrons and holes to describe the photogenerated potential for driving the cell reaction; this potential is equated to the difference between the quasi-Fermi levels for electrons and holes at the semiconductor-electrolyte interface. The splitting of electron and hole quasi-Fermi levels is dependent upon light intensity and, hence, this model predicts a threshold light intensity before the cell reaction occurs. A free energy balance based on this model would be $({}_nE_F^* - {}_pE_F^*) = \frac{\Delta G}{nF} + \eta_c' + \eta_a' + iR$, where ${}_nE_F^*$ and ${}_pE_F^*$ are the quasi-Fermi levels at the interface for electrons and holes, respectively; η_a' is the difference between ${}_nE_F^*$ and the O_2/H_2O redox potential; and η_c' is the difference between ${}_pE_F^*$ and the H^+/H_2 redox potential.

The quasi-Fermi level concept depends upon local statistical detailed balance between the creation and annihilation of electrons and holes. It has been argued by Williams and Nozik [29] that this detailed balance is not a good approximation for photoelectrolysis, as evidenced by the very high quantum efficiencies (0.8 to 1.0) which are experimentally observed [16,31].

It has been proposed [29] that charge transfer across the semi-conductor-electrolyte interface is highly irreversible, and analogous in some respects, to photoemission into a vacuum. On this basis, it is believed [29] that the charge transfer process is better approximated as individual events. The energy balance of Figure 3 would then correspond to a description of the energetics of these individual events. In the Williams and Nozik model, the semiconductor-electrolyte interface is treated as a heterojunction, and the band properties of liquid water are taken into account (see Figure 4). The injection of holes from the photo-excited n-type semiconductor anode is then described as a tunneling process which is fast compared to electron-hole recombination or to thermal relaxation. Thus, "hot" holes can be also injected into the electrolyte [29].

While the Gerischer model [26] predicts a light intensity threshold for photoelectrolysis, the irreversible photoinjection model [29] predicts no threshold intensity. Experimental results with TiO_2 [30,31] over a light intensity range of about 1×10^{-3} mw/cm^2 to 4×10^5 mw/cm^2 show no threshold intensity and a linear photocurrent response with intensity. Furthermore, studies [31] of the temperature dependence of photoelectrolysis show no temperature effects on the action spectra (2500Å to 4200Å) from 25°C to 80°C. These results appear to support the irreversible photoinjection model. However, because of the wide band gap of TiO_2, the threshold light intensity predicted by the Gerischer model may be too low to be readily observed [26].

Other models [32-36] have also been proposed for the energetics of photoelectrolysis. In general, the theory of photoelectrolysis is still in a very rudimentary state, and further research is required to establish a valid theory.

2. p/n cells

p/n cells can be either homotype or heterotype, depending whether the two semiconductor electrodes are the same (differing only in the type of dopant present), or whether the two electrodes are different materials. The latter case is the most interesting, and the energetics of this type of cell is depicted in Figure 5 [10]. The important feature of this configuration is that the electron affinities of the two semiconductors are different. This produces an enhancement of the available electron-hole pair potential for the net electrolyte reaction when both electrodes are illuminated. As seen from the energy balance presented at the bottom of Figure 5, the net electron-hole pair potential available at the electrode surfaces is the sum of the semiconductor band gaps minus the difference in their flat-band potentials and minus the Fermi level terms, ΔE_F, for each semiconductor.

The sum of the band bending values produced at each semiconductor is equal to the difference between their flat-band potentials. The amount of potential enhancement is maximized if the difference in flat-band potentials is minimized. The minimum difference in flat-band potentials is determined by the minimum band bending required in each semiconductor electrode to produce efficient charge carrier separation. The total band bending present in the two electrodes is independent of light intensity. However, the distribution of the total band bending between the electrodes is dependent upon light intensity and the carrier densities of each electrode.

In the p/n cell, two photons must be absorbed (one in each electrode) to produce one net electron-hole pair for the cell reaction. This electron-hole pair consists of the minority hole and minority electron from the n-type and p-type electrodes, respectively, and it has a potential energy greater than that available from the absorption of one photon. The majority electron and hole recombine at the ohmic contacts.

An important possible advantage of the p/n cell is that for a given cell reaction it may allow the use of smaller band gap semiconductors. Since the maximum photocurrent available from sunlight increases rapidly with decreasing band gap [37-39] this could produce higher conversion efficiencies. This possibility is discussed in further detail below.

The enhanced electron-hole pair potential available from p/n cells can eliminate the bias required for Schottky-cells when U_{fb} is below the H^+/H_2 redox potential. This was shown for the first time [10,40] with a n-TiO_2/p-GaP cell; photoelectrolysis of water into H_2 and O_2 was achieved at zero bias with simulated sunlight [10]. Further studies have been made with n-TiO_2/p-CdTe, n-$SrTiO_3$/p-CdTe, and n-$SrTiO_3$/p-GaP cells [41, 42].

Conversion Efficiency

A rigorous calculation of the maximum possible theoretical conversion efficiency for photoelectrolysis cells has not yet been made. Although certain aspects of the calculation are similar to those for photovoltaic cells, there are sufficient basic differences between photoelectrolysis cells and photovoltaic cells to require a separate, independent treatment of the efficiency calculation for photoelectrolysis cells [9].

The primary factor governing the maximum efficiency for photoelectrolysis is the minimum semiconductor band gap(s) which can be used and still drive the desired electrolyte reaction. The efficiency, η, for the photoelectrolysis of water into H_2 and O_2 is 1.23 I_{ss}/P_{in}, where I_{ss} is the short circuit current flowing in the cell, and P_{in} is the input optical power (it is assumed here that the bias requirements are zero; finite bias (E_B) reduces the efficiency by $E_B I_{ss}/P_{in}$). The maximum possible value for the photocurrent (I_g) from a semiconductor with a band gap, E_g, which is exposed to solar radiation, has been published by several authors [37-39]. At air mass I (sun at Zenith on clear day, at sea level), I_g saturates at about 80 ma/cm^2 for $E_g \leq 0.4$eV; I_g values of about 50 ma/cm^2, 25 ma/cm^2, and 12 ma/cm^2 are obtained at E_g values of 1.0, 1.5, and 2.0 eV, respectively. Hence, smaller band gaps produce larger maximum photocurrents.

The actual photocurrent, I_{ss}, will depend upon the "squareness" of the current-voltage characteristic and the quantum efficiency of the cell [9]. The former factor is the ratio of the short circuit current to the saturated photocurrent obtained with bias. Both of these factors can be accounted for by defining a current factor, α, which is equal to I_{ss}/I_g. A model has not yet been established which permits the calculation of α from fundamental principles and properties.

For photovoltaic cells, the efficiency is given by $\eta_{PV} = E_g I_g (CF)(VF)/P_{in}$, where CF and VF are the current factor and voltage factor, respectively. The former is related to the squareness of the current-voltage curve for the photovoltaic cell, and the latter is equal to the ratio of the open circuit voltage to the semiconductor band gap. Both factors decrease with decreasing band gap, which leads to a maximum η_{PV} of about 25% at $E_g = 1.4$eV [37].

The basic difference between a photovoltaic cell and a photoelectrolysis cell is that in the former a definite, well-defined external photovoltage is generated. In photoelectrolysis, an external photovoltage is not normally produced (cell operating under short circuit conditions); if an external photovoltage is produced by introducing a load in the external circuit between the electrodes, the magnitude of this external photovoltage is not related at all to the photopotential developed for the net electrolyte reaction. In fact, the external photovoltage can be varied over a wide range by simply changing the relative initial O_2 and H_2 concentration in the cell, thereby changing the Fermi level of the electrolyte and the band bending of the semiconductor electrode in the dark [9].

In a simple analysis, Gerischer [26,28] equates the maximum conversion efficiency for photoelectrolysis to that for photovoltaic cells times the factor $1.23/E_g$. Making the substitution for η_{PV} indicated above, this leads to $\eta = 1.23 I_g(CF)(VF)/P_{in}$. Thus, the Gerischer analysis suggests that $\alpha = (CF)(VF)$. However, the open circuit photovoltage (V_{oc}) is not a meaningful variable in photoelectrolysis; hence the voltage factor, VF, is not applicable to the efficiency analysis (VF = V_{OC}/E_g). In other words, the Gerischer analysis assumes that the optimum current-voltage curve for a semiconductor of a given band gap will be equivalent regardless of whether the semiconductor is used in a photovoltaic cell or in a photoelectrolysis cell. This assumption is in serious question since the significance of V_{OC} is so different for the two cases. Furthermore, other factors affecting CF and VF, such as the dark saturation current, are also quite different for photovoltaic cells and photoelectrolysis cells.

For photoelectrolysis, the best case for maximum theoretical efficiency is believed to be the heterotype p/n cell [10]. Hence, the minimum sum of band gaps depends on the minimum permissible values of V_B, η_a, and η_c. Crude approximations concerning these values [9] and of α, yield an estimate of about 25% for the maximum possible efficiency. This is to be compared to the value of about 16% for the maximum efficiency of water splitting by a photovoltaic cell in series with a conventional electrolysis cell [9].

Further research must be done to establish the efficiency limitations more rigorously, and to determine whether the optimum values of E_g, V_B, U_{fb}, η_a, η_c, as well as the required stability can be achieved in semiconductor materials. These values are roughly estimated to be: $E_g \sim 2.0eV$, η_a (or η_c) $< 0.4V$, and $V_B < 0.3V$ for Schottky-type cells, and $E_g(n) + E_g(p) \sim 2.3eV$, $\eta_a + \eta_c \sim 0.6V$, and $V_B(n) + V_B(p) \sim 0.4V$ for heterotype p/n cells [9]. Furthermore, the flat-band potentials of the semiconductor electrodes must bear certain relationshps with respect to the electrolyte redox couples, as discussed earlier, in order for the cells to operate at zero bias.

The maximum experimental conversion efficiency with sunllght for the group of materials studied to date (see Table I) is in the range of 0.5 to 1.0%.

SEMICONDUCTOR ELECTRODE STABILITY

The photogenerated holes and electrons in semiconductor electrodes are generally characterized by strong oxidizing and reducing potentials, respectively. Instead of being injected into the electrolyte to drive redox reactions, these holes and electrons may oxidize or reduce the semiconductor itself, and cause decomposition. This possibility is a serious problem for practical photoelectrochemical devices, since photodecomposition of the electrode leads to inoperability or to short electrode lifetimes.

A simple model of electrode stability has been presented by Gerischer [26-27] and by Bard & Wrighton [43]. The redox potential of the oxidative and reductive decomposition reactions are calculated and put on an energy level scale like the one shown in Figure 1. The relative positions of the decomposition reactions are compared with those of the desired redox reactions in the electrolyte and with the positions of the semiconductor valence and conduction band edges. Absolute thermodynamic stability of the electrode is assured if the redox potential of the oxidative decomposition reaction of the semiconductor lies below (has a more positive value on the SCE scale) the valence band edge, and if the redox potential of the reductive decomposition reaction lies above (has a more negative value on the SCE scale) the conduction band edge. This situation does not exist in any of the semiconductors studied to date. More typically, one or both of the redox potentials of the semiconductor oxidative and reductive decomposition reactions lie within the band gap, and hence become thermodynamically possible. Electrode stability then depends upon the competition between thermodynamically possible semiconductor decomposition reactions and thermodynamically possible redox reactions in the electrolyte. This competition is governed by the relative kinetics of the two possible types of reactions.

For TiO_2, the redox potential for the oxidation of water to O_2 is more negative than that for the oxidation of TiO_2 to O_2; therefore, the former couple lies further above the valence band edge than the latter couple. This makes the water oxidation reaction more thermodynamically favored than the TiO_2 oxidation reaction. This means that at equilibrium, the conversion of water to oxygen would be greater than the conversion of the TiO_2 oxidation reaction. However, equilibrium is generally never attained here, and kinetic factors predominate. It is very difficult to predict kinetic behavior, and one must rely on experimental data. For TiO_2, it has been found that water oxidation proceeds nearly exclusively rather than TiO_2 oxidative decomposition. However, some evidence of TiO_2 degradation does exist [44,45], especially in acidic solutions.

In cases where the redox potentials of the electrode decomposition reactions are more thermodynamically favored than the electrolyte redox reactions (oxidative decomposition potential more negative, reductive decomposition potential more positive, than the corresponding electrolyte redox reactions), the products of the electrolyte redox reactions have sufficient potential to drive the electrode decomposition reactions [43]. Hence, this situation usually results in electrode instability; ZnO [26-27], Cu_2O [26-27], and CdS [26-27,43] semiconductor electrodes are unstable in simple aqueous electrolytes, and represent examples of this situation.

It appears that the more thermodynamically favored redox reactions also become kinetically favored so that these reactions
predominate. The origin of this effect has been generally attributed [46,26-27] to the existence of surface states [47-50]
within the semiconductor band gap. These surface states allow
photogenerated minority holes in n-type electrodes to rise to
the highest redox level, where efficient, isoenergetic hole
transfer can then occur; an inverse process occurs with p-type
electrodes. This effect has been used to stabilize semiconductor electrodes by establishing a redox couple in the electrolyte with a redox potential more negative than the oxidative decomposition potential (or more positive than the reductive decomposition potential), such that this electrolyte redox reaction occurs preferentially to the decomposition reaction, and
scavenges the photogenerated minority carriers [26-27,43]. However, this stabilization technique can only be used for electrochemical photovoltaic cells, and not for photoelectrolysis cells.

ELECTRODE MATERIALS AND CONFIGURATIONS

Photochemical Diodes

The elimination of bias requirements for photoelectrolysis cells
through the use of p/n-type systems leads to a very interesting
configurational variation. This configuration has been labeled
"photochemical diodes" [51], and comprises photoelectrolysis
cells which are collapsed into monolithic particles containing
no external wires. The photochemical diodes can be Schottky-
type or p/n type, and in one simple form respectively consist of
a sandwich of either a semiconductor and a metal, or an n-type
and a p-type semiconductor, connected together through ohmic contacts.

To generate the relevant redox reactions, photochemical diodes
are simply immersed in an appropriate electrolyte and the semiconductor face(s) illuminated. The size of a photochemical
diode is arbitrary,provided that there is sufficient optical absorption, and that a space charge layer exists which is consistent with efficient charge carrier separation. Within these constraints, the particle size may approach colloidal, or perhaps
macromolecular dimensions, and the diodes may possibly operate
as a colloidal dispersion or as a solute in solution, thereby
greatly simplifying the design and operation of photoelectrochemical reactors using sunlight.

A comparison of the energy level schemes for p/n heterotype
photochemical diodes and biological photosynthesis reveals very
interesting analogies [9,51]. Both systems require the absorption of two photons to produce one useful electron-hole pair.
The potential energy of this electron-hole pair is enhanced so

that chemical reactions can be driven which require energies greater than that available from one photon. Furthermore, the n-type semiconductor is analogous to Photosystem II, the p-type semiconductor is analogous to Photosystem I. and the recombination of majority carriers at the ohmic contacts is analogous to the recombination of the excited electron from Excited Pigment II with the photogenerated hole in Photosystem I [see Figure 6].

Semiconductor Materials

Approximately two dozen semiconductors have been studied as electrodes, and they are listed in Table I. All of the n-type materials studied thus far are oxides since they are generally more stable against photoelectrochemical oxidation than non-oxides.

The bias required for each electrode in order to generate H_2 in a Schottky-type cell is approximately equal to the difference between the flat-band potential and the H^+/H_2 redox potential. For oxides, this difference is approximately constant with changing pH since the adsorbed ions are H^+ and OH^- and the drop across the Helmholtz layer is given by Equation 3. Semiconductor anodes which operate with zero bias (U_{fb} above the H^+/H_2 level) are $SrTiO_3$, $KTaO_3$, Ta_2O_5, and ZrO_2. All the n-type semiconductors in Table I are reported to be stable except Bi_2O_3, PbO, and V_2O_5.

A few p-type materials have been examined as cathodes for photoelectrolysis cells (Table 1). The relative stability of p type semiconductors against reductive decomposition is generally greater than the stability of n-type semiconductors against oxidative decomposition. Non-oxide semiconductors with small band gaps, such as p-Si and p-GaP, appear to be stable when used as cathodes. However, the flat-band potentials of these materials with respect to the H_2O/O_2 redox level are unfavorable, and large bias voltages must be applied. For zero bias operation in p-type Schottky-type cells, U_{fb} must be below (more positive than) the H_2O/O_2 redox level.

Coated Semiconductors

One approach to the problem of obtaining stable and efficient semiconductor electrodes has been to overcoat unstable small band gap semiconductors with thin films of stable wide band gap semiconductors or with thin metal films. The former approach has been studied [74-75] with TiO_2 deposited (by CVD) on n-Si, n-CdS, n-GaAs, n-GaP, and n-InP. Additional studies have been made [76] of n-GaAs and n-GaAlAs overcoated with RF-sputtered films of TiO_2, SnO_2, Nb_2O_3, Al_2O_3, and Si_3N_4, and n-GaAs coated with MoO_3 [77].

The general problem with this approach is that films which are thick enough ($\gtrsim$500Å) to provide protection against decomposition do not permit the transport of photogenerated minority carriers from the small band gap semiconductor into the electrolyte. Films thinner than about 400Å allow diffusion of electrolyte species to the inner semiconductor, or they do not provide a continuous coating [76].

An electrode comprising a thick TiO_2 film deposited over a silicon p-n junction solar cell has been successfully shown [78] to produce photoelectrolysis without an external bias. Here, the silicon solar cell photovoltage is in series with the photoelectrolytic cell and provides the necessary bias for H_2O splitting required by the TiO_2 anode.

It has been claimed by Nakato et.al. [79,80] that a 300Å layer of gold on n-GaP prevents the photodecomposition of n-GaP, and permits the photo-oxidation of water to oxygen. However, this claim has been disputed by Wilson et.al. [81-82]; this thickness does provide protection of the n-GaP against decomposition.

Nakato et.al. [59] also claim that very small amounts of Pt, Pd, Ni, and Cu deposited on p-GaP act as catalysts for H_2 evolution. The overvoltage is reduced at the cathode, leading to H_2 evolution at potentials more positive by 0.2 to 0.3V compared to the base p-GaP electrode.

Dye sensitization

Efforts have been made to increase the visible absorptivity of stable, wide band gap semiconductors, and hence improve their conversion efficiency, by means of dye sensitization techniques [83-93]. The dye molecules are adsorbed on the semiconductor surface; when they become excited by the absorption of visible light, charge transfer may occur from the excited state of the adsorbed molecule to the semiconductor energy bands. The direction of the charge transfer (electron injection or hole injection) depends upon the relative positions of the energy levels of the excited dye and the semiconductor.

Successful sensitization of TiO_2 electrodes to visible light has been demonstrated using alizarin dyes [78], polypyridineruthenium (II) complexes [89], and rose bengal [90]. ZnO has been successfully sensitized with rhodamine B [85,88], rose bengal [87,93], and alizarin dyes [84]; CdS has been sensitized with rhodamine B [92]. The general problems with this approach are: (1) adsorbed layers which are sufficiently thick to absorb appreciable amounts of incident light inhibit the charge transport process because of the low carrier mobility

and low electrical conductivity of the dyes; and (2) the dyes
are not sufficiently stable and/or are not regenerated during
the oxidation-reduction cycles.

SUMMARY

Photoelectrochemistry is based on the unique properties of the
semiconductor-electrolyte junction. The energetics and kinetics of photo-induced charge transfer reactions across semiconductor-electrolyte interfaces, as well as other important aspects of the junction potential barrier, have been studied by
many researchers since 1954. However, beginning about 1970,
Honda & coworkers [1] were the first to point out the potential application of photoelectrochemical systems for solar
energy conversion and storage. This work has stimulated a
large international research effort, and since 1975 over 150
papers have been published on photoelectrochemical systems for
energy conversion.

Two classes of photoelectrochemical devices are possible:
photoelectrolysis cells and electrochemical photovoltaic cells.
In the former, optical energy is converted into chemical energy,
while in the latter optical energy is converted into electrical
energy. Critical semiconductor electrode parameters for both
systems are the flat-band potential and the band-gap. The
latter governs the absorption characteristics and hence the
ultimate conversion efficiency by establishing the external
bias requirements for photoelectrolysis cells and the maximum
attainable photovoltage from electrochemical photovoltaic cells.

Progress has been made in predicting flat-band potentials from
atomic properties of the constituent atoms of the semiconductor
electrode. Experimental techniques for measuring flat-band potentials have been critically examined and are better understood.

Semiconductor surface states may control the kinetics of charge
transfer across the semiconductor-electrolyte interface, and
they may play a major role in determining the photoelectrochemical stability of semiconductor electrodes.

Electrode stability is a major problem for photoelectrochemical
devices. For photoelectrolysis cells, n-type semiconductors
which have been found to be stable against oxidative decomposition during O_2 evolution are limited thus far to wide band
gap oxides ($E_g > 2.2eV$), and this results in low conversion efficiency (maximum reported values with sunlight are about 0.5 to
1.0%). Smaller band gap p-type semiconductors appear to be
stable against reductive decomposition, but large external bias
voltages are required for photoelectrolysis.

Efforts to improve the efficiency of photoelectrolysis cells include the use of bi-layered electrode systems, and dye sensitization techniques.

A new and interesting configuration for photoelectrolysis cells, comprising monolithic structures containing no external wires, has been described and named photochemical diodes. These photochemical diodes are simply immersed in electrolyte and illuminated to drive the electrolyte reactions.

A rigorous and detailed theory for photoelectrolysis has not yet been derived, and controversy exists concerning the energetics, mechanism, and upper limit conversion efficiency for photoelectrolysis. However, it is generally accepted that for the splitting of water into H_2 and O_2 using sunlight, the maximum possible efficiency for an optimum photoelectrolysis cell is significantly greater than that obtained with an optimum solar cell (solid state or liquid junction) in series with a conventional electrolysis cell.

The major advantages of photoelectrochemical devices are the great ease with which the semiconductor-electrolyte junction is formed (the semiconductor is simply immersed into the electrolyte), and the fact that thin, polycrystalline, semiconductor films can be used without serious loss of conversion efficiency compared to single crystal semiconductors.

Although significant advances have been made both in the basic understanding of photoelectrochemical devices and in the development of systems with good conversion efficiency and stability, much additional research and development must be done before photoelectrochemical systems can be seriously considered for practical solar energy conversion schemes.

REFERENCES

1. a) Fujishima, A., Honda, K. 1972. Nature 238:37; b) 1971. Kogyo Kagaku Zasshi (J. Chem. Soc. Japan). 74:355; c) 1971. Bull. Chem. Soc. Japan. 44:1148; d) 1970. J. Institute of Industrial Science, Univ. of Tokyo. 22:478.

2. Nozik, A.J. 1978. Annual Reviews Physical Chem., Vol. 29 (in press).

3. Nozik, A.J., 1976. Proc. First World Hydrogen Energy Conf., Vol. II 5B-31, Miami Beach, Fla.

4. Gerischer, H. 1975. Electroanal. Chem. and Interfacial Electrochem. 58:263.

5. Lohmann, F. 1967. Z. Naturforsch. 22a:843.

6. Butler, M., Ginley, D.S. 1978. J. Electrochem. Soc. 125:228.

7. Sze, S.M. 1969. Physics of Semiconductor Devices, pp. 25-38. New York:John Wiley. 812 pp.

8. Gomes, W.P., Cardon, F. 1977. See Ref. 9, pp. 120-31.

9. Nozik, A.J. 1977. In Semiconductor Liquid-Junction Solar Cells, Proceedings of Conference on Electrochemistry and Physics of Semiconductor-Liquid Interfaces Under Illumination, Airlie, Va., ed. A. Heller, Electrochemical Soc., Princeton, N.J., pp. 272-289.

10. Nozik, A.J. 1976. Appl. Phys. Lett. 29:150.

11. Ohnishi, T., Nakato, Y., Tsubomura, H. 1975. Ber. Bunsenges. Phys. Chem. 79:523.

12. Mavroides, J.G., Kafalos, J.A., Kolesar, D.F. 1976. Appl. Phys. Lett. 28:241.

13. Tomkiewicz, M., Woodall, J.M. 1977. Science 196:4293.

14. Nozik, A.J. 1975. Nature 257:383.

15. Wrighton, M.S., Ginley, D.S., Wolczanski, P.T., Ellis, A.B., Morse, D.L., Linz, A. 1975. Proc. Nat. Acad. Sci. U.S.A. 72:1518.

16. Mavroides, J.G., Tcherner, D.I., Kafalas, J.A., Kolesar, D.F. 1975. Mat. Res. Bull. 10:1023.

17. Watanabe, T., Fujishima, A., Honda, K. 1976. Bull. Chem. Soc. Japan 49:355.

18. Wrighton, M.S., Ellis, A.B., Wolczanski, P.T., Morse, D.L., Abrahamson, H.B., Ginley, D.S. 1976. J. Am. Chem. Soc. 98:2774.

19. Nasby, R.D., Quinn, R.K. 1976. Mat. Res. Bull. 11:985.

20. Ellis, A.B., Kaiser, S.W., Wrighton, M.S. 1976. J. Phys. Chem. 80:1325.

21. Clechet, P., Martin, J., Olier, R., Vallouy, C. 1976. C.R. Acad. Sci. Ser. C 282:887.

22. Nozik, A.J. 1977. J. Cryst. Growth 39:200.

23. Thurmond, C.D., 1975. J. Electrochem. Soc. 122:1133.

24. Gerischer, H. Private communication.

25. Kung, H.H., Jarrett, H.S., Private communication.

26. Gerischer, H. 1977. See Ref. 9, pp. 1-19.

27. Gerischer, H. 1977. J. Electroanal. Chem. 82:133.

28. Gerischer, H. 1976. In Solar Power and Fuels - Proceedings First International Conf. on the Photochemical Conversion and Storage of Solar Energy, ed. J.R. Bolton, New York:Academic Press.

29. Williams, F., Nozik, A.J. 1978. Nature. 271:137.

30. Bocarsly, A.B., Bolts, J.M., Cummins, P.G., Wrighton, M.S. 1977. Appl. Phys. Lett. 31:568.

31. Chance, R.R., Nozik, A.J., to be published.

32. Bockris, J. O'M., Uosaki, K. 1977. See Ref. 9, pp. 315-330.

33. Schwerzel, R.E., Brooman, E.W., Craig, R.A., Wood, V.E. 1977. See Ref. 9, pp. 293-314.

34. Bockris, J. O'M., Khan, V.M., Uosaki, K. 1976. J. Res. Inst. for Cat., Hokkaido Univ. 24:1.

35. Beni, G. 1977. See Ref. 9, pp. 108-119.

36. Mattis, D.C. 1977. See Ref. 9, pp. 99-107.

37. Wolf, M., 1960. Proc. IRE 48:1246.

38. Loferski, J.J., 1956. J. Appl. Phys. 27:777.

39. Prince, M.B. 1955. J. Appl. Phys. 26:534.

40. Yoneyama, H., Sakamoto, H., Tamura, H. 1975. Electrochim. Acta 20:341.

41. Ohashi, K., McCann, J., Bockris, J.O'M. 1977. Int. J. Energy Res. 1:259.

42. Ohashi, K., McCann, J., Bockris, J. O'M. 1977. Nature 266:610.

43. Bard, A.J., Wrighton, M.S. 1977. J. Electrochem. Soc. 124:1706.

44. Harris, L.A., Wilson, R.H. 1976. J. Electrochem. Soc. 123:1010.

45. Harris, L.A., Cross, D.R., Gerstner, M.E. 1977. See Ref. 9, pp. 157-171.

46. Memming, R. 1977. See Ref. 9, pp. 38-53.

47. Tomkiewicz, M. 1977. See Ref. 9, pp. 92-98.

48. Mavroides, J.G. 1977. See Ref. 9, pp. 84-91.

49. Frank, S.N., Bard, A.J. 1975. J. Am. Chem. Soc. 97:7427.

50. Laser, D., Bard, A.J. 1976. J. Electrochem. Soc. 123: 1828.

51. Nozik, A.J. 1977. Appl. Phys. Letts. 30:567.

52. Kung, H.H., Jarrett, H.S., Sleight, A.W., Ferretti, A. 1977. J. Appl. Phys. 48:2463.

53. Bolts, J.M., Wrighton, M.S. 1976. J. Phys. Chem. 80:2641.

54. Mollers, F., Memming, R. 1972. Ber. Bunsenges. Phys. Chem. 76:469.

55. Wrighton, M.S., Morse, D.L., Ellis, A.B., Ginley, D.S., Abrahamson, H.B. 1976. J. Am. Chem. Soc. 98:44.

56. Kennedy, J.H., Frese, K.W. 1976. J. Electrochem. Soc. 123:1683.

57. Schleich, D.M., Derrington, C., Godek, W., Weisberg, D., Wold, A. 1977. Mat. Res. Bull. 12:321.

58. Bockris, J. O'M., Uosaki, K. 1977. J. Electrochem. Soc. 124:1348.

59. Nakato, Y., Tonomura, S., Tsubomura, H. 1976. Ber. Bunsenges, Phys. Chem. 80:1289.

60. Ginley, D.S., Butler, M.A. 1977. J. Appl. Phys. 48:2019.

61. Butler, M.A., Nasby, R.D., Quinn, R.K. 1976. Solid State Comm. 19:1011.

62. Hardee, K.L., Bard, A.J. 1977. J. Electrochem. Soc. 124: 215.

63. Hodes, G., Cohen, D., Manassen, J. 1976. Nature 260:312.

64. Derrington, C.E., Godek, W.S., Castro, C.H., Wold, A. 1978. Inorgan. Chem. In Press.

65. Derrington, C.E., Castro, C.A., Godek, W., Sanchez, R.L., Wold, A. 1977. See Ref. 54, pp. 254-60.

66. Butler, M.A., Ginley, D.S., Eibschutz, M. 1977. J. Appl. Phys. 48:3070.

67. Ohashi, K., Uosaki, K., Bockris, J. O'M. 1977. Int. J. Energy Res. 1:25

68. Tamura, H., Yoneyama, H., Iwakwa, C., Sakamoto, H. J. Electroanal. Soc. 80:357.

69. Hardee, K.L., Bard, A.J. 1976. J. Electrochem. Soc. 123: 1024.

70. Quinn, R.K., Nasby, R.D., Baughman, R.J. Mat. Res. Bull. 11:1011.

71. Yeh, L.S., Hackerman, N. 1977. J. Electrochem. Soc. 124: 833.

72. Mayumi, S., Iwakwa, C., Yoneyama, H., Tamura, H. 1976. J. Electrochem. Soc. Japan 44:339.

73. Candea, R.M., Kostner, M., Goodman, R., Hickok, J. 1976. J. Appl. Phys. 47:2724.

74. Kohl, P.A., Frank, S.W., Bard, A.J. 1977. J. Electrochem. Soc. 124:225.

75. Bockris, J. O'M., Uosaki, K. 1976. Energy 1:95.

76. Tomkiewicz, M., Woodall, J. 1977. J. Electrochem. Soc. 124:1436.

77. Gourgaud, S., Elliott, D. 1977. J. Electrochem. Soc. 124:102.

78. Morisaki, H., Watanabe, T., Iwase, M., Yazawa, K. 1976. Appl. Phys. Letts. 29:338.

79. Nakato, Y., Abe, K., Tsubomura, H. 1976. Ber. Bunsenges. Phys. Chem. 80:1002.

80. Nakato, Y., Ohinishi, T., Tsubomura, H., 1975. Chem. Letts. pp. 883-886.

81. Wilson, R.H., Harris, L.A., Gerstner, M.E. 1977. J. Electrochem. Soc. 124:1233.

82. Harris, L.A., Gerstner, M.E., Wilson, R.H. 1977. J. Electrochem. Soc. 124:1511.

83. Fujishima, A., Honda, K. 1974. Nippon Shashin Gakkaishi 37:303.

84. Matsumura, M., Nomura, Y., Tsubomura, H. 1976. Bull. Chem. Soc. Japan 49:1409.

85. Osa, T., Fujihira, M. 1976. Nature 264:349.

86. Watanabe, T., Fujishima, A., Tatsushi, O., Honda, K. 1976. Bull. Chem. Soc. Japan 49:8.

87. Tsubomura, H., Matsumura, M., Nakatoni, K., Nomura, Y. 1977. See Ref. 54, pp. 178-185.

88. Fujihira, M., Ohishi, N., Osa, T. 1977. Nature 268:226.

89. Clark, W.D.,., Sutin, N. 1977. J. Am. Chem. Soc. 99:4676.

90. Spitler, M.T., Calvin, M. 1977. J. Chem. Phys. 66:4294.

91. Fujishima, A., Iwase, T., Honda, K. 1976. J. Am. Chem. Soc. 98:1625.

92. Watanabe, T., Fujishima, A., Honda, K. 1975. Ber. Bunsenges. Phys. Chem. 79:1213.

93. Tsubomura, H., Matsumura, M., Nomura, Y., Amamiya, T. 1976. Nature 261:402.

Table 1 Summary of semiconductors studied as photoelectrolysis electrodes

Semiconductor Electrode[a]	Band Gap (eV)	Flat-Band Potential[b] (Volts vs SCE)	pH	References[c]
$n-ZrO_2$	5.0	-2.0	13.3	21,52
$n-Ta_2O_5$	4.0	-1.4	13.3	21,52
$n-SnO_2$	3.5	-0.7	13.0	46,53,54,55
$n-KTaO_3$	3.5	-1.3	13.0	20
$n-SrTiO_3$	3.4	-1.4	13.0	12,17,18,48,52
$n-Nb_2O_5$	3.4	-1.0	13.3	21
$n-BaTiO_3$	3.3	-0.8	13.6	19,52,56,57
$n-TiO_2$	3.0	-1.0	13.0	9,10,14-16,21, 40,41,44,52,53, 56,62,68
$p-SiC$	3.0	+1.1	14.0	58
$n-V_2O_5$	2.8	+0.8	7.0	62
$n-Bi_2O_3$	2.8	-0.1	9.0	62
$n-PbO$	2.8	-0.3	9.0	62
$n-FeTiO_3$	2.8	+0.1	14.0	60
$n-WO_3$	2.7	+0.2	7.0	21,61-63
$n-WO_{3-x}F_x$	2.7	-	-	64,65
$n-YFeO_3$	2.6	-0.5	14.0	66
$n-Pb_2Ti_{1.5}W_{0.5}O_{6.5}$	2.4	-0.6	13.3	52
$p-ZnTe$	2.3	-1.0	14.0	58,67
$n-PbFe_{12}O_{19}$	2.3	+1.0	13.3	52
$p-GaP$	2.3	+0.8	7.0	10,13,40-42,59,68

Table 1 Cont.

n-CdFe$_2$O$_4$	2.3	-0.2	13.3	52
n-Fe$_2$O$_3$	2.2	-0.1	9.0	69-71
n-CdO	2.2	-0.2	13.3	52
n-Hg$_2$Nb$_2$O$_7$	1.8	-0.1	13.3	52
n-Hg$_2$Ta$_2$O$_7$	1.8	-0.2	13.3	52
p-CuO	1.7	+0.3	7.0	62
p-InP	1.3	-0.1	1.0	58,72
p-CdTe	1.4	-	-	58,67
p-GaAs	1.4	0.2	9.2	67
p-Si	1.1	-0.3	1.0	73

[a] Maximum experimental conversion efficiency with sunlight for this group of materials is between 0.5 to 1.0%. Semiconductor electrodes having band gaps >3eV are limited by low solar absorptivity; those with band gaps >3eV have been limited by large bias requirements (unfavorable flat-band potentials).

[b] Values decrease approximately 0.06 volts per unit increase in pH. Values vs (SCE) are converted to NHE scale by adding +0.24 volts.

[c] Only specific studies on photoelectrolysis listed.

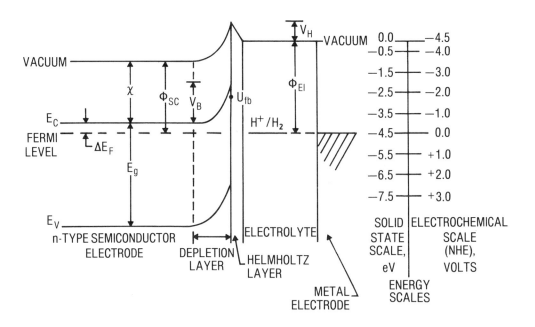

Figure 1. Energy level diagram for semiconductor-electrolyte junction showing the relationships between the electrolyte redox couple (H^+/H_2), the Helmholtz layer potential drop (V_H), and the semiconductor band gap (E_g), electron affinity (χ), work function (ϕ_{sc}), band bending (V_B), and flat-band potential (U_{fb}). The electrochemical and solid state energy scales are shown for comparison. ϕ_{El} is the electrolyte work function.

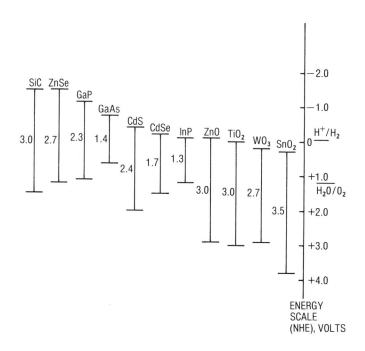

Figure 2. Position of valence and conduction band edges for
several semiconductors in contact with aqueous
electrolyte at pH=1.0. The position of the H^+/H_2
and H_2O/O_2 redox couples are indicated at the right

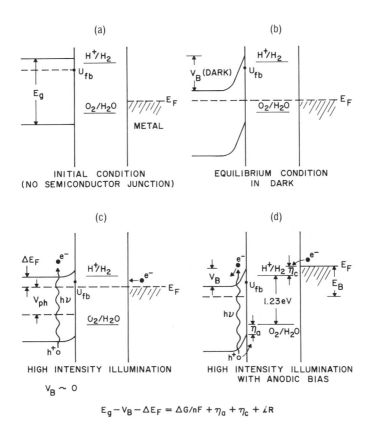

Figure 3.   Sequence of energy level diagrams
for Schottky-type photoelectroly-
sis cell from the initial condition
to the final condition of photoelectrolysis with bias.

1244

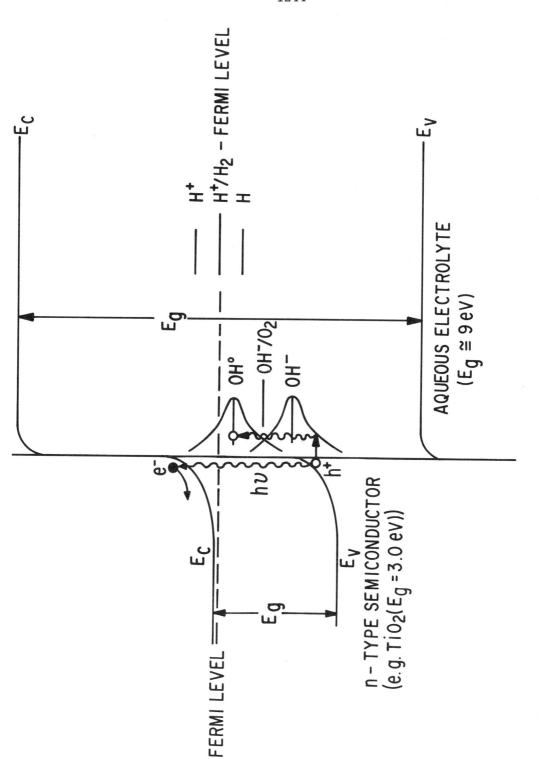

Figure 4.   Energy band diagram of semiconductor-electrolyte
junction showing band states of water and its ex-

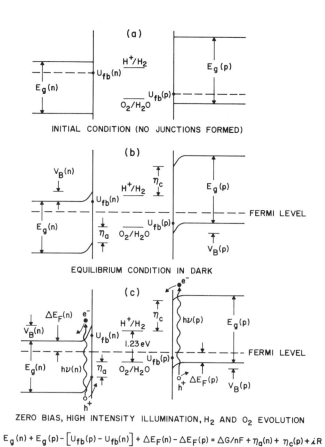

Figure 5.   Sequence of energy level diagrams for p/n photo-electrolysis cells.

ELECTRON FLOW IN PHOTOSYNTHESIS

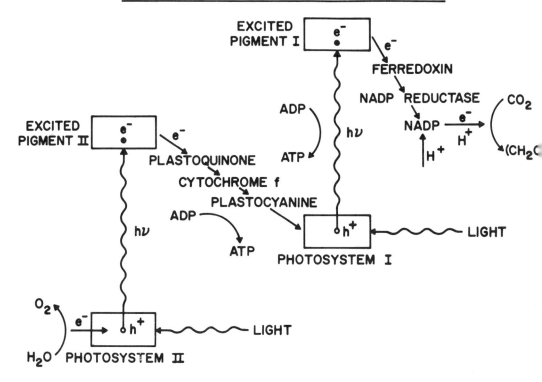

ELECTRON FLOW IN p-n PHOTOCHEMICAL DIODES

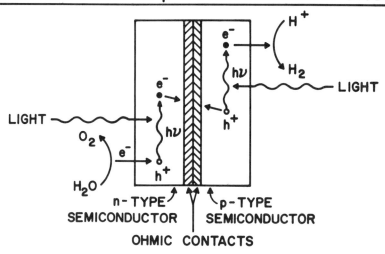

Figure 6.   Comparison of electron flow in photosynthesis with that for p/n photochemical diodes.

PHOTOELECTROCHEMICAL GENERATION
OF HYDROGEN WITH HYBRID ELECTRODES

K. Yazawa and H. Morisaki
The University of Electro-Communications
Chofu-shi, Tokyo, Japan

ABSTRACT

A novel semiconductor electrode structure is described which enables the stable photoelectrolysis of water in a single cell without any auxiliary power source. A TiO_2 thin film is fabricated on a Si solar cell either by chemical vapor deposition or by thermal oxidation of a sputtered Ti film in vacuum. This structure of electrodes has several advantages: (1) TiO_2 film is anodically biased by the solar cell, and the efficiency of its catalytic action to decompose water is greatly improved. (2) Photons with energy higher than 3.0 eV of the solar radiation are absorbed in the TiO_2 film and those with energies between 1.2 and 3.0 eV are used in the solar cell. (3) The electrode is free from the corrosion problem.

INTRODUCTION

Photoelectrolysis of water with a catalytic semiconductor electrode was first reported on a TiO_2 electrode by Fujishima and Honda [1]. It is expected that the process will provide a prospective technique for the hydrogen production with solar energy. One of the restrictions in selecting the semiconducting materials for this purpose is its chemical stability. Rather few materials are known to be free from photo-corrosion, TiO_2 being among the rest. It is also known that the anodic charge transfer at the interface between TiO_2 crystal and water proceeds very efficiently because of the help of its surface states [2]. In spite of these favorable properties there is a difficulty of low conversion efficiency in photoelectrolysis with this material. The energy band gap of TiO_2 (3.0 eV) is too large for efficient utilization of the solar spectrum, while the minimum free energy required for the dissociation of water is only 1.23 eV. So Fujishima and Honda were only successful by resorting to a differential cell structure.

It is known that the quantum efficiency of the photoelectrolysis of water with TiO_2 is greatly improved when anode bias is applied to the TiO_2 solar cell hybrid electrodes in the hydrogen production with the solar energy [3]. In the present paper we are going to report on the experimental results of the photoelectrolysis of water with the electrodes.

HYBRID ELECTRODES

A schematic view of the TiO_2-Si solar cell hybrid electrode is shown in Fig. 1. The significance of the use of the hybrid electrode (hereafter referred to as HE) is as follows:

(1) The efficiency of the photoelectrolysis of water is improved as the TiO_2 electrode is biased anodically against the counterelectrode by the electromotive of the solar cell.

(2) The solar spectrum per unit area is utilized efficiently in HE, because the TiO_2 film absorbs only photons with energy higher than 3.0 eV of the incident radiation and transmits the rest to the Si solar cell, which absorbs those with energies between 1.2 and 3.0 eV.

(3) The electrode is free from the often encountered corrosion problem, being protected on its surroundings with the stable TiO_2 film.

The substrates were p^+/n Si solar cells purchased from Nippon Electric Co., Japan. In the preliminary experiments TiO_2 films were prepared by chemical vapor deposition (CVD) with the following reaction:

$$TiCl_4 + 2CO_2 + 2H_2 \rightarrow TiO_2 + 2CO + 3HCl \qquad (1)$$

The optimum substrate temperature for the reaction was about 750° C, and the deposition time was about 1 hr to form a 5000 Å thick film. The TiO_2 films as deposited were semiconducting with the resistivity being typically 1-10 cm. Later, we employed the following CVD reaction as the substitute for (1):

$$TiCl_4 + 2H_2O \rightarrow TiO_2 + 4HCl \qquad (2)$$

The requisite substrate temperature with this reaction was 300° C, and the speed of the film growth was higher than that with the former reaction. The TiO_2 films as deposited were again semiconducting with resistivities of the order of 10^3 Ω-cm. The purpose of the alteration of the CVD reaction from (1) to (2) was to avoid the probable thermal degradation of the Si solar cell.

Thermal oxidation of a sputtered Ti film was also tried in fabricating the TiO_2 film. It has been shown that a TiO_2 film with excellent photoelectrolytic capability can be prepared by proper heat treatment of a metallic Ti plate in vacuum (10^{-2} - 10^{-3} Torr.) [4]. We found that TiO_2 films can also be obtained by heat treatment of sputtered Ti films in vacuum, and their physical properties depends upon the fabricating conditions in a similar way as in the case of metallic Ti plates. We had the best results with HE prepared by a heat treatment of 20 min. at 700° C of a sputtered Ti film with 2500 Å thickness. A Cu wire was connected with Ag past to the back electrode of each solar cell substrate. The entire electrode except the TiO_2 surface was then coated with insulating epoxy resin. To see the effect of solar cells on the electrode properties of photoelectrolysis, we also fabricated a standard TiO_2 electrode (SE) prepared by the same method on n-Si substrates.

OPERATION OF HYBRID ELECTRODES

Both electrodes for the photoelectrolysis (HE or SE and Pt) and a standard

calomel electrode (SCE) were immersed into electrolyte contained in a glass
vessel. The electrolyte used was 0.1N NaOH. The surface of HE (or SE) was
illuminated by a 500W Xe lamp through a quartz window of the vessel. A
monochromator was inserted between the window and the lamp when required.
Photoelectrolysis of water by HE was demonstrated under the illumination of
the Xe lamp. To achieve maximum efficiency, the HE and Pt electrodes were
shorted together through an external lead wire. When the surface of HE
was illuminated, H_2 and O_2 gas evolutions were observed at the Pt and HE
respectively. We also illuminated HE by sunlight, and obtained an energy
conversion efficiency of about 0.1% for the production of hydrogen. Thus
the photoelectrolytic cell with HE is quite simple. It requires neither a
differential cell (two compartment cell) structure, nor an external power
source.

To study the fundamental behavior of HE as an electrode for photoelectrolysis,
current-voltage characteristics were measured of the electrode under illumination of the Xe lamp by applying variable dc bias against the counter-
electrode. Typical results are shown in Fig. 2. The electrode potentials
were measured against SCE, and the curves 1 and 2 stand for the characteristics of Pt and HE, respectively. Curve 2 should be compared with curve
3 in the figure, that is a typical characteristic of SE. It is noted that
the Pt potential (curve 1) exhibits the normal value for H_2 evolution above
the critical bias ($\doteq$ 1.0 V), whereas the current increase above +0.6 V vs.
SCE corresponds to O_2 evolution, the overvoltage for the O_2 evolution being
about 0.39 V (the normal potential is +0.21 V vs. SCE at pH = 13). A small
amount of current observed in the potential range between 0 and -1 V vs.
SCE may result from the photogalvanic mode operation of the cell as reported
by Mavroides et al. [5], since the current increases remarkably when O_2 gas
is bubbled through the solution.

The characteristics of SE prepared either by CVD or by thermal oxidation of
the sputtered Ti film are quite similar to those of single crystal TiO_2,
however the photocurrent is somewhat lower. In our preliminary experiment
the photocurrent of SE with an Al doped TiO_2 film showed even higher photocurrent than a single crystal, suggesting future possibilities.

The anodic current accompanied by O_2 evolution starts to increase at about
-0.7 V and tends to saturate above +0.4 V, the saturation current being proportional to the light intensity. For H_2 evolution to occur on the Pt electrode without external bias, the potential for the O_2 evolution on TiO_2 must
be shifted to the negative side of the H_2 potential on the Pt electrode. In
the case of the SE-Pt system the above condition is not satisfied, and H_2
gas is not evolved.

Comparing curves 2 and 3 in Fig. 2, we noted that the current-vs.-potential
characteristic of HE was shifted by about -0.7 V from that of the TiO_2 electrode. The shift of the potential originates from the electromotive force
generated by the Si solar cell, apparently indicating that there was no
undesirable potential barrier at the interface between the Si solar cell and
the TiO_2 film. Fig. 3 shows the characteristics of a HE prepared by the
CVD method with reaction (1). We noted that the electromotive force of the

Si solar cell decreased with increasing current in this case. We also heat-treated a Si solar cell at 700°C for 1 hr in a hydrogen atmosphere to find that the solar cell was degraded. This was the reason why we altered the chemical reaction for CVE, as mentioned already.

We further tried another CVD reaction with low substrate temperature:

$$Ti(OC_3H_7)_4 + 2H_2O \rightarrow TiO_2 + 4C_3H_7OH \tag{3}$$

With this reaction too, we had satisfactory results. Thus the limitation of the temperature of the substrate solar cells in preparing HE seems to be essential.

CONCLUSION

It was found that a TiO_2 solar cell hybrid electrode works successfully as the anode for the photoelectrolysis of water. The energy conversion efficiency is low at present, although hydrogen was produced in a single cell without any auxiliary power supply. We hope it may be improved by proper selection of electrode materials and optimizing the device design (e.g. impurity profile).

REFERENCES

1. A. Fujishima and K. Honda, "Electro-chemical Photolysis of Water at a Semiconductor Electrode", Nature 238, 37 (1972).

2. H. Morisaki, M. Hariya, and K. Yazawa, "Anomalous Photoresponse of n-TiO_2 Electrode in a photoelectrochemical Cell", Appl. Phys. Lett. 30, 7-9 (1977).

3. H. Morisaki, T. Watanabe, M. Iwase, and K. Yazawa, "Photoelectrolysis of Water with TiO_2 covered Solar Cell Electrodes", Appl. Phys. Lett. 29, 338-340 (1976).

4. K. Yazawa, H. Kamogawa, and H. Morisaki, "Semiconducting TiO_2 Films for Photoelectrolysis of Water", submitted for publication in International Journal of Hydrogen Energy.

5. J. G. Mavroides, J. A. Kafalas, and D. F. Kolesar, "Photoelectrolysis of Water in Cells with $SrTiO_3$ Anodes", Appl. Phys. Lett. 28, 241-243 (1976).

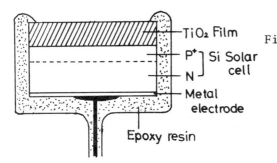

Fig.1.

Schematic view of TiO$_2$-Si solar cell hybrid electrode.

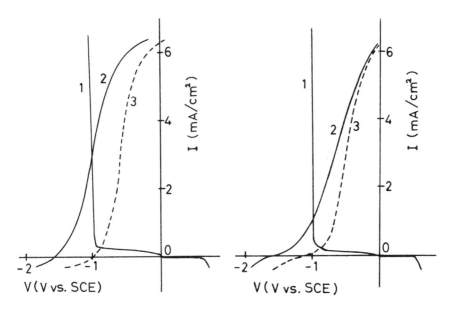

Fig.2. Current-vs.-potential characteristics of electrodes in 0.1N NaOH solution. (1) Pt. (2) hybrid electrode (HE) (3) TiO$_2$ electrode (SE)

Fig.3. Current-vs-potential characteristics of a hybrid electrode prepared by the high temperature CVD method.

PHOTOCHEMICAL PRODUCTION OF HYDROGEN FROM WATER

E. Broda
Institute of Physical Chemistry
University, Vienna, Austria

ABSTRACT

The energy flux in sunlight is 40 000 kW per head of the world population. Theoretically much of this energy can be used to photolyze water, in presence of a sensitizer, to H_2 (and O_2) for a hydrogen economy. The main difficulty in a homogeneous medium is the back-reaction of the primary products. According to the "membrane principle", the reducing and the oxidizing primary products are released on opposite sides of asymmetric membranes, and so prevented from back-reacting. In essence, this is the mechanism of the photosynthetic machinery in plants and bacteria. This therefore serves as an example in the artificial construction of suitable asymmetric "vectorial" membranes. Relatively small areas of photolytic collectors, e.g. in tropical deserts, could cover the energy needs of large populations through hydrogen.

INTRODUCTION

The bulk of the technical hydrogen now comes from limited, fossil, raw materials, namely, coal and oil. However, alternative processes are available or under development that are based on an unlimited raw material, water. On utilization in combustion this hydrogen returns to the form of water so that the cycle is closed. Water requires for the splitting into $H_2 + 0.5 O_2$ 56.7 kcal (237 kJ) in the liquid and 68.3 kcal (285 kJ) in the gas form, per mole. This dissociation energy can come either from a limited, i.e., exhaustible, or from an inexhaustible, regenerative, source.

An old alternative process is electrolysis. This process is technically simple, but is bound to remain expensive. The conversion of heat, e.g. from nuclear reactors, into electric energy is burdened with the huge loss due to the Carnot factor, in practice hardly less than 60%. Moreover, the yield in technical electrolysis is far removed from the theoretical value.

A more recent group of processes is called "thermochemical". The energy source is always heat from nuclear reactors. More or less complicated cycles of reactions involving inorganic

1253

reactants in the end lead to the production of hydrogen. Many different cycles have been proposed. Formally, the totality of the participating reactants can be considered as a catalytic system for water splitting. Again a Carnot factor cannot be avoided.

The process for the production of hydrogen from water to be discussed here is photochemical. The dissociation energy of 1 mole H_2O (237 kJ) equals the energy of 1 mole (einstein) of green light. Thus thermodynamically there is no obstacle to the "photolysis" of 1 mole H_2O by 1 quantum of green, blue or violet light, and with 2 quanta of yellow or red light the process could also be carried out. The photolysis of water may be possible at room, or similar, temperature and is altogether expected to be a clean, nonhazardous and convenient process. Because of the high temperature to be ascribed thermodynamically to sunlight, even to diffuse sunlight, the Carnot loss is not large. The existing or envisaged processes for the photolysis of water are biotic or abiotic.

The possible scope of water photolysis can be judged from the flux of light energy that comes, uninterruptedly, to the Earth: $1,7.10^{14}$ kW, corresponding to 40 000 kW per head of the world population. Any light, whether direct or diffuse, is useful photochemically. While the light flux varies during the day and the year and also according to the weather, this is not a decisive drawback in a process where the product (primarily, hydrogen) is stored quite easily.

In a "solar hydrogen", in contrast to a "nuclear hydrogen" economomy, no additional energy is introduced into the biosphere, and this is therefore not heated up. It is true that large-scale photochemical hydrogen factories might influence the local albedo, but this is a second-order effect, and measures can be taken against it.

HYDROGEN FROM PLANT PHOTOSYNTHESIS

In photosynthesis, the plants extract about 10^{11} tons of carbon per year from the air, twenty times more than is returned now to the air in the combustion of fossil fuels. Several authors have recently discussed the global effects of photosynthesis [1,2].

About 40 years ago it has been established by the genius from the Netherlands, the microbiologist C.B. van Niel [3], that the essence of plant photosynthesis consists precisely in the splitting of water. The energy of visible light is captured by plant pigments and transferred to an enzyme system for the

photolysis of H_2O. Now normally the hydrogen from the water
does not appear as the gas, which would from the standpoint
of the plant be useless. The hydrogen - more precisely: its
electron - rather serves, in some reactive form, for the reductive assimilation of the CO_2. Van Niel's theory has super-
seded older hypotheses, according to which the action of light
consists in the direct photochemical splitting of CO_2.

It has later become known through the work of D.I.Arnon [4]
that the primary reductant, which subsequently enters the dark
reactions of photosynthesis, consists of the reduced form of
ferredoxin, a well-defined iron-containing protein. The standard electrochemical redox potential of ferredoxin is about
-0.4 volts and consequently coincides with the potential of
the couple H_2/H_2O. Thus from a thermodynamical point of view
the achievement of the plant in reducing ferredoxin is as
great as if it had produced free hydrogen. The cycle of dark
reactions has been elucidated by M. Calvin [5].

(In the analysis of the mechanisms of plant photosynthesis,
comparison with the photosynthetic bacteria is of extra-
ordinary value [1]. These bacteria are the ancestors of the
plants. Their most direct descendants among the plants are
the blue-green algae, recently often called "blue bacteria".
The photosynthetic machinery of the bacteria is simpler and
less efficient than that of the plants, including the blue-
green algae. This fact is expressed in the fundamental result established by H. Molisch that bacteria never release O_2.
They cannot break the bond between hydrogen and oxygen, and
so they require "easier" source of hydrogen, e.g. H_2S. Indeed
many photosynthetic bacteria are intolerant of oxygen and are
strict anaerobes. Nevertheless, the broad features of bacterial photosynthesis are similar to those of plant photosynthesis.)

It has been found by H. Gaffron [6,7] that in certain artificial conditions plants produce H_2 as a product of their
photosynthesis. Larger amounts can be made by photosynthetic
bacteria. The release of hydrogen is explained by the presence
in such organisms of a hydrogenase. This enzyme transfers an
electron, as soon as this has been lifted to a high negative
potential by the energy of light, to an ion of hydrogen. In
the plants, the hydrogenase normally is latent, and therefore
must be induced. The hydrogenase also is rather inefficient,
and it is easily poisoned by oxygen, itself a product of photosynthesis. Therefore plants are unlikely ever to serve directly as a major source of hydrogen through photosynthesis [7].

To improve yields, components of plants have been combined with more strongly acting hydrogenases of bacterial origin [9]. However, such hydrogenases are also oxygen-sensitive, and the hydrogen yields are still modest. Moreover, such systems must be carefully prepared from living organisms. They are expensive and deteriorate quickly.

ENERGY FARMING

Another approach to hydrogen production through biotic processes is "energy farming" [10,11]. The biomass of productive plants, e.g., alfalfa, poplar trees or sugar cane, can be subjected to hydrogen fermentation. However, yields of hydrogen would not be high; methane fermentation gives more fuel. Practically complete conversion of biomass to hydrogen (plus CO_2) could be obtained by steam reforming, similar to the well-known process for the production of hydrogen on the basis of coal:

$$(CH_2O) + H_2O_g = CO_2 + 2 H_2; \quad \Delta G_o = -12 \text{ kJ}$$

(CH_2O) indicates unit quantity of carbohydrate, the main component of biomass - not formaldehyde!

However, even assuming that this process were economical, the fundamental obstacle would remain that any possible farming area could better be used to make food, feedstuff or technical raw materials rather than hydrogen. This would apply also if blue-green or green algae, as many hope, could be grown economically in tanks or basins [12,13] so that no, or practically no, area would be diverted from ordinary agriculture. In this case, too, biomass would best be used otherwise. In the consideration of energy farming it must furthermore not be forgotten that high yields require heavy investment of fossil energy through fertilizers, pesticides and transport [14].

What remains of energy farming is the possibility to produce hydrogen (or methane) from waste like straw, corn cobs, sawdust, agricultural or town effluents [15]. Important though such processes certainly are, no independent increase in output is possible. The amount of waste remains linked to agricultural or silvicultural production and to population size.

WATER PHOTOLYSIS WITH MEMBRANES

Single- and double-phase systems

One approach to direct water-splitting is to let light act on water or a single-phase aqueous solution (an abiotic system) in presence of a sensitizer, also known as a photocatalyst. The sensitizer would play the role of chlorophyll A in the plants. The energy of the light would be transferred to the water and used for its photolysis. Such single-phase systems have so far not been successful [16]. Whenever hydrogen was observed, its concentration was stationary and small. Moreover, in most cases only UV light (typically, 254 nm) proved active. Such light does not occur naturally at ground level.

The basic difficulty in the experiments with the homogeneous aqueous solutions consists in the appearance, at close distances, of both reducing and oxidizing primary products. These "complementary" products may or may not be free radicals. The products rapidly meet by diffusion, back react with each other to water, and disappear ("geminate reaction"). This has also so far been the experience with solutions of complexes of transition metals, e.g. Fe or Ru [18]. Such complexes are hoped to provide separate sites for subsequent reaction steps in photolysis.

In double-phase systems it is in principle possible to separate spatially the complementary primary products, and so to prevent, partly or wholly, their recombination and mutual annihilation. Systems with one liquid and one solid bulk phase have been studied [19]. Often the solid is a semiconductor. In some cases an external potential was applied, and the energy of light was used only to supplement the electric energy invested (photoassisted electrolysis).

Here we shall rather concentrate on heteroabiotic systems consisting of two liquid phases ("compartments") [20]. The phases obviously must be separated by a more or less permanent membrane, consisting of solids and or liquids, which may contain the sensitizer and catalysts. The energy transducing machinery is distinguished by an asymmetric structure and is arranged in such a way that some of the primary products of photolysis appear in one liquid compartment, the other (complementary) products in the other liquid compartment. In this way, a "vectorial" rather than a "scalar" reaction takes place; the membrane can simply be called a "vectorial membrane".

The membrane principle

Vectorial membranes have never as yet been constructed by man in a planned way. But Nature has "learned" their construction and utilization soon after life arose, 3-4 gigayears (10^9 years) ago, although at first the membranes were not photosynthetic [1]. No living cell exists now that does not make use of this "membrane principle", as it might be called. While the importance of membranes in bioenergetics has long been known, the contribution of P. Mitchell [21] should be singled out, who in his "chemiosmotic" theory gave a central role to asymmetric biomembranes.

Let it first be recalled that already the enzyme molecules, indispensable components of even the most ancient and primitive bacteria, carry out vectorial reactions. Each enzyme molecule has at least one reactive site, which accepts reactant(s) coming from some (not all) directions, and releases product(s) into some (not all) directions. However, with dissolved enzymes the vectoriality is obscured by the disorientation of the enzyme molecules within the solution. This applies, e.g. to the enzymes of lactic acid or alcoholic fermentation. One could speak of "microvectoriality". Vectoriality emerges more clearly as soon as enzyme molecules are positioned in an ordered way within permanent structures, i.e. in membranes ("macrovectoriality").

Presumably the first macrovectorial system in the evolution of organisms was that for "active transport". This is defined as the energy-driven transport across membranes, notably the membranes surrounding the cells. Such systems are more or less specific for particular substrates, and can transport them even against large gradients of the electrochemical potential. This is defined as

$$\Delta \bar{\mu} = RT\ln(a_2/a_1) + nEF$$

where a_1 and a_2 are the activities inside and outside, n the charge of an ion (if it is an ion), E the electric potential difference, and F the faraday. For example, bacteria generally transport K ions actively inward, and H ions outward. Active transport is by definition endergonic, i.e. it requires free energy.

The most ancient bacteria are not photosynthetic, but derive their energy from fermentation processes. However, with the energy of sunlight the bacteria tapped later in evolution a tremendous, additional, source of energy. Experience shows that the photosystems of the bacteria (and also of higher organisms) are without exception contained exclusively within membranes. The asymmetry of the membranes that served and serves active transport has also been exploited to ensure the vectoriality of the primary processes in photosynthesis: the separation of the complementary products.

(The similarity of the working principles of photosynthesis in all organisms, bacteria and plants, and also the similarity of the chemical constituents of the photosystems leave hardly any doubt that all photosynthesizers are monophyletic, i.e. that in evolution photosynthesis arose only once. All later, and more efficient, photosynthetic organisms decended from earlier, and more primitive, photosynthetic organisms. Yet fairly recently one important exception has become known: Halobacterium halobium [22], a photosynthetic bacterium living in brines, has a qualitatively different kind of photosystem and apparently arose independently and fairly late [23]. But this organism likewise carries its photosystem within the cell membrane.)

In the simplest plants, the "prokaryotic" blue-green algae, the photosynthetic, water-splitting, membranes are usually contained as a network within the protoplasm. In the higher ("eukaryotic") plants such networks are restricted to specialized, intracellular, "organelles", the chloroplasts. Otherwise the photosynthetic machineries in lower and higher plants are nearly identical. Now, how do these photosynthetic membranes ("thylakoid" membranes) work ?

Two photosystems in succession

Before answering this question, an important feature of plant photosynthesis [24,25] must be mentioned. As pointed out, the synthesis of (CH_2O) from CO_2 and H_2O (by plants) requires far more energy than from CO_2 and H_2S (by bacteria). The figures are 112 and 12 kcal (480 and 50 kJ) per mole, respectively. To overcome the enormous endergonicity of water splitting, the plants are found to apply 2 light quanta for each reduction equivalent (electron) that is transferred from water to ferredoxin, and therefore, in physiological conditions, to CO_2. According to the "Z scheme" put forward first by Hill and Bendall [26] each electron is lifted first (in "photosystem 2") by the energy of 1 quantum to a primary acceptor. Some non-photosynthetic ("dark", i.e., exergonic) steps follow, after which the electron is present within a different kind of molecule, and this molecule then serves as electron donor for "photosystem 1". As sensitizers, both systems contain chlorophyll A, but in different forms. Ferredoxin is reduced only through the action of photosystem 1.

An ordered sequence of 2 different, collaborating, photoreactions in different places is again an achievement of Nature
that man has not yet equalled. Now what are the primary products, the two couples of complementary compounds, in the two
subsequent photoreactions? Though the investigation of the
photosynthetic machinery of the plants is still in flux, we
already have broad knowledge of the primary (and also of the
subsequent) reaction steps.

The first photostep in the plants leads to the reaction

$$H_2O + plastoquinone = 0.5\ O_2 + plastohydroquinone$$

and the second photostep to the reaction

$$plastocyanine\ (red) + ferredoxin\ (ox) =$$
$$plastocyanin\ (ox) + ferredoxin\ (red)$$

In either case, the first-named product is formed at the inside of the thylakoid membrane. The reactions are here
written down in a purely formal way, and transient, very
short-lived, intermediates are not taken into account. Each
of the two photosteps requires an energy per electron of
about 77 kJ, corresponding to light of about 1500 nm. For the
release of each molecule of O_2 and for the concomitant reduction of each atom of carbon to the level of carbohydrate
the lifting of 4 electrons is needed, each by 2 steps.

Photolysis with visible light

It has been said that the photolysis of water cannot be
carried out with visible light because the primary step, the
breaking of a O-H bond, needs too much energy: 460 kJ per
mole, corresponding to 1 einstein of UV light (260 nm). This
is wrong. In fact, Nature has demonstrated to us that light
around 700 nm (more precisely: about 700 in photosystem 1,
and 680 in photosystem 2) is sufficient, i.e. long-wave
visible light. One reason is that Nature has avoided unduly
strong endergonic steps, notably the uncompensated breaking
of O-H bonds. Expressed in terms of energy per mole, the
limiting energies for the two photosteps mentioned before are
only near 170 kJ.

Indeed the energy balance in the natural system is even more
favourable than has appeared so far. Ferredoxin (red) and O_2
are not the only products of electron flow from the first
donor, H_2O, to the last acceptor, ferredoxin (ox). The electron flow is enzymatically coupled to the simultaneous endergonic synthesis of adenosine triphosphate (ATP) from its
precursors adenosine diphosphate (ADP) and orthophosphate[27].

The ATP is needed, along with ferredoxin (red), for the assimilation of the CO_2. It also serves other essential purposes of the living cell.

(Photosynthetic bacteria can utilize light of even longer wave length, namely, infrared light. The mechanistic analysis, from the point of view of photochemistry, is most interesting, but cannot be undertaken here).

The energy yield in the overall process of plant photosynthesis can be estimated by comparing the light energy absorbed with the energy available in the back-conversion of the product into the starting materials, H_2O and CO_2. In sufficient approximation, this is measured as the heat of combustion of the biomass. In optimum conditions with microscopic algae, the energy yield has the astonishingly high value of 33%. By optimum conditions is meant not only that the plants are healthy and active, but also that light without excess energy is applied, i.e. red light. With larger quanta (short-wave light) the quantum yield remains the same, but part of the energy of each quantum is wasted.

Artificial photolytic membranes

The basic task in the artificial photolysis of water is, then, the construction of photoactive vectorial membranes that release the complementary products to opposite sides. These products may be H_2 and O_2 directly or at least products of the photoredox reaction that in subsequent dark steps give theses gases. For instance one could think of reduced ferredoxin as an intermediate stage on the way to H_2 if a hydrogenase-active catalyst were present. Possibly such artificial vectorial membranes can be constructed ab initio on the basis of our knowledge of physical chemistry. Alternatively, we may learn from the plants and be "inspired" by them [28,29, 30]. For the purpose it is necessary not only to understand photosynthesis in depth, but its essential features must also be distinguished from those that have been dragged along from an evolutionary past. After all, the plants were not designed and optimized at a green table, but they could evolve only from their ancestors. This evolution needed advance in small steps, as is true for evolution in general.

"Structured systems" [20] have been built that contain micelles or microscopic vesicles. The insides are separate compartments. In a few cases, notably by Henglein and his colleagues [31, 32, 33], it has been observed that quantum yields in micellar systems may be far higher than in homogeneous solution. Obviously the spontaneously formed membranes are asymmetric and geminate recombination is, because of the com-

partmentation, inhibited. Work has also been started with
"black" lipid bilayer membranes ("BLM") [34] and with mono-
layer assemblies [35]. A further possibility are macroscopic
synthetic systems, e.g. on the basis of polymer membranes.
They may perhaps at the beginning be difficult to approach,
but from a technical point of view such systems may in the
end have the highest feasibility.

ESTIMATE OF TECHNICAL POSSIBILITIES

The ultimate goal is a massive contribution of water photolysis to a hydrogen economy. According to theory, the rate of
hydrogen production would, within wide limits, be just proportional to light flux. The hydrogen would be collected at
the rate made possible by sunshine at the given time and
place, stored as a gas or in the form of organic fuels, e.g.
methanol or hydrocarbons, and transported to the consumption
centres.

Current estimates for mean light energy fluxes through all
time are 110 watts/m² in England or Southern Sweden, 185 in
the USA, and 250 in the Sahara, with 300 in a few ultra-hot
places. These differences are not excessive, and as soon as
photolysis works at all, it will have to be considered not
only for hot, but also for temperate climates. Of course,
few large contiguous areas are free in Europe or the USA, but
even so such areas may be found, e.g. in Arizona. Moreover,
decentralisation of the hydrogen factories would be an
advantage from more than one point of view.

Nevertheless, the utilization of tropical deserts would be
especially attractive. To paraphrase Kurt Mendelssohn, the
Sahara is, by common consent, immensely large and immensely
useless. Ocean water would feed the hydrogen factories;
there is no obvious reason why salt water should not directly
be used as a photolyte. Nor would the quantity be prohibi-
tive. Be it mentioned that of the water consumed in the
irrigation of plants 999 parts are lost by transpiration and
only 1 part is used for photolysis.

While the light collecting areas will be provided best by
developing countries, the know-how and the equipment will
for a long time have to be supplied by the developed countries.
Clearly socio-political problems will arise that ought to be
studied in time, but problems there will be whatever the
energy sources of the future! At some future time, photolytic hydrogen factories might also be built at sea.

For a rough estimate, we assume an energy flux, infrared included, of 250 watts/m^2, and an energy yield in the form of
hydrogen of 10% of the theory. With these assumptions made,
hydrogen energy is obtained at a rate of 25 000 kW per km^2
(18 t H$_2$/day). With a consumption of 1 kW of thermal energy
per head, as is typical of large parts of the world now,
about 40 km^2 would be needed per million people. Even with
10 kW per head, about the present world record value, only
400 km^2 per million would be required. This is to be compared with the area of the Sahara of some 10 million km^2.

Here it has not been assumed that the radiation is concentrated (focussed) or that the photoactive, collecting, area
is enclined towards the sun. Though in either way savings in
photoactive area can be made, no total economy of area would
be expected, and presumably investment per unit photoactive
area would go up considerably.

It may be premature to guess the economics. However, we can
approach the question from the back end. We allow a price of
hydrogen, per unit calorific (joulific?) value, equal to that
of crude oil, and capital service (interest and depreciation)
of 10% per annum. No operating costs are taken into account,
as they are entirely unknown, but probably relatively modest.
On this basis, one would arrive at a break-even line with an
investment of 20 million dollars/km^2, or 20 dollars/m^2,
accessory equipment (piping, etc.) included.

In spite of the fact that the basis of these estimates may
still appear optimistic, it will not be easy to come down to
the break-even line given. But as the price of fossil fuel
will undoubtedly go up steeply in the future, the break-even
line must rise correspondingly. Moreover, for the comparison
with hydrogen from thermochemical (nuclear) cycles, many, not
only economic, factors will have to be taken into account.

REFERENCES

1. E. Broda, The Evolution of the Bioenergetic Processes,
      Pergamon Press, Oxford 1975
2. P. Böger, Photosynthese in globaler Sicht, Naturwiss.
      Rundsch. 28, 429-435 (1975)
3. C.B. van Niel, The Present Status of the Comparative Study
      of Photosynthesis, Ann.Rev.Plant.Physiol. 13, 1-21
      (1962)
4. D.I. Arnon, The Light Reactions of Photosynthesis, Proc.
      Nat.Acad.Sci.,Wash. 68, 2883-2892 (1971)

5. J.A. Bassham and M. Calvin, The Path of Carbon in Photosynthesis, Prentice-Hall, Englewood Cliffs, 1957
6. H. Gaffron and J. Rubin, Fermentative and Photochemical
   Production of Hydrogen in Algae, J.Gen.Physiol. 26,
   219-240 (1943)
7. N.I. Bishop, M. Frick and L.W. Jones, in [8],p.3-22
8. A. Mitsui, S. Miyachi, A. San Pietro and S. Tamura, Eds.,
   Biological Solar Energy Conversion, Acad.Press,
   New York 1977
9. D. Hoffmann, R. Thauer and A. Trebst, Photosynthetic
   Hydrogen Evolution by Spinach Chloroplasts Coupled
   to a Clostridium Hydrogenase, Z.Naturforsch. 32 c,
   257-266 (1977)
10. M. Calvin, Photosynthesis as a Resource for Energy and
    Materials, Amer.Scient. 64, 270-278 (1976)
11. M. Calvin, Photosynthesis as a Resource for Energy and
    Materials, Photochem. Photobiol. 23, 425-444 (1976)
12. J.C. Goldman and J.H. Ryther, Mass Production of Algae:
    Bioengineering Aspects, in [8], p. 367-378
13. W.J. Oswald and J.R. Benemann, A Critical Analysis of
    Bioconversion with Microalgae, in [8], p. 379-396
14. D. Pimentel, L.E. Hurd, A.C. Bellotti, M.J. Forster,
    I.N. Oka, O.D. Sholes and R.J. Whitman, Food Pro-
    duction and the Energy Crisis, Science 182, 443-449
    (1973)
15. J.R. Benemann, J.C. Weissman, B.L. Koopman and W.J.
    Oswald, Energy Production by Microbial Photosynthesis,
    Nature 268, 19-23 (1977)
16. B. Holmström, Photoinduced Redox Processes: Production
    of Hydrogen and Other Fuels, in [17], p.IV, 1-62
17. S. Claesson and L. Engström, Solar Energy - Photochemical
    Conversion and Storage, National Swedish Board for
    Energy Development, Stockholm 1977
18. R. Larsson, Photochemical Production of Hydrogen Cata-
    lyzed by Transition Metal Complexes, in [17], p.V,
    1-41
19. L. Tegnér, Photoelectrochemical Production of Electricity,
    in [17], p.III, 1-58
20. M. Almgren, Photochemical Solar Energy Conversion in
    Organized Systems, in [17], p.VII, 1-88
21. P. Mitchell, Chemiosmotic Coupling and Energy Trans-
    duction, Bodmin (England) 1968
22. D. Oesterhelt, R. Gottschlich, R. Hartmann, H. Michel and
    G. Wagner, Light Energy Conversion in Halobacteria, in:
    B.A. Haddock and W.A. Hamilton, Eds., Microbial Ener-
    getics, Cambridge Univ.Press 1977, p. 333-349
23. E. Broda, Die Evolution der bioenergetischen Prozesse,
    Biologie in uns.Zeit 7, 33-40 (1977)

24. R.K. Clayton, Photosynthesis: How Light is Converted to Chemical Energy, Addison-Wesley, Reading, Mass., 1974

25. H.T. Witt, Biophysikalische Primärvorgänge in der Photosynthesemembran, Naturwiss. 63, 23-27 (1976)

26. R. Hill and F. Bendall, Function of the Two Cytochrome Components in Chloroplasts, Nature 186, 136-139 (1960)

27. D.I. Arnon and R.K. Chain, Ferredoxin-catalyzed Photophosphorylation, Plant Cell Physiol. Special Issue 129-147 (1977)

28. E. Broda, Großtechnische Nutzbarmachung der Sonnenenergie durch Photochemie, Naturwiss. Rundsch. 28, 365-372 (1975)

29. E. Broda, Solar Power - The Photochemical Alternative. Bull. Atom. Sci. 32 (3) 49-52 (1976)

30. E. Broda, Photochemical Hydrogen Production Through Solar Radiation by Means of the Membrane Principle, UNESCO/WMO Solar Energy Symposium, Geneva 1976

31. S.A. Alkaitis, M. Grätzel and A. Henglein, Laser Photoionization of Phenothiazine in Micellar Solution, Ber. Bunsenges. physik. Chem. 79, 541-546 (1975)

32. A. Henglein, Estimated Distribution Functions of Electronic Redox Levels and the Rate of Chemical Reactions of Excess Electrons in Dielectric Liquids, Ber. Bunsenges. physik. Chem. 79, 475-480 (1975)

33. A.J. Frank, M. Grätzel, A. Henglein and E. Janata, The Influence of the Interface Potential on the Reaction of Hydrated Electrons and Neutral Radicals with Acceptors in Micelles, Ber. Bunsenges. physik. Chem. 80, 547-551 (1976)

34. H.T. Tien, Electronic Processes and Electronic Aspects of Bilayer Lipid Membranes, Photochem. Photobiol. 24, 97-102 (1976)

35. O. Inacker, H. Kuhn, D. Möbius and G. Debuch, Manipulation in Molecular Dimensions, Z.Physik.Chem. N.F. 101, 337-360 (1976)

BIO-SOLAR HYDROGEN PRODUCTION

A. Mitsui
School of Marine and Atmospheric Science
University of Miami
4600 Rickenbacker Causeway, Miami, Florida, U.S.A.

ABSTRACT

The subject of biological hydrogen photoproduction is reviewed, with an update of recent developments. Special emphasis is given to the research on hydrogen photoproduction in salt water systems presently underway in the laboratory at the University of Miami Marine School.

The overall aims of the paper are to: 1) discuss the advantages of biological hydrogen photoproduction as a source of fuel and biomass; 2) examine the key areas of research in this field, and 3) evaluate the economic potentials and feasibility of such systems.

INTRODUCTION

The world's present level of technology and standard of living has been attained through the expenditure of vast amounts of fossil fuel energy. Over the past half century, oil has become the most important source of energy and synthetic materials. Consequently the petrochemical industry has become the economic base of our society. The recent energy crisis has awakened us to the fact that this base will have to undergo major changes if our present material state is to be perpetuated. Coal, natural gas, and nuclear power have been suggested as the future solutions to the energy problem [e.g., 25, 94]. However, all three of these alternatives involve serious environmental hazards and/or face eventual depletion.

The ideal solution to this problem would be the development of a renewable energy source which is non-polluting. Hydrogen meets both these criteria [e.g., 120, 155-157]. While hydrogen has long been recognized as an efficient fuel, a sophisticated technology has not been developed for its production and use on a commercial scale. An important consideration in the development of such a technology is that energy is required for hydrogen formation. This stems from the fact that in the natural environment almost all hydrogen is found in bound form, as in the case of water, H_2O. This paper will primarily deal with various techniques for hydrogen production, with special consideration given to the use of solar radiation as the source of energy for the process.

Non-Biological Hydrogen Production [e.g., 25, 70, 97, 154]

As stated in the introduction, the major factor in hydrogen production is the requirement for input energy. The methods being developed for hydrogen evolution can therefore be conveniently subdivided into the three types; (1) those using heat as the driving force, i.e., coal gasification and thermochemical reaction, (2) those using electricity, i.e., electrolysis, and (3) those using direct light energy, i.e., photolysis and biophotolysis.

In terms of the maximum preservation of natural resources one of the most attractive methods of hydrogen production is photolysis, since it is based on the direct utilization of solar energy. At the same time, however, photolysis techniques are among the newest and least developed areas of research.

Biological Hydrogen Production

The capability of many photosynthetic microorganisms to produce hydrogen gas has been recognized for many years. The process itself is directly linked to the light dependent photosynthetic pathway. Therefore, the application of biological hydrogen evolution to the production of fuel would result in a new solar energy based technology. Actually there are numerous advantages to this approach to hydrogen production [106], including the following:
(1) The system could be operated at low physiological temperatures (i.e., 10-40°C) as opposed to the normally high temperatures required for chemical or physical production of hydrogen (400 to 1000+$^{\circ}$K).
(2) The only major input into the system would be solar energy and a hydrogen donor, possibly water (salt water).
(3) The production of hydrogen would not involve the evolution of pollutants, as in the case of fossil fuel refineries.
(4) The fuel produced, H_2 gas, would be clean burning (i.e., yielding H_2O).
(5) The system would readily lend itself to a potentially beneficial and profitable program of multiple utilization, including the production of food for human and animal consumption, compounds required by methane producing bacteria, and commercially usable chemical byproducts.

A. A Brief History of Biological Hydrogen Production Research

The photoproduction of hydrogen by intact cells was first observed in algae by Gaffron and Rubin in 1942 [34] and in photosynthetic bacteria by Gest and Kamen in 1949 [40, 41]. After these pioneering experiments, hydrogen photoproduction by intact cells was extensively studied by several laboratories [7, 8, 11, 14-16, 20-23, 27, 33, 34, 36-44, 47-49, 53, 55-58, 61, 63-66, 69, 71-74, 77, 80, 81, 84, 87, 90, 95, 96, 101, 103-108, 112, 117, 118, 121, 122, 127, 131, 132, 134-136, 138-144, 151, 152, 159, 160, 162].

In recent years many new avenues of hydrogen photoproduction and solar energy bioconversion research have been proposed, including applied aspects [5, 17, 18, 24, 29, 35, 45, 52, 59, 74, 80, 93, 100, 106, 110-112, 129].

Through these and other experiments the mechanism of hydrogen evolution is starting to be elucidated. Basically, solar radiation is absorbed by some form of pigment, for example, the light harvesting chlorophyll protein. This energy is transmitted to the electron through the reaction center chlorophyll protein. The source of the electron is some electron donor compound (example, water). This energized electron is the key to the process of hydrogen gas evolution. The electrons for the H_2 production step follow essentially the same pathway as those leading to carbon fixation and other associated functions [see 99, 100]. The actual shunting of electrons from the photosynthetic carbon fixation pathway can occur just after ferredoxin (or another primary electron acceptor) to proton can take place in three different ways [see 104, 112]. The transfer could occur through: (1) an ATP independent hydrogenase catalyst; (2) an ATP dependent nitrogenase system; or (3) NADP and hydrogenase mediated system (or analogous systems). Of course, all species of algae and photosynthetic bacteria do not contain all three hydrogen evolution systems. Energetically very little is known about the relative efficiency of each system. The key element in the process of H_2 production is solar energy input. Solar radiation drives the photosynthetic system by supplying the energy for the transfer of electrons from the donor compound to proton, and for ATP formation through photophosphorylation. This donor substance varies from species to species. Photosynthetic bacteria require an electron donor such as an organic compound [41, 43] or a sulfur compound [95]. However, algae may utilize water as their electron donor [14-16, 33, 55, 56, 63, 64, 72, 138-144]. Figure 1 illustrates major differences between algal and photosynthetic bacterial systems.

However, it should be pointed out that the question of electron donors of algae has been a very controversial issue. Many hypotheses have been presented along with experimental data. For example, King et al. [75], and Jones and Bishop [61] recently have shown that hydrogen production in some algal species may not be directly linked to the photolysis of water. Rather they suggest that some organic compound serves as the electron donor in such systems. Hydrogen production by heterocystous blue green algae appear to require organic electron donors. On the other hand, Bishop et al. [15] have demonstrated that in Scenedesmus and other green algae, water serves as the primary substrate for hydrogen and oxygen production. Whether the electron donor of algal systems is water or organic substances could depend on genetic groups, culture conditions, and physicochemical experimental conditions. Energetically there seem to be some basic differences between photosynthetic bacteria and algae. The minimum solar energy input for these two systems is quite different [see 112]. It should also be pointed out that algal biophotolysis systems usually need two light reactions and bacterial hydrogen production systems may require only one.

Another issue of great importance in the study of hydrogen production is the question of oxygen inhibition. In many species of algae oxygen inhibits hydrogen production by inactivating the major catalysts of the reaction, hydrogenase and/or nitrogenase. Recently, several researchers have shown that some blue green algae are not subject to oxygen inhibition. They have hypothesized that this resistance stems from the presence of a specialized cell type, or heterocyst, which is impermeable to oxygen, and contains the essential photosystem I components and nitrogenase [11, 137]. The electron donor for the reaction is organic and is provided by adjacent normal vegetative cells [61, 152].

The oxygen inhibition problem faced by other algae and photosynthetic bacteria is being worked on by numerous researchers. Intensive studies of the structural and functional properties of hydrogenases and nitrogenases are going on in several laboratories. For example, Krasna and his coworkers have investigated the hydrogenases of a wide range of microorganisms [54, 83, 85]. Attempts are being made to find oxygen insensitive forms of hydrogenase and nitrogenase.

In addition to intact cell experiments with hydrogen photoproduction, several research groups have initiated studies with cell-free systems. During 1961-1963, my coworkers and I [3, 109, 113, see also 100] succeeded in constructing a working cell-free hydrogen photoproduction system. This cell-free system consisted of chloroplasts of plant leaves or lamellae of blue green algae, methyl viologen or ferredoxin, and bacterial hydrogenase (bacterial hydrogenase was isolated from Desulfovibrio, Chromatium and Clostridium). Cysteine or ascorbate was the electron donor of the system. This research was quickly followed by a demonstration that hydrogen could be produced from water [163]. More recently Abeles [1] successfully promoted hydrogen photoproduction by a cell-free green algal system. From the standpoint of energy production, earlier experiments were reexamined. In 1972-1973, Krampitz [82] demonstrated cell-free photoproduction of hydrogen using an NADP mediated system in the presence of chloroplasts, ferredoxin and bacterial hydrogenase. Ben-Amotz and Gibbs [9], Benemann et al. [10] and Rao et al. [130] reported cell-free photoproduction of hydrogen using a system similar to that described above (chloroplast-ferredoxin-bacterial hydrogenase). Ben-Amotz and Gibbs used dithiothreitol as an electron donor. Rao et al. used chloroplasts which were stabilized by glutaraldehyde.

Numerous other studies of hydrogen production through biochemical and chemical pathways have been proposed and researched. These studies have been reviewed in other literature sources [17, 24, 26, 120].

Another part of the cell free work involves attempts to stabilize hydrogen production through microencapsulation and enzyme immobilization [12, 76, 125, 166]. It has been shown that hydrogenase enzyme immobilized onto glass beads exhibit less oxygen sensitivity. Yagi and his coworkers are intensively researching specialized cell-free techniques involving immobilized hydrogenases [165, 166]. Basic research of the molecular mechanisms of hydrogen and oxygen evolution is being carried out in laboratories such as Dr. B. Kok's and Dr. E. Greenbaum's [51].

In addition, genetic techniques are being explored in order to obtain algal and bacterial mutants with favorable hydrogen and oxygen production characteristics. This type of research is being carried out in several laboratories in the world, for example Drs. Togasaki and San Pietro's [96] and Dr. Bishop's laboratories [16, 96].

Research into all areas of biological hydrogen production is expanding rapidly. Numerous laboratories throughout the world are initiating research in this area. Among those engaging in these new efforts are the laboratories of Drs. Bothe [21, 22] and Boger in West Germany, Drs. Gogotov et al. [47, 48] in the USSR, Dr. Bachofen in Switzerland, Dr. D. Hall [2, 52, 130] of England, Dr. T. Bockris [117] of Australia, Dr. Borda [18] of Austria, and Dr. Packer [125, 151, 152] of the U.S.A.

B. Hydrogen Production by Photosynthetic Marine Organisms

As the history of biological hydrogen production reveals, many new approaches to this area of research are being undertaken [5, 17, 24, 29, 35, 45, 52, 59, 80, 93, 100, 106]. Our approach at the University of Miami's Marine School is based on the belief that there are undiscovered organisms which exhibit special characteristics suited for application. The methodology used to attack this problem and our results to date will be described below.

The first step was to survey the tropical marine environment for species with high hydrogen producing capabilities. Most of the early experiments with intact-cell systems utilized freshwater species of algae and photosynthetic bacteria. However, the marine environment contains many species of photosynthetic bacteria and algae which are capable of the same function.

The advantages of using marine photosynthetic microorganisms in mass cultures are many. The most apparent of these is the availability of salt water. Since many regions of the world suffer from lack of fresh water [115], the ability to use salt water becomes an important factor. Secondly, salt water, in its natural state, abounds with many nutrients (including CO_2, magnesium, sulfate, potassium) essential to the growth of photosynthetic organisms. In tropical and subtropical marine environments photosynthetic organisms (bacteria and algae) can be found in abundance year round. Since the tropical environment is well endowed with a wide diversity of marine algae and bacterial species it is an ideal location for the study of genetically different hydrogen photoproduction systems. We felt that a survey of hydrogen-producing tropical marine algae and photosynthetic bacteria would reveal species with exceptionally high hydrogen photoproducing capabilities. In addition, utilizing marine species would permit the use of salt water as a hydrogen donor. The impending shortage of fresh water in many areas of the world made this a key advantage. Other key elements of the study included: (1) isolation and identification of individual species from mixed samples, (2) screening for H_2 production capabilities, (3) determination of culture conditions which maximize H_2 production and growth, and (4) the selection of a hydrogen donor which would be both efficient and inexpensive.

In accordance with our experimental plan we surveyed numerous locations in the tropical Atlantic Ocean for hydrogen producing blue-green algae and photosynthetic bacteria. Pure cultures of many of the collected organisms were established.

Using these isolated stains as a working base we performed screening experiments for long term (3 to 7 days) hydrogen production capability. This approach diverged from the more common short term (e.g., 1-15 minute) hydrogen production experiments performed using hydrogen electrodes or manometry. We chose this "long term" method because it made it possible to discuss the commercially important question of stability and quantity of H_2 production. Most previous experiments had shown that hydrogen photoproduction lasts only a very short time (a few minutes to 1 hour), unless the experimental gas phase is kept free of hydrogen gas (and in most cases oxygen) by the continuous flow of an inert gas, ex. argon.

In the first set of experiments several strains of blue green algae were shown to produce H_2 gas in the dark. Approximately one half of the tested species produced hydrogen in the dark [86]. The fact that so many strains of tropical marine blue green algae produced H_2 in the dark coincided with the results of Ben Amotz et al. [8] with temperate and cold water green and red marine algae. This led to the observation that a wide diversity of marine species contain H_2 producing mechanisms, probably including some form of hydrogenase (or nitrogenase). However, the rates of dark hydrogen production were relatively low. Light hydrogen production capabilities of marine blue-green algae were then tested. Three different types of hydrogen production in the light were observed [86, 87, 112].

(1) In the first type, hydrogen was produced in the dark and not in the light both with and without oxygen trapping agents (chromous chloride).

(2) In some cases hydrogen production occurred in the dark at moderate rates and accelerated in the light in the presence of an oxygen trapping agent. This acceleration depended on the given light intensities. Without an oxygen trapping agent, hydrogen produced in the dark was consumed photochemically.

(3) In others, hydrogen was produced in the light with and without an oxygen trapping agent, but not in the dark.

Most of the strains produced hydrogen at rates of 0.002-0.1 m moles H_2 per mg chlorophyll per day or 0.002-1.9 μ moles H_2 per mg dry weight per day [86]. However, one marine blue-green algal strain (Miami BG7) exhibited high light-dependent production rates. In subsequent experiments [112] the culture conditions of this strain were altered (i.e., minus combined nitrogen) yielding exceptionally high H_2 production rates. Under the new conditions, Miami BG7 produced 230 μ moles of hydrogen per mg chlorophyll per hour, or 1 ml of H_2 gas per ml reaction mixture within a three-day illumination period. Furthermore, the H_2 gas was continually produced at approximately the same rate throughout the 5-7 day period (based on later tests).

The results of these experiments indicate that it is feasible to conceive of using hydrogen photoproduction by algae as a source of fuel. In terms of gross hydrogen production it is possible to make an estimate of the amount of gas which could be made available for commercial use. Based on our experiments with Miami BG7 and to calculate what size culture would be needed to supply the electrical demands of an average sized house (2 bedrooms) during the summer months (peak use period) in southern Florida (i.e., 1000 kwhr/month, estimate by Florida Power and Light (see Fig. 2). Using the same production rates as above, a culture 1 m deep and having a total area of $(8 m)^2$ or $(26 feet)^2$ would be adequate to meet these needs (see Fig. 2).

It is obvious that these figures are highly tentative and involve many simplifying assumptions. Nevertheless they do provide some indication of the great potential usefulness of this approach. Accordingly, a three-year project was recently initiated to study hydrogen production pathways of Miami BG7 and the environmental conditions which increase and stabilize production rates.

ECONOMY OF BIOLOGICAL HYDROGEN PRODUCTION

A. The Problem of Solar Energy Bioconversion Efficiency

The problem which has plagued researchers in their attempt to practically apply the above mechanisms of H_2 production is the question of efficiency.

Since the production of hydrogen gas is an offshoot of the photosynthetic pathway, the solar energy conversion efficiency is directly related to the metabolic demands of the plant cells being used in the system.

This, in large part, is why the theoretical maximum of 30% photosynthetic solar energy conversion efficiency [80, 93. 129] is never observed in nature. As such, the problem of increasing bioconversion efficiency involves a complex of variables associated with important considerations: (1) metabolic potential--the metabolic potential of plants is associated with their photosynthetic capability, which in turn determines their H_2-producing potential. Different species exhibit different metabolic potentials. Even within the same species the attainment of this potential is dependent on the existence of proper environmental conditions (i.e., sunlight, temperature, salinity, nutrient availability); (2) competing pathways for electrons, there are numerous pathways which the excited electron from photosystem I (PS I) can take. These include CO_2 fixation, O_2 reduction, cyclic electron flow, N_2 fixation, nitrite reduction or hydrogen production [81, 100, 101]. Under natural conditions the percentage of electrons flowing to each of these pathways will depend on the metabolic state of the cell; (3) oxygen inactivation of hydrogenase and nitrogenase-- in the natural environment, O_2 produced during the photolysis of water will, at certain concentrations, inhibit the activity of some types of hydrogenase and nitrogenase enzymes which catalyze the H_2-producing reaction. This phenomenon, where it occurs, reduces the net yield of H_2 gas; (4) rate limiting reactions--certain steps in the transfer of electrons to H_2 production, especially electron flow between photosystem I and photosystem II, and hydrogenase and nitrogenase catalysis, are rate limiting (see 112).

There have been two major new directions which research has taken in order to solve these problems and thereby enhance solar energy conversion: the manipulation of living cells and the development of cell-free production systems.

Living Cell H_2 Production

Our approach to increasing hydrogen photoproduction capability in living systems involves several steps [101, 112]. These include: (1) survey and selection of proper organisms, (2) development of a method of efficient removal of oxidizing agents such as oxygen, (3) physical and chemical treatment of cells, (4) use of specific metabolic inhibitors, (5) searching for oxygen insensitive hydrogenase and nitrogenase enzymes, (6) genetically altering cells in order to increase the efficiency of hydrogen production, and (7) finding appropriate electron donors for the process (see Fig. 1).

Cell Free H_2 Production (100)

Another extension of intact cell H_2 photoproduction research would be the use of a cell free approach. It is well known that natural, intact cell biochemical systems operate under a network of checks and balances. This system helps to maintain organisms within a state of homeostasis (equilibrium). This precludes excessive buildup of any end product, and it also sets an upper limit to the rates of hydrogen production and nitrogen fixation (as well as other biochemical reactions). One of the most efficient ways of removing these restraints would be to isolate the hydrogen producing system from its cellular environment, creating a cell free system.

The intact cell and cell free approaches are bound together in two respects: (1) by their mutual use of the algal and photosynthetic bacterial photosynthetic production mechanisms, and (2) by the similarities in experimental methodologies used to study both approaches.

The initial step is to collect, isolate and culture different species (genetic groups) of marine algae and photosynthetic bacteria, which will serve as a source for the "pool of potential components" for the cell-free system. The next step is to acquire information about the structure (e.g., X-ray crystallography) and function of these components so that "selection processes" can be carried out.

The next steps in the process include "stabilization" (e.g., prevention of chlorophyll protein photobleaching), immobilization, and the alteration of the molecular structure of components.

Stabilization is one of the most important factors in successful cell-free system projects, primarily because components of cell-free H_2 photoproduction systems are more unstable than their intact cell counterparts.

Finally, the selected components can be combined or arranged on a membrane or column to make the reaction system. At this stage tests must be made to determine the physicochemical conditions, solar energy intensities, and the electron and H^+ donors, which together yield the highest H_2 gas production rates.

This cell free approach, while being initially more complicated than the intact cell approach, may in the long run yield simple and stable hydrogen production systems.

B. Profit Margin and the Multiple Utilization Approach

In the development of any new technology one of the most important considerations must be economic potential. This has led to pessimism amongst some researchers over the future of biological hydrogen production systems. It must be remembered, however, that many new enterprises have encountered and successfully overcome such barriers. In order to resolve these problems considerable effort and money will have to be devoted to research on ways to minimize construction and maintenance costs of production systems, while at the same time maximizing profit.

Profit is important since the eventual cooperation of private enterprise in these ventures is highly desirable if not indispensable. In many cases the development of a biological hydrogen production system designed solely for hydrogen evolution may not be profitable enough to justify its existence. It will be necessary to implement a program for the multiple utilization of harvested materials. Many microorganismal and plant species would probably lend themselves to the production of several economically valuable substances. Some raw plant materials could be subdivided and refined to yield food, feed, medicine, and fuel (through such processes as fermentation). These types of multiple utilization procedures have already been successfully applied with several land agriculture projects. There is no reason why they could not be applied to hydrogen production systems. This multiple utilization potential includes:

Biomass Production--Food and Feed

Photosynthetic bioconversion mechanisms are associated with the production of many valuable byproducts, one of the most important being potential sources of food and feed. For many years algal and marine plant products have received regionally limited use as food items in Asia, Northern Europe, the Pacific islands, and in some areas of South America. Macroalgae such as Porphyra and Laminaria have a high commercial value in Japan and nearby regions. As discussed above, considerable effort has been devoted to the cultivation of these algae as a food source, especially in Asian countries such as Japan. In the western world brown algae have traditionally been valued as fertilizers and animal feed [60].

Over the past three decades, the cultivation of several types of unicellular green and blue green algae as a protein source has been undertaken. Fresh water and brackish species of the photosynthetic microorganisms Chlorella, Scenedesmus, Spirulina, and Rhodopseudomonas have also been cultured for use as a protein source in aquaculture, as feed additives, and as food additives [30, 78, 79, 128, 148-150, 153].

Mass cultures (after use in hydrogen production), could serve as a protein source for human or animal consumption or for aquaculture (fish, shellfish, shrimp, crabs). Microorganismal carbohydrates, lipids, minerals, and vitamins could also be utilized.

Many algae contain large quantities of vitamin A, the B group vitamins (B_1, B_2, B_{12}, niacin, pantothenic acid, and folic acid) and vitamins C and D. In certain species of seaweeds and unicellular algae, vitamin content is as rich and varied as that of meat [67, 68, 92].

Numerous minerals are also provided by algae. Sodium, potassium, calcium, magnesium, and iron may constitute up to 15-25% by dry weight in certain algae [92, 116]. These inorganic components play an important role in preventing blood acidosis [145].

Fertilizer

The technology for producing high crop yields depends on large amounts of fertilizers, especially combined nitrogen compounds such as ammonia. The synthesis of ammonia by industrial methods spends non-renewable fossil fuels and requires a considerable input of energy for the conversion process. These energy costs are also carried over in the production of animal protein. The study and application of biological nitrogen fixation and the control of denitrification would reduce the requirements for industrially produced fertilizer.

Applications to Aquaculture (101)

One of the factors which determines the success or failure of ventures in aquaculture is the proper choice of primary food sources. Green algae and diatoms have already been successfully employed, as the base of the food chain, to feed zooplankton, which in turn are utilized by the larval and adult stages of the organisms being exploited (i.e., crustaceans, shellfish and fish). Dr. Kobayashi's laboratory in Japan has successfully used freshwater or soil photosynthetic bacteria for aquaculture purposes. However, cultured photosynthetic bacteria and blue-green algae have yet to be tested for their applicability as food in aquaculture. The question might arise as to why marine photosynthetic bacteria and blue green algae should be used instead of presently developed food sources. There are several advantages which favor the former as a food resource: (1) the cell walls of photosynthetic bacteria and blue green algae are much softer and easier to digest than those of green algal species, (2) some marine photosynthetic bacteria and blue-green algae have a high nutritional value as mentioned above, and (3) the growth rate of some marine photosynthetic bacteria and blue-green algae is considerably faster than many of the presently used food organisms. Mass cultures of H_2-producing algal strains could also provide food for the aquaculture of shrimp, crabs, shellfish, and fish (either directly or through the culture of zooplankton).

Methane Production

Several years ago, Oswald and his coworkers reported that methane production using sewage and algae might be economically feasible [123]. More recently Wilcox, North and their colleagues have studied the possibility of using kelp as a substrate for methane production [119, 164]. Some other photosynthetic strains may prove to be an economical source of carbohydrate material for bacteria-mediated methane production.

Medicine

The surveying and research of metabolically active substances produced by algae has received some attention in the past few years. However, it remains a relatively little-studied area for investigation. New metabolically active substances may therefore be found in many of these as yet relatively unexplored species.

Several algae are already known to produce substances of potential medical importance. The anthelmintics L-kainic acid and L-allokainic acid have been obtained from the red alga Digenia simplex [114]; domoic acid has also been obtained from Chondria armata [28, 147]. Laminine, obtained from the brown alga Laminaria, has been shown to be effective as an antihypertensive [124, 146]. There are also indications that sodium alginate prevents gastroenteric absorption of radioactive strontium [133]. The compounds sodium laminarin sulfate and fucoidin could be potentially useful as blood coagulants [158]. In addition, several algae are known to produce antibiotic substances. Survey and further study of marine algae will probably lead to the discovery of other medically useful compounds.

Chemicals

There is an enormous range of chemical substances produced by algae and plants which are of interest to man. In a project in which there is a potential for developing systems which will continuously produce large quantities of material, it is important to explore ways of exploiting this resource from a chemical point of view.

Algal polysaccharides such as agar, carrageenan, and furcellaran from red algae and algin from brown algae are important algal products presently in commercial use. These substances are used in numerous industries as emulsifying agents, gelling agents, stabilizers, suspension agents, and thickeners. Cultivation of algae to provide these substances could be combined with their utilization for other purposes. For example, mass production of Macrocystis (Ocean Farm Project) will supply algin as the main product and also fertilizer and feed as byproducts [91, 164].

Sugar alcohols represent another potentially important algal product. The accumulation of these compounds appears to be a common feature among algae. Some brown algae are known to accumulate mannitol to levels as high as 50% of the dry weight [98]. Halophilic species of the unicellular green alga <u>Dunaliella</u> have been found to store glycerol under conditions of high salinity [50]. When cells of this alga are suspended in 1.5 M NaCl solution, an intracellular glycerol concentration of 2 M can be reached [6, 19].

Relatively few surveys of the chemical and biological properties of photosynthetic microorganisms have been conducted. Further studies on the chemical composition, biochemistry, and physiology of these organisms will undoubtedly reveal new chemical resources of industrial value.

CONCLUSION

In this paper the potential of hydrogen as a fuel has been described. Particular emphasis has been placed on the importance of utilizing solar radiation as the principal energy source for production of hydrogen. There is, however, some opposition to the immediate development of technologies, based on the grounds that coal and nuclear energy are more reasonable alternatives. There is certainly no doubt that the latter forms of energy will dominate the fuel economy over the next few decades because of the existence of pre-adapted utilization technologies.

In the long run, however, there are drawbacks to the use of coal, natural gas, and nuclear power. The most serious of these are health and safety considerations. First of all, the combustion of coal involves the evolution of numerous toxic and environmentally hazardous substances. In addition to these inherently dangerous substances the production of large quantities of CO_2 presents a long term risk to the biosphere because of atmospheric heating, or the "green house" effect. Even if these problems could be resolved, the fact remains that coal and natural gas were not renewable and will eventually meet the same fate as oil.

Nuclear energy, though renewable (through breeder reactors), is accompanied by the most serious potential dangers of any fuel. The production of large quantities of nuclear waste could lead to future health dangers of catastrophic proportions.

Solar energy, especially as it relates to hydrogen production (i.e., hydrogen photoproduction) represents one of the most viable substitutes for coal and nuclear power. Both of the major elements in this type of fuel technology would be renewable and pollution free. The barrier which now stands in the way of widespread application of hydrogen photoproduction is the lack of an advanced production, storage and utilization technology. This problem could be resolved by a strong research and development effort in this area.

ACKNOWLEDGMENT

The author would like to thank Mr. S. Kumazawa, Mr. E. Phlips and Ms. J. Radway for the preparation of the manuscript.

This material is based upon research supported by the U.S. National Science Foundation under Grant no. AER 75-11171 A01 and AER 77-11545. Any opinions, findings, and conclusions or recommendations expressed in this publication are those of the author and do not necessarily reflect the views of the National Science Foundation.

REFERENCES

1. Abeles, F.B. 1964. Cell free hydrogenase from Chlamydomonas. Plant Physiol. 39: 169.

2. Adams, M.W.W. and D.O. Hall. 1977. Isolation of the membrane-bound hydrogenase from Rhodospirillum rubrum. Biochem. Biophys. Res. Commun. 77: 730.

3. Arnon, D.I., A. Mitsui and A. Paneque. 1961. Photoproduction of hydrogen gas coupled with photosynthetic photophosphorylation. Science 134: 1425.

4. Barr, W.J. and F.A. Parker. 1976. The introduction of methanol as a new fuel into United States economy. Foundation for Ocean Research, San Diego.

5. Beck, R.W. 1976. Workshop on solar energy for nitrogen fixation and hydrogen production. University of Tennessee, Knoxville.

6. Ben-Amotz, A. and M. Avron. 1973. The role of glycerol in the osmotic regulation of the halophilic alga Dunaliella parva. Plant Physiol. 51: 875.

7. Ben-Amotz, and M. Gibbs. 1974. H_2 photoevolution and photoreduction by algae. Abstracts of the Annual Meeting of the American Society of Plant Physiologists. p. 27.

8. Ben-Amotz, A., D.L. Erbes, M.A. Riederer-Henderson, D.G. Peavey and M. Gibbs. 1975. H_2 metabolism in photosynthetic organisms. I. Dark H_2 evolution and uptake by algae and mosses. Plant Physiol. 56: 72.

9. Ben-Amotz, A. and M. Gibbs. 1975. H_2 metabolism in photosynthetic organisms. II. Light-dependent H_2 evolution by preparations from Chlamydomonas, Scenedesmus and spinach. Biochem. Biophys. Res. Comm. 64: 355.

10. Benemann, J.R., J.A. Berenson, N.O. Kaplan and M.D. Kamen. 1973. Hydrogen evolution by a chloroplast-ferrodoxin-hydrogenase system. Proc. Natl. Acad. Sci. U.S.A. 70: 2317.

11. Benemann, J.R. and N.M. Weare. 1974. Hydrogen evolution by nitrogen fixing Anabaena cylindrica cultures. Science 184: 174.

12. Berenson, J.A. and J.R. Benemann 1977. Immobilization of hydrogenase and ferrodoxins on glass beads. FEBS Letters 76: 105.

13. Bishop, N.I. 1966. Partial reactions of photosynthesis and photoreduction. Ann. Rev. Plant Physiol. 17: 185.

14. Bishop, N.I. and H. Gaffron. 1963. On the interrelation of the mechanisms for oxygen and hydrogen evolution in adapted algae. In: Photosynthetic Mechanisms of Green Plants. NAS-NRC Publication 1145. p. 441.

15. Bishop, N.I. and L.W. Jones. 1976. Aspects of the mechanisms of photohydrogen metabolism in green and blue-green algae. Abstracts of U.S.-Japan Cooperative Science Seminar on Biological Solar Energy Conversion, University of Miami, Miami. p. 16.

16. Bishop, N.I., M. Frick and L.W. Jones. 1977. Photohydrogen production in green algae: Water serves as the primary substrate for hydrogen and oxygen production. In: Biological Solar Energy Conversion (Eds.) A. Mitsui et al. Academic Press, New York.

17. Bolton, J.R. (Ed). 1976. Abstracts of International Conference on the Photochemical Conversion and Storage of Solar Energy.

18. Borda, E. 1976. Solar power, the photochemical alternative. The Bulletin of the Atomic Scientists 32: 48.

19. Borowitzka, L.J. and A.D. Brown. 1974. The salt relation to marine and halophilic species of the unicellular green alga Dunaliella. The role of glycerol as a compatible solute. Arch. Mikrobiol. 96: 37.

20. Bose, S.K. and H. Gest. 1962. Hydrogenase and light stimulated electron transfer reactions in photosynthetic bacteria. Nature 195: 1168.

21. Bothe, H. J. Tennigkeit and G. Eisbrenner. 1977. The hydrogenase-nitrogenase relationship in the blue-green alga Anabaena cylindrica Planta 133: 237.

22. Bothe, H., J. Tennigkeit and G. Eisbrenner. 1977. The utilization of molecular hydrogen by the blue-green alga Anabaena cylindrica. Arch Mikrobiol. 114: 43.

23. Bregoff, H.M. and M.D. Kamen. 1952. Studies on the metabolism of photosynthetic bacteria. XIV. Quantitative relations between malate dissimilation, photoproduction of hydrogen and nitrogen metabolism in Rhodospirillum rubrum. Arch. Biochem. Biophys. 36: 202.

24. Calvin, M. 1974. Solar energy by photosynthesis. Science 184: 375.

25. C.E.Q., E.R.D.A., E.P.A., F.E.A., F.P.C., D.I., N.S.F. 1975. Energy alternatives: a comparative analysis. The Science and Public Policy Program. University of Oklahoma, Norman, Oklahoma.

26. Cooper, S.R. and M. Calvin. 1974. Solar energy by photosynthesis: manganese complex photolysis. Science 185: 376.

27. Daday, A., R.A. Platz and G.D. Smith. 1977. Anaerobic and aerobic hydrogen gas formation by the blue-green alga Anabaena cylindrica. Appl. Environ. Microbiol. 34: 478.

28. Daigo, K. 1959. Studies on the constituents of Chondria armata. J. Pharm. Soc. Japan 79: 350.

29. Donovan, P., W. Woodward, F.H. Morse and L.O. Herwig. 1972. An assessment of solar energy as a national energy resource. NSF/NASA Solar Energy Panel.

30. Durand-Chastel, H. and M.D. Silve. 1977. The Spirulina algae. Euro. Sem. Biol. Solar Energy Conversion Syst., Grenoble-Autrans. p. 1-15.

31. Energy Research and Development Administration. 1976. Hydrogen Fuels: A. Bibliography. Office of Public Affairs, Technical Information Center. Washington, D.C.

32. Feigenblum, E. and Alvin I. Krasna. 1970. Solubilization and properties of the hydrogenase of Chromatium. Biochem. Biophys. Acta 198: 157.

33. Frenkel, A.W. 1952. Hydrogen evolution by the flagellate green alga Chlamydomonas moewusii. Arch. Biochem. Biophys. 38: 219.

34. Gaffron, H. and J. Rubin. 1942. Fermentative and photochemical production of hydrogen in algae. J. Gen. Physiol. 26: 219.

35. Gainer, J.L. (Ed). 1976. Enzyme Technology and Renewable Resources. University of Virginia and National Science Foundation (RANN).

36. Gest, H. 1951. Metabolic patterns in photosynthetic bacteria. Bacteriol. Rev. 15: 183.

37. Gest, H. 1963. Metabolic aspects of bacterial photosynthesis. In: Bacterial Photosynthesis. (Eds) H. Gest, A. San Pietro and L.P. Vernon. The Antioch Press, Yellow Springs. p. 129.

38. Gest, H. 1971. Energy conversion and generation of reducing power in bacterial photosynthesis. Advances in Microbial Physiology 7:243.

39. Gest, H., J. Judis and H.D. Peck, Jr. 1956. Reduction of molecular nitrogen and relationships with photosynthesis and hydrogen metabolism. In: Inorganic Nitrogen Metabolism. (Eds) W.D. McElroy and B. Glass. Johns Hopkins Press, Baltimore. p. 298.

40. Gest, H. and M.D. Kamen. 1949. Studies on the metabolism of photosynthetic bacteria. IV. Photochemical production of molecular hydrogen by growing cultures of photosynthetic bacteria. J. Bacteriol. 58: 239.

41. Gest, H. and M.D. Kamen. 1949. Photoproduction of molecular hydrogen by Rhodospirillum rubrum. Science 109: 558.

42. Gest, H. and M.D. Kamen. 1960. The photosynthetic bacteria. In: Encyclopedia of Plant Physiology. (Ed) W. Ruhland. Vol. 2. Springer Verlag, Berlin. p. 568.

43. Gest, H., M.D. Kamen and H.M. Bregoff. 1950. Studies on the metabolism of photosynthetic bacteria. V. Photoproduction of hydrogen and nitrogen fixation by Rhodospirillum rubrum. J. Biol. Chem. 182: 153.

44. Gest, H., J.G. Ormerod and K.S. Ormerod. 1962. Photometabolism of Rhodospirillum rubrum: Light-dependent dissimilation of organic compounds to carbon dioxide and molecular hydrogen by an anaerobic citric acid cycle. Arch. Biochem. Biophys. 97: 21.

45. Gibbs, M., A. Hollaender, B. Kok, L.O. Krampitz and A. San Pietro (Organizers) 1973. Proceeding of the workshop on bio-solar conversion. NSF-RANN.

46. Gitlitz, P.H. and A. I. Krasna. 1975. Structural and catalytic properties of hydrogenase from Chromatium. Biochemistry 14: 2561.

47. Gogotov, I.N. and A.V. Kosyak. 1976. Hydrogen metabolism in Anabaena variabilis in the dark. Mikrobiologiya 45: 586.

48. Gogotov, I.N., A.V. Kosyak and A.N. Krupenko. 1976. Formation of hydrogen by cyanobacteria Anabaena variabilis in the presence of light. Mikrobiologiya 45: 941.

49. Gray, C.T. and H. Gest. 1965. Biological formation of molecular hydrogen. Science 148: 186.

50. Graigie, J.S. and J. McLachlan. 1964. Glycerol as a photosynthetic product in Dunaliella tertiolecta Butcher. Can. J. Bot. 42: 777

51. Greenbaum, E. 1977. The molecular mechanisms of photosynthetic hydrogen and oxygen production. In: Biological Solar Energy Conversion (Eds.) A. Mitsui et al. Academic Press, New York. p. 101.

52. Hall, D.O. 1976. Photobiological energy conversion. FEBS Letters 64: 6.

53. Hartman, H. and A.I. Krasna. 1963. Studies on the "Adaptation" of hydrogenase in Scenedesmus. J. Biol. Chem. 238: 749.

54. Hartman, H., and A.I. Krasna. 1964. Properties of the hydrogenase of Scenedesmus. Biochem. Biophys. Acta. 92: 52.

55. Healey, F.P. 1970. The mechanism of hydrogen evolution by Chlamydomonas moewusii. Plant Physiol. 45: 153.

56. Healey, F.P. 1970. Hydrogen evolution by several algae. Planta 91: 220.

57. Hillmer, P. and H. Gest. 1977. H_2 metabolism in the photosynthetic bacterium Rhodopseudomonas capsulata: H_2 production by growing cultures. J. Bacteriol. 129: 724.

58. Hillmer, P. and H. Gest. 1977. H_2 metabolism in the photosynthetic bacterium _Rhodopseudomonas capsulata_: Production and utilization of H_2 by resting cells. J. Bacteriol. 129: 732.

59. Hollaender, A., K.J. Monty, R.M. Pearlstein, F. Schmidt-Bleek, W.T. Snyder, and E. Volkin (Eds). 1972. An inquiry into biological energy conversion. Gatlinburg. NSF-RANN.

60. Jensen, A. 1971. The nutritive value of seaweed meal for domestic animals. In: Proc. 7th Int. Seaweed Symp. (Ed.) K. Nisizawa et al. John Wiley & Sons, New York. p. 7.

61. Jones, L.W. and N.I. Bishop. 1976. Simultaneous measurement of oxygen and hydrogen exchange from the blue-green alga _Anabaena_. Plant Physiol. 57: 659.

62. Kakuno, T., N.O. Kaplan and M.D. Kamen. 1977. Chromatium hydrogenase. Proc. Natl. Acad. Sci. USA 74: 861.

63. Kaltwasser, H. and H. Gaffron. 1964. Effects of carbon dioxide and glucose on photohydrogen production in _Scenedesmus_. Plant Physiol. 39: xiii.

64. Kaltwasser, H., T.S. Stuart and H. Gaffron. 1969. Light-dependent hydrogen evolution by _Scenedesmus_. Planta 89: 309.

65. Kamen, M.D. and H. Gest. 1949. Evidence for a nitrogenase system in the photosynthetic bacteria. Science 109: 560.

66. Kamen, M.D. and H. Gest. 1952. Serendipic aspects of recent nutritional research in bacterial photosynthesis. In: Phosphorus metabolism. Vol. II. (Eds) W.D. McElory and B. Glass. The Johns Hopkins Press, Baltimore. p. 507.

67. Kanazawa, A. and D. Kakimoto. 1958. Studies on the vitamins of seaweeds. 1. Folic acid and folinic acid. Bull. Jap. Soc. Sci. Fish. 24: 573.

68. Kanazawa, A. 1961. Studies on the vitamin B - complex in marine algae. 1. On vitamin contents. Mem. Fac. Fish. Kagoshima Univ. 10: 38.

69. Kelley, B.C., C.M. Meyer, C. Gandy and P.M. Vignais. 1977. Hydrogen recycling by _Rhodopseudomonas capsulata_. FEBS Letters 81: 281.

70. Kelley, J.H. and E.A. Laumann. 1975. Hydrogen Tomorrow. Jet Propulsion Laboratory, California Institute of Technology, Pasadena, California.

71. Kessler, E. 1962. Hydrogenase und H_2 Stoffwechsel bei Algen. Deut. Bot. Ges. (N.F.) 1: 92.

72. Kessler, E. and H. Maifarth. 1960. Vorkommen und Leistungstahigkeit von Hydrogenase bei linigen Grunalgen. Arch. Mikrobiol. 37: 215.

73. Kessler, E. 1974. Hydrogenase, photoreduction and anaerobic growth. In: Algal Physiology and Biochemistry. (Ed) W.D.P. Stewart. Blackwell Scientific Publication, Oxford. p. 456.

74. Kessler, E. 1976. Microbial production and utilization of gases. Akademie der Wissenschaften zu Gottingen, p. 247.

75. King, D., D.L. Erbes, A. Ben-Amotz and M. Gibbs. 1977. The mechanism of hydrogen photoevolution in photosynthetic organisms. In: Biological Solar Energy Conversion (Eds.) A. Mitsui et al. Academic Press, New York. p. 69.

76. Kitajima, M. and W.L. Butler. 1976. Microencapsulation of chloroplast particles. Plant Physiol. 57: 746.

77. Klein, U. and A. Betz. 1978. Induced protein synthesis during the adaptation to H_2 production in Chlamydomonas moewusii. Physiol. Plant. 42: 1.

78. Kobayashi, M. et al. 1969. Sewage purification by photosynthetic bacteria and its use as a fish feed. Bull. Jap. Soc. Sci. Fish. 35: 1021.

79. Kobayashi, M., K. Mochida and A. Okuda. 1967. The amino acid composition of photosynthetic bacterial cells. Bull. Jap. Soc. Sci. Fish. 33: 657.

80. Kok, B. 1973. Photosynthesis. In: Proceedings of the Workshop on Bio-Solar Conversion. (Ed) M. Gibbs, A. Hollaender, B. Kok, L.O. Krampitz, and A. San Pietro. NSF-RANN Report. p. 22.

81. Kok, B., C.F. Fowler, H.H. Hardt, and R.J. Radmer. 1976. Biological solar energy conversion: approaches to overcome yield stability and product limitations. In: Enzyme Technology and Renewable Resources. (Ed) J.L. Gainer. University of Virginia and NSF-RANN and Oral Presentation by B. Kok.

82. Krampitz, L.O. 1973. Hydrogen production by photosynthesis and hydrogenase activity. NSF-RANN Report No. HA1, HA3, HA5. N-73-013. Biophotolysis of water. NSF-RANN Report No. HA2. N-73-014.

83. Krasna, A.I. and D. Rittenberg. 1956. A comparison of the hydrogenase activities of different microorganisms. Proc. Natl. Acad. Sci. 42: 180.

84. Krasna, A.I. 1976. Bioconversion of solar energy. In: Enzyme Technology and Renewable Resources. (Ed.) J. L. Gainer. University of Virginia NSF-RANN. p. 61.

85. Krasna, A.I. 1977. Catalytic and structural properties of the enzyme hydrogenase and its role in biophotolysis of water. In: Biological Solar Energy Conversion (Eds.) A. Mitsui et al. Academic Press, New York. p. 53.

86. Kumazawa, S. and A. Mitsui. 1978; Kumazawa, S., S. Barciela and A. Mitsui. 1978. Manuscripts in preparation.

87. Kumazawa, S., J. Frank, H.R. Skjoldal and A. Mitsui. 1976. Hydrogen production by tropical marine blue-green algae and photosynthetic bacteria. Abstract of the Annual Meeting of the American Society of Plant Physiologists. p. 61.

88. Kurita, S. K. Toyoda, T. Endo, N. Mochizuki, M. Honya and T. Onami. 1977. Hydrogen photoproduction from water. In: Biological Solar Energy Conversion. (Eds.) A. Mitsui et al. Academic Press, New York. p. 87.

89. Kwei-Hwang Lee, J. and M. Stiller. 1967. Hydrogenase activity in cell-free preparations of Chlorella. Biochem. Biophys. Acta. 132: 503.

90. Lambert, G.R. and G.D. Smith. 1977. Hydrogen formation by marine blue-green algae. FEBS Letters 83: 159.

91. Leese, T.M. 1975. Ocean food and energy farm kelp product conversion. (manuscript) Presented to 141st Ann. Meeting Amer. Assoc. Adv. Sci., New York.

92. Levring, T., H.A. Hoppe, O.J. Schmid (Eds.). 1969. Marine algae. Cram, de Gruyter and Co., Hamburg.

93. Lien, S. and A. San Pietro. 1975. An inquiry into biophotolysis of water to produce hydrogen. Indiana University and NSF.

94. Livingston, R.S. and B. McNeill (Eds). 1975. Beyond Petroleum. Stanford University, Stanford.

95. Losada, M., M. Nozaki and D.I. Arnon. 1961. Photoproduction of molecular hydrogen from thiosulfate by Chromatium cells. In: Light and Life. (Eds) W.D. McElroy and B. Glass. Johns Hopkins University Press, Baltimore. p. 570.

96. McBride, A.C., S. Lien, R.K. Togasaki and A. San Pietro. 1977. Mutational analysis of Chlamydomonas reinhardi: Application to biological solar energy conversion. In: Biological Solar Energy Conversion. (Eds.) A. Mitsui et al. Academic PRess, New York. p. 77.

97. McGraw-Hill Encyclopedia of Energy. 1976. McGraw-Hill, Inc. 785 pp.

98. Meeuse, B.J.D. 1962. Storage products. In: Physiology and Biochemistry of Algae. (Ed.) R.A. Lewin. Academic Press, New York. p. 289.

99. Mitsui, A. 1967. Physiological role of algal ferredoxin: Relation to photoproduction of hydrogen gas, photoreduction of NADP, photoreduction of nitrite, photofixation of nitrogen and photophosphorylation. In: Studies of mechanism in photosynthesis I. (Ed) A. Takamiya. Tokyo University, Tokyo, Japan. p. 53. (In Japanese).

100. Mitsui, A. 1975. Utilization of solar energy for hydrogen production by cell free system of photosynthetic organisms. In: Hydrogen Energy, Part A (Ed) T.N. Veziroglu, Plenum Publishing Co. , New York. p. 309.

101. Mitsui, A. 1975. Photoproduction of hydrogen via microbial and biochemical processes. In: Proceedings of Symposium-Course "Hydrogen Energy Fundamentals." (Ed) T.N. Veziroglu, University of Miami. S-2, 31.

102. Mitsui, A. 1975. Multiple utilization of tropical and subtropical marine photosynthetic organisms. In: The Proceedings of the Third International Ocean Development Conference. Seino Printing Co. 3: 11.

103. Mitsui, A. 1975. Photoproduction of hydrogen via photosynthetic processes. In: Proceeding of US-Japan Joint Seminar, "Key Technologies for the Hydrogen Energy System." (Ed) T. Ohta. Yokohama National University. p. 75. (Revised form of the symposium proceeding "Hydrogen Energy Fundamentals").

104. Mitsui, A. 1976. Bioconversion of solar energy in salt water photosynthetic hydrogen production system. In: Proceedings of the First World Hydrogen Energy Conference. (Ed) T.N. Veziroglu, University of Miami. 2: 4B-77.

105. Mitsui, A. 1976. Solar energy bioconversion by marine blue-green algae. In: Abstracts of the International Conference on the Photochemical Conversion and Storage of Solar Energy A2-3. Middlesex College, The University of Western Ontario.

106. Mitsui, A. 1976. Long-range concepts: Application of photosynthetic hydrogen production and nitrogen fixation research. In: Proceedings of a Conference on Capturing the Sun Through Bioconversion. The Washington Metropolitan Studies. Washington, D.C., p. 653.

107. Mitsui, A. 1976. A survey of hydrogen producing photosynthetic organism in sub-tropical and tropical marine environments (Abstract). In: Enzyme Technology and Renewable Resources. (Ed) J.L. Gainer. University of Virginia and NSF-RANN. p. 39.

108. Mitsui, A. 1976. A survey of hydrogen producing photosynthetic organisms in tropical and subtropical marine environments. NSF/RANN Annual Report.

109. Mitsui, A. and D.I. Arnon. 1962. Photoproduction of hydrogen gas by isolated chloroplasts in relation to cyclic and non-cyclic electron flow. Plant Physiol. 37S: IV.

110. Mitsui, A., S. Miyachi, A. San Pietro, S. Tamura (Organizers). 1976. Abstracts of U.S.-Japan Cooperative Science Seminar on Biological Solar Energy Conversion, RSMAS, University of Miami, Miami.

111. Mitsui, A., S. Miyachi, A. San Pietro, S. Tamura (Eds.). 1977. Biological Solar Energy Conversion. Academic Press, New York.

1286

112. Mitsui, A. and S. Kumazawa. 1977. Hydrogen production by marine photosynthetic organisms as a potential energy resource. In: Biological Solar Enrgy Conversion. (Eds.) A. Mitsui et al. Academic Press, New York. p. 23.

113. Mitsui, A., A. Paneque and D.I. Arnon. 1962. Photoreduction of methylviologen by isolated chloroplasts. (Manuscript) Photoproduction of hydrogen by chloroplast-methylviologen-hydrogenase system. (Manuscript) Photoproduction of hydrogen and cyclic and non-cyclic phosphorylation by chromatophores of blue-green algae. Quoted in following review papers by D.I. Arnon: Photosynthetic Mechanisms of Green Plants. (Eds) B. Kok and A.T. Jagendorf. Pub. No. 1145, Washington, D.C., NAS-NRC. p. 195. (1962) and Science 194: 1460 (1965).

114. Murakami, S., T. Takemoto and Z. Shimizu. 1953. Studies on the effective principles of Digenea simplex. J. Pharm. Soc. Japan 73: 1026.

115. NAS. 1973. Water scarcity may limit use of western coal. Science. 181: 525.

116. Naylor, J. 1976. Production, trade and utilization of seaweeds and seaweed products. FAO Fish. Tech. Paper No. 159.

117. Neil, G., D.J.D. Nicholas, J. O'M. Bockris and J.F. McCann. 1976. The photosynthetic production of hydrogen. International J. of Hydrogen Energy. 1: 45.

118. Newton, J.W. 1976. Photoproduction of molecular hydrogen by a plant-algal symbiotic system. Science 191: 559.

119. North, W.J. 1977. Possibilities of biomass from the ocean: the marine farm project. In: Biological Solar Energy Conversion. (Eds.) A. Mitsui et al. Academic Press, New York. p. 347.

120. Ohta, T. (Ed). 1975. U.S.-Japan Joint Seminar, Key Technologies for Hydrogen Energy System. U.S. National Science Foundation and Japan Society for Promotion of Science. Yokohama National University, Yokohama.

121. Ormerod, J.G. and H. Gest. 1962. Hydrogen photosynthesis and alternative metabolic pathways in photosynthetic bacteria. Bacteriol. Rev. 26: 51.

122. Ormerod, J.A., K.S. Ormerod and H. Gest. 1961. Light-dependent utilization of organic compounds and photoproduction of molecular hydrogen by photosynthetic bacteria: relationships with nitrogen metabolisms. Arch. Biochem. Biophys. 94: 449.

123. Oswald, W.J. and G.G. Golueke. 1960. Biological transformation of solar energy. Adv. in Applied Microbiol. 2: 223.

124. Ozawa H., Y. Gomi and I. Otsuki. 1967. Pharmacological studies on laminine monocitrate. J. Pharm. Soc. Japan 87: 935.

125. Packer, L. 1976. Problems in the stabilization of the in vitro photochemical activity of chloroplasts used for H_2 production. FEBS Letters 64: 17.

126. Pearson, W.H. and W.J. Darby. 1961. Protein nutrition. Ann. Rev. Biochem. 30: 325.

127. Peters, G.A., W.R. Evans and R.E. Toia, Jr. 1976. Azolla-Anabaena azollae relationship. IV. Photosynthetically driven, nitrogenase-catalyzed H_2 production. Plant Physiol 58: 118.

128. Pirie, N.W. The Spirulina algae. In: Food protein sources. (Ed.) N.W. Pirie. Cambridge Univ. Press, Cambridge. p. 33.

129. Radmer, R. and B. Kok. 1977. Photosynthesis: limited yields, unlimited dreams. BioScience 27: 599.

130. Rao, K.K., L. Rosa and D.O. Hall. 1976. Prolonged production of hydrogen gas by a chloroplast biocatalytic system. Biochem. Biophys. Res. Comm. 68: 21.

131. Schick, J. J. 1971. Interrelationship of nitrogen fixation, hydrogen evolution and photoreduction in Rhodospirillum rubrum. Arch. Mikrobiol. 75: 102.

132. San Pietro, A. and R. K. Togasaki. 1976. Bio-solar conversion: Search for algal hydrogenase with greater oxygen resistance. In: Enzyme Technology and Renewable Resources. (Ed) J.L. Gainer. University of Virginia and NSF-RANN. p. 45.

133. Skoryna, S.C., K.C. Hong and Y. Tanaka. 1971. The effects of enzymatic degradation products of alginates on intestinal absorption of radiostrontium. In: Proc. 7th Int. Seaweed Symp. (Eds.) K. Nisizawa et al. John Wiley and Sons, New York. p. 605.

134. Spruit, C.J.P. 1954. Photoproduction of hydrogen and oxygen in Chlorella. In: Proceedings of First International Photobiology Congress. Amsterdam. p. 323.

135. Spruit, C.J.P. 1958. Simultaneous photoproduction of hydrogen and oxygen by Chlorella. Meded. Landbouwhogesch. Wageningen 58: 1.

136. Spruit, C.J.P. 1962. Photoreduction and anaerobiosis. In: Physiology and Biochemistry of Algae. (Ed.) R.A. Lewin. Academic Press, New York and London. p. 47.

137. Stewart, W.R.D., A. Haystead and H.W. Pearson. 1969. Nitrogenase activity in heterocysts of blue-green algae. Nature 224: 226.

138. Stuart, T.S. 1971. Hydrogen production by photosystem I of Scenedesmus: Effect of heat and salicylaldoxime on electron transport and photophosphorylation. Planta 96: 81.

139. Stuart, T.S. and H. Gaffron. 1971. The kinetics of hydrogen photoproduction by adapted algae. Planta 100: 228.

140. Stuart, T.S. and H. Gaffron. 1972. The mechanism of hydrogen photoproduction by several algae. I. The effect of inhibitors of photophosphorylation. Planta 106: 91.

141. Stuart, T.S. and H. Gaffron. 1972. The mechanism of hydrogen photoproduction by several algae. II. The contribution of photosystem II. Planta 106: 101.

142. Stuart, T.S. and H. Gaffron. 1972. The gas exchange of hydrogen-adapted algae as followed by mass spectrometry. Plant Physiol. 50: 130.

143. Stuart, T.S., E.W. Herold, Jr. and H. Gaffron. 1972. A simple combination mass spectrometer inlet and oxygen electrode chamber for sampling gases dissolved in liquids. Anal. Biochem. 46: 91.

144. Stuart, T.S. and H. Kaltwasser. 1970. Photoproduction of hydrogen by photosystem I of Scenedesmus. Planta 91: 302.

145. Takagi, M. 1975. Seaweeds as medicine. In: Advance of phycology in Japan. (Eds.) J. Tokida and H. Hirose. Dr. W. Funk Publ., The Hague. p. 321.

146. Takemoto, T., K. Daigo and N. Takagi. 1964. Studies on the hypotensive constituents of marine algae. J. Pharm. Soc. Japan 84: 1176.

147. Takemoto, K. Daigo, Y. Kondo and Y. Kondo. 1966. Studies on the constituents of Chondria armata. 8. On the structure of domoic acid. J. Pharm. Soc. Japan. 86: 874.

148. Tamiya, H. 1955. Growing Chlorella for food and feed. Proceedings of world symposium on applied solar energy. Phoenix, Arizona. p. 231.

149. Tamiya, H. 1957. Mass culture of algae. Ann. Rev. Plant Physiol. 8: 309.

150. Tamiya, H. 1975. Green micro-algae. In: Food protein sources. (Ed.) N.W. Pirie. Cambridge Univ. Press, Cambridge. p. 35.

151. Tel-Or, E., L.W. Lujik and L. Packer. 1977. An inducible hydrogenase in cyanobacteria enhances N_2 fixation. FEBS Letters 78: 49.

152. Tel-Or, E., L.W. Lujik and L. Packer. 1978. Hydrogenase in N_2 fixing cyanobacteria. Arch. Bichem. Biophys. 185: 185.

153. Tsukada, O., T. Kawahara and S. Miyachi. 1977. Mass culture of Chlorella in Asian countries. In: Biological Solar Energy Conversion (Eds.) A. Mitsui et al. Academic Press, New York. p. 363.

154. Van Norstrand's Scientific Encyclopedia. 1975. Fifth Edition. Van Norstrand Reinhold Company. 237 pp.

155. Veziroglu, T.N. (Ed.). 1975. Symposium Proceedings: Hydrogen Energy Fundamentals. University of Miami, Miami.

156. Veziroglu, T.N. (Ed.). 1975. Hydrogen Energy, Parts A and B. Plenum Press, New York.

157. Veziroglu, T.N. (Ed.). 1976. Conference Proceedings of the 1st World Hydrogen Energy Conference. University of Miami, Miami.

158. Volesky, B., J.E. Zajic and E. Knettig. 1970. Algal products. In: Properties and products of algae. (Ed.) J E. Zajic. Plenum Press, New York. p. 49.

159. Wang, R., R.P. Healy and J. Myers. 1971. Amperometric measurement of hydrogen evolution in Chlamydomonas. Plant Physiol. 48:108.

160. Ward, M.A. 1970a. Whole cell and cell-free hydrogenases of algae. Phytochemistry 9:259.

161. Ward, M.A. 1970b. Adaptation of hydrogenase in cell-free preparations from Chlamydomonas. Phytochemistry 9: 267.

162. Weissmann, J.C and J.R. Benemann. 1977. Hydrogen production by nitrogen-starved cultures of Anabaena cylindrica. Appl. Environ. Microbiol. 33: 123.

163. Whatley, F.R. and B.R. Grant. 1963. Photoreduction of methyl viologen by spinach chloroplasts. Federation Proceedings. p. 227.

164. Wilcox, H.A. 1975. The ocean food and energy farm project. In: Proc. 3rd Int. Ocean Dev. Conf. Seino Printing Co., Tokyo. Vol. 3. p. 43.

165. Yagi, T. 1976. Separation of hydrogenase-catalyzed hydrogen-evolution system from electron-donating system by means of enzymic electric cell techniques. Proc. Natl. Acad. Sci. U.S.A. 73:2947.

166. Yagi, T. 1977. Use of an enzymic electric cell and immobilized hydrogenase in the study of the biophotolysis of water to produce hydrogen. In: Biological Solar Energy Conversion. (Eds.) A. Mitsui et al. Academic Press, New York. p. 61.

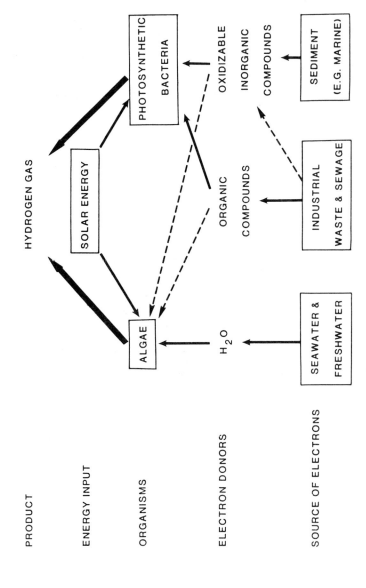

Figure 1. Electron and hydrogen donors for the photoproduction of hydrogen in algal and bacterial systems

DOMESTIC REQUIREMENT

Energy Consumption in Average South Florida House (2 Bedrooms)

During Peak Use Month (Summer)

Estimated by Florida Power and Light

1000 KWatt-Hr / Month

$= 2.9 \times 10^4$ Kcal / Day

(1 mole H_2 $\equiv$ 57 Kcal)

$= 500$ moles H_2 / Day

$= 11$ m^3 H_2 / Day

Experimental

0.17 m^3 H_2 Produced / m^3 Algal Suspension / Solar Day

Size of Algal Hydrogen Production System

11 m^3 H_2 $\div$ 0.17 m^3 H_2

$= 8$ m $\times$ 8 m $\times$ 1 m (depth)

$= 11$ m $\times$ 11 m $\times$ 0.5 m (depth)

Figure 2.   Algal hydrogen production system applied to domestic needs.

ATTEMPTS TO PRODUCE HYDROGEN BY COUPLING HYDROGENASE AND CHLOROPLAST PHOTOSYSTEMS

T. Yagi and H. Ochiai
Department of Chemistry, School of Education,
Shizuoka University, Shizuoka 422, Japan, and
Laboratory of Biochemistry, College of Agriculture,
Shimane University, Matsue 690, Japan

ABSTRACT

All the energy available to biological systems depends on the process of photosynthesis which converts solar energy to chemical energy stored in carbohydrates and other organic molecules. It cannot be expected, however, that the conversion efficiency shall be much improved by usual agricultural techniques, since the process is dependent on the complex mechanisms of plant cells which have not fully been elucidated. At the same time, the photosynthetic products are by no means versatile energy sources. The present research aims at finding out an efficient water–biophotolytic system, i. e., a convenient hydrogen-producing system by coupling chloroplast photosystems and bacterial hydrogenase with an electron carrier between them, so that we are prepared for biological solar energy conversion to meet future needs of energy supply.

Main difficulties include the photoinactivation of chloroplast photosystems, oxygen-sensitivity of bacterial hydrogenase, and autoxidation of the reduced form of electron carriers. To overcome these difficulties, the following attempts were made. Factors affecting the stabilities of bacterial hydrogenase and chloroplasts were studied, and conditions to immobilize them were examined. The immobilized hydrogenase thus prepared did not loose activity appreciably upon storage and repeated uses. The immobilized chloroplast films revealed marked stability in both photosystems, and acted as a photo-anode upon illumination of visible light. By coupling immobilized chloroplast and hydrogenase in the presence of methylviologen as an electron carrier and ascorbate and 2,6-dichlorophenol-indophenol as an electron donor, hydrogen photoproduction was observed. Possibility of separating chloroplast-dependent electron-donating system from hydrogenase-catalyzed hydrogen-producing system by electrochemical as well as mechanical means was discussed. Also the use of non-biological or semi-biological system as auxiliary components of artificial hydrogen photoproducing systems was discussed.

INTRODUCTION

Large scale consumption of energy stored in fossil fuels resulted in increasing pollution, and coming shortage of these energy sources will be unavoidable. It will, therefore, be necessary to seek for new energy sources. At present, the solar radiation is the only unexhausted and non-polluting energy sources available, and the photosynthesis is the only process which converts the solar energy to chemical energy stored in carbohydrates and other bio-organic compounds in large scale. Since the conversion efficiency of this process is limited by complex regulation mechanisms of plant cells, and the photosynthetic products are not versatile energy sources, it would be advisable to develop a new artificial system which efficiently converts solar energy to chemical energy in the form of non-polluting fuel molecules such as hydrogen gas.

The phenomenon of biological hydrogen production upon illumination, now is known as photoproduction of hydrogen, was first described on green alga, Scenedesmus, by Gaffron & Rubin in 1942 [1]. Since then, this phenomenon has been observed with many other green algae which contain hydrogenase [2-15]. The mechanism of hydrogen photoproduction, now is being believed, includes essential contribution of photosystem I [6-8, 14] and algal hydrogenase with electron mediator between them, and of auxiliary contribution of photosystem II [5, 12]. Recent paper by Bishop et al. [14], however, suggested essential contribution of photosystem II. The electrons are believed to be supplied from endogenous organic compounds which have been accumulated during photosynthesis [1, 9, 10], but possibility remains that water is the primary electron donor for hydrogen photoproduction [8, 14]. There is no contribution of photophosphorylation to the hydrogen photoevolution, since the phenomenon was not inhibited by uncoupling agents [8, 14]. The mechanism of algal hydrogen photoproduction can be illustrated as shown in eq. 1.

$$
\begin{array}{c}
O_2 \\
\uparrow \\
H_2O
\end{array}
\xrightarrow{\text{light} \downarrow\downarrow\downarrow}
\text{photosystem II}
\xrightarrow{}
\underset{e^-}{\diagup}
\text{photosystem I}
\xrightarrow{\text{light} \downarrow\downarrow\downarrow}
\begin{array}{c}
H_2 \\
\uparrow \\
H^+
\end{array}
\qquad \text{eq. 1}
$$

Similar phenomenon was observed with photosynthetic bacteria, Rhodospirillum rubrum by Gest & Kamen in 1949 [16, 17]. The mechanism of bacterial hydrogen photoproduction is, however, different from that of algal one. ATP is produced by photophosphorylation [18-20], and is supplying energy for bacterial nitrogenase which is catalyzing hydrogen evolution. The electrons are supplied from endogenous organic compounds [19, 20].

$$2\,H^+ + 2\,e^- + 2\,ATP \rightarrow H_2 + 2\,ADP + 2\,Pi \qquad\qquad \text{eq. 2}$$

The first attempt to mimic algal hydrogen photoproduction was reported by Arnon, Mitsui, & Paneque [21]. They successfully produced hydrogen upon illuminating an artificial system containing spinach chloroplasts and bacterial hydrogenase with cysteine as an electron donor. This artificial system, thus composed of the right half part of the electron transporting system shown in eq. 1. Recent hydrogen photoproducing system reported by Mitsui [22] contained, in addition to above-mentioned system, 2, 6-dichlorophenolindophenol as an electron mediator between cysteine and photoelectron transport system, methylviologen as an electron carrier between photosystem I and hydrogenase, and N-(3, 4-dichlorophenyl)-N', N'-dimethylurea (DCMU*) as an inhibitor of electron flow from photosystem II to photosystem I. Therefore the electrons flew efficiently from cysteine to methylviologen, and then to protons to produce hydrogen by the catalytic action of bacterial hydrogenase.

$$2 H^+ + 2 \text{ methylviologen}^+ \rightarrow H_2 + 2 \text{ methylviologen}^{2+} \qquad \text{eq. 3}$$

The addition of DCMU effectively prevented the production of oxygen gas which blocks hydrogen production either by inhibiting hydrogenase itself or by reoxidizing the reduced electron carrier formed during illumination of chloroplast photosystems.

Much progress has been made in recent years on biological solar energy conversion [14, 15, 22-34]. Ben-Amotz & Gibbs succeeded in hydrogen photo-production by illuminating spinach chloroplast-algal hydrogenase-methylviologen system with dithiothreitol as an electron donor in the presence of DCMU [26, 33]. Kitajima & Butler attempted to stabilize and to improve artificial hydrogen photoproducing system by microencapsulating both of chloroplasts and bacterial hydrogenase [27]. Ascorbate was donating electrons to the electron phototransfer system via 2, 6-dichlorophenolindophenol.

In these artificial hydrogen photoproducing systems, electrons supplied to hydrogenase system were derived from the added electron donors, and not from water. Therefore the operation of these systems did not result in net conversion of light energy to hydrogen energy, since exogenous electron donor molecules had to be consumed. Therefore only a half way to the final goal of biophotolysis of water has been achieved.

Benemann et al. succeeded in producing hydrogen gas by illuminating a mix-ture containing spinach chloroplasts, clostridial hydrogenase [EC 1.12.7.1] and clostridial ferredoxin, without exogenously added electron donors [22], thus proved the biophotolysis of water to be possible. Their system, however, had to be supplied with glucose and glucose oxidase [EC 1.1.3.4] to trap photo-produced oxygen gas. Without this oxygen scavenger, the rate of hydrogen photoproduction decreased rapidly due to oxidative inactivation of hydrogenase.

---

* Abbreviations. DCMU: N-(3, 4-dichlorophenyl)-N', N'-dimethylurea, PVA-105: polyvinyl alcohol (average degree of polymerization, 550), PVA-117: polyvinyl alcohol (av. d. p., 1750), and Tris: 2-amino-2-hydroxymethyl-1,3-propanediol.

Rao, Rosa, & Hall [28] stabilized spinach chloroplasts by cross-linking them with glutaraldehyde, and coupled them with bacterial hydrogenase to photo-produce hydrogen.   In this case also, the addition of glucose and glucose oxidase as oxygen scavenger was essential for the long-lasting photoproduction of hydrogen.   This means that chemical energy stored in glucose molecules are being consumed during the process of hydrogen photoproduction, and that the conversion efficiency of light energy to hydrogen energy is very low. Since our final goal is biophotolysis of water to produce hydrogen to meet future needs of energy supply, simultaneous production of hydrogen and oxygen gases is desirable.   Therefore, it will be necessary to solve the problems of how to protect hydrogenase from oxidative inactivation, how to prevent autoxidation of electron donors for hydrogenase, and how to stabilize chloro-plast photosystems.

To solve these problems, some proposals have been made [23-33].   One is to find stable hydrogenase and chloroplasts which are resistant to inactivation, or to stabilize hydrogenase and chloroplasts by immobilization technique.   At the same time, synthesis of non-autoxidizable new electron carrier will be desirable.   Another is to separate hydrogenase-catalyzed hydrogen-evolution system from electron-donating system which produces oxygen by means of enzymic electric cell technique [29].   This technique will be promising if we have some devices to transfer photoexcited electrons efficiently to electrode. Mechanical separation of hydrogen-producing system from electron-donating system [34] also deserves considerations.

In this paper, we shall present our recent research on stabilizing hydrogenase and chloroplasts by immobilization technique, and our attempts to produce hydrogen by coupling stabilized hydrogenase and chloroplasts.

MATERIALS AND METHODS

Hydrogenase [hydrogen : ferricytochrome c_3 oxidoreductase, EC 1.12.2.1] was solubilized and purified from particulate fraction of Desulfovibrio vulgaris, strain Miyazaki, as described by Yagi et al. [35].   Cytochrome c_3 was puri-fied from soluble fraction of desulfovibrio sonicate as described by Yagi & Maruyama [36].   Broken chloroplast of type C was prepared from spinach leaves by routine procedure in 0.4 M sucrose-0.05 M Tris-0.01 M NaCl buffer at pH 7.2 [37], and used for studies on immobilization and photoproduction of hydrogen. Chlorophyll content was determined by the method of Mac Kinney [38]. Methylviologen was a product of British Drug Houses.   Methylviologen-phosphate stock solution is a mixture of 1 volume of 0.04 M methylviologen and 3 volumes of 0.2 M phosphate buffer of pH 7.0.   PVA-105 and PVA-117 were generous gifts from Kuraray Co., Osaka.

Activity of hydrogenase expressed in units was assayed and monitored by enzymic electric cell method [39]. According to this method, the short–circuit current of the cell represents the activity of hydrogenase. The principle of this assay method is formulated as eq. 4-6.

Anode :

$$H_2 + 2\,methylviologen^{2+} \rightarrow 2\,H^+ + 2\,methylviologen^+ \qquad \text{eq. 4}$$
$$2\,methylviologen^+ \rightarrow 2\,methylviologen^{2+} + 2\,e^- \text{ (to electrode)} \qquad \text{eq. 5}$$

Cathode :

$$Hg_2Cl_2 + 2\,e^- \text{ (from electrode)} \rightarrow 2\,Cl^- + 2\,Hg \qquad \text{eq. 6}$$

The activity of photosystem I was assayed by measuring the rate of oxygen absorption with a dissolved–oxygen–meter (YSI 54 ARC) in accordance with the method of Epel & Neumann [40] with slight modification [41, 42]. The activity of photosystem II was assayed by measuring the rate of photoreduction of 2, 6-dichlorophenolindophenol [42, 43]. The activity of photophosphorylation was estimated according to the procedure of Asada et al. [44]. Production of hydrogen was measured by conventional manometric technique.

EXPERIMENTAL RESULTS

Improvement of Stability of Desulfovibrio Hydrogenase in the Aqueous State

It is important to obtain fundamental knowledge as to the factors influencing the stability of hydrogenase in the aqueous state. In general, enzymes are stabilized by their substrates, so the influence of cytochrome c_3 on the stability of hydrogenase was tested. To the anode room of the enzymic electric cell [39], a reaction mixture (3.0 ml) containing hydrogenase (85 milliunits at 30°), various amount of cytochrome c_3, and 0.5 ml of methylviologen–phosphate stock solution was added. Oxygen-free hydrogen gas was bubbled through the reaction mixture kept at 50°. After electrolytic reduction of the anode room, the short-circuit current of the cell was monitored. Figure 1 shows the time-course curves of the activity of hydrogenase at 50° expressed in terms of the current. As shown in this figure, the activities of native hydrogenase calculated by extrapolating the activity curves to time 0, were not influenced by the addition of cytochrome c_3, but the enzyme was protected from thermal inactivation in the presence of cytochrome c_3. The concentration of cytochrome c_3 which endowed the enzyme with half maximum stability, i. e., K_S, at 50° was 3×10^{-7} M. It was reported earlier [45] that hydrogenase had two K_S values for cytochrome c_3. One of them, K_{1S}, is for the first cytochrome c_3 molecule which was assumed to stabilize the enzyme, and not replaceable by methylviologen. The other, K_{2S}, is for the second cytochrome c_3 molecule which acted as an electron carrier for hydrogenase reaction and replaceable by methylviologen. The K_S value for cytochrome c_3 estimated by its stabilizing effect at 50° agrees well with that of K_{1S} at 30°.

Immobilization of Hydrogenase in Polyacrylamide Gel

At first, conditions for preparing immobilized hydrogenase were examined. The immobilized hydrogenase reported in our previous paper [32] retained activity on storage, there was tendency for the enzyme to leak out of gel, especially after repeated uses. Use of double amount of the cross-linker molecule, N,N'-methylenebisacrylamide, did not prevent leakage of the enzyme, but resulted in decreased activity. It seems advisable to use another cross-linker molecule to improve the stability of the gel, and this was tried. To a solution containing 750 mg of acrylamide, 40 mg of N,N'-methylenebisacrylamide, 200 units of hydrogenase and 28 nmoles of cytochrome c_3 in 5.0 ml of 0.014 M phosphate buffer (pH 7.0), 0.5 ml of 5 % N,N,N',N'-tetramethylethylenediamine and 0.5 ml of freshly prepared 1 % ammonium peroxodisulfate solution was added in this order, and the mixture was gently mixed, sucked into a plastic straw (inner diameter ; 4.5 mm), and was left to stand for gelation. The gel was then squeezed out of the straw, and the resulting gel was cut into pieces of 70 mg weight. Hydrogenase gel pieces without cytochrome c_3 was also prepared.

Hydrogenase gel thus prepared was stored in 0.02 M phosphate buffer (pH 7.0), and at intervals, the activity of the gel was measured by the enzymic electric cell method. The results are shown in Figure 2. Hydrogenase gel prepared in the presence of glutaraldehyde had enough stability, and there was no room to be improved by the addition of cytochrome c_3. The extent of leakage of the enzyme from the gel was also estimated by measuring the residual current of the cell after the gel had been removed from the cell. In no case, the residual current exceeded $1\,\mu A$, even after repeated uses of the gel.

Activity of Hydrogenase Gel after Repeated Uses

For the biological conversion system of solar energy to work efficiently, it is mandatory for the components of the conversion system to be used continuously and repeatedly without loss of activity, or else, more energy have to be consumed to isolate these components than that obtainable by solar energy conversion. Therefore, the activity of the gel after repeated uses was tested.

The activity of hydrogenase gel was monitored for about 100 minutes in the enzymic electric cell, then the gel was removed from the cell, and was stored in 0.02 M phosphate buffer (pH 7.0) at 5°. No precaution was made to eliminate atmospheric oxygen. The activity of the stored gel was again assayed and monitored, and this was repeated several times. The results are also shown in Figure 2. This shows that the immobilized hydrogenase gel can be used continuously and repeatedly without much loss of activity.

Immobilization of Hydrogenase in Polyvinyl Alcohol Film

Hydrogenase gel described in the preceding section had characteristic properties suitable for use in biological solar energy conversion system, but the
activity yield was not high due to limited surface area of the gel.   Therefore
the immobilized hydrogenase preparation of larger surface area is desirable.
For this purpose, entrapping the enzyme in polyvinyl alcohol film was tried
as suggested by Ochiai et al. [46].   An 0.8 ml portion of hydrogenase (80 units)
containing 23 nmoles of cytochrome c_3 was added to 5.5 ml of 18 % PVA-117
containing 0.02 M phosphate buffer (pH 7.0), and the mixture was gently mixed,
spread on a glass plate, and dried in vacuo over phosphorus pentoxide.   The
activity of the resulting rosy film was assayed in the enzymic electric cell,
and was found to retain full activity of the starting material.   Unfortunately
this film was too loose for the enzyme, and all the activity went out to the
reaction mixture after single use.   This film may be useful as ready-to-use
film of hydrogenase which can be stored without much precaution.

Attempts to Immobilize Chloroplasts in Polyacrylamide Gel

Another member of an artificial hydrogen photoproducing system is chloroplast.
This has two photosystems, I and II, of which, photosystem II is more labile.
Various kinds of treatment including aging decrease the activity of photosystem
II.   At first, the effects of immobilizing reagents were tested.   Neither of
acrylamide, N, N'-methylenebisacrylamide, 3-dimethylaminopropionitrile, nor
potassium peroxodisulfate at their concentrations used in gelation had inhibitory
effects on chloroplast photosystem II, but gelation upon mixing them with
chloroplasts resulted in complete loss of activity of photosystem II [41, 42].
This suggests that radical reaction is injurious to chloroplast photosystem II.
Employing another polymerization technique called redox polymerization [47]
with bovine serum albumin as stabilizing reagent, considerable stabilization
of chloroplast photosystems were achieved [41, 42], but the immobilization of
chloroplasts in polyvinyl alcohol film gave more satisfactory results as shown
in the following section.

Immobilization of Chloroplasts in Polyvinyl Alcohol Film

In most literatures on chloroplast preparations, polyol compounds such as
sucrose or sorbitol was used as stabilizing reagents.   Polyols are also known
experimentally to stabilize many other enzymes.   Polyvinyl alcohol is a member
of polyols, and is known to polymerize to form film under dehydrating conditions.   Therefore, test was made whether polyvinyl alcohol had similar
protecting effect as sucrose on chloroplast photosystems.   Chloroplast suspension was prepared in 0.05 M Tris-0.01 M NaCl buffer (pH 7.2) containing
3 % PVA-105, and the activities of photosystems I and II, and of cyclic
photophosphorylation were compared with those of chloroplast suspension

prepared in 0.4 M sucrose–0.05 M Tris–0.01 M NaCl buffer (pH 7.2). The activities of both of photosystems were essentially the same in both of them either just after the preparation or after 24 hours. These results suggested that polyvinyl alcohol had protecting effect similar to that of sucrose, and can be used as a substitute for sucrose [46].

Freshly prepared chloroplasts containing 3 mg of chlorophyll were suspended in 18 ml of 0.05 M Tris–0.01 M NaCl buffer (pH 7.2) containing 18 % PVA-117 and 1 % bovine serum albumin. The resulting sticky suspension was spread on a glass plate, and dried in vacuo over phosphorus pentoxide to obtain film [46]. The recovery of the activities of both photosystems were nearly quantitative, and that of the activity of photoreduction of methylviologen with water as an electron donor [40] was 29 %. The activities of both photosystems were maintained several weeks without appreciable loss. On the other hand, the activity yield of cyclic photophosphorylation was only 8 % of that of freshly prepared chloroplast suspension. This does not mean that the immobilization technique was inadequate for use in hydrogen production system, since photophosphorylation is known not to participate in photoproduction of hydrogen in living algae.

One side of the immobilized chloroplast film thus prepared was coated with metal by vacuum evaporation, connected to lead wire, and the open surface of the metal was covered with insulator by means of adhesive. By this way, a chloroplast electrode was prepared. The electric cell was then constructed with this chloroplast electrode and a counter electrode made of untreated metal as described by Ochiai et al. [42]. Upon illumination, the chloroplast electrode acted as a photoanode, and photocurrent of as much as 10 μA was observed [48].

A Preliminary System for Artificial Hydrogen Photoproduction

In a Warburg vessel, a reaction mixture containing 0.3 ml of 0.02 M ascorbate, 0.3 ml of 2.5×10^{-4} M 2,6-dichlorophenolindophenol, 0.3 ml of 0.04 M methylviologen, 0.1 ml of 3×10^{-4} M alcoholic DCMU solution, 0.1 ml of 0.2 M glucose, 0.1 ml of glucose oxidase (10 units), and immobilized chloroplast film containing 3 mg of chlorophyll in 0.02 M Tris buffer with or without hydrogenase (40 units) was added as described by Rao, Rosa, & Hall [28]. The final volume of the reaction mixture was 3.0 ml, and pH was 7.0. Gas phase was replaced by nitrogen. After 30 minutes' preincubation, 3 fluorescent lamps were turn on, and the change in gas volume was recorded at 25°. The intensity of light was about 5000 lux. The results are shown in Figure 3. As shown, gas volume increased in the presence of hydrogenase. The production of hydrogen gas lasted for longer than 4 hours, suggesting the stability of the immobilized chloroplast film.

DISCUSSION

An artificial system for hydrogen photoproduction is consisted of chloroplasts, bacterial hydrogenase, and an electron carrier between them. In our present research, much progress has been made to obtain stabilized hydrogenase and chloroplast preparations, and succeeded in photoproducing hydrogen by coupling stabilized components in the presence of an exogenously added electron donor. In order to achieve true solar energy conversion, water should supply electrons to reduce protons with concomitant production of oxygen gas. In living plants, electrons are transferred via ferredoxin to Calvin cycle, where ferredoxin is protected from oxygen by organized structure inside the chloroplast, but not in artificial hydrogen photoproducing system, where the carrier is ready to be oxidized by oxygen.

One approach to solve this problem is the synthesis of a non-autoxidizable electron carrier of low redox potential which specifically accepts electrons from photosystem I. Krampitz [49] suggested the possibility to synthesize non-autoxidizable viologens, but this does not seem to be successful.

Another approach is the separation of hydrogenase-catalyzed hydrogen-evolution system from chloroplast-dependent electron donating system either electrochemically as suggested by Yagi [29], or mechanically as suggested by Shin [34]. In order for electrochemical separation system to work, it is necessary to take out photoexcited electrons from chloroplasts to the electrode without being reoxidized by oxygen (Figure 4). Since the immobilized chloroplast film works as a photoanode, it may be possible to construct a photoelectric cell which produce hydrogen by biophotolysis of water.

Mechanical separation of chloroplast-dependent system from hydrogen-producing system may be achieved by coupling two columns as shown in Figure 5 [34]. The first column is packed with chloroplast, ferredoxin, and ferredoxin-NADP$^+$ reductase in their immobilized state. The second column is packed with ferredoxin-NADP$^+$ reductase, ferredoxin, and hydrogenase, also in their immobilized state. The circulating medium contains an NADP$^+$/NADPH couple. Upon illumination, water will be docomposed to produce oxygen, and NADP$^+$ will be reduced to NADPH. NADPH is not autoxidizable, and will be separated from oxygen, and transferred to the second column, where it will undergo reoxidation to form NADP$^+$ and hydrogen gas by the combined catalytic actions of immobilized enzymes. Immobilization of ferredoxin-NADP$^+$ reductase and ferredoxin has already been published [50, 51].

Turning our view point to non-biological systems, some examples of photocells capable of decomposing water have been deviced. In 1972, Fujishima & Honda announced that n-type semiconductor, TiO_2, could work as a photoanode, and take up electrons from water to produce oxygen upon illumination [52]. They constructed an electrochemical photocell by coupling this photoanode with platinum electrode. Upon illumination, electrons were taken up from water and transferred to the cathode to reduce protons; oxygen and hydrogen being the products at the anode and cathode, respectively. It could have been true photolysis of water, if it had not been for help of alkali in the anode and acid in the cathode. Decomposition of water had mainly been driven by light energy, but contribution of neutralization energy of alkali and acid should not be neglected.

Watanabe et al. [53] reported a new photocell composed of another n-type semiconductor, $SrTiO_3$, as a photoanode and platinum electrode as a cathode. The mechanism of water photolysis by this cell was similar to that of the TiO_2 - Pt photocell. As this cell worked at neutrality in both electrodes, true photolysis of water was achieved. Unfortunately, the photoanode is excitable only by near ultraviolget radiation, which comprises only a minor fraction of total solar energy reaching the surface of the earth.

Unless the percentage of energy conversion is much improved, excitation of electrons of water ($E_0' = +0.8$ V at pH 7.0) to the level of proton ($E_0' = -0.42$ V at pH 7.0) is not feasible by absorbing single photon of visible light, but will be possible by absorbing two photons of visible light consecutively as in the case of chloroplast photosystems. Therefore coupling of a photoanode with a photocathode will be advisable in constructing photocells. A p-type semiconductor, GaP, may be used as a photocathode, because electrolytic hydrogen production in GaP electrode was known to take place at more positive potential than on platinum electrode under illumination [54].

Recently, Fujita reported that platinum electrode coated with zinc tetraphenylporphyrin worked as a photocathode, with reversible photopotential [55]. A preliminary study to construct photoelectric cell by coupling a photoanode and a photocathode with bacterial hydrogenase is in progress [56].

Another attempt is to construct a semi-biological water-photolytic cell. Tributsh & Calvin [57] reported the photoanodic character of chlorophyll molecules deposited on semiconductor, and observed the spectral dependence of the photocurrents to be dependent on absorption spectrum of the adsorbed molecules. Takahashi et al. [58] constructed a photocell by coupling a photoanode composed of chlorophyll and anthrahydroquinone in hexacyanoferrate(II) solution, and a photocathode composed of chlorophyll and naphthoquinone in NAD^+ solution. Upon illumination of both electrodes, NAD^+ was photoreduced to NADH, and hexacyanoferrate(II) was photooxidized to hexacyanoferrate(III). Photoelectromotive force of the cell during illumination was 0.25 V. These examples of non-biological and semi-biological photoelectrodes will partly substitute some labile components of biological photocells. A coupling of natural and artificial energy-conversion systems described above will lead to developing an efficient system of biological solar energy conversion system, so that hydrogen and oxygen will be photoproduced to meet future needs of energy supply.

We are indebted to Miss Yoko Mizukami, Miss Teruyo Minami, Mr Mutsumi Fujioka, and Miss Keiko Tsuji for their excellent technical assistances. This work was supported in part by a grant (No. 211112, 1977) from the Ministry of Education, Science and Culture of Japan.

REFERENCES

1. Gaffron, H., & Rubin, J. (1942) J. Gen. Physiol. 26, 219-240
2. Damaschke, K. (1957) Z. Naturf. 12b, 441-443
3. Damaschke, K., & Lübke, M. (1958) Z. Naturf. 13b, 54-55
4. Syrett, P. J., & Wong, H. A. (1963) Biochem. J. 89, 308-315
5. Healey, F. P. (1970) Planta 91, 220-226
6. Healey, F. P. (1970) Plant Physiol. 45, 153-159
7. Stuart, T. S., & Kaltwasser, H. (1970) Planta 91, 302-313
8. Stuart, T. S. (1971) Planta 96, 81-92
9. Stuart, T. S., & Gaffron, H. (1971) Planta 100, 228-243
10. Stuart, T. S., & Gaffron, H. (1971) Plant Physiol. 50, 136-140
11. Stuart, T. S., & Gaffron, H. (1972) Planta 106, 91-100
12. Stuart, T. S., & Gaffron, H. (1972) Planta 106, 101-112
13. Kessler, E. (1974) Bot. Monogr. 10, 456-473
14. Bishop, N. I., Frick, M., & Jones, L. W. (1977) in A. Mitsui, S. Miyachi, A. San Pietro, & S. Tamura (ed.) Biological Solar Energy Conversion, Academic Press, New York, pp. 3-22
15. McBride, A. C., Lien, S., Togasaki, R. K., & San Pietro, A. (1977) ibid. 77-86
16. Gest, H., & Kamen, M. D. (1949) Science 109, 558-559
17. Gest, H., & Kamen, M. D. (1949) J. Bacteriol. 58, 239-245
18. Arnon, D. I., Losada, M., Nozaki, M., & Tagawa, K. (1961) Nature 190, 601-606
19. Bose, S. K., & Gest, H. (1962) Nature 195, 1168-1171
20. Gest, H. (1972) Adv. Microbiol. Physiol. 7, 243-282
21. Arnon, D. I., Mitsui, A., & Paneque, A. (1961) Science 134, 1425
22. Benemann, J. R., Berenson, J. A., Kaplan, N. O., & Kamen, M. D. (1973) Proc. Nat. Acad. Sci. USA 70, 2317-2320
23. Gibbs, M., Hollaender, A., Kok, B., Krampitz, L. O., & San Pietro, A. (Organizers) (1973) Proc. Workshop Bio-Solar Conversion, NSF-RANN
24. Lien, S., & San Pietro, A. (1975) An Inquiry into Biophotolysis of Water to Produce Hydrogen, Indiana University and NSF
25. Mitsui, A. (1975) in T. N. Veziroglu (ed.) Hydrogen Energy, Part A. Plenum Publ. Corp., New York, pp. 309-316
26. Ben-Amotz, A., & Gibbs, M. (1975) Biochem. Biophys. Res. Commun. 64, 355-359

27. Kitajima, M., & Butler, W. L. (1976) Plant Physiol. 57, 746-750
28. Rao, K.K., Rosa, L., & Hall, D.O. (1976) Biochem. Biophys. Res.Commun. 68, 21-28
29. Yagi, T., (1976) Proc. Nat. Acad. Sci. USA 73, 2947-2949
30. Mitsui, A. (1977) in A. Mitsui, S. Miyachi, A. San Pietro, & S. Tamura (ed.) Biological Solar Energy Conversion, Academic Press, New York, pp. 23-51
31. Krasna, A. I. (1977) ibid. 53-60
32. Yagi, T. (1977) ibid. 61-68
33. King, D., Erbes, D. L., Ben-Amotz, A., & Gibbs, M. (1977) ibid. 69-75
34. Shin, M. (1977) in S. Tamura (ed.) Development of Biological Productivity (Rep. submitted to Ministry of Educ., Sci., Culture of Japan, 1976) p. 36
35. Yagi, T., Kimura, K., Daidoji, H., Sakai, F., Tamura, S., & Inokuchi, H. (1976) J. Biochem. 79, 661-671
36. Yagi, T., & Maruyama, K. (1971) Biochim. Biophys. Acta 243, 214-224
37. Hall, D. O. (1972) Nature, New Biology 235, 125-126
38. Mac Kinney, G. (1941) J. Biol. Chem. 140, 315-322
39. Yagi, T., Goto, M., Nakano, K., Kimura, K., & Inokuchi. H. (1975) J. Biochem. 78, 443-454
40. Epel, B. L., & Neumann, J. (1973) Biochim. Biophys. Acta 325, 520-529
41. Ochiai, H., Shibata, H., Matsuo, T., Hashinokuchi, K., & Yukawa, M. (1977) Agr. Biol. Chem. 41, 721-722
42. Ochiai, H., Shibata, H., Matsuo, T., Hashinokuchi, K., & Yukawa, M. (1978) Nogei Kagaku Kaishi 52, 31-36 (in Japanese)
43. Shibata, H., & Ochiai, H. (1973) Agr. Biol. Chem. 37, 471-476
44. Asada, K., Takahashi, M., & Urano, M. (1972) Anal. Biochem. 48, 311-315
45. Yagi, T. (1970) J. Biochem. 68, 649-657
46. Ochiai, H., Shibata, H., Matsuo, T., Hashinokuchi, K., & Inamura, I. (1978) Agr. Biol. Chem. 42, 683-685
47. Kern, W., (1948) Makromol. Chem. 1, 209-228
48. Ochiai, H., & Honda, K. (1978) to be published
49. Krampitz, L. O. (1973) U.S. Nat. Tech. Inform. Serv. (Chem.Abst. 81, 101997h)
50. Shin, M., & Oshino, R. (1978) J. Biochem. 83, 357-361
51. Shin, M. (1978) in S. Tamura (ed.) Development of Biological Productivity (Rep. submitted to Ministry of Educ., Sci., Culture of Japan, 1977) in press
52. Honda, K., & Fujishima, A. (1972) Nature 238, 37-38
53. Watanabe, T., Fujishima, A., & Honda, K. (1976) Bull. Chem. Soc. Japan 49, 355-358
54. Memming, R., & Schwandt, S. (1968) Electrochim. Acta 13, 1299-1310
55. Fujita, Y. (1978) in S. Tamura (ed.) Development of Biological Productivity (Rep. submitted to Ministry of Educ., Sci., Culture of Japan, 1977) in press
56. Sakata, T. private communication
57. Tributsh, H., & Calvin, M. (1971) Photochem. Photobiol. 14, 95-112
58. Takahashi, F., & Kikuchi, R. (1976) Biochim. Biophys. Acta 430, 490-500

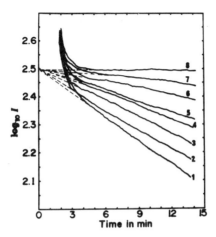

Figure 1. Effect of cytochrome c_3 on the activity of hydrogenase. A reaction mixture (3.0 ml) containing 85 milliU of hydrogenase, 0.5 ml of methylviologen-phosphate stock solution, and cytochrome c_3 was placed in the anode room of the cell. After electrolytic reduction, the current (in μA) of the cell was recorded at 50°. Concentration of cytochrome c_3 in μM was: 1) 0, 2) 0.06, 3) 0.12, 4) 0.24, 5) 0.37, 6) 0.62, 7) 1.43, and 8) 4.28.

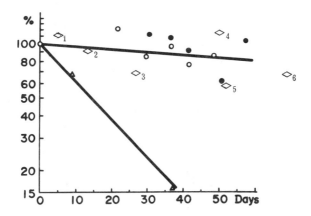

Figure 2. Change of activity of hydrogenase gel upon storage and after repeated uses. Three kinds of hydrogenase gel were stored in 0.02 M phosphate buffer (pH 7.0) at 4°, and their activities were measured at intervals. The activities relative to those of freshly prepared samples were plotted. O: gel prepared in the presence of cytochrome c_3, ●: gel prepared in the absence of cytochrome c_3, and △: gel without glutaraldehyde. The relative activity of reused gel was shown with numerals indicating the times of reuses.

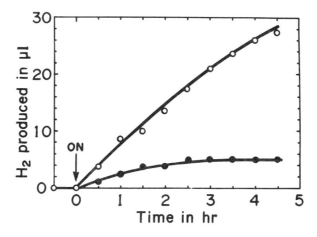

Figure 3. Hydrogen photoevolution from immobilized chloroplast-hydrogenase-methylviologen system. The experimental conditions are described in the text. O : in the presence of hydrogenase, and ● : in its absence.

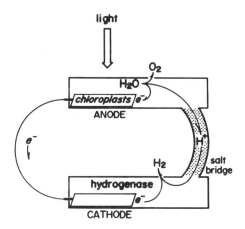

Figure 4. Electrochemical separation of hydrogen-evolution system from oxygen-evolution system.

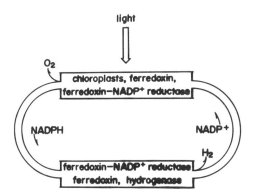

Figure 5. Mechanical separation of hydrogen-evolution system from oxygen-evolution system. Cited from ref. [34] with slight modification.

THERMOELECTROCHEMICAL CYCLES FOR POWER AND HYDROGEN PRODUCTION*

Meyer Steinberg
Process Sciences Division
Department of Energy and Environment
Brookhaven National Laboratory
Upton, New York 11973
U.S.A.

ABSTRACT

Based on electrochemical mechanisms, a new power cycle for converting thermal to electrical energy is presented. The cycle is referred to as the thermo-electrochemical (TEC) power cycle. The general principle involves combining the electrochemical decomposition of a compound to its elements at a condition where the free energy change is low with the recombination of the same or a different compound from its elements at a condition where the free energy change is high. The difference in free energies gives a net difference in electromotive force which results in a net power output for the system. The power cycle combines the operation of an electrolyzer at a high temperature low emf condition, with a fuel cell at a low temperature high emf condition. The thermal energy is used to provide the high level heat in the electrolyzer while the low level heat is rejected in the fuel cell. Heat is thus converted to DC electricity. Ideal power cycle efficiencies are equal to the Carnot efficiency for non-condensing systems. The principles are illustrated with a H_2-O_2-H_2O system and with a H_2-O_2-Cl_2-HCl system. In the H_2-O_2-H_2O system, utilizing heat at 727°C (1000°K) in the electrolytic cell and rejecting heat at 25°C (298°K) in the fuel cell, results in a net ideal emf difference of 0.187 volts with a Carnot cycle efficiency of 70%. The system could be directed towards application of nuclear heat from high temperature gas cooled reactors or at still higher temperatures from coal, oil, or gas fired plants where efficiencies could reach over 80%. Overvoltage and cell IR losses will reduce these ideal efficiencies, however, the potential for improvement over existing cycles appears to be very high. A feasible system, minimizing overvoltage losses in the electrolytic cells appears to be the operation of a cycle in which water is first electrolyzed at 1000°K (727°C) to H_2 and O_2. The H_2 is then fed to a fuel cell where it combines with Cl_2 at 298°K (25°C) forming aqueous HCl. The aqueous HCl is oxidized with the O_2 from the electrolyzer back to water and Cl_2, thus closing the cycle. The net cell voltage is calculated to be 0.362 volts and efficiencies approaching the Rankine cycle efficiency of 47.2% could be obtained. Other cycles are described. The benefits of the TEC cycle is that it produces electricity from thermal energy with no moving parts and at moderate pressures which should make this system highly reliable, and less costly than conventional power conversion cycles. TEC may have especially suitable applications for the production of electrometals, (Al and Cu), electrochemicals and synthetic fuels because of the direct use of the DC current generated in the cycle. It is recommended that these systems be further developed and evaluated.

INTRODUCTION

The conversion of thermal energy to electrical power is a subject of much study in the present energy crisis situation. Regardless of the energy source, be it fossil, nuclear, solar, or geothermal, part, if not all of the energy always appears as heat (enthalpy) to be utilized in a power producing cycle. Thermodynamically, these cycles are usually Carnot cycle limited, in that the efficiency of energy conversion is determined by the upper energy supply temperature limits and the lower temperature at which energy is rejected to the environment ($\varepsilon_c = (T_1-T_2)/T_1$). Furthermore, because of the physical properties of the working fluid, which is usually water, carbon dioxide or helium, the power cycle efficiency is further reduced in the usual Rankine (condensing steam cycles) and Brayton (gas recompression) cycles. Thus for thermal energy supplied at $1000°K$ and reject heat at $300°K$, Carnot thermal to electrical (work) cycles of 70% should be realized whereas Rankine and Brayton cycles yield no more than about 42% overall efficiency in large power plants. Some of the inefficiency can be attributed to the mechanical and electrical generating equipment however, this is usually not a significant fraction.

Attempts to improve thermal to electrical power cycle efficiency are being made by going to open cycle systems in such devices as MHD, fuel cells or gas turbines. With the exception of the fuel cell, these open cycles, however, are limited to the upper parts of the thermal energy spectrum and must be used in conjunction with secondary power cycles (topping and bottoming cycles) which are usually Rankine and Brayton limited. We have been intrigued for a number of years by the possibility of a closed electrochemical power conversion cycle. The electrochemical cell itself converts electrical to chemical energy and vice-versa under ideal conditions in a reversible manner. That is, thermodynamically, the free energy of the chemical reaction is converted to electrical energy

$$\Delta F = nfe$$

n = no. of chemical equivalent
f = Faradays constant
e = emf or voltage potential of the cell
ΔF = the free energy charge of the chemical reaction

When electrical energy is converted to chemical energy in a cell, the process is called electrolysis and the equipment is usually called an electrolyzer. A large electrochemical industry has played a significant role in the economy of the country for over a century in producing basic chemicals (hydrogen peroxide, chlor-alkali, etc.) and metallurgical materials (copper, aluminum, etc.). When chemical energy is converted to electrical energy, the process and equipment is usually called a fuel cell. This process has mainly been practiced in the laboratory and only recently has the technology been emerging in industry. Probably the main reason for a disinterest in its use is because the Rankine and Brayton cycles were economical based on low-cost fuels and that now when fuel costs are skyrocketting, the country is realizing the need for higher efficiency devices.

A property of an electrochemical cell is that it is controlled by the second
law of thermodynamics in than not all the chemical energy (enthalpy) can be
converted to electrical (or work) energy and that the chemical energy appears
both as electrical energy and thermal energy in the process. Stating this
thermodynamically

$$\Delta F = \Delta H - T\Delta S$$

$$\text{or } \Delta H = \Delta F + T\Delta S$$

ΔF = free energy of the chemical reaction, Kcal/mol
ΔH = enthalpy of the chemical reaction, Kcal/mol
 T = absolute temperature at which cell is operating, $T^{o}K$
ΔS = entropy of the chemical reaction, Kcal/mol-^{o}K
$T\Delta S$ = the unavailable thermal energy for conversion to
     electrical (or work), Kcal/mol

Thus in an electrolyzer, supplying electrical energy to the cell (ΔF), means
that chemical energy is released (ΔH), and that an additional amount of
energy ($T\Delta S$), is required to make the process work. In the reverse
situation when chemical energy is converted to electrical in a fuel cell,
the chemical energy (ΔH) appears as electrical energy (ΔF) and thermal energy
($T\Delta S$).

Normally as the temperature of an electrolyzer is raised, the emf or electrical potential (ΔF) decreases while the ΔH remains fairly constant.
However, this depends strongly on the type of chemical reaction taking
place. As an example, we can take the electrolysis or the decomposition of
water (H_2O) in an electrolyzer. At 298.2^{o}K (25°C) the free energy for
$H_2O_{(g)}$ decomposition to hydrogen and oxygen, $\Delta F_{298.2}$ = +54.6 Kcal/mol which
is equivalent to a reversible potential of 1.19 volts. For liquid water to
gaseous H_2 and O_2, the reversible potential is 1.23 volts, however, at higher
temperatures water only exists as a gas and, therefore, the lower potential
is assumed in this study. The enthalpy of decomposition of gaseous H_2O is
+57.9 Kcal/mol. As the temperature increases ΔH increases very slightly,
however, ΔF decreases markedly and thus $T\Delta S$ increases markedly. Table 1
indicates the range of values and the corresponding reversible potentials.
For a liquid water cell, $H_2O_{(aq)} = H_{2(g)} + 1/2O_{2(g)}$, ΔF = +56.69 Kcal/mol
and the reversible potential is 1.225 volts.

The data in Table 1 indicates that ideally when electrical energy is fed to
an electrolytic water decomposition cell, an additional amount of energy in
the form of thermal energy ($T\Delta S$) must be fed to or absorbed by the cell,
otherwise, the cell will tend to lose heat and the temperature will drop.
As the temperature level of the cell increases, the electrical potential
needed to drive the reaction (ΔF) decreases and more thermal energy must be
fed to the cell to maintain the temperature. Thus at 298.2^{o}K (25°C) the
fraction of the total energy (ΔH) needed by the cell in terms of thermal
energy is 5.5% and at 2000°K (1727°C) it is 46.5%. At approximately 4200°K,
ΔF is 0 and the fraction of thermal energy needed to decompose water is
100% and no electrical energy is required. Thus, as the cell energy is

raised more thermal energy is used directly to produce chemical energy.

In reality there are inefficiencies in electrolytic cell operations which tend to modify the above ideal processes. These losses are a function of the current density and temperature. These are the IR resistance losses in the electrolyte and the electrode over-potential, especially the oxygen overvoltage. Thus where an ideal voltage of at $298.2^{\circ}K$ is 1.23 volts, the inefficiencies will increase the voltage generally to 1.8 volts or more; the main increase usually being due to the oxygen electrode overvoltage which amounts to about 0.5 volts. In a real cell, these inefficiencies degrade the electrical energy to thermal energy and thus goes to supply the $T\Delta S$ thermal energy needed by the cell. One, therefore, does not observe a lowering in temperature as indicated by the thermodynamics of an ideal adiabatic cell operating at $298.2^{\circ}K$. However, as the temperature is raised, the overvoltage effects decreases and the cell operates closer to the thermodynamically ideal situation. It is estimated that at about $1000^{\circ}K$ ($727^{\circ}C$) the overvoltage should be reduced to less than 0.1 volt (100 milli-volts), so that the water electrolysis cell voltage required for operation would be $0.998 + 0.100$ or approximately 1.1 volts.

H_2-O_2 TEC Power Cycle

Based on the principle that voltage decreases with increasing temperature, a power cycle can be devised using an electrolytic cell operating at elevated temperature and a fuel cell at a lower temperature. Ideally, the cell voltage produced at the lower temperature is higher than the voltage pro-duced at the higher temperature, thus the electrolytic cell can be powered by the fuel cell. Referring to Figure 1, a thermal power source, can heat the electrolytic cell thus supplying the $T\Delta S$ requirement. Power from the fuel cell is fed back to the electrolytic cell (ΔF) and hydrogen and oxygen is produced (ΔH). The H_2 and O_2 is heat exchanged with the returning cooler water vapor from the fuel cell and combined in the fuel cell to produce electrical power. External cooling water is used to maintain the fuel cell temperature near ambient or at a fixed temperature lower than the electrolyzer cell. Part of the fuel cell power is sent back to the electro-lytic cell which is at a higher potential than the electrolytic cell. The difference in cell voltage ($\Delta F_f + \Delta F_e$) is then the net potential which produces the power from the cycle. Note that the thermodynamic sign con-vention is preserved here; a negative sign meaning energy evolved from the reaction and a positive sign meaning energy added to the system. The cycle efficiency is:

$$\varepsilon = \frac{\Delta F_f + \Delta F_e}{T_e \Delta S_e}$$

Subscript f refers to fuel cell and subscript e to electrolyzer

Theoretically generating and utilizing heat at $1000^{\circ}K$ in the electrolytic cell and rejecting heat at $298^{\circ}K$ ($25^{\circ}C$) in the fuel cell gives a Carnot efficiency of 70%. In Table 2, we calculate the ideal cycle based on the thermodynamic data given in Table 1. The values are slightly lower than the ideal Carnot, probably due to the precision of the thermodynamic data but are nevertheless very close. Figure 2 plots the cell voltage for the H_2-O_2 system and Figure 3 shows the T-S diagram for the TEC cycle.

The values in Table 2 assume that the fuel cell operates totally in the vapor phase. Due to the physical properties of water, this means operating at a pressure below the vapor pressure of water, which at 25°C is 18 mm Hg. Subatmospheric operation can be avoided by operating above 100°C which would decrease the efficiency somewhat and allow operation at 1 atm. If a liquid water phase is allowed to form in the fuel cell, then the heat of vaporization of water is extracted and must be added to the water before entering the electrolyzer. It should be noted that the heat capacity of the H_2 and $1/2O_2$ gas is about equal to the heat capacity of H_2O gas so that a heat balance between gaseous water and gaseous H_2 and O_2 can be maintained.

However, if the heat of condensation of water is lost in this fuel cell at lower temperatures, it must then be made up by addition of heat of vaporization (ΔH_v). The efficiency of the cycle can then be expressed by the following

$$\varepsilon = \frac{\Delta F_f + \Delta F_e}{T_e \Delta S_e + \Delta H_v}$$

Values of the condensing cycle, consisting of electrolyzing water gas at high temperatures and operating a H_2-O_2 fuel cell at 25°C forming liquid water, are given in Table 2A. It is noted that the net voltages are higher than that of the noncondensing system and cycle efficiencies exceeding Rankine (40%) at temperatures above 700°C can be obtained. However, there are overvoltage problems.

The problem is that the H_2-O_2 power cycle suffers because of the overvoltage inefficiencies which exist in the H_2-O_2 cells. For example, if there is a 0.1 volt energy loss in the electrolyzer at 1000°K, the cell voltage required is 0.998 + 0.100 = 1.098 volts. At the lower temperature fuel cell operation (298.2°K) the overvoltage goes as high as 0.500 volts so that the fuel cell voltage = 1.185 -0.500 = 0.685 volts which is clearly lower than the required electrolytic cell voltage and thus the cycle is inoperative. Another way of looking at the operability of the electrolytic cell/fuel-cell combination as a power producer is that if the sum of the cell overvoltages and IR losses are greater than the net ideal thermodynamic potential, then the cell will not work. It is possible that for electrolytic cell temperatures greater than 2000°K cell voltages will be less than 0.700 volts and the low temperature fuel cell could then supply the necessary energy. However, this temperature presents a formidable materials problem for nuclear fission reactors and limits the heat source to fossil fueled power plants, or in the future to fusion reactors. Another approach is to select an electrochemical system that would have low electrode overvoltage inefficiencies. Such a system is the hydrogen-chlorine reaction system.

H_2-Cl_2 TEC Power Cycle

Applying the same principles to the H_2-Cl_2 system the basic thermodynamic data are presented in Table 3. Even at 25°C, the overvoltage potential of the chlorine electrode is very small in the order of 0.050 to 0.100 volts, (50 to 100 mv).

One can determine from Table 3 that for the H_2-Cl_2 system, the ΔF for HCl decomposition increases so that the relative stability of HCl increases with temperatures. This means that the reversible cell voltage increases with temperature and that, for net power production, the fuel cell (recombining H_2 + Cl_2 + HCl) must operate at the higher temperature and the electrolytic cell (decomposition of HCl to H_2 + Cl_2) must operate at a lower temperature. This is distinctly possible, and data for the ideal net voltage output cycle efficiencies are shown in Table 4.

It is noted from Table 4 that the net ideal voltage for the power cycle is only in the order of 0.020 to 0.118 volts (or 20 to 118 millivolts), up to temperatures of 2000°K (1726.8°C). The precision of the thermodynamic data is also in question here. This means that if the total overvoltage losses exceed 100 millivolts the power cycle will not operate. It is possible that the cell overvoltages can be less than 100 millivolts, however, the margin of safety makes this power cycle unrealistic. The only point that can be made here is that power cycles can be designed either for an electrolytic-cell fuelcell mode or vice-versa, depending on the thermodynamics of the system. Also energy storage at isothermal conditions in the H_2-Cl_2 system looks to be more attractive than in the H_2-O_2 system, because of the lower overvoltages. A system which appears more realistic is the H_2-O_2-Cl_2 power cycle described in the next section.

H_2-O_2-Cl_2 TEC Power Cycle

From the above, the optimal electrothermal system should have the following criteria:

    1. The electrochemical reactions should have a large temperature coefficient.

    2. The electrode overvoltages should be a minimum.

    3. Materials should be universally available and materials of construction materials should not be severe.

    4. The cycle should operate at temperatures reasonably attainable with nuclear fission reactors in addition to fossil energy thermal plants.

A system which has these characteristics and relies on two interconnected electrochemical reactions is outlined in the following sequence of reaction steps.

I. High temperature electrolytic decomposition of water; the $T_e \Delta S_e$ thermal energy is supplied by a high temperature power source (fission fusion or fossil).

$$H_2O_{(g)} = H_{2(g)} + 1/2O_{2(g)}$$

II. Oxidation of HCl with oxygen generated from the electrolytic cells (I) thus regenerating Cl_2

$$2HCl_{(g)} + 1/2O_{2(g)} = H_2O_{(g)} + Cl_{2(g)}$$

III. Low temperature fuel cell recombination of Cl_2 from II with H_2 from I and generating power for I, thus completing the power cycle.

$$H_{2(g)} + Cl_{2(g)} = 2HCl_{(g)}$$

If we examine the plot the data in Table 1 for H_2O decomposition and Table 3 for HCl decomposition as is done in Figure 2, we find that at a temperature below 875^oK (600^oC) the cell voltage for HCl is lower than for H_2O, whereas above this temperature HCl has a higher cell potential than H_2O. We can thus design power cycles for either side of this nodal point. However, we want to work the H_2O cell at a high temperature to reduce the oxygen electrode overvoltage. Therefore, we design for the right side of the figure and operate the HCl fuel cell at high temperatures, feeding power back to a water decomposition cell at the same high temperature, or at any temperature where the cell overvoltage is minimized. It should be noted here that this combined system can also operate ideally, isothermally to produce power at high thermal efficiencies especially in cascade operation. The high temperature electrolytic decomposition of water reduces the overvoltage to the order of 0.1 volts. The overvoltage for H_2-Cl_2 fuel cell reaction is not more than 0.1 volt and probably less, so that the total overvoltage losses should be low, probably < 0.2 volt. From Figure 2, in order to maintain a spread of 0.2 V between the HCl cell and the H_2O cell, temperatures higher than 1440^oK (1167^oC) or if the HCl cell is maintained at 298.2^oK, the H_2O cell must operate above 1700^oK (1427^oC). These temperatures would leave out the possibility of using nuclear fission reactors which have temperature limitations presently in the order of 1000^oK (727^oC). High temperature reactors including fusion reactors, in the future, designed to generate to 1200^oC and above could be considered here. For fossil fuel fired plants, temperatures to 2000^oK (1727^oC) are available. It should be noted that development of high temperature electrochemical and fuel cells are assumed in this analysis.

A more practical solution is to run the H_2-Cl_2 fuel cell for HCl formation in the aqueous phase.

IV. $$1/2H_{2(g)} + 1/2Cl_{2(g)} = HCl_{(aq)}$$

The thermodynamic values at $298.2^\circ K$ are

$$\Delta H_{298.2} = -39.85 \text{ Kcal/mol}$$

$$\Delta F_{298.2} = -31.33 \text{ Kcal/mol}$$

$$T\Delta S_{298.2} = -8.52 \text{ Kcal/mol}$$

For this reaction, the cell voltage is then $31.33/23.06 = 1.36$ volts at $25^\circ C$. Combining this with water decomposition at $1000^\circ K$ gives a net voltage of $1.36-1.0 = 0.36$ volts well above the cell overvoltage potentials.

The power cycle depends upon the oxidation of HCl for regeneration of chlorine. For the gas phase reaction II,

$$2HCl_{(g)} + 1/2O_{2(g)} = H_2O_{(g)} + Cl_{2(g)}$$

The thermodynamic values at $298.2^\circ K$ are as follows:

$$\Delta H_{298.2} = -6.84 \text{ Kcal/mol}$$

$$\Delta F_{298.2} = -4.54 \text{ Kcal/mol}$$

$$T\Delta S_{298.2} = -2.30 \text{ Kcal/mol}$$

The gas phase reaction is thus exothermic requiring no heat. However, the more practical system as mentioned above is the regeneration of chlorine from the aqueous phase.

V. $\quad HCl_{(aq)} + 1/4O_{2(g)} = 1/2H_2O_{(g)} + 1/2Cl_{2(g)}$

$$\Delta H_{298.2} = +10.95 \text{ Kcal/mol}$$

$$\Delta F_{298.2} = +4.01 \text{ Kcal/mol}$$

$$T\Delta S_{298.2} = +6.94 \text{ Kcal/mol}$$

The net reaction is endothermic thus requiring energy input. The process for regeneration of Cl_2 from HCl has been known in the past as the Decon Process and more recently has been further developed under the name Kel-Chlor Process [1]. The process usually operates at about 300°C for both the anhydrous and aqueous system and at pressures of from 1 to 15 atm.

The cycle efficiency of the three part power process is then as follows:

$$\varepsilon = \frac{\Delta F_f + \Delta F_e}{T_e \Delta S_e + \Delta H_{HCl_{(aq)}}}$$

Table 5 lists the net cell potential for a range of temperatures for the H_2-O_2-Cl_2.

Table 5 indicates that over 47% of the energy can be converted to electrical energy at temperatures above about 700°C. This cycle efficiency is actually a Rankine cycle because of the condensed hydrochloric acid. The overvoltage and IR losses should be low in this system under these conditions. The system is unique in that there are no moving parts as pertains in today's tubro-generator system. The system can actually operate at low pressure (atmospheric up to a few atmospheres pressure, as desired). Also if there is a low level heat source available, for example, geothermal or solar energy, this can supply the heat of vaporization of the aqueous hydrochloric acid and thus enhance the higher temperature thermal efficiency. Cascading cycles may also be applicable. The system also lends itself to integration with the use of hydrogen and oxygen for the production of synthetic fuel and feedstocks and for energy storage. Because of the direct production of DC electrical power, the thermoelectrochemical (TEC) power cycle may be especially suited for the production of electrochemicals and the electro-winning of metals such as aluminum and copper. DC power can be directly used in these operations without power conditioning and conversion equipment.

The application of higher temperature electrolytic decomposition (electro-lyzers) combined with low temperature recombination (fuel cells) in a power cycle has been mentioned previously [2] but has not been further pursued. There may be electrochemical systems other than those described above which may prove to have better electrochemical characteristics. Anodic depolar-izers such as SO_2 may have some interesting possibilities in reducing electrode overvoltages [3] and allowing cyclical power process applications [4].

REFERENCES

1. C. P. van Dijk and W. C. Schreener, Chem. Eng. Progress 69 (4) 57-63 (1973).

2. M. S. S. Hsu, W. E. Morrow, Jr., J. B. Goodenough, 10th Intersociety Energy Conversion Engineering Conference (Newark, Delaware) Proceedings, Paper 759085, 555-63 (1975).

3. W. Juda and D. M. Moulton, Chem. Eng. Progress 63 (4) 59-60 (1967).

4. G. H. Farbman and L. E. Bricker, 10th Intersociety Energy Conversion Engineering Conference (Newark, Delaware) Proceedings, Paper 759178, 1199-1204 (1975).

* The submitted manuscript has been authored under contract EY-76-C-02-0016 with the U.S. Department of Energy. Accordingly, the U.S. Govern-ment retains a nonexclusive, royalty-free license to publish or reproduce the published form of this contribution, or allow others to do so, for U.S. Government purposes.

TABLE 1

ENTHALPY, FREE ENERGY UNAVAILABLE ENERGY (TΔS)
REVERSIBLE CELL POTENTIALS FOR THE
ELECTROLYTIC DECOMPOSITION OF WATER

$$H_2O_{(g)} = 1/2H_{2(g)} + 1/2O_{2(g)}$$

t°C	T°K	ΔH Kcal/mol	ΔF Kcal/mol	Reversible emf-volts	TΔS Kcal/mol	Fraction addition of thermal energy TΔS/ΔH
25.0	298.2°K	+57.8	+54.64	1.185	+3.16	0.055
226.8	500	+58.27	+52.36	1.135	+5.91	0.101
726.8	1000	+59.21	+46.03	0.998	+13.18	0.228
1226.8	1500	+59.84	+39.26	0.851	+20.58	0.344
1726.8	2000	+60.26	+32.31	0.700	+27.95	0.464

TABLE 2

IDEAL EFFICIENCY FOR
H₂-O₂ THERMOELECTROCHEMICAL POWER CYCLE
GASEOUS WATER FUEL CELL

| Electrolytic cell | | | Fuel Cell | | | Ideal Cycle Efficiency,% | Carnot Eff.,% |
| Temp. | Elec. Energy | Heat | Temp. | Elec. Energy | Net Cell | $\frac{\Delta F_f + \Delta F_e}{T_e \Delta S_e}$ | $\frac{T_e - T_f}{T_e}$ |
$t\,^{\circ}C$ — T_e	ΔF_e	$T_e \Delta S_e$	T_f	ΔF_f	Volt.		
25.0 — 298.2	+54.64	+3.16	298.2	-54.64	0.000	0.0%	0.0%
226.8 — 500	+52.36	+5.91	"	"	0.049	38.6	40.4
726.8 — 1000	+46.03	+13.18	"	"	0.187	65.3	70.2
1226.8 — 1500	+39.26	+20.58	"	"	0.333	74.7	80.1
1726.8 — 2000	+32.31	+27.95	"	"	0.485	80.1	85.1

TABLE 2A

IDEAL EFFICIENCY FOR H₂-O₂
THERMOELECTROCHEMICAL POWER CYCLES
LIQUID WATER FUEL CELL SYSTEM

Electrolytic Cell H₂O(g)=H₂(g) + 1/2O₂(g) = H₂O(ℓ)

Temp. $t_e\,^{\circ}C$	Elec. Energy ΔF_e	Heat $T_e \Delta S_e$	Fuel Cell Temp. $t_f\,^{\circ}C$	Elec. Energy ΔF_f	Net Cell Volt.	Ideal Cycle Eff.,%	Carnot Eff.,%
25.0	54.64	3.16	25	56.69[1]	0.0444	15.0	0
226.8	52.36	5.91	"	"	0.094	26.4	40.4
726.8	46.03	13.18	"	"	0.230	45.0	70.2
1226.8	39.26	20.58	"	"	0.378	56.1	80.1
1726.8	32.31	27.95	"	"	0.529	63.4	85.1

1) Cell voltage = 1.229.

TABLE 3

ENTHALPY, FREE ENERGY, UNAVAILABLE ENERGY (TΔS)
AND REVERSIBLE CELL POTENTIALS FOR THE ELECTROLYTIC
DECOMPOSITION OF HYDROGEN CHLORIDE

$$HCl_{(g)} = 1/2 H_{2(g)} + 1/2 Cl_{2(g)}$$

$t°C$	$T°K$	ΔH Kcal/mol	ΔF Kcal/mol	TΔS Kcal/mol	Reversible emf-volts
25	298.2	+22.06	+22.78	-0.72	0.988
226.8	500	+23.22	+23.22	-1.01	1.006
728.8	1000	+22.56	+24.09	-1.53	1.044
1226.8	1500	+22.74	+24.81	-2.07	1.076
1726.8	2000	+22.84	+25.49	-2.65	1.106

TABLE 4

IDEAL EFFICIENCY FOR H_2-Cl_2
THERMOELECTROCHEMICAL POWER CYCLE

	Fuel Cell		Elec. Energy		Electrolytic Cell	Energy				
$t°C$	T_f	$ΔH_f$ Kcal/mol	$ΔF_f$ Kcal/mol	$TΔS_f$ Kcal/mol	T_e	$ΔF_e$	Net ΔF Kcal/mol	Ideal Net Volt.	Ideal Cycle Eff., %	Carnot Eff., %
25.0	298	22.06	-22.78	+0.72	29.82	+22.78	0.00	0.000	-0.0	0.0
226.8	500	22.21	-23.22	+1.01	"	"	-0.44	0.020	-44.0	40.4
726.8	1000	22.56	-24.09	+1.53	"	"	-1.31	0.056	-85.6	70.2
1226.8	1500	22.74	-24.81	+2.07	"	"	-2.03	0.088	-98.1	80.1
1726.8	2000	22.84	-25.49	+2.65	"	"	-2.71	0.118	-102.3	85.1

TABLE 5

IDEAL EFFICIENCY FOR H_2-O_2-HCl(aq)
THERMOELECTROTHERMAL POWER CYCLE

WATER ELECTROLYZERS AT HIGH TEMPERATURE AND
AQUEOUS HYDROCHLORIC ACID AND FUEL CELL OPERATION AT LOW TEMPERATURE

Intermediate Reaction:
$HCl_{(aq)} + 1/2O_{2(g)} = 1/2H_2O_{(g)} + 1/2Cl_{2(g)}$; $\Delta H_{298.2} = +10.95$

| Electrolytic Cell $H_2O_{(g)} = H_{2(g)} + 1/2O_{2(g)}$ | | | | | Fuel Cell $1/2H_{2(g)} + 1/2Cl_{(g)} = 2HCl_{(aq)}$ | | | | | | |
t°C	T°K	ΔF_e	$T\Delta S_e$	Cell Volt	t°C	T°K	ΔF	Net ΔF	Net Ideal Volt.	Ideal Cycle Eff.,%	Carnot Eff., %
25.0	298.2	54.64	3.16	1.185	25	298.2	-31.33 [1]	4.01	0.174	32.0	0.4
226.8	500	52.36	5.91	1.135	"	"	"	5.15	0.223	37.0	40.4
726.8	1000	46.03	13.18	0.998	"	"	"	8.32	0.360	47.4	70.2
1226.8	1500	39.26	20.58	0.851	"	"	"	11.70	0.507	55.1	80.1
1726.8	2000	32.31	27.95	0.700	"	"	"	15.18	0.658	60.9	85.1

1) Cell volt = 1.359

1322

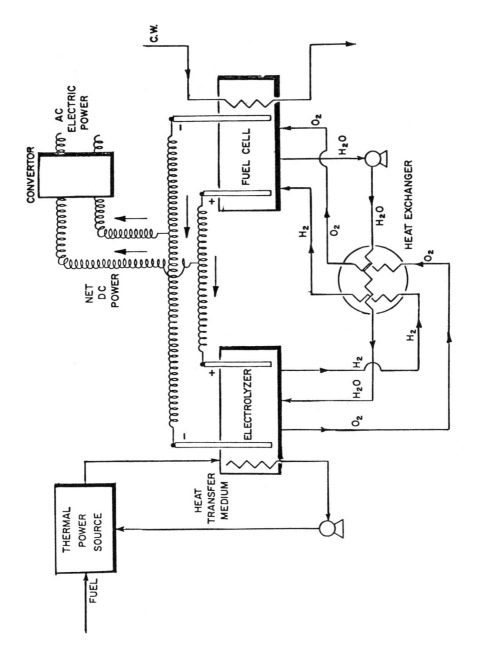

Figure 1. H_2-O_2 Thermoelectrochemical Power Cycle

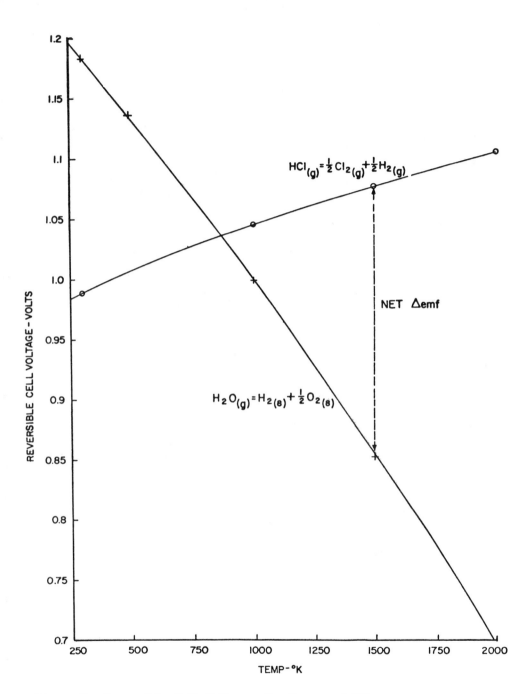

Figure 2.   Reversible Cell Voltages for Water and HCl

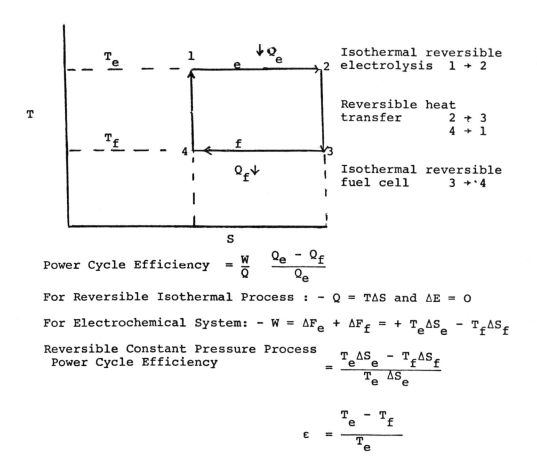

Power Cycle Efficiency $= \dfrac{W}{Q} \quad \dfrac{Q_e - Q_f}{Q_e}$

For Reversible Isothermal Process : $- Q = T\Delta S$ and $\Delta E = 0$

For Electrochemical System: $- W = \Delta F_e + \Delta F_f = + T_e \Delta S_e - T_f \Delta S_f$

Reversible Constant Pressure Process
  Power Cycle Efficiency $\qquad = \dfrac{T_e \Delta S_e - T_f \Delta S_f}{T_e \ \Delta S_e}$

$$\varepsilon = \dfrac{T_e - T_f}{T_e}$$

Figure 3. T-S Diagram from Thermoelectrochemical (TEC) Power Cycle

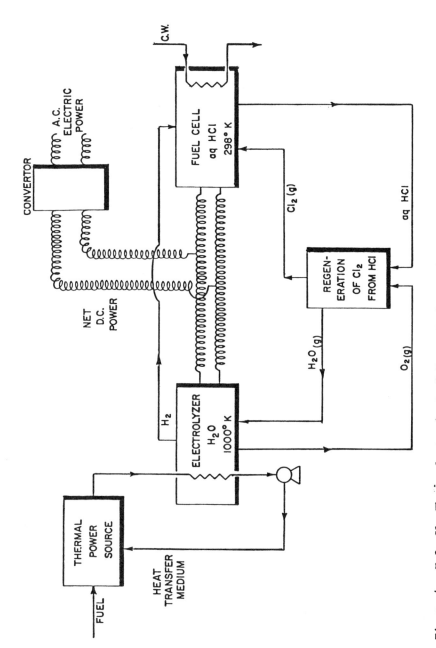

Figure 4. H_2O_2–Cl_2 Thermoelectrochemical Power Cycle

TECHNICAL SESSION 6

TRANSMISSION
AND
DISTRIBUTION

GAS DISTRIBUTION EQUIPMENT
IN HYDROGEN SERVICE,
PRELIMINARY FINDINGS

J. B. Pangborn, D. G. Johnson, and W. J. Jasionowski
Institute of Gas Technology
IIT CENTER, 3424 S. State Street
Chicago, Illinois 60616, U.S.A.

ABSTRACT

The Institute of Gas Technology is conducting an experimental program to
identify problem areas that could occur with the use of conventional
natural gas distribution equipment in hydrogen service. Funded by the
U. S. Department of Energy, there are 15 manufacturers and gas distribution
companies that are participating by loaning or donating equipment and
services. Three operational model test loops have been constructed: a
Residential/Commercial loop with smaller sized components, an Industrial
loop with larger sized components, and a small test loop for special
(created) leakage tests. We are measuring flow rates and energy delivery,
and leakage rates, and we will note apparent problems in materials
compatibility. Baseline data on natural gas operation have been taken for
the Residential/Commercial loop and the Industrial loop, and both of these
loops are now operating on pure hydrogen.

INTRODUCTION

This program is part of a multi-year effort to supply needed information
about hydrogen delivery in natural gas distribution equipment. The overall program will identify operating, safety, and materials problems associated with the use of hydrogen in conventional distribution systems. One
of the major incentives behind (nonfossil-based) hydrogen as a future
supplement and eventual replacement for natural gas is the expectation
that the existing gas delivery system can be used without major modifications. Natural gas is a major portion of the energy usage in the United
States, now accounting for 19.8 Quads (19.8 X 10^{15} Btu) or 25.4% of our energy
consumption. The incentive for supplemental fuel gases (including
hydrogen), to enable continued delivery of energy as a fuel gas,
is primarily financial. The embedded capital investment in the gas distribution industry in the U.S. now exceeds 20 billion dollars. This includes
over 650,000 miles of distribution mains which carry gas to about 45,000,000
customers. Replacement of this distribution system with another would cost
many times this investment. Furthermore, most equipment and lines now
being installed are expected to last 50 years or longer. If it is practical
to carry hydrogen safely in this existing gas distribution equipment, then
hydrogen is indeed an attractive form for energy delivery in the future.

If moderate problems are identified now, then we have sufficient time to define and develop alternative operating procedures. If, however, serious problems are found, then other alternatives (besides hydrogen) must be weighed against major system modifications.

CONSTRUCTION OF TEST LOOPS

After site facility preparation, IGT proceeded during 1977 to construct three model test loops using equipment loaned or donated by 15 manufacturers and gas utility companies. The 15 companies that are program participants (along with the DOE) are listed in the Appendix of this paper. All construction for the test loops was done in accordance with gas industry procedures and requirements. In fact, a field construction welder from Northern Illinois Gas Company welded the necessary steel joints with standard field equipment. All subassemblies were leak tested prior to and after integration into a total assembly or test loop.

RESIDENTIAL/COMMERCIAL TEST LOOP

The Residential/Commercial Test Loop consists of four subloops and a bypass. The pipeline materials of construction are 1) steel, 2) copper, 3) plastic (high-molecular-weight, high-density polyethylene), and 4) cast iron. This model contains components and equipment normally installed in typical residential and/or commercial service. Figure 1 is a photograph of the completed Residential/Commercial Test Loop and Figure 2 is a simplified diagram of this loop with its major equipment components. In operation, a single-stage compressor feeds 15 SCF/min of natural gas or hydrogen to the model at a pressure between 60 and 65 psig. The compressed gases pass through an aftercooler to reduce gas temperature to ambient and then through a surge tank to dampen pulsations. A regulator reduces the pressure to about 50 to 60 psig, simulating pressures in the distribution mains and the service lines to commercial or residential consumers. In the case of the cast-iron subloop, the pressure is reduced to 6 inches water column by another pressure regulator. At the simulated building line, which is the termination of the service line, the distribution pressure (50 to 60 psig) is reduced further by a service regulator to 6 inches water column. The gases then pass through a gas meter to the inlet of the compressor and are recompressed and recycled. Flow is controlled and proportioned through the subloops with valves and the bypass.

INDUSTRIAL TEST LOOP

The Industrial Test Loop consists of one loop and a bypass with steel pipeline material. This model contains components and equipment normally installed in typical industrial service. Figure 3 is a photograph of the Industrial model, and Figure 4 is a simplified diagram of this loop with its major equipment components. In operation, a two-stage compressor feeds 15 SCF/min of natural gas or hydrogen to the model at a pressure between 170 and 175 psig. The compressed gases pass through an after-cooler to reduce gas temperatures to ambient and then through a surge tank to dampen pulsations. A line regulator installation reduces the pressure to 60 psig, simulating pressures in the distribution main. Another regulator downstream, in series, reduces the pressure further to 8 psig, simulating the operating service pressures of industrial components. The gases then pass through several industrial gas meters (for example, dia-phragm, rotary, or turbine) connected in series, to the inlet of the compressor, and are recompressed and recycled. Flow is controlled and proportioned with valves and the bypass.

SAFETY TEST LOOP

The Safety Test Loop consists of one loop with a leak zone and a bypass. The leak zone provides a space for testing and defining problems associ-ated with mechanical or corrosion leaks, leak clamps, and ruptures. The pipeline material is steel except at the leak zone. Figure 5 is a photo-graph of the Safety Test model and Figure 6 is a simplified diagram of this loop with its major equipment components. In operation, a single-stage compressor feeds 10 to 15 SCF/min of natural gas or hydrogen to the Safety Test model at a pressure between 50 and 60 psig. The compressed gases pass through an aftercooler to reduce gas temperatures to ambient and then through a surge tank to dampen pulsations. The gases then pass through the experimental setup in the leak zone to the inlet of the com-pressor and are recompressed and recycled. If excess leakage is a problem, the gas flow will terminate in the leak zone and the gases will be vented to the outdoors.

SPECIAL MEASUREMENT PROCEDURES

The three reciprocating piston compressors were installed to recycle and compress natural gas and/or hydrogen to the operating design conditions of the model test loops. An initial 40-hour test was performed with the Residential/Commercial and Industrial models at design conditions using nitrogen gas. The systems operated satisfactorily, and leak rates of the components and model were characterized. During this operation it was again assured that the systems and the individual components, fittings, and connections met industry requirements for leak-tightness. The Residential/Commercial loop and the Industrial loop were then operated for about two weeks to gain base-line data on flow and leakage with natural gas. Measurements with hydrogen are now under way for both loops.

As the gases pass through various regulators or stations in the two major loops, the operating pressures are reduced from the feeder main (150 to 200 psi), to the distribution main (50 to 60 psi), and to the point of service (5 to 10 psi for Industrial and 6 to 10 inches water column for

Residential/Commercial). Meters are installed either in series or in
controlled subloops so that comparative flow measurement data (on natural
gas and hydrogen) can be taken. At a simulated building line, the gases
are filtered, recompressed, and recycled to the loops. The Residential/
Commercial and the Industrial models are designed to operate continuously,
whereas the Safety Test model will operate intermittently. Using local
pressure and temperature measurements, all gas flows are reduced to gas
industry standard cubic feet (SCF, 60°F, 1 atm).

Eleven components (couplings, unions, a pressure regulator, and a flow
meter) were enclosed with sheet metal or Plexiglas enclosures to monitor
and compare leakage of natural gas and hydrogen from these specific components. Figure 7 is a photograph of a pressure regulator sealed in a
Plexiglas enclosure. Volumetric displacement and gas analysis (by gas
chromatography or mass spectrograph) were the methods selected to measure
the leak rates of these components.

Total system leakage is being determined by measuring makeup gas additions
to the high-pressure side of the test loops (after the compressor). The
makeup gas quantities are being determined by pressure decay from calibrated cylinders.

PRELIMINARY OBSERVATIONS AND RESULTS

At this writing (March 1978), the Residential/Commercial Test Loop has
experienced 6.5 weeks of operation with hydrogen, while the Industrial Test
Loop has experienced more hydrogen time — about 12.5 weeks. We will be
operating for several more months before the equipment and components will
be examined for physical, chemical, or metallurgical effects due to the
hydrogen exposure. (Each individual component has a duplicate, stored in
original condition, for later comparisons with the hydrogen-exposed components.) So far, no materials problems or incompatibilities in the distribution equipment are evident. However, both the Residential/Commercial
model and the Industrial model have suffered compressor gas leakage problems. The compressors are reciprocating piston machines with Teflon rings
and Teflon dynamic seals for oil-free operation in natural gas and/or hydrogen
service. (They were selected over diaphragm models on the basis of costs
and delivery time.) Both compressors have been serviced by the supplier to
replace the dynamic seals and the head gaskets. These repairs were found to
be necessary with both natural gas and hydrogen operation to keep compressor
leakage small relative to the very small leakage rates observed for the
distribution loops and components.

GAS FLOW AND ENERGY DELIVERY

The high heating value of hydrogen is 325 Btu/SCF (1.21×10^7 J/m^3); whereas natural gas heating values are commonly in the range of 950 to 1100

Btu/SCF (3.54×10^7 J/m^3 to 4.10×10^7 J/m^3). Thus, the flow of hydrogen must be increased by a factor of about 2.9 to 3.4 (relative to natural gas under the same temperature and pressure conditions) to deliver energy at a rate equivalent to that of natural gas. A previous study considered the gas flow equations for laminar, partially turbulent, and turbulent flow hydrogen and natural gas in pipes. Some conclusions of that study are defined below: [1]

> "Usually, the natural gas flow in distribution mains is partially turbulent, but conditions of laminar and fully turbulent flow occur. If hydrogen is to be delivered in this or a future system built for natural gas service, certain operating and procedural changes are to be anticipated. If volumetric flow rates for hydrogen were about 325% (on the average) of those for natural gas, an equivalent amount of energy would be delivered. However, if the pipes and operating pressures are unchanged:
>
> - For Laminar flow, the volumetric flow rate of hydrogen will be about 130% of that of natural gas and the delivered energy will be only 40% of that of natural gas.
>
> - For partially turbulent flow, the volumetric flow rate of hydrogen will be about 260% of that of natural gas and the delivered energy will be 80% of that of natural gas.
>
> - For turbulent flow, the volumetric flow rate of hydrogen will be about 280% of that of natural gas and the delivered energy will be 85% of that of natural gas.
>
> - For all categories of flow, the Reynolds number for hydrogen will be one-half of or less than that of natural gas, and the designed-for categories of flow might be downgraded, e.g., partially turbulent to laminar, in some instances.
>
> To deliver the same quantity of energy, the hydrogen gas density is best increased by increasing the operating pressures of the pipelines."

In our experiments with the model distribution loops we have observed:

- For the Residential/Commercial Distribution Loop operating on hydrogen, the volumetric flow rate is about 300% that of natural gas. This percentage factor is for the entire loop operating under the same overall pressure drop (compressor output to compressor intake) for hydrogen and for natural gas. Our natural gas (Peoples Gas Light

& Coke Company, Chicago) has a heating value of 1008
to 1016 Btu/SCF. Hence, the energy delivery as hydrogen
is about 97% that of natural gas under the same operating conditions for this model.

- For the Industrial Distribution Loop operating on hydrogen, the volumetric flow rate is about 245% that of
natural gas. Hence, the energy delivery as hydrogen
is 80% that of natural gas under the same operating
conditions for this model.

Comparative flow meter readings in the Industrial Loop are presented in Table 1 for tests on natural gas and hydrogen (534-hour
cumulative test on natural gas and 935-hour and 1170-hour
cumulative tests on hydrogen). All meters are of a different
brand (manufacturer): the mean value of the average flow
rates is 237.5 CF/hr for natural gas, 555.9 CF/hr for hydrogen
(Test 1), and 597.9 CF/hr for hydrogen (Test 2).

GAS LEAKAGE: HYDROGEN VERSUS NATURAL GAS

In order to determine the leakage of a test loop, the compressor leakage
must be substracted from the combined system leakage.

Figure 8 presents example cumulative leakage data for the Industrial Loop
in a 500-hour test with circulating natural gas. This base-line test
shows three notable aspects of system leakage:

1. Compressor leakage apparently decreased with time (possibly as rings
    and seals "wore in").

2. The model loop leakage rate is quite small, (about 0.07 SCF/hr).

3. Loop leakage is relatively constant and no significant new leakage
    appears to be developing with time.

When switched to hydrogen, the Industrial Loop exhibited a volumetric
leakage rate more than three times higher than for natural gas. Also,
the compressor dynamic seals began to leak progressively (but gradually)
more. Leakage rates for most of the components are not unproportionately
large for hydrogen versus natural gas. Exceptions are pipe couplings
with rubber seals (various rubber compositions) and leakage through valve
stem packings. Although the relative leakage rates are yet to be quantified with precision, preliminary indications are that rubber coupling
seals and valve stem seals are exhibiting hydrogen leakage rates that are
four to five times higher than were observed for natural gas. The
Residential/Commercial Loop has a compressor that leaks more. Subsequently,
we are still developing the base-line natural gas and hydrogen data that
is necessary for valid comparisons. The previous study cited above for

gas flow and energy delivery also considered leakage (diffusional and that through an orifice). The conclusions of that (mathematical) study are listed below:

"Leakage ratio $\dfrac{H_2}{CH_4}$ = 2.83 (diffusion)

Leakage ratio $\dfrac{H_2}{CH_4}$ = 2.54 (laminar flow through orifice)

Although leaks measured volumetrically will be 2.5 to 2.8 times greater for hydrogen (versus those for natural gas for the same pressure drop), the energy loss would be less. If for hydrogen delivery the operating pressure is increased (by 20%), the energy loss (through a leak) should be about equivalent."

In the Industrial Loop experiments, a leakage rate comparison study showed that with natural gas, the overall loop leakage rate (under continuous flow conditions) was measured to be 0.070 SCF/hr. Under the same operating conditions, but with hydrogen (see Figure 9), the overall Industrial Loop leakage rate was 0.227 SCF/hr. Therefore, the observed volumetric leakage ratio is 0.227 $\div$ 0.070 = 3.24. In terms of energy loss via leakage, the ratio is 3.24 $(\dfrac{325}{1012})$ = 1.04 (energy lost as hydrogen to energy lost as natural gas).

BIBLIOGRAPHY

[1] Pangborn, J. B., Scott, M. I. and Sharer, J. C., "Technical Prospects for Commercial and Residential Distribution and Utilization of Hydrogen." Paper presented at the 1st World Hydrogen Energy Meeting, Miami Beach, Florida, March 1976.

APPENDIX

This program is being conducted with the support of the U.S. Department of Energy (DOE) under contract EY-76-C-02-2907. The program technical writer for DOE is Mr. Ray Hagler of the Jet Propulsion Laboratory, Pasadena, California. Industrial companies participating in this program through the donation or loan of equipment and services are listed below. We wish to express our gratitude for their interest and support.

The Peoples Gas Light & Coke Company
Chicago, Illinois

Northern Illinois Gas Company
Aurora, Illinois

American Meter Division of Singer
Philadelphia, Pennsylvania

Municipal and Utility Division of
Rockwell International
Pittsburgh, Pennsylvania

Kerotest Manufacturing Corporation
Pittsburgh, Pennsylvania

Mueller Company
Decatur, Illinois

The Sprague Meter Company
Bridgeport, Connecticut

Dresser Measurement Division of
Dresser Industries, Inc.
Houston, Texas

Phillips Products Company, Inc.
Dallas, Texas

Fisher Control Company
Marshalltown, Iowa

Dresser Manufacturing Division
Dresser Industries, Inc.
Bradford, Pennsylvania

E.I. duPont de Nemours and Company
Wilmington, Delaware

Republic Steel Corporation
Cleveland, Ohio

Rego Company
Chicago, Illinois

Flow Control Division of
Rockwell International
Pittsburgh, Pennsylvania

Table 1. COMPARATIVE FLOW METER READING IN NATURAL GAS AND HYDROGEN OPERATIONS

Meter Type, Capacity	Natural Gas (534 hours) $*$ Mean Flow Rate = 237.5 CF/hr $*$ Sample Standard Deviation = 4.3		Hydrogen Test 1 (935 hours) 555.9 CF/hr 14.1		Hydrogen Test 2 (1170 hours) 597.9 CF/hr 12.7	
	Average Flow Rate, CF/hr	% Deviation From Mean	Average Flow Rate, CF/hr	% Deviation From Mean	Average Flow Rate, CF/hr	% Deviation From Mean
Turbine, 4000 CF/hr	233.5	-1.7	533.4	-4.0	582.6	-2.6
Diaphragm #1 1000 CF/hr	242.8	+2.2	570.7	+2.7	611.4	+2.3
Diaphragm #2 1000 CF/hr	235.7	-0.8	545.6	-1.9	590.0	-1.3
Diaphragm #3 1000 CF/hr	232.7	-2.0	542.2	-2.5	584.5	-2.2
Rotary 3000 CF/hr	238.8	+0.5	565.1	+1.7	605.6	+1.3

$*$ Based on the three diaphragm meters and the rotary meter.

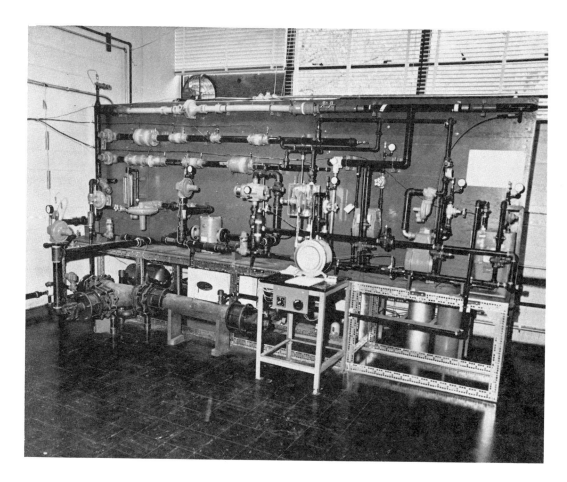

Figure 1.   Residential/Commercial Test Loop

1339

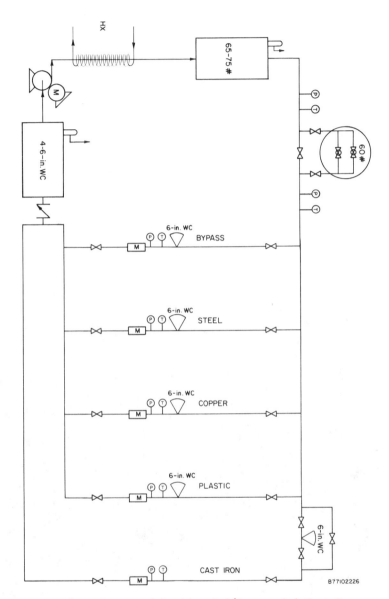

Figure 2.   Diagram of Residential/Commercial Test Loop

Figure 3. Industrial Test Loop

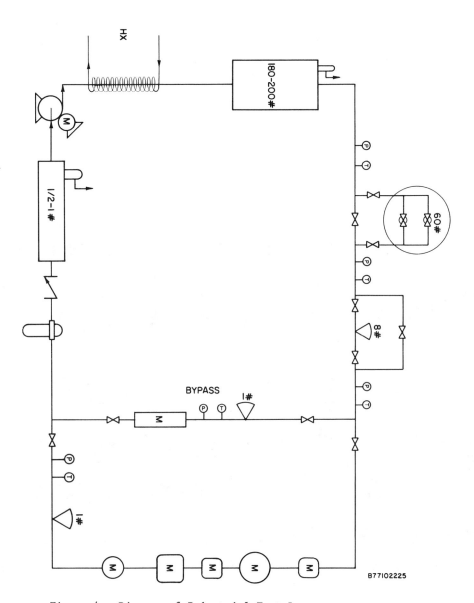

Figure 4. Diagram of Industrial Test Loop

Figure 5. Safety Test Loop

1343

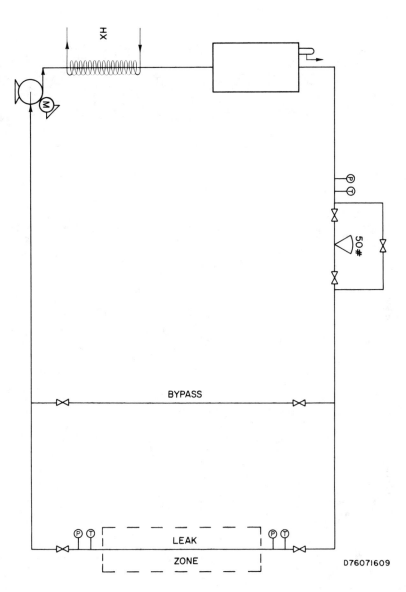

Figure 6.   Diagram of Safety Test Loop

Figure 7.   Pressure Regulator in Enclosure

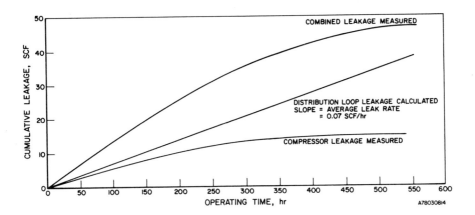

Figure 8.   Baseline Natural Gas Leakage for Industrial Distribution Test Loop

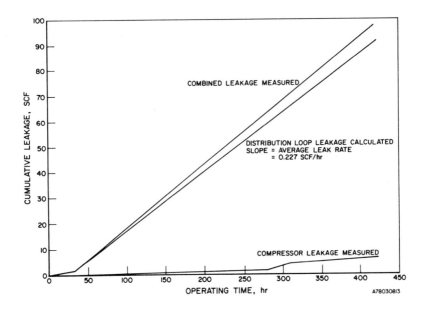

Figure 9.   Leakage Characteristics of the Industrial Test Loop, Hydrogen Operations

ANALYSIS OF THE POTENTIAL TRANSMISSION
OF HYDROGEN BY PIPELINE IN SWITZERLAND

E. Anderson, J. Davies, M. Kornmann, G. Capitaine
Battelle
Geneva Research Centre
Geneva, Switzerland

ABSTRACT

The potential transmission of hydrogen by pipeline has been
studied, particular attention being paid to the consequence of
the specific situations in Switzerland (forecast energy demand
and breakdown by fuel types, relatively short transport distances, existence of large hydraulic facilities, etc.). Although
transmission costs are higher for hydrogen than for natural gas,
for pipeline operation at throughputs significantly below
maximum capacity, as is the case in Switzerland, there would
be currently no capacity problems and transmission costs would
be virtually the same if hydrogen were transported.

Analysis of available data on the specific problems of materials
and standards of pipeline construction, with particular regard
to environmental hydrogen cracking, shows that slow crack
growth from existing cracks under low cycle fatigue due to
daily pressure fluctuation, if no special precautions are taken, may
lead to problems during the pipe lifetime. Fracture propagation
should be less of a problem with hydrogen than with natural gas.

INTRODUCTION

With the increasing interest in utilising hydrogen as an energy
vector, many studies have been carried out on the technical and
economic feasibility of transporting hydrogen gas by pipeline
[1-7]. Although the results of these studies are important in
the overall evaluation of the concept, their often general
nature does not allow one to draw conclusions on detailed design
for individual situations and countries.

The present study was undertaken specifically for Switzerland,
so that in assessing hydrogen pipelines, current natural gas
pipeline practice is taken into account.

SWISS NATURAL GAS PIPELINE NETWORK

Figure 1 presents the natural gas transmission network in
Switzerland as of 1977. Details of each pipeline are given in
Table 1. The main characteristics of the network are its rela-
tively recent construction, transmission pressure in the range

1347

64-70 bars with branch lines at 6-25 bars, and the use of relatively low strength steels. Except for the Transit Gas line,
electrical resistance welded or seamless pipes (given the
relatively small diameter) have been mainly employed, although
Gaz Nat installed spiral welded pipe in its overland sector;
seamless pipe was installed in the Lake of Geneva.

Design factors vary between 1.7 for open country and 2.5 near
inhabited areas.

The few instances of pipeline failure that have been reported
occured during hydrostatic testing of ERW pipes to between 90%
and 100% SMYS and were due to defects in the longitudinal weld.

FUTURE ENERGY AND GAS DEMAND IN SWITZERLAND

In order to be able to evaluate the economic feasibility of
transporting hydrogen by pipeline in Switzerland it was necessary to have an estimate of the energy demand evolution. This
is one of the subjects studied by a Swiss Federal Commission
on the Total Energy Concept (GEK*). The following actual figures
for 1975 and their anticipated evolution to 1985 for the energy
demand and distribution [8] have been furnished:

	1975		1985	
	Tcal	%	Tcal	%
new forms			1 000	0.5
wood	1 995	1.3	3 500	1.7
coal	2 288	1.5	4 000	1.9
electricity	27 753	18.4	41 900	20.0
gas	5 009	3.3	19 000	9.0
oil	114 107	75.5	140 600	66.9
TOTAL	151 152	100.0	210 000	100.0

If gas achieves a 9% share by 1985, it may be possible
for the gas share to rise as high as 20% by 2000. This would
not be out of line with existing experience in such countries
as Belgium, UK and W. Germany.

If a total energy demand in the region of 250 000 Tcal per year
is postulated for the end of the century, we find that the
total gas demand could be of the order of 50 000 Tcal per year.

---

*) Eidgenössische Kommission für die Gesamtenergiekonzeption

That this is relatively small by current standards can be judged
by the fact that the Transit Gas pipeline, which transports
Dutch natural gas across Switzerland to Italy, currently
supplies roughly 60 000 Tcal per year, has a capacity of
100 000 Tcal per year and yet is only 36" in diameter. Even
if all the gas demand for Switzerland were to be satisfied by
hydrogen, a single large diameter pipeline would be sufficient
to ensure the transportation.

GAS PIPELINE TRANSPORT CHARACTERISTICS

In order to assess the interest of transporting hydrogen a
computer programme, enabling the various input parameters to be
varied easily, was set up to calculate the design parameters of
suitable pipelines and to estimate the total transportation cost.
The programme was used to calculate the pipeline transport of
hydrogen and methane and any desired mixture of the two gases.
Basic assumptions in the calculations shown are that the total
transport distance is 200 km and that two compressor stations
are needed to ensure transportation. A peak load factor of 2.0
is assumed i.e. the pipeline is dimensioned so as to be able
to transmit at the peak twice the average daily throughput
derived from the annual throughput quoted.

Optimum pipeline diameter for a given energy throughput

One possible comparison between a methane and a hydrogen pipeline is to consider the optimum diameter for a constant energy
supply. The energy content of methane is taken as 5.01×10^7
joules/kg (8 566 Kcal/m^3) and that of hydrogen as 12.10×10^7
joules/kg (2 598 Kcal/m^3). Thus, more than three times as much
hydrogen by volume as methane must be transported to satisfy
a given energy demand.

Figures 2 and 3 show the variation in cost of kWh transported for
energy throughputs of 2000 and 50 000 Tcal per year respectively.
It may be noted that the larger throughput corresponds to the
maximum estimated total gas/hydrogen demand in Switzerland by
the year 2000 while the former figure corresponds roughly to the
existing annual throughput in the main individual pipelines in
Switzerland. The figures show two intermediate levels of
hydrogen: methane mixture (40% and 80% hydrogen by volume).

The conclusions here are as follows:

a) The optimum diameters and corresponding costs in the two
    cases are as follows:

Total energy/year	Diameters (m)		Transport cost (cents/kWh)	
	methane	hydrogen	methane	hydrogen
2 000 Tcal	0.30	0.35	0.48	0.57
50 000 Tcal	1.10	1.40	0.08	0.12

b) The difference in optimum transport cost is less than 0.1 centimes/kWh, which may be compared with a current natural gas price in Switzerland (according to demand level) of 3.5-9.4 centimes/thermie or roughly 3.0-8.0 centimes/kWh [8].

c) Even the maximum projected total gas energy demand for Switzerland (50 000 Tcal/year) may be transported as hydrogen in a single pipe of 1.40 metre diameter.

d) As found in all types of pipeline transport, there is a significant reduction in transport costs as throughput increases from 2 000 Tcal to 50 000 Tcal per year.

Varying throughput of different gases in fixed pipe diameter

The above discussion of pipeline transport costs deals with the design of a single pipeline and compares a specific natural gas line with a specific hydrogen line.

Of interest in practice will be the effect on transport cost of passing increasing amounts of hydrogen through a pipeline optimized for the transport of natural gas. This case is shown in Figure 4 which gives the transport cost for methane and hydrogen transport through a single 0.3 m diameter pipe as a function of increasing energy throughput. The following conclusions may be drawn:

a) The optimum cost point for a hydrogen pipeline is at a lower energy throughput than for a methane pipeline (1 800 Tcal/year compared with 2 800 Tcal/year). At the respective optimum points the costs are 0.64 centimes per kWh for hydrogen and 0.39 centimes per kWh for methane.

b) In practice, a methane pipeline will not be operating near its optimum point. In the present case, if the methane line is operating at around 50% of capacity (say 1 500 Tcal/year i.e. a typical throughput for a pipeline of this size in existing Swiss conditions), the line may not only be converted to hydrogen without any further compressor stations being necessary, but at this operating point the transport

cost difference is only 0.61 centimes/kWh compared with
0.69 centimes/kWh. Furthermore, if the methane line is
operating at 1/3 capacity or less (e.g. 1 000 Tcal/year)
there will be no transport cost differences at all between
the line operating with methane or hydrogen.

c) One significant result of using a pipeline for hydrogen
    rather than methane transmission is that the ultimate ca-
    pacity of the line will be reduced - in the present case
    from roughly 3 200 to 2 500 Tcal/year. This could in
    practice be countered either by ensuring that hydrogen flow
    was more even (i.e. with a lower peak demand) or by adding
    a further compressor station to the hydrogen line.

One final point may be made with respect to the gas transport
situation in Switzerland. Due to the supply conditions and the
short transport distance, compressor stations are not in fact
necessary (the only exception is for gas received at Basle
from France where the pressure needs to be raised from 25-40
bars to 70 bars for transmission). If in a future hydrogen
production plant the hydrogen is supplied at a high pressure
(60-70 bars), then it is probable that no intermediate compressor stations would be needed either for hydrogen transmission,
and pipeline transport costs would then be virtually the same
as for natural gas (the pipeline capacity would, however, be
reduced).

Summarizing the comparison of pipeline transport of natural gas
and hydrogen, it was found that:

-   for a given energy throughput, the optimum pipe diameter for
    hydrogen transport will be 15-25% greater than the optimum
    diameter for natural gas,

-   at the optimum points the transport costs for hydrogen
    would be 25-50% higher than for natural gas,

-   for a given pipeline diameter, however, optimum hydrogen
    transport costs would be 60-70% higher than the optimum
    natural gas costs,

-   a given pipeline, without any changes in operating condi-
    tions, would have an energy transport capacity 30% greater
    when transporting natural gas than if transporting hydrogen.

ENVIRONMENTAL HYDROGEN CRACKING

Review of data on external hydrogen cracking of pipeline steels

From the beginning, it has been apparent that materials problems could be significant in a hydrogen pipeline. It is often stated that hydrogen embrittlement is not a problem for low strength steels and examples such as hydrogen storage bottles, existing hydrogen pipelines and hydrogen storage tanks are given. However, the working conditions of these structures are different from those of a hydrogen energy transmission pipeline: in storage bottles, the stresses are low and there is no weld; the number of fillings during the working life is low and there is no fatigue. Concerning hydrogen storage tanks, the stresses are higher and 70 accidents have been observed in Europe over the last fifteen years. The specific case of hydrogen pipelines is studied later.

Information and data concerning environmental hydrogen embrittlement of pipeline steels (e.g. API 5LX 48, X52, X60, X65) have been reviewed with a view to apply them to an actual Swiss pipeline network.

Studies of hydrogen environmental embrittlement of lower strength steels are rare because it was thought for a long time that hydrogen had no effect when the elastic limit is below about 56 kg/mm^2 (80 000 psi), and because most of the environmental embrittlement studies have been concerned with aeronautical and space applications where such steels are little used.

The influence of hydrogen on the different properties of C-Mn ferritic-pearlitic steels can be summed up as follows:

tensile strength: no marked influence of hydrogen;

yield strength: no major influence of hydrogen [9]. This value is used in the design of the pipeline;

ductility at rupture: the ductility decreases with the square root of the pressure up to a certain pressure above which an increase has no further influence. Figure 5 gives some data for different steels [9-10]. A minimum ductility is normally specified for a pipeline (e.g. 18%).

reduction in area: decreases in the same way as the ductility. The preceeding results have not been obtained specifically on pipeline steels but on carbon ferrite-pearlite steels of lower strength;

notch sensitivity: the notch sensitivity is very high under
hydrogen pressure. This sensitivity increases with crack acuity
up to a maximum stress intensity. For pipeline steels, Table 2
shows the reduced ductility obtained on notched samples [11].

slow growth of cracks: when the elastic limit of the steel is
lower than 56 kg/mm^2 (80 ksi), no measurable slow crack growth
has been observed till now by different authors [12,13,14].
Blunting of the crack, and sometimes branching, is observed on
the other hand. However, these tests have been carried only at
room temperature and not on high strength pipeline steels. On
higher strength steels, slow crack growth has been observed
when the stress intensity factor is higher than a certain value,
K_{TH}, function of the steel and the H_2 pressure (Figure 6) [15].
Of course, this K_{TH} is lower than K_{IC}, critical stress intensity
factor.

fatigue crack growth: recently, results [14-16] have been obtained concerning the crack growth in fatigue. A crack grows
much faster in hydrogen than in air. This rate can be 10-20
times higher at frequency of 1 Hz. The rate, da/dn is a function
of the difference of the stress intensities ΔK and increases
with pressure up to 20-70 atmospheres and then is independent
of the pressure. The main results are given on Figure 7 [17].
A recent publication [18] shows the existence of a threshold
around 15 MN/m$^{3/2}$ at 1 Hz frequency (Figure 6). Increasing
hydrogen pressure is seen to increase da/dn and decrease the
apparent ΔK threshold. The growth rate, da/dn is four times
higher at low frequency (0.001 Hz) than at higher frequency
(1 Hz) and the threshold, if any, is much lower. It looks also
to be possible to slow the crack growth by mixing the pure gas
with impurities such as O_2, CO_2, sulfur compounds ..., but
mixing with methane alone does not slow the growth.

Consequences for a hydrogen pipeline

It can be concluded that, if a pipeline is perfect, without any
defect, it will be able to carry hydrogen with no additional
problems compared with natural gas. However, a pipeline is never
defect-free and the defects of most concern for a hydrogen
pipeline are hard spots and cracks. Hard spots result from a
partial quenching of the steel during rolling or welding. Cracks
develop near slag inclusions, at the root of the welds, at
inadequate joint penetrations, etc. The next chapter deals with
the influence of these problems on pipeline safety.

TECHNICAL FEASIBILITY OF A HYDROGEN PIPELINE

The main problem in deciding whether a high pressure hydrogen pipeline is technically feasible or not is that no practical experience exists in other countries and very little experimental testing has been carried out on conventional pipe materials and welds. The few results that exist have at various times been used to conclude both that no problems will exist in practice as well as that testing and defect detection limits will have to be pushed to their limits in order to allow hydrogen to be transported.

Defect tolerance

Figure 8 shows the size of critical longitudinal defects (superficial and through thickness cracks) which are of greatest significance to performance in a typical pipeline case: external diameter 306 mm (12"), thickness: 4.5 mm, steel X60, pressure 70 bars.

The data were calculated using the theory of Kiefner et al. [19] which, although applying specifically to external superficial defects, can also be used for superficial internal defects. These results are slightly different from those of Swisher [20] due to different basic hypotheses. At a through crack, there is a leak; when the stress intensity is high enough, fracture propagation occurs. When a superficial crack is considered, two possibilities can occur: at a critical stress intensity, this crack can propagate through the thickness giving a leak; then the crack stops or propagates over a long distance. On the Figure 8, it is possible to differentiate both cases according to the position of the through crack curve and the superficial crack one.

Slow growth

Where there is a hard spot, a defect is able to grow slowly in hydrogen up to this critical dimension when the size of the hard spot is greater than the critical dimension. The non-dangerous defect must thus be so small that it does not propagate. For example, a C-Mn steel ($\sim$ X52) can have a K_{IC} value of 150 $MN/m^{3/2}$, and the critical length of a full thickness crack is 6.2 cm. On a hard spot, the steel is quenched with a hardness of about 400 HV, and its tensile strength is about 200 000 psi. K_{IC} can be 50 $MN/m^{3/2}$, corresponding to a critical through thickness defect length of 2.2 cm. K_{TH} is lower and could be of the order of 12 $MN/m^{3/2}$ at 70 bars with a critical crack length of only 0.3 cm. Of course, the actual values are not known, but these figures have been measured on a quenched low alloy steel AISI 340 at the same strength level (Figure 6).

Fatigue crack growth

The stress variation in a pipeline is very slow (for instance one cycle/day depending on gas consumption) but the ΔP may be considerable.

Normally, there is no significant crack growth by fatigue in a natural gas pipeline, but rather rapid growth has been observed under corrosion fatigue conditions. Thus, it is important to estimate the possibility of crack growth under fatigue in hydrogen gas.

A calculation of the fatigue crack growth enables the determination of the initial size of a defect which may become critical. This can be carried out by integration of the growth rate with K given by Keifner et al. [19]. The most interesting is to calculate the evolution of a superficial crack in order to check whether a leak or a fracture is obtained. An estimate has already been made on a simpler, but perhaps less realistic case: the fatigue growth of a through crack. This preliminary work allows an estimate to be made of the importance of the problem of fatigue crack growth in hydrogen pipelines.

For a pipeline made of X60 steel, diameter 0.3 m, thickness 4.5 mm, with a working life of 25 years, a pressure variation of 45% and taking a crack growth rate measured under hydrogen at low frequency, preliminary estimates show that a longitudinal through-thickness crack longer than 1.5 mm is able to grow up to the critical size; as an example of a shorter crack, a longitudinal through-thickness crack of 1.2 mm length will grow to 1.21 mm under natural gas but up to 7.5 mm under pure hydrogen. This defect is critical for a toughness lower than 40 $MN/m^{3/2}$.

Defects in actual pipelines

Although all pipeline codes state that pipes and welds that contain cracks and crack-like defects should not be passed for operation, the limits of detection of current NDT methods mean that pipelines do get installed wich contain small "injurious defects". Also, the API 5LX and Swiss Codes* allow some defects

---

* Richtlinien für Plannung and Bau von Rohrleitungsanlagen zur Beförderung Flüssiger und Gasförmigen Brenn- und Triebstoffe, Eidg. Rohrleitungsinspektorat, 1977.

to be passed which could be considered potentially dangerous
under fatigue conditions. For example, it is possible to have
a pipeline with undercuts/defects of the following dimensions
in the longitudinal and circumferential welds:

- maximum depth 0.8 mm and length one-half the wall thickness,

- maximum depth 0.4 mm, any length.

Also some other defects may be as deep as 12.5% wall thickness.
In the case of pipes installed in Switzerland this can correspond to defects of between 0.4 and 1.0 mm in depth. There is
no experimental indication that such real defects will grow
in hydrogen but their dimensions are of the order of those
theoretical critical cracks presented in the previous section.
Thus fatigue crack growth could be a problem.

The other dangerous defects, hard spots, may reach a hardness
as high as 400 HV on the body of a pipe. By comparison with the
preceeding results, it is very important to avoid hard spots
during manufacturing and to reduce the maximum tolerable hardness. If this is not possible, then eventually a sophisticated
non destructive testing procedure must be used to detect such
structures.

Fracture propagation

Pipeline puncturing due to impact from bulldozers, etc. is one
of the main causes of pipeline accidents and even with natural
gas these can be very serious. Whether hydrogen or hydrogen/
methane mixtures would be worse from this point of view is not
evident, but one eventually positive aspect would be that
fracture propagation may be less of a problem than with natural
gas [21] . Material specifications may be drawn up to preclude
both brittle and ductile propagating fractures in conventional
API 5LX60 and X65 steels. The newer, high toughness X60-X70
grades still pose a problem from the point of view of a satisfactory correlation of material characteristics with ductile
propagation, but this does not apply to the current Swiss pipeline system and probably will not in the future since conventional API 5LX steels can be used to provide entire satisfaction
from the technical and economic points of view.

As far as hydrogen is concerned, current impact specifications
on the ductile-brittle transition temperature employed for
natural gas should readily preclude brittle fractures, especially since, in addition, the high speed of decompression in
hydrogen (Figure 9) means that, as with crude oil, no long
sustained fracture should occur since the stress at the crack
tip is released [22] .

In the case of ductile fracture the higher decompression wave speed may lead to the situation where cracks that can propagate in natural gas pipelines will be arrested in hydrogen. Thus, the specification of a minimum impact energy level* should readily preclude ductile crack propagation in conventional pipeline steels. No particular need for high toughness steels is foreseen for Swiss conditions.

RELEVANT EXPERIENCE ON PIPELINES

Hydrogen at a pressure of 18 bars has been transported now for about 40 years in the Hüls pipeline network with apparently no major problems [23]. Given the relatively low stress level in the pipe, crack growth as described above would not be expected to occur but any incidents such as accidents, leaks, etc. which could result in hydrogen gas escape have not led to reported explosions. Several other hydrogen pipelines are reported in the world [20]: they are either low pressure, as at Hüls, made of austenitic steel or very small diameter. One high pressure pipeline constructed of low alloy steel was shut down because of leaks in welds.

Equally, many years of experience of the transmission and mainly distribution of town gas containing often considerable quantities of hydrogen was not considered a more dangerous operation than transmitting natural gas. One advantage of town gas was the presence of traces of water vapour and carbon monoxide which are efficient inhibitors of hydrogen environmental cracking, and the water vapour also assisted in reducing leaks at mechanical joints. The feasibility of adding trace quantities of inhibitors to hydrogen or hydrogen/methane mixtures should be seriously examined if insurmountable technical and economic problems are foreseen for the pure gases. Currently in the USA the study phase of hydrogen as a potential energy vector is over and practical problems involved are now starting to be examined in two test loops:

- transmission loop at SANDIA, Livermore

- distribution loop at IGT, Chicago.

Parallel studies in Europe where pipe materials specifications are different would seem to be indicated in order to build up the know-how necessary to arrive at practical solutions.

---

*) Impact energy measured in a Charpy V-notch test in the region of 100% shear fracture.

SITUATION IN SWITZERLAND

Specifically for the case of installing a new pipeline in
Switzerland for hydrogen transport, the pipe material specifications and installation controls would certainly have to be
strengthened, mainly with respect to the control of cracks,
crack-like defects and hard spots. Experimental confirmation
is certainly required so that irrealistic limits are not placed
on materials and controls that may lead to a false picture of
pipeline transport of hydrogen.

For the conversion of the present natural gas network to hydrogen it would be necessary to control the pipeline quality by the
passage of an inspection survey pig, several commercial versions
of which exist or are under development for the detection of
cracks, wall thinning, etc. [24]. The limits of detection of
such pigs are still the question of some discussion, but for
one version specially developed for longitudinal, internal
defects, cracks of 80 mm length and 2 mm depth have been detected [25]. This might not be a fine enough resolution for hydrogen
pipelines.
In all events, the critical aspect in any pipeline utilised for
hydrogen transmission will be slow crack growth due to fatigue.
The incidence of high cycle fatigue being negligible, it is
the daily variation in pressure that must be controlled to a
higher degree than with natural gas. If this is not strictly
controlled then the initial defect detection limit will have to
be very low. The smaller the pressure variations, the larger
the initial crack that can be tolerated and hence the less
strict need be the detection.

Recommendations for a Swiss hydrogen pipeline would thus include:

- pipe material supplied to reinforced API 5LX and DIN specifications especially with respect to defect detection,

- pipeline installation to reinforced API 1104 specifications with respect to hard spots and crack-like defects,

- operating pressure fluctuations as low as possible,

- periodic testing by inspection survey pigs in order to
  detect cracks before they break through the wall or become
  critical.

CONCLUSIONS

The present study shows that pipeline transmission of hydrogen under optimum conditions in new, full-capacity pipelines will be 25-50% more expensive than natural gas transmission. However, for a given pipeline diameter and for pipeline operation at throughputs significantly below maximum capacity, which is the current situation for national gas pipelines in Switzerland, there would be no capacity problems and transport costs would be virtually the same if hydrogen were transported.

Also, current and future forecast levels of gas demand in Switzerland should pose no problems of pipe dimensions. Even the maximum likely national hydrogen demand could be accomodated within a single large diameter pipeline.

The major technical problem with transmission of hydrogen gas at high pressure is the possibility of slow fatigue crack growth from existing cracks or crack-like defects in the pipe body or weld. The presence of hard spots makes this situation worse. Conditions particularly conducive to crack growth are low cycle fatigue - daily variation of the pressure.

Because of the nature of the decomppression in hydrogen gas, a leak-before-break situation should occur and no large-scale fractures are to be expected, as with oil pipelines. Thus, the philosophy of safety in a hydrogen pipeline should concentrate on the crack growth from existing defects. However, it is evident that, before conversion of a natural gas pipeline or the design and installation of a new hydrogen pipeline, experimental results on crack growth of defects in pipes and welds are required together with more accurate calculations of crack growth from a variety of defects. Such results are imperative to the realistic analysis of the feasibility of transmitting large quantities of hydrogen by pipeline.

If it is confirmed that fatigue is a major problem, possible technical solutions include:

- an accurate control and regulation of the hydrogen pressure,
- use of inhibitors to slow crack growth,
- more accurate detection of the defects before the pipeline is put into operation,
- periodic testing of the installed pipeline.

ACKNOWLEDGEMENTS

The authors wish to thank Dr. R. Tenne for assistance with the crack growth calculations and the Fonds national suisse de la recherche scientifique for sponsoring the programme.

REFERENCES

1. D.P. Gregory, A Hydrogen Energy System, Chapter 4, AGA, August (1972)

2. G. Beghi et al., Hydrogen Economy Energy (THEME) Conference, Miami, March (1974)

3. R. Reynolds and W. Slager, ibid.

4. A. Konopka and J. Wurm, 9th Intersociety Energy Conversion Conference, San Francisco, August (1974)

5. P. Hampson et al., Hydrogen Energy Fundamentals Symposium, Miami, March (1975)

6. J. Hollenberg, ibid.

7. G. Donat and J. Colonna, AIM, Liège, November (1976)

8. GFK Intermediate Report, May (1976)

9. W. Hoffmann and W. Rauls, Welding Research Supplement 225-S, May (1965)

10. A.W. Thompson and I.M. Bernstein, Int. J. of Hydrogen Energy, 2, (1977) 163-173

11. E.E. Fletcher and A.R. Elsea, private communication

12. A.W. Loginow and E.H. Phelps, Trans ASME, J. of Eng. for Industry, February (1975), 274

13. E. Irving and E.A. Almond, Metals Technology, February (1977) 115

14. H.G. Nelson, Effects of Hydrogen on Behaviour of Materials, Edit. Bernstein, September 7-11 (1975), 602

15. R.J. Walter and W.T. Chandler, Influence of Gaseous Hydrogen on Metals, Final Report, prepared for NASA, NASA Cr-124410, October (1973)

16. R.J. Walter and W.T. Chandler, Effects of Hydrogen on Behaviour of Materials, Edit. Bernstein, September 7-11 (1975) 273

17. S.L. Robinson, Hydrogen Compatibility of Structural Materials, SAND 76-8255, December (1976)

18. H.G. Nelson, preprint, November (1977)

19. J.F. Kiefner, W.A. Maxey, R.J. Eiber and A.R. Duffy, ASTM Special Technical Publication 536 (1972), 461

20. J.H. Swisher, Effects of Hydrogen on Behaviour of Materials, Edit. Bernstein, September 7-11 (1975), 558.

21. 5th AGA Symposium on Line Pipe Research, Houston, November (1974), Catalogue No. L 30174

22. J.E. Hood, Int. J. Pres. Ves. and Piping (2) (1974), 165

23. C. Isting, Conference on Hydrogen and its Perspectives, A.I.M., Liège, November (1976)

24. For example, W. Flamank, Pipes and Pipelines International, June 1974, 27

25. A. Kittel, Erdöl und Kohle, Erdgas, Petrochimie, Compendium 1974/75, 865

TABLE 1 - HIGH PRESSURE NATURAL GAS PIPELINES IN SWITZERLAND

	Diameter mm	Length km	Wall thickness mm	Steel type*	Design pressure bars	Year of construction
Transit Gas	914.4	71.6	12.10	X60	67.5	
	863.6	72.0	12.70	X60	70.0	
		14.0	14.27	X60	"	1972-74
		7.8	15.88	X60	"	
		0.8	17.48	X60	"	
		1.9	19.05	X60	"	
		168.1				
Swissgas	406	54.2	7.1	St 53.7	70	
		43.0	"	-	"	
	356	49.1	6.3	St 53.7	"	1973-74
	324	93.0	-	-	"	
	273	3.5	5.0	St 47.5	"	
		1.5				
		244.3				
EGZ	273	13.8		St 47.7	70	
	273	2.9		St 47.7	25	1974
	224	4.4		St 24.7	25	
		21.3				
GVO Ostschweiz	273	32.9		St 34.7		
	219	95.7				1968-69
	168	33.8		38.7		
	114	41.7		43.7		
		204.1				
Thayngen-Schlieren	273	53.5		47.7	64	1969
	168	5.2		53.7		
		58.7				
		310.8				
GVM,original network	324	5.8		St 34.7		
	273	33.2		42.7	50	
	219	134.9		47.7	64	1966-67
	168	3.8		53.7	64	
	114	4.0			64	
		181.7				
Hochrhein	324	30.3	5.6	St 53.7	70	1973-74
GASNAT	324	5.5	5.6	St 53.7	70	
	273	13.5	5.0	St 53.7	70	
	324	32.0	5.6	St 53.7	70	
	273	11.7	5.0	St 38.7	25	1973-74
	273	1.3	5.0	St 53.7	70	
	219	1.0	5.0	St 38.7	25	
Lake of Geneva	273	78.5	10.0	St 53.7	70	1973-74
	273	20.0	10.0	St 60.7	70	
		163.5				

* Steel mostly according to DIN 17 172

TABLE 2 - EFFECTS OF TESTING AS-RECEIVED (UNCHARGED) SAMPLES OF THE THREE STEELS IN A HIGH-
PRESSURE HYDROGEN ENVIRONMENT AT ROOM TEMPERATURE ON THE REDUCTION IN AREA
OBTAINED WITH NOTCHED TENSILE SPECIMENS (a,b), AFTER FLETCHER AND ELSEA [11]

Specimen number	Test environment	Test pressure psig	Reduction in area percent	Change in reduction in area by hydrogen environment percent
		X52 STEEL		
X-9-N	Argon	2,000	3.9	
X-13-N	Argon	2,000	6.8	
Avg.			5.4	-.-
X-10-N	Hydrogen(c)	2,000	1.5	
X-11-N	Hydrogen	2,000	1.5	
X-12-N	Hydrogen	2,000	1.9	
Avg.			1.6	-70
		X65 STEEL		
A-9-N	Argon	2,000	2.9	
A-13-N	Argon	2,000	4.5	
Avg.			3.7	-.-
A-10-N	Hydrogen	2,000	2.0	
A-11-N	Hydrogen	2,000	1.1	
A-12-N	Hydrogen	2,000	1.7	
Avg.			1.6	-57
		JALLOY S-100 STEEL		
3-9-N	Argon	2,000	9.4	
3-13-N	Argon	2,000	6.8	
Avg.			8.1	-.-
3-10-N	Hydrogen	2,000	5.9	
3-11-N	Hydrogen	2,000	4.2	
3-12-N	Hydrogen	2,000	5.8	
Avg.			5.3	-35

(a) Other specimens were tested in argon at the same pressure and in the same equipment for
comparison

(b) These specimens were tested with a constant rate of load application, which resulted in
a very short time for plastic deformation to occur prior to fracture

(c) Ordinary tank hydrogen was used without further purification for all specimens in this
table that were tested in a hydrogen environment

Gasversorgung der Schweiz

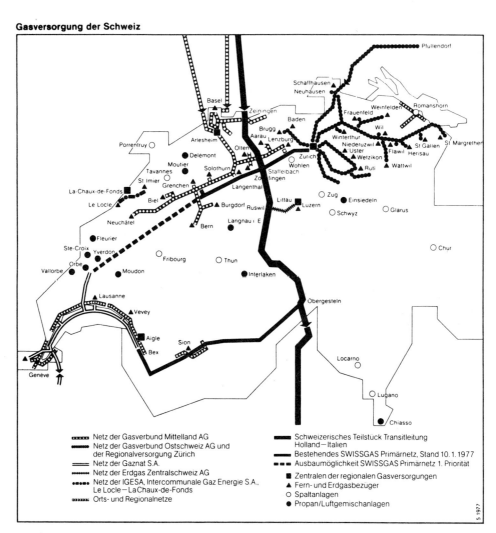

FIGURE 1 - NATURAL GAS NETWORK IN SWITZERLAND

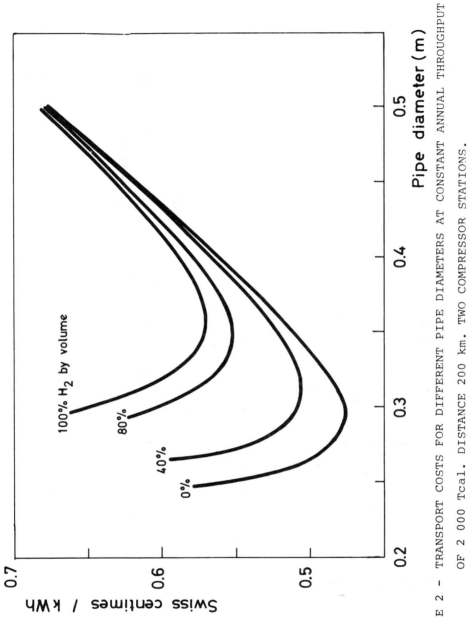

FIGURE 2 - TRANSPORT COSTS FOR DIFFERENT PIPE DIAMETERS AT CONSTANT ANNUAL THROUGHPUT OF 2 000 Tcal. DISTANCE 200 km. TWO COMPRESSOR STATIONS.

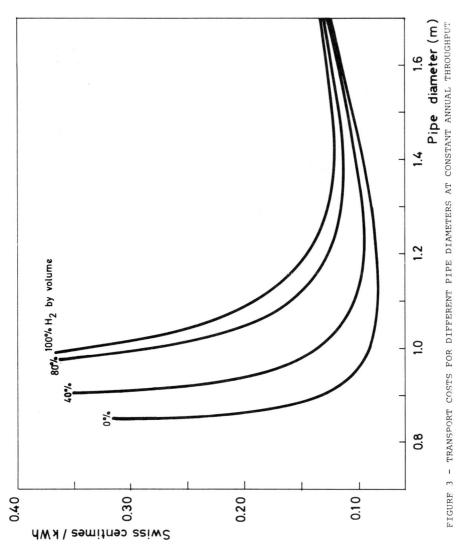

FIGURE 3 - TRANSPORT COSTS FOR DIFFERENT PIPE DIAMETERS AT CONSTANT ANNUAL THROUGHPUT
OF 50 000 Tcal. DISTANCE 200 km. TWO COMPRESSOR STATIONS.

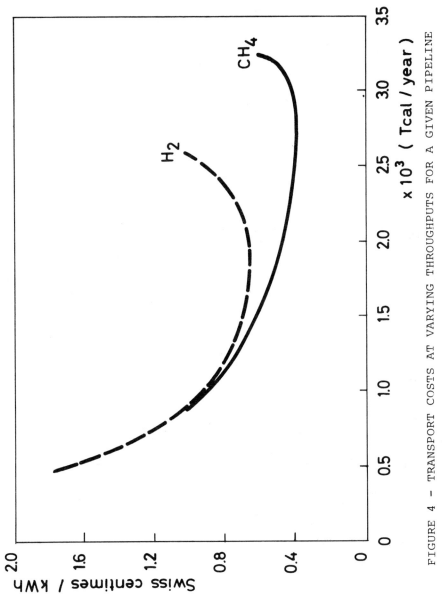

FIGURE 4 - TRANSPORT COSTS AT VARYING THROUGHPUTS FOR A GIVEN PIPELINE

(DIAMETER 0.3 m). DISTANCE 200 km. TWO COMPRESSOR STATIONS.

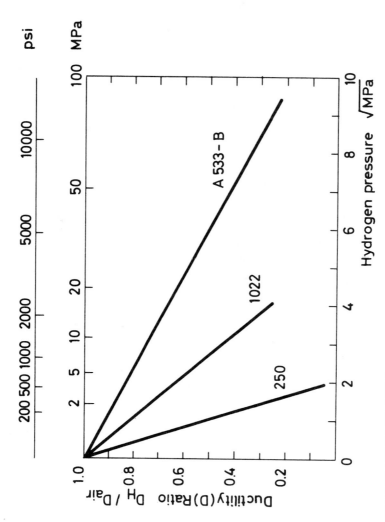

FIGURE 5 - DUCTILITY RATIO OF THREE STEELS AS A FUNCTION OF HYDROGEN GAS PRESSURE IN WHICH THE TEST WAS CONDUCTED. 250 = MARAGING STEEL, YIELD STRENGTH 1 725 MN/m² (250 ksi); 1022 = Mn CARBON STEEL, ULTIMATE STRENGTH 500 MN/m³; A533-B = LOW ALLOY STEEL (Ni, Mo), YIELD STRENGTH 485 MN/m², ULTIMATE STRENGTH 690 MN/m². AFTER HOFFMANN AND RAULS [8].

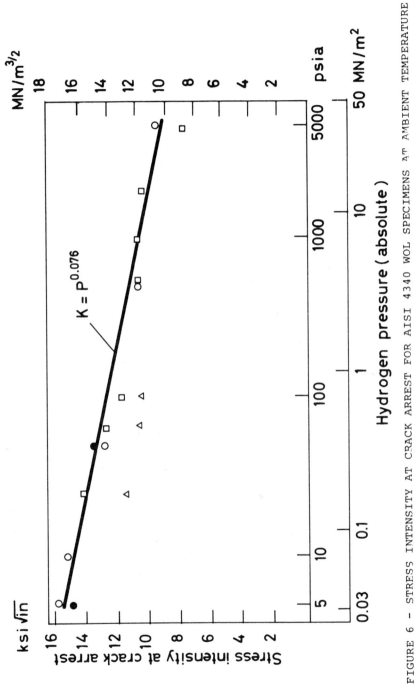

FIGURE 6 - STRESS INTENSITY AT CRACK ARREST FOR AISI 4340 WOL SPECIMENS AT AMBIENT TEMPERATURE AS A FUNCTION OF HYDROGEN PRESSURE. AFTER WALTER AND CHANDLER [15].

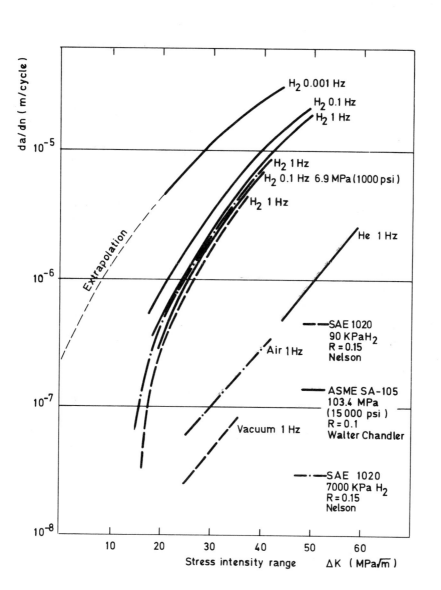

FIGURE 7 – COLLECTED CYCLIC FLAW GROWTH DATA FOR MANGANESE-CARBON MILD STEELS IN HYDROGEN GAS ENVIRONMENTS.

SAE 1020 = 0.2% C CARBON-Mn STEEL, ULTIMATE STRENGTH 500 MN/m^2

SAE 105 = 0.2% C CARBON-Mn STEEL, ULTIMATE STRENGTH 462 MN/m^2, YIELD STRENGTH 269 MN/m^3.

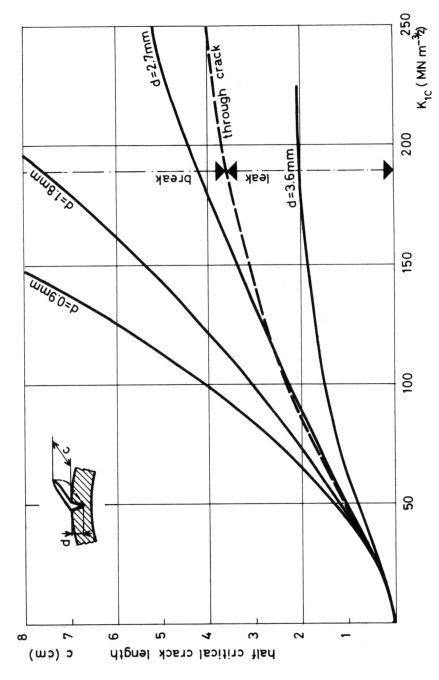

FIGURE 3 - CRITICAL CRACK LENGTH, 2C, AS A FUNCTION OF STRESS INTENSITY FACTOR FOR A TYPICAL PIPELINE IN SWITZERLAND

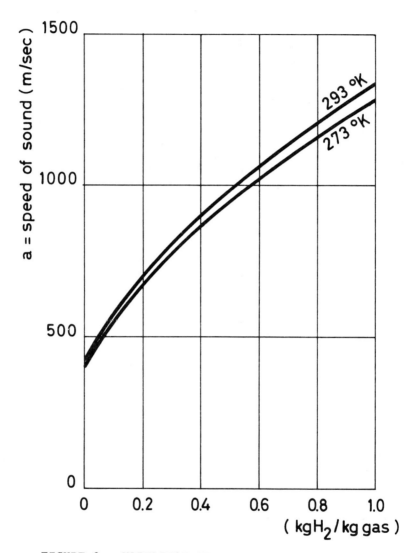

FIGURE 9 - VARIATION OF SPEED OF SOUND IN HYDROGEN:
NATURAL GAS MIXTURES.

MIXTURE PROPERTIES FOR HYDROGEN SUPPLEMENTATION OF NATURAL GAS

N.R. Baker[*]
and
W.D. Van Vorst
University of California
Los Angeles, California, U.S.A.

ABSTRACT

It has been proposed that in the future, manufactured gases, such as substitute natural gas (SNG) or hydrogen, will be required to supplement and eventually replace a rapidly diminishing natural gas resource. Such a proposal
requires that both methane and hydrogen, in varying amounts, be transmitted,
distributed, and utilized in the existing natural gas supply network. The
differences in the thermophysical and combustion properties of hydrogen and
methane are, in general, rather significant. This paper discusses the prediction of properties of hydrogen-methane mixtures, and presents the results
of computation of Joule-Thomson coefficients, flammability limits, quenching
distances and diameters, minimum ignition energies and flame stability of
mixtures of several compositions.

INTRODUCTION

As the natural gas reserves of the world are depleted, it will become necessary to either supplement the gas supplies with a manufactured gas, or replace the energy currently supplied by the gas network with another energy
form such as electricity. From at least a cost viewpoint, it seems that a
transition to a manufactured gas system is more likely. It is unclear at
this stage whether this replacement gas for residential and commercial use
would be a substitute natural gas (SNG), which is primarily methane with hydrogen added, or a complete hydrogen system. SNG implementation seems more
reasonable in the near-term future, while hydrogen might dominate sometime
in the next century.

If the natural gas supplies are supplemented with a manufactured gas, several
consequences become apparent. First, the natural gas resource obviously is
extended by whatever fraction of the pipeline gas is replaced by the manufactured gas. Second, more time is allowed for the transition to a completed
manufactured gas system.

Assuming the production of SNG is dependent upon the world's supply of hydrocarbons such as coal or oil, it seems that the SNG would have to be supplemented eventually and then replaced by hydrogen. If the gaseous fuel of the
future is in fact hydrogen, the experience and technology developed during
the transition would be invaluable. Whatever the scenario of the future, such
a proposal requires that both methane and hydrogen, in varying amounts, be present simultaneously in the gaseous fuel supply network. This implies the need
to know properties of the system: methane-hydrogen.

_____

[*] Present address - Institute of Gas Technology, Chicago, Illinois 60616.

JOULE-THOMSON COEFFICIENTS

When gas expands in a throttling process from a high pressure to a lower one, the temperature of the gas will change; this phenomenon is called the Joule-Thomson effect. For a particular gas, at a specified initial temperature and pressure, there is an associated Joule-Thomson coefficient which defines both the magnitude and direction of this temperature change.

Experimentally, a set of isenthalpic curves can be plotted on a temperature versus pressure diagram denoting the change in temperature for the given final and initial pressures. The slope of these curves is called the Joule-Thomson coefficient and is defined as:

$$\mu = (\frac{\partial T}{\partial P})_h \tag{1}$$

When this slope is negative, the gas temperature will rise upon expansion; when the slope is positive, the gas will cool upon expansion. If all the isenthalpic curves are connected at the point where their slope equals zero, an "inversion curve" is formed, denoting a region where throttling causes cooling and a region where it causes heating. The point where this curve intersects the temperature axis (i.e., at $P = 0$) is called the maximum inversion temperature. A gas having a temperature greater than this maximum inversion temperature will always show an increase in temperature during throttling. Below this temperature, the equilibrium state of the gas may fall inside the region of cooling and the gas will cool upon expansion.

In this respect, hydrogen and methane are very different. Hydrogen has a maximum inversion temperature of 202 K (−96°F) while methane has a value close to 950 K (1250°F). Thus, in the temperature region of interest here, ambient, hydrogen will rise in temperature upon expansion while the temperature of methane will fall during throttling. Since there is an abundance of pressure-reducing equipment in the natural gas delivery system now in use, the supplementation or complete substitution of hydrogen for methane will cause an increase in the temperature of the equipment itself. The importance of this effect lies in its influence on components made of materials such as rubber or plastics which are present. For this reason, Joule-Thomson coefficients were found for hydrogen-methane mixtures in order to determine the magnitude of this effect.

From classical thermodynamics,

$$dh = Tds + vdP \tag{2}$$

$$= [C_p dT - T(\frac{\partial v}{\partial T})_p dP] + vdP \tag{3}$$

or

$$dT = \frac{1}{C_p}[T(\frac{\partial v}{\partial T})_P - v]dP + \frac{1}{C_p}dh \tag{4}$$

For the isenthalpic process of expansion

$$(\frac{dT}{dP})_h = \mu = \frac{T^2}{C_p}[\frac{\partial (v/T)}{\partial T}]_p \tag{5}$$

The virial equation of state for a gas in terms of pressure is

$$\frac{Pv}{R_o T} = 1 + \frac{B(T)P}{R_o T} + \frac{C(T)P^2}{R_o^2 T^2} + \dots \tag{6}$$

or

$$\frac{v}{T} = \frac{R_o}{P} + \frac{B(T)}{T} + \dots \tag{7}$$

where B is the second virial coefficient, C is the third, etc. The derivative with respect to temperature is

$$\left(\frac{\partial(v/T)}{\partial T}\right)_p = \frac{1}{T}\frac{dB(T)}{dT} - \frac{B(T)}{T^2} + \dots \tag{8}$$

Therefore,

$$\mu = \frac{T^2}{C_p}\left[\frac{(v/T)}{T}\right]_p = \frac{1}{C_p}\left[T\frac{dB(T)}{dT} - B(T) + \dots\right] \tag{9}$$

For an ideal, perfect gas μ is equal to zero, since all virial coefficients are equal to zero. If the above equation is evaluated for a real gas at low pressures, C_p can be replaced by C_p^o, the molar heat capacity at zero pressure (identical to an ideal gas). Normally, C_p is also a function of pressure for a real gas. Keeping the first two terms of the expansion, yields

$$\mu^o = \frac{1}{C_p^o}\left[T\frac{dB(T)}{dT} - B(T)\right] \tag{10}$$

The dimensionless variables and function to be used for this analysis are:

$$B^* = B(T)/b_o \tag{11}$$

$$b_o = \frac{2}{3}\pi N_A \sigma^3 \tag{12}$$

$$T^* = \kappa T/\varepsilon \tag{13}$$

where N_A is Avogadro's number, and κ is the Boltzman constant. σ and ε are taken from the Lennard-Jones potential function for the modified hard-sphere models of statistical thermodynamics and their values are shown in Table 1. Simply, b_o is a molecular volume term, and ε the minimum energy of the potential function. Thus, the zero pressure Joule-Thomson coefficient equation becomes

$$\mu^o = \frac{2}{3}\frac{\pi N_A^3}{C_p^o}\left[T^*\frac{dB^*(T^*)}{dT^*} - B^*(T^*)\right] \tag{14}$$

For ease of notation

$$B_1^* = T^*\frac{dB^*(T^*)}{dT^*} \tag{15}$$

so

$$\mu^o = \frac{2}{3} \frac{\pi N_A \sigma^3}{C_p^{\,o}} [B_1^* - B^*] \tag{16}$$

The function $B_1^* - B^*$ (a function of T^* only), has been tabulated in Reference 1. For a binary mixture of gases 1 and 2, the equation

$$\mu_{mix}^o = \frac{x_1^2 (C_p^{\,o} \mu^o)_1 + x_2^2 (C_p^{\,o} \mu^o)_2 + 2x_1 x_2 (b_o)_{12} (B_1^* - B^*)_{12}}{x_1 (C_p^{\,o})_1 + x_2 (C_p^{\,o})_2} \tag{17}$$

is used where

$$x_i = \text{mole fraction of species i}$$

$$(b_o)_{12} = \frac{2}{3} \pi N_A \sigma_{12}^3 \tag{18}$$

$$\sigma_{12} = \frac{(\sigma_1 + \sigma_2)}{2} \tag{19}$$

and

$$\varepsilon_{12} = (\varepsilon_1 \varepsilon_2)^{\frac{1}{2}} \tag{20}$$

is used for finding T^*_{12} and thus $(B_1^* - B^*)_{12}$.

The equation for the Joule-Thomson coefficient was evaluated for mixtures of hydrogen and methane with temperatures ranging from 150 K to 1100 K; results are shown in Figure 1. It can be seen that although, at temperatures above 200 K, hydrogen is heated upon expansion, the coefficient is small (a 200 atmosphere pressure drop at an initial temperature of 300 K raises the temperature less than 6 K). For methane, however, the effect is much greater and opposite; at 300 K, a 200 atmosphere pressure drop causes an 85 K drop in temperature.

The coefficient changes almost linearly with changing mixture composition. Therefore, it seems that the use of any mixture of hydrogen and methane should present few new problems when used in pressure reducing equipment because of temperature effects.

It should also be possible to separate such mixtures by cooling and liquefaction of the methane by use of throttling with existing equipment since the effect of hydrogen is so small in comparison.

FLAMMABILITY LIMITS

The upper limit of flammability is that mixture of fuel and air in which a flame will just propagate; an increase in the fuel concentration makes the mixture non-flammable. Conversely, the lower limit has just enough fuel to sustain flame propagation. The limits are a function of the fuel, gas temperature, pressure, and the direction of propagation.

Upward propagation increases the range of flammable mixtures because of the effect of convection assisting the propagation. For example, in methane-air mixtures, the downward propagation concentration limits are 5.8% and 13.6% in air while for upward propagation the limits are 5.3% and 13.9%. The behavior of hydrogen is unusual. The upper limit, 74%, is independent of the propagation direction. Moreover, hydrogen has two lower limits. The limit for normal coherent flame propagation is 9.0%. The limit for a non-coherent flame, in which not all the fuel is consumed, and the flame propagates as a series of distinct globules, is 4.0%.[2]

The flammable range is widened by increasing the temperature of the initial mixture. The effect of pressure on the limits, except at very low or very high pressures, is small. Since the work contained herein deals with essentially atmospheric pressures, the effect, if any exists, was ignored.

Many natural gas appliances use primary mixtures which are not flammable. That is, the mixture of fuel and air supplied to the burner ports will not sustain a flame until secondary air is entrained outside of the burner itself. Since the upper limits of methane and hydrogen are very different, and the use of non-flammable primary mixtures is an important part of natural gas utilization, calculations were performed to determine the change in limits with increasing hydrogen concentration in the fuel gases.

For calculation of the limits of mixtures of the two fuel gases with air, Le Chatelier's equation may be used:

$$\text{Mixture Limit} = \frac{100}{\sum\limits_{i}^{N} \dfrac{\text{mole fraction of fuel } i}{\text{limit of } i \text{ in air}}} \qquad (21)$$

Previous work has shown that for complex mixtures of hydrogen, methane, and carbon monoxide, Le Chatelier's equation yields results with good correlation to experimental data.[3]

The data of White [4] were used to compute the increase in the limits as the initial mixture temperature is increased. The data were used to form a linear relation by the least squares method. The equations are:

$$\text{Hydrogen lower limit} = 4.0 - (T-300)/121.61 \qquad (22)$$

$$\text{Hydrogen upper limit} = 74.0 + (T-300)/37.85 \qquad (23)$$

$$\text{Methane lower limit} = 5.3 - (T-300)/245.79 \qquad (24)$$

$$\text{Methane upper limit} = 13.9 + (T-300)/104.71 \qquad (25)$$

The limits for upward propagation were used, with the non-coherent flame limit for hydrogen. Thus, the limits computed are the widest possible.

As shown in Figure 2, additions of hydrogen to methane do not produce significant increases in the upper limit. An 85% hydrogen-15% methane fuel mix is required before the limit is extended to a value midway between the limits of the pure gases in air.

At ambient temperature, the lower limits of the two gases are very close, but at elevated temperatures, the difference is increased. However, hydrogen addition causes the opposite effect as that for upper limit: the limit decreases rapidly with hydrogen addition. Yet, the closeness of the two limit values decreases this importance.

With regard to the more important upper limit, it seems that the addition of limited amounts of hydrogen to methane should not cause much change in the limit of flammability. This conclusion should be viewed from both a safety and burner operation standpoint.

CALCULATION OF QUENCHING DIAMETERS

The quenching, or extinguishment, of flames due to the presence of a solid surface may be described by two different mechanisms; thermal and diffusion. The basic premise of thermal theories is that a flame is quenched by the heat lost to the surface being in excess of that produced by the combustion of fuel in the flame. The diffusional theories have the wall acting as a sink for the destruction of various intermediate chain reaction species necessary to complete the reaction of the fuel and the oxidant. In essence, the reaction between the two can only proceed as far as the production of the intermediate species, a critical number of which are destroyed at the surface. Thus, propagation is stopped and the flame is quenched.

In general, neither theory reproduces known data more accurately than the other. However, the thermal theories are superior in one respect; they require less detailed information and assumptions. For example, in the various thermal theories, the required information consists of values for thermal conductivity, specific heat, density, flame speed and temperature, and, in one theory, an activation energy. All of these are known and readily found in the literature. The diffusional theories may require the above, plus information on the order of reaction, existence and concentration of the various intermediate species, and their diffusional rates through other species to the surface.[5,6] This data may not be known for the fuel-oxidant combination in question and can only be approximated.

Because of these reservations, only the thermal theories were investigated in order to produce values for the quenching distance, i.e., that separation of surfaces through which no flame of a given mixture composition, pressure and temperature can pass. If the fuel is specified and at atmospheric pressure, there exists an opening through which no flame can pass, regardless of the fuel concentration; this distance then is only a function of the initial mixture temperature. It is defined as the minimum quenching distance and has a unique value for each fuel gas.

The actual magnitude of the minimum opening depends upon the geometry of the surface. The separation of parallel plates is designated the quenching distance, while the diameter of a cylindrical opening is called the quenching diameter. Because of these geometric differences, the values are not the same; the quenching diameter is always greater than the quenching distance.

It is fortunate that the minimum quenching distance occurs at, or very near, the stoichiometric concentrations of fuel and oxidant; thus little accuracy is lost by computing the value at stoichiometric.

Knowledge of the quenching distance is an important part in the design of both combustion devices using a premixed fuel-air combination and safety devices used to prevent the spread of a flame from one area to another.

The method described by Mayer [7] was chosen for use in calculating the quenching diameters of hydrogen-methane mixtures because of its simplicity of assumptions and its accuracy in reproducing known experimental data as discussed later.

Two variables were used for calculating the mixture properties: mole fractions of hydrogen and methane in the fuel mix, and the initial mixture temperature. Fuel-air mixtures were always evaluated at stoichiometric.

Thermal conductivity (k), viscosity, specific heat and density data for the pure gases were correlated by computer program into polynomials as a function of mixture temperature. Mixture rules for thermal conductivity and viscosity were based on methods of statistical thermodynamics as shown in the work of Wilke [8] and Lindsay and Bromley [9].

The flame speed, S_u, for stoichiometric hydrogen-air [10] and methane-air[11] mixtures were expressed by the equations:

$$\text{Hydrogen:} \quad S_u = 1.011 \times 10^{-4} \, (T_u)^{1.721} \, (\text{m/sec}) \qquad (26)$$

$$\text{Methane:} \quad S_u = 5.416 \times 10^{-6} \, (T_u)^{1.936} \, (\text{m/sec}) \qquad (27)$$

where T_u = initial mixture temperature.

Adiabatic flame temperatures, T_f^a, were found by using linear regression analysis on existing data[2,12], yielding:

$$\text{Hydrogen:} \quad T_f^a = 1490 \, (T_u)^{0.0827} \, (^\circ K) \qquad (28)$$

$$\text{Methane:} \quad T_f^a = 2190.0 + 0.5084 \, (T_u - 300)(^\circ K) \qquad (29)$$

It was found that a linear function reproduced the data better, for methane, than a power function.

Le Chatelier's law for mixtures was used to calculate both flame speed and temperature for the fuel mixtures and has proved quite accurate in reproducing known data for both applications.[13,14]

Data for the activation energies were taken from the work of Kaskan[15] for cooled flames on porous metal burners. For stoichiometric mixtures with air,

$$E_{H_2} = 1.966 \text{ X } 10^8 \text{ J/kmol} \tag{30}$$

$$E_{CH_4} = 2.7196 \text{ X } 10^8 \text{ J/kmol} \tag{31}$$

For mixtures of the two fuels, the activation energy was taken on a mole fraction weighted basis.

Calculations were performed to determine the value of n in Mayer's equations. It was found that a value of 2.5 yielded good results for methane-air, but produced errors for hydrogen. Known data were correlated for both and produced:

$$n_{CH_4} = 2.485 \tag{32}$$

$$n_{H_2} = 2.368 \tag{33}$$

Using n = 2.5 for hydrogen gave an error increase of 10% for the quenching diameter, while for methane the error increase was only 1%. For mixtures of the two, a mole fraction weighted average correlated very well (within 0.5%).

Calculated values of quenching diameter, d_o, as a function of fuel mixture and initial temperature (assumed to equal the quenching surface temperature) are shown in Figure 3.

QUENCHING DISTANCES AND MINIMUM IGNITION ENERGIES

Berlad and Potter,[6] using one of the diffusional quenching mechanisms, developed a method of predicting the relationship between various quenching surface geometries. This relationship should be independent of the quenching mechanism used in the derivation. The two geometries of interest are the quenching distance (between parallel plates) and the quenching diameter (of a cylindrical tube). It was found that:

$$d_p = \frac{\sqrt{12}}{32} d_o \tag{34}$$

where d_p is the quenching distance.

This relationship is necessary to calculate the minimum ignition energy. Williams[16] states, "When a combustible mixture is ignited by heating a plane slab of the gas, it is found that the amount of energy (per unit surface area of the slab) added to the gas must exceed a definite minimum value for ignition to occur." On the basis of theoretical and experimental studies, a rule may be put forth as, "Ignition will occur only if enough energy is added to the gas to heat a slab about as thick as a steadily propagating adiabatic laminar flame to the adiabatic flame temperature."

Note that this gas slab thickness has already been identified as the quenching distance and is discussed in the work of Lewis and von Elbe[17,18].
Slab thicknesses less than d_p will not produce ignition, and the slab temperature will then merely decay to the ambient temperature by heat conduction.[19]

The ignition criterion may, therefore, be stated as:

$$H_i = \frac{d_p^2 k}{S_u^a}(T_f^a - T_u) = \frac{12}{32}\frac{d_o^2 k \ (T_f^a - T_u)}{S_u^a} \tag{35}$$

where H_i is the minimum ignition energy (Joules) and the superscript a denotes adiabatic. Results of calculations are shown in Figure 4. Note that
the shape of the curve is very similar to the quenching curves.

It should be noted that the minimum ignition energy is more applicable to
spark ignition processes and may be important only in relation to the replacement of standing pilot flames by automatic ignition devices.

Discussion of Results

There is considerable disagreement in published values for quenching distances, as shown in Table 2 for methane-air mixtures.

While the last entry in the above table contains more fuel than for a stoichiometric mixture, it is still very close to the minimum value due to the
nature of the "flatness" of curve at this point.

Table 3 compares results of this work with experimental values and those
calculated using the method of Lewis and von Elbe for identical mixture
properties. Mixtures are stoichiometric.

Similarly, the minimum ignition energy values are compared in Table 4.

The considerable errors associated with calculating minimum ignition energies
must be related to the inherently simple ignition criterion. Furthermore,
Lewis and von Elbe used a sperical ignition source for their model, as opposed to William's flat slab. As a result, there is a factor of π multiplied
in Lewis and von Elbe's final result. In general, however, calculations of
minimum ignition energies agree in order of magnitude, and exhibit reasonable influence by common factors such as pressure, temperature, and fuel
concentration.

It seems, therefore, that the theory of Mayer, with a little modification,
produces satisfactory agreement with experimental data. It has the advantage, over other theories, of being simple in nature, being easy to calculate, and requiring input values which are known to exist and need not be
inferred or estimated.

FLAME STABILITY

A flame becomes stabilized on a burner port at points in the gas stream where the normal flame speed equals the velocity of the gas stream. In general, this occurs at or near the boundary of the gas stream (e.g., the port wall).

Experiments with flames stabilized at the ends of long tubes have shown[22] that a certain quantity, the boundary velocity gradient, describes the tendency of that particular flame to either flashback into the tube or blow off the end of the tube and be extinguished. An equation for the boundary velocity gradient may be found using techniques of fluid mechanics. If the flow in the tube is laminar, the velocity distribution is parabolic and given by Poiseuille's equation:

$$U = \frac{-\Delta P}{4\mu L} (R^2 - r^2) \tag{36}$$

where R is the tube radius, r the radius to any point in the flow, μ the viscosity coefficient, ΔP the pressure drop over pipe length L, and U is the gas velocity at r.

The boundary velocity gradient is —

$$g = \frac{-\partial U}{\partial r}\bigg|_{r = R} = \frac{-\Delta P R}{2\mu L} \tag{37}$$

While this equation may be used in the above form, it can be simplified considerably. The volumetric flow V, is —

$$V = \int_{o}^{R} 2\pi U r \, dr = \frac{-\pi \Delta P R^4}{8 L \mu} \tag{38}$$

and therefore —

$$g = \frac{4V}{\pi R^3} \tag{39}$$

Figure 5 shows a typical stability diagram, in this case for methane.[*] There are three regions of interest: the areas of blowoff, flashback, and stable flame. Any point on either the blowoff of flashback curve is called the critical boundary velocity gradient and signifies that the instability occurs at precisely this condition. Above the blowoff curve or below the flashback curve denotes flame extinguishment or flame flashback into the primary mixture, respectively. The area in between denotes a stable flame situated at the burner port outlet, the desired condition.

---

[*] In flame stability diagrams, it is common to use a fuel concentration, F, defined as:

$$\frac{\% \text{ fuel in fuel-air mixtures}}{\% \text{ fuel in a stoichiometric fuel-air mixture}}$$

Note that this is not equal to the equivalence ratio.

The experimental data on flame stability is obtained on single long (40 to 100 tube diameters) upright tubes standing in free air at room temperature. Most burners in common usage satisfy none of the above conditions; the burner has many short, inclined ports and each port flame is influenced by the presence of its neighboring ports.

The above reservations would seem to affect the blowoff curve more than the flashback curve. Inclining the port introduces buoyancy distortion of the flame. This, in effect, will "tilt" the flame, allowing the dilution of the primary mixture by ambient air, and so the flame may blow off at a boundary velocity gradient less than the critical value. The flashback phenomena deals primarily with the port conditions, not flame conditions, and should be less susceptible to flame tilt.

The use of short ports is not as serious a problem as first considered. Wilson [23] showed that sharp-edged, short ports with non-steady laminar flow produced identical flame stability gradients as do long cylindrical tubes with steady laminar flow. The presence of other nearby ports will influence the flame and may cause the flames to partially coalesce. This is very commonly seen. But as with port inclination, this should affect the blowoff values but change the flashback gradients little.

The final reservation on the use of data obtained from a single port applied to a multiport burner is that not all ports are identical in effective diameter because of casting, drilling or lint and oil build-up differences and each port may not have identical flows and mixtures. These differences will affect the critical boundary velocity gradients for both flashback and blowoff.

The differences in combustion properties between methane (Figure 5) and hydrogen (Figure 6) are very obvious when the two are compared. In fact, the two are at the extremes; all other fuel gases in use (C_3H_8, C_2H_4, CO, C_3H_6, C_6H_6) have flame stability curves which lie in between those of hydrogen and methane.[22] In general, it can be seen that hydrogen flames are more susceptible to flashback but much less susceptible to blowoff than are methane flames.

CALCULATION OF CRITICAL BOUNDARY VELOCITY GRADIENTS

Some data exist for mixtures of hydrogen and methane. Interpolation between these known mixtures is difficult because of the size of the intervals (e.g., 7%, 26%, 46%, 70%, 84%, 94% hydrogen in methane). Therefore, Van Krevelen and Chermin's [24] method, later modified, of generalizing flame stability data was used.

Figure 7 is typical of the results obtained by using the above method and equations in reproducing known data on flashback. The correlation was excellent at fuel lean mixtures (F < 1) but very poor at richer fuel mixtures. The reason for this discrepancy is that the original theory [24] assumes that the flashback curve is symmetrical. In reality, this is essentially true for hydrocarbons, and the theory works well for those mixtures. However, by referring back to Figure 7 it can be seen that hydrogen's flashback curve is not symmetrical.

Upon examination of this discrepancy, it was shown that the asymmetrical
nature of the hydrogen flashback curve is compounded by the theory's assumption of a linearity existing in the manner the methane curve "widens" to
form the hydrogen curve. This error is especially evident at fuel-rich
conditions.

Using correlations developed from the existing experimental data, the method
of Van Krevelen and Chermin was modified to reflect the above changes. The
authors carried out calculations for mixtures of hydrogen and methane, by
10% increments. Figure 8 shows the results for the calculated critical
boundary velocity gradients for flashback and Figure 9 shows the results
for blowoff.

Evidently, as hydrogen is added to methane, the flashback curve changes
little in shape or form. The peak of the curve increases with hydrogen
content, but the right-hand branch of the parabola remains identical with
that for pure methane. The phenomenon is similar in nature to that seen
for the upper flammability limit for hydrogen-methane mixtures. The limited
effect of hydrogen addition is more clearly shown in Figures 10 and 11.
High concentrations of hydrogen are required to effect significant changes
in either the maximum critical boundary velocity gradient or its corresponding fuel gas concentration.

BURNER OPERATION

As may be inferred from the preceeding discussions, it does not seem possible to predict the maximum permissible concentration of hydrogen in methane within the constraint of acceptable burner operation. Each type and
configuration of burner in use today may have to be tested in order to determine such a limit.

Research at the British Gas Corporation [25] has indicated that this limit
may occur during start-up of a burner, as opposed to the steady-stage operation assumed in this analysis. Due to fluctuations in both mixture velocity and fuel concentration at any one port, flashback can occur during
burner initiation, but may not occur during steady-stage operation. If the
hydrogen addition limit is caused by such an action, the problem may be
solved by either better fuel and flow distribution (which involves modification or replacement of all burners in use) or some method of achieving
steady-state operation, bypassing the initial instabilities, such as de-
laying burner ignition.

CONCLUSIONS

The work presented in this paper indicates that the supplementing of methane
by hydrogen, in the fuel supply, is possible. No prediction is made on the
level of supplementation permissible. However, the following conclusions may
be made:

1. Little problem should be encountered with Joule-Thomson heating of the
    gas mixtures.

2. Hydrogen has an upper flammability limit of 530% greater than methane's. However, the addition of even 50% hydrogen increases the limit less than 70%. Thus, the upper limit is not a strong function of hydrogen concentration below an equimolar mixture.

3. Minimum quenching diameters and distances are strongly influenced by even small hydrogen additions.

4. The maximum critical boundary velocity gradient for hydrogen flash-back is 26 times greater than for methane. A mixture of 85% hydrogen-15% methane is required to effect half that increase. Few problems with blowoff should exist as the hydrogen concentration is increased.

REFERENCES

1. Hirschfelder, J. O., C. F. Curtiss, and R. B. Bird. Molecular Theory of Gases and Liquids, Wiley, New York, 1954.

2. Drell, I. L. and F. E. Belles. "Survey of Hydrogen Combustion Properties," Lewis Flight Propulsion Laboratory, Cleveland, Ohio, NACA Report 1383, April 1957.

3. Coward, H. F. and C. W. Jones. "Limits of Flammability of Gases and Vapors," U. S. Department of the Interior, Washington, D.C., U. S. Bureau of Mines Bulletin 503, 1952.

4. White, Albert G. "Limits for the Propagation of Flame in Inflammable Gas-Air Mixtures. III - The Effect of Temperature on the Limits," Journal of the Chemical Society (London) Transactions, Vo. CXXVII, Part I, pp. 672-684, 1925.

5. Potter, A. E. and A. L. Berlad. "A Relation Between Burning Velocity and Quenching Distance," Lewis Flight Propulsion Laboratory, Cleveland, Ohio, NACA Technical Note 3882, November 1956.

6. Berlad, A. L. and A. E. Potter. "Prediction of the Quenching Effect of Various Surface Geometries," Fifth Symposium (International) on Combustion, University of Pittsburg, Pittsburg, Pennsylvania, August 1954, Reinhold, New York, 1955, pp 728-735.

7. Mayer, E. "A Theory of Flame Propagation Limits Due to Heat Loss," Combustion and Flame, Vol. 1, No. 4, pp. 438-452, March 1957.

8. Wilke, C. R. "Diffusional Properties of Multicomponent Gases," Chemical Engineering Progress, Vol. 46, Nc. 2, pp. 95-104, February, 1950.

9. Lindsay, A. L. and L. A. Bromley. "Thermal Conductivity of Gas Mixtures," Industrial Engineering Chemistry, Vol. 42, pp. 1508-1511, 1950.

10. Heimel, S. "Effect of Initial Mixture Temperature on Burning Velocity of Hydrogen-Air Mixtures with Preheating and Simulated Preburning," Lewis Flight Propulsion Laboratory, Cleveland, Ohio, NACA Technical Note 4156, 1957.

11. Dugger, G.L. "Effect of Initial Mixture Temperature on Flame Speed of Methane-Air, Propane-Air, and Ethylene-Air Mixtures," Lewis Flight Propulsion Laboratory, Cleveland, Ohio, NACA Technical Note 2374, May 1951.

12. American Gas Association. Gas Engineers Handbook, Industrial Press, New York, 1974.

13. Jones, G. W., B. Lewis, and H. Seaman. "The Flame Temperatures of Mixtures of Methane-Oxygen, Methane-Hydrogen, and Methane-Acetylene with Air," Journal of the American Chemical Society, Vol. 53, No. 11, pp. 3992-4001, November 1931.

14. Payman, W. "The Propagation of Flame in Complex Gaseous Mixtures. Part II. The Uniform Movement of Flame in Mixtures of Air with the Paraffin Hydrocarbons," Journal of the Chemical Society (London), Vol. 115, pp. 1446-1453, 1919.

15. Kaskan, W. E. "The Dependence of Flame Temperature on Mass Burning Velocity," Sixth Symposium (International) on Combustion, Yale University, New Haven, Connecticut, August 1956, Reinhold, New York, 1956, pp. 134-142.

16. Williams, F. A. Combustion Theory, Addison-Wesley, Reading, Massachusetts, 1965.

17. von Elbe, G. and B. Lewis. "Theory of Ignition, Quenching and Stabilization of Flames of Nonturbulent Gas Mixtures," Third Symposium on Combustion, Flame and Explosion Phenomena, Madison, Wisconsin, September 1948, Williams and Wilkins, Baltimore, Maryland, 1949, pp. 68-79.

18. Lewis, B. and G. von Elbe. "Ignition of Explosive Gas Mixtures by Electric Sparks. II. Theory of the Propagation of Flame from an Instantaneous Point Source of Ignition," Journal of Chemical Physics, Vol. 15, No. 11, pp. 803-808, November 1947.

19. Spalding, D. B. "Approximate Solutions of Transient and Two-Dimensional Flame Phenomena: Constant-Enthalpy Flames," Proceedings of the Royal Society of London, Vol. A245, No. 1242, pp. 352-372, June 1958.

20. Lewis, B. and G. von Elbe. Combusion, Flames and Explosions of Gases, 2nd ed., Academic Press, New York, 1961.

21. Harris, M. E., J. Grumer, G. von Elbe, and B. Lewis. "Burning Velocities, Quenching, and Stability Data on Nonturbulent Flames of Methane and Propane with Oxygen and Nitrogen," Third Symposium on Combustion, Flame and Explosion Phenomena, Madison, Wisconsin, September 1948, Williams and Wilkins, Baltimore, Maryland, 1949, pp. 80-89.

22. Grumer, J., M. E. Harris, and W. E. Rowe. "Fundamental Flashback, Blowoff, and YellowTip Limits of Fuel Gas-Air Mixtures," U. S. Department of the Interior, Washington, D.C., U. S. Bureau of Mines Report of Investigation 5225, 1956.

23. Wilson, C.W. "Lifting and Flowoff of Flames from Short Cylindrical Burner Ports," Industrial Engineering Chemistry, Vol. 44, pp. 2937-2942, 1952.

24. Van Krevelen, D. W. and H. A. G. Chermin. "Generalized Flame Stability Diagram for the Prediction of Interchangeability of Gases," Seventh Symposium on Combustion, London and Oxford, England, August 1958, Butterworths, London, 1959, pp. 358-368.

25. Harris, J. A. and J. R. Wilson. "Utilisation-The Burning Issue," Institution of Gas Engineers, London, Communication 949, 1974.

TABLE 1. FORCE CONSTANTS FOR HYDROGEN AND METHANE
(Values averaged from Reference 1)

Gas	$\sigma(x\ 10^{-8}cm)$	b_o(cc/mol)	ε/K (K)
CH_4	3.806	69.565	146.08
H_2	2.919	31.369	31.18

TABLE 2. MEASURED QUENCHING DISTANCES OF METHANE
T = 300 K, p = 1 X 10^5 N/m^2 (1 atm)

$\emptyset$	Quenching Distance (cm)	Reference
1.0	0.21	20
1.0	0.25	21
1.058	0.28	17

TABLE 3. COMPARISON OF CALCULATED AND EXPERIMENTAL
QUENCHING DISTANCES AND DIAMETERS, STOICHIOMETRIC MIXTURES
T = 300 K, p = 1 X 10^5 N/m^2 (1 atm)

Value/Fuel	Quenching Distance d_p (cm)		Quenching Diameter d_o (cm)	
	CH_4	H_2	CH_4	H_2
Experimental	0.21-0.28[1]	0.064[2]	0.335[3]	--
Calculated				
a. This work	0.2205[4]	0.069[4]	0.36[5]	0.113[5]
b. Lewis & von Elbe	0.1836[6]	--	0.383[6]	--

1. See references in TABLE 2.
2. Reference 2.
3. Reference 21.
4. $d_p = (\frac{12}{32})^{\frac{1}{2}} d_o$ (Reference 6).
5. Modified method of Reference 7.
6. Method and data ($\emptyset$ = 1.058) of References 17 and 18.

TABLE 4. COMPARISON OF CALCULATED AND EXPERIMENTAL
MINIMUM IGNITION ENERGIES, STOICHIOMETRIC MIXTURES
T = 300 K, p = 1 X 10^5 N/m^2 (1 atm)

Value/Fuel	Minimum Ignition Energy (mJ)	
	CH_4	H_2
Experimental	0.3[1]	0.0188[1]
Calculated		
a. This work	0.625[2]	0.0263[2]
b. Lewis & von Elbe	3.47[1]	0.0419[1]

1. Reference 20.
2. Equation 35 (References 6, 7, and 16).

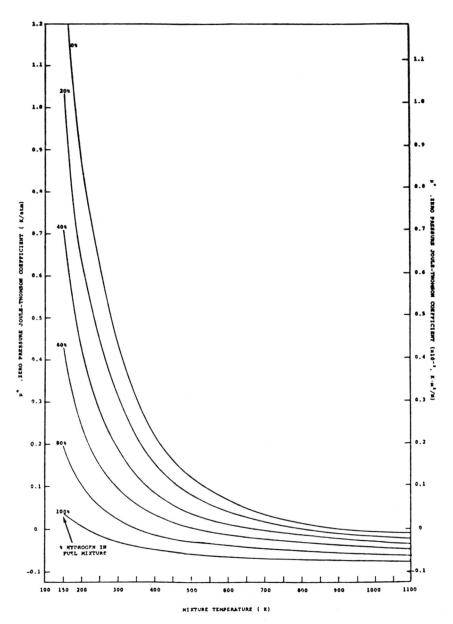

Figure 1.   Calculated Zero-Pressure Joule-Thomson Coefficient
For Hydrogen-Methane Mixtures
(SI Units on Right Scale, Conventional Units on Left Scale)

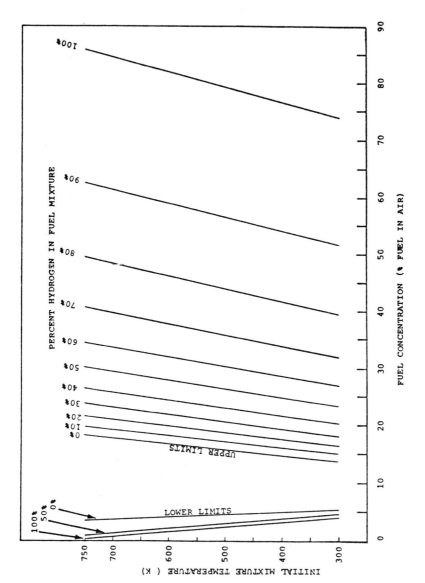

Figure 2. Calculated Upper and Lower Flammability Concentration Limits for Hydrogen-Methane Mixtures as a Function of Initial Gas Mixture Temperature. Direction of Propagation is Upward. Pressure = 1 X 10^5 N/m^2.

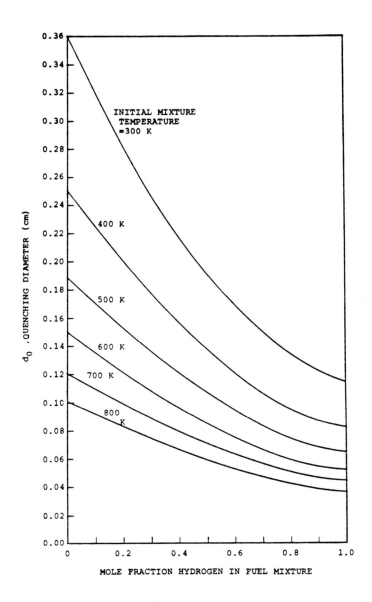

Figure 3. Calculated Quenching Diameters for Hydrogen-Methane Mixtures as a Function of Initial Mixture-Surface Temperature. Stoichiometric Mixtures With Air, Pressure = 1×10^5 N/m^2 (1 atm).

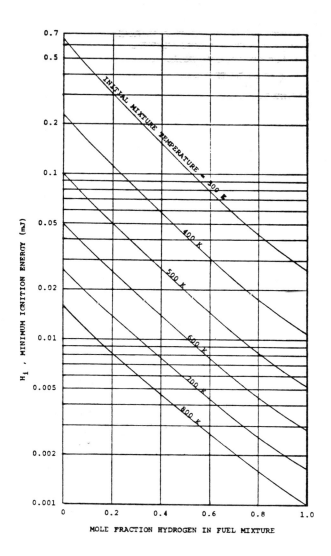

Figure 4. Calculated Minimum Ignition Energies for Hydrogen-Methane Mixtures as a Function of Initial Mixture Temperature Stoichiometric Mixtures With Air, Pressure = 1×10^5 N/m^2 (1 atm).

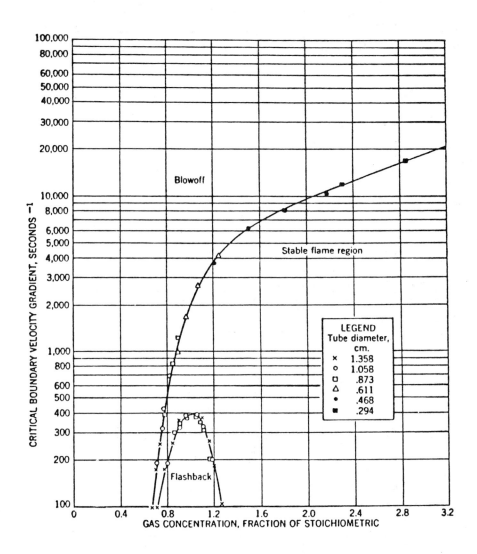

Figure 5.   Flame Stability Diagram for Methane.
Gas Temperature = 300 K, Pressure = 1×10^5 N/m^2
(From Grumer et al., Reference 22)

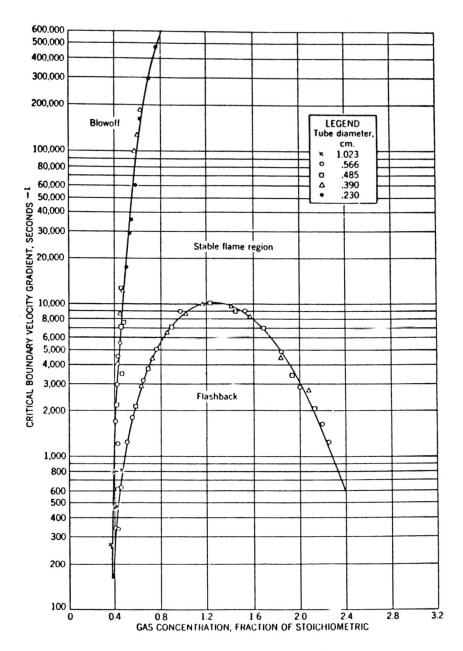

Figure 6. Flame Stability Diagram for Hydrogen (Plus 0.3% Oxygen)
Gas Temperature = 300 K, Pressure = 1×10^5 N/m^2
(From Grumer et al., Reference 22)

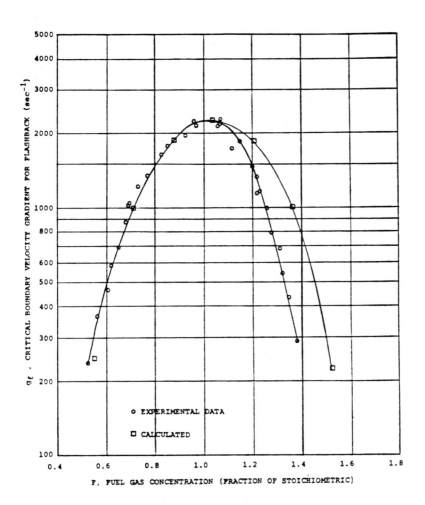

Figure 7. Comparison of Experimental (Grumer et al.) and Calculated
(Method of Van Krevelen and Chermin) Critical Boundary Velocity Gradient
For Flashback. Fuel Mixture of 70.7% Hydrogen-29.3% Methane.
Gas Temperature = 300 K, Pressure = 1 X 10⁵ N/m² (1 atm)

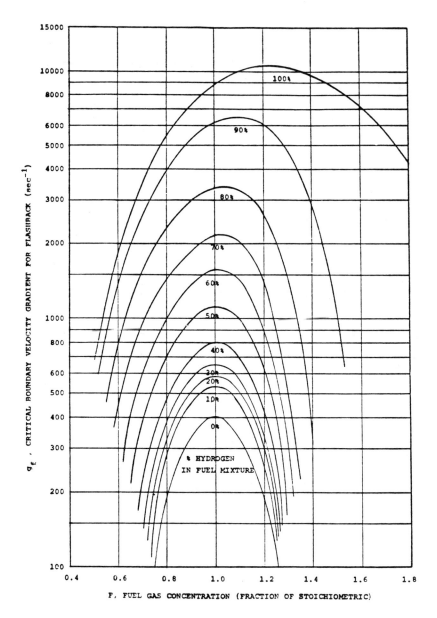

Figure 8.   Calculated Critical Boundary Velocity Gradients for Flashback
of Hydrogen-Methane Mixtures With Air.
Gas Temperature = 300 K, Pressure = 1×10^5 N/m^2 (1 atm)

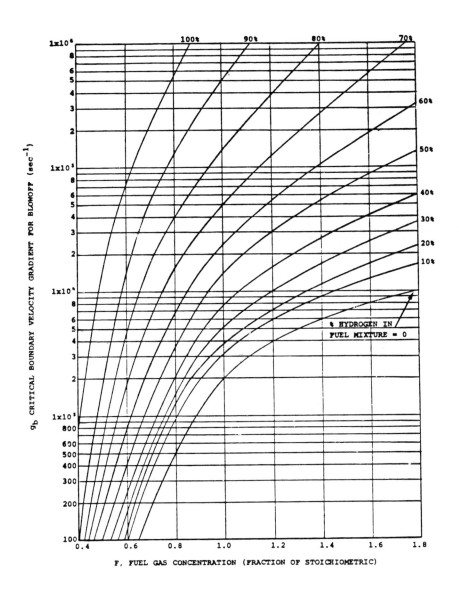

Figure 9. Calculated Critical Boundary Velocity Gradients for Blowoff of Hydrogen-Methane Mixtures With Air. Gas Temperature = 300 K, Pressure = 1×10^5 N/m^2 (1 atm)

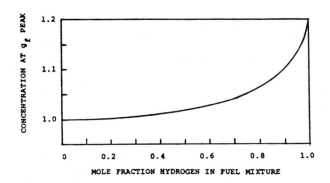

Figure 10. Calculated Fuel Gas Concentration (Fraction of Stoichiometric)
Corresponding to Peak Value of Flashback Curve for Hydrogen-Methane Mixtures.
Gas Temperature = 300 K, Pressure = 1×10^5 N/m^2 (1 atm)

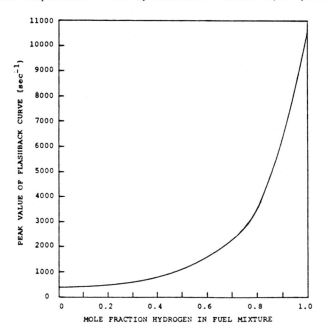

Figure 11. Calculated Maximum Critical Boundary Velocity Gradients for
Flashback of Hydrogen-Methane Mixtures
Gas Temperature = 300 K, Pressure = 1×10^5 N/m^2 (1 atm)

SMALL SCALE AMMONIA PRODUCTION AS A MEANS FOR HYDROGEN STORAGE

JP. Jourdan and R. Roguenant
Rhône-Poulenc Industries
92 408 Courbevoie, France

ABSTRACT

Solar electricity can be seasonally stored via the process chain : electrolysis - hydrogen conversion to ammonia - storage of ammonia - ammonia decomposition to hydrogen - fuel cell. The overall electric efficiency of the
chain would be about 53 % for a large and continuous power input (500 MW =
1500 t/d ammonia). For solar power stations the scale will be limited and
the efficiency will drop to about 36 %, based on the data from an existing,
non-optimized, 75 t/d ammonia plant which is described in detail.

Solar power stations based on this concept would be well suited to serve various energy needs of isolated regions all year round : electricity, gas, ammonia as fertilizer and as a portable, easy to store, back-up source of hydrogen.

The electricity and fertilizer needs of a population in the range of 10 to
50 thousand could be met by a 100 MWe peak solar power station coupled to a
75 t/d ammonia plant, the corresponding investment being in the range 1000 -
8000 $/ton of oil equivalent per year, depending on the climate and the
energy used. Specific examples are given within these ranges.

INTRODUCTION

Hydrogen gas will be an ideal energy carrier wherever a large network of gas
pipe-lines exists, coupled with natural reservoirs such as aquifers or depleted gas or coal fields, or cavities leached out from salt domes. These
conditions assume a fairly dense population over a large area and special
geological formations. Where they are not met, a condensed phase will have
to be used to move and store energy : ammonia (NH_3), liquid hydrogen (LH_2),
liquid methane (LNG), methanol or synthetic hydrocarbons.

Such liquid energy carriers will be available from energy rich regions or
from OTEC floating plants to populations having easy access to sea ports,
directly or via pipe-lines. But in many instances, imports will not be possible for geographical or political reasons, or will be strictly limited :
in such cases solar energy may become the main source of energy, either directly or via the atmosphere (wind, precipitations), and will have to be
stored to ensure a steady and safe supply over the year.

1401

ADVANTAGES OF AMMONIA AS AN ENERGY CARRIER

TABLE I lists some of the major characteristics of various liquid energy carriers. As can be seen from it, NH_3 is very similar to propane in terms of vapor pressure and very similar to methanol in terms of heat of combustion and specific gravity.

Compared to liquid hydrogen (LH_2) or liquid natural gas (LNG), NH_3 has the obvious advantage of not requiring sophisticated cryogenic technology or a large power consumption for refrigerating the storage tanks.

Compared to any of the carbon containing products, NH_3 has the important advantage of not requiring any carbon source for its synthesis. Carbon dioxide (CO_2) can be recovered from lime kilns or cement furnaces in limited quantities or can be extracted from the atmosphere at a high cost. Biomass is an excellent carbon source but cannot be grown efficiently without consuming large amounts of water (about 1000 m3 per ton of oil equivalent), which may not be available. In contrast, nitrogen is available everywhere in the air at a concentration 2000 times that of CO_2. Nitrogen can be efficiently separated from the air using well proven, clean and automated techniques.

Another advantage of NH_3 over carbon containing products is its easier conversion back to hydrogen for use in fuel cells or combustion. Instead of steam reforming followed by shift conversion of carbon monoxide and extraction of CO_2 (a poison for fuel cells using an alkaline medium), NH_3 requires only a cracking and washing. Ammonia crackers have been commercially used since 50 years to produce a 75 % hydrogen mixture with nitrogen. Hydrogen can be separated from such a mixture by differential permeation, but this is not necessary for the vast majority of energy uses.

Cracked NH_3 burns cleanly : no carbon monoxide or dioxide, no particulate matter and no more nitrogen oxides than with hydrocarbons. Use of ammonia as an engine fuel has been reported in the litterature [1].

One further, obvious advantage of ammonia is its value as a fertilizer.

HANDLING OF AMMONIA

Handling of liquid ammonia at room temperature, much like propane, is very common, bulk or in cylinders, but precautions must be taken because a ruptured line or tank will cause the formation of a heavy, toxic fog. Therefore it is recommended to store large quantities as a refrigerated liquid under atmospheric pressure. In case the tank is located near a town, it should be surrounded by a safety containment : such a tank located near Rouen has a capacity of 20,000 t (it is interesting to note that the energy content of this tank is equivalent to the annual consumption of electricity of a population of 20,000, based on the average French rate, including industrial needs).

Transportation of ammonia by tank trucks is obviously very common, but pipe-lines offer an interesting possibility : the cost of transportating energy by pipe-lines should be much lower for NH3 than for natural gas or hydrogen, respectively 1/2 and 1/3 (2).

STORAGE OF ELECTRICITY VIA AMMONIA

The concept of storing electricity via the following process chain :

- water electrolysis to hydrogen
- ammonia synthesis from hydrogen and nitrogen (separated from the air)
- ammonia storage
- ammonia cracking back to hydrogen
- fuel cells

is not a new one. A recent article by G.L. Dugger, E.J. Francis and W.H. Avery (3) contains an excellent description of this concept as applied to ocean thermal energy conversion (OTEC). Land-based solar power stations could also use the scheme, as will be discussed here, but they will have to overcome two drawbacks as compared to OTEC : non-continuous power supply and a much smaller scale, resulting in lower efficiencies.

In order to first delineate the potential of the system on a large scale, the data given in the above-mentioned reference were used to assess the follo-wing case.

Large scale seasonal storage of electricity

In this example, a 1000 MWe power station (hydroelectric or nuclear) is as-sumed to have 50 % excess capacity for four months a year, used to produce and store 180,000 t/year ammonia. A 500 MWe fuel cell peaking power plant uses the ammonia as fuel. TABLE II lists the relevant data on specific con-sumptions and investment for each section. The overall electricity to elec-tricity efficiency resulting from these data is 53 % and the total estimated capital cost in 1975 dollars is $ 215 million ($ 77/GJe = 0,21 $/kWhe).

TABLE III gives the resultant cost price of peak electricity recovered from the system as ranging from 10 to 24 $/GJ.

As can be seen, the penalty incurred in storing electricity over long pe-riods of time is heavy, even in this case where very favourable data were assumed. These data correspond to technologies that should be developed by 1990 but are not fully available yet.

Another point must be made clear : diurnal storage will not be accomplished via the ammonia route since it is impractical to shut down and restart the ammonia plant every day. Some demand adjustment is possible, but hydrogen storage should preferrably be used for short term storage since it allows a higher overall efficiency.

Seasonal storage of solar electricity

Land-based solar power stations will generally need some sort of seasonal energy storage to make use of excess power generated in summer (or in winter where irrigation and/or air conditioning are the main uses for power). The difficulty of seasonal storage is compounded here by the small capacity of the individual solar stations.

A peak power of 100 MWe has been assumed for the reference case presented in this paper. Such a capacity can result from interconnecting a number of smaller power stations in the same region. While 100 MWe central receiver power plants are seriously proposed, it is doubtful that any larger ones will ever be built.

The output of a 100 MWe station, averaged over the day, corresponds to a 75 t/d ammonia plant, at best. This capacity was taken for the reference station. It is a very small size plant, now considered quite uneconomical and inefficient.

The other components of the storage system (electrolysis, fuel cells) are modular by nature and suffer no significant penalty from the small scale.

The utilization of the solar power plant in terms of equivalent peak hours of operation per year can very between wide limits depending on the climate : in northern Europe it could be as low as 800, and in deserts it could approach 3000. It will be designated here by "h".

The electricity sold by the power station may have one of three immediate origins :

- solar power plant directly
- fuel cells burning hydrogen from the short term gas storage
- fuel cells burning cracked ammonia.

For simplicity the second one will not be considered here. The fraction of electricity made from ammonia, which will be designated by "f", may then vary over a wide range :

- f = 1 would mean that the power station is used only as a back-up for other power stations not equipped with seasonal storage
- f = 0.5 could mean that all electricity consumed during the bad season originates from stored ammonia.

Reference solar power station

Fig. 1 is a block diagram of the reference power station showing the various sections, their capacities and corresponding estimated investment. Fig. 2 lists the basic efficiencies used in the various steps.

The 100 MWe peak power plant itself can be of any type but it is actually envisioned as fixed arrays of thin-film photovoltaic solar cells with a minimum of power conditioning to fit the requirements of the electrolysis plant. The assumed cost of $ 0.5/peak watt, including site preparation, array installation, interconnection and power conditioning, would correspond to a 1990 technology.

The water electrolysis plant is sized to use the entire peak power available from the power plant. Generally part of the power would be sent to the grid directly, resulting in a lower average current density on the electrodes and a higher than nominal efficiency. Although a number of suitable electrolytic cells will be developed by 1990, the solid polymer electrolyte (SPE) system in development at General Electric (4) was chosen as an example. Eighteen 5 MW - output standard, automated SPE modules would be required and could actually be installed within the field of photovoltaic arrays so as to reduce the electric losses and cost of interconnection. Pure water would be fed from the central station : a maximum of 200 m3/day is required, and, although it might be cheaper to recover and seasonally store condensate from the fuel cells, a reverse osmosis purification system is included.

Hydrogen gas could be produced directly under the 200 bar pressure neccessary for daily storage. Otherwise a compressor handling 25,000 Nm3 has to be included between electrolysis and gas storage.

Eight hours of equivalent peak power is provided as H_2 gas storage to make it possible for the ammonia synthesis plant to operate 24 hours a day at approximately constant throughput. This was found to be cheaper than oversizing the ammonia plant. Two types of hydrogen storage were considered : standard 200 bar, 10 Nm3 special steel cylinders, and a hydride system. Although the latter offer the attractive possibility of using the heat of reaction from the NH_3 synthesis to desorb the stored hydrogen, the former system was chosen as being well known, for simplicity of estimating. In practice, however, it is felt that a better system would be available by 1990, as the present one is really cumbersome (19,000 cylinders) and heavy (1340 t).

The 75 t/d ammonia synthesis plant, including the air separation, is similar to the one described later which is an existing, well-proven plant in operation on the Saint-Auban site of Rhône-Poulenc Industries. It must be stressed here that this is not an optimized plant : it is very energy inefficient but rugged and relatively cheap to build. Wet cooling towers were assumed as an energy sink.

A 10,000 t NH_3 storage tank is provided, of the refrigerated (-34°C), atmospheric pressure type. Its capacity could be adjusted within a wide range without materially changing the overall investment, because this type of equipment benefits from large economies of scale.

Refrigeration should consume less than 1 % of the yearly net electrical output of the power station in a desert type of environment.

Thermal cracking of ammonia has been practised since 50 years to commercially prepare H_2/N_2 mixtures on a small scale. Automatic systems are sold for this purpose, but they are generally electrically heated. Cracking is the inverse of the synthesis reaction and is therefore an equilibrium-limited, catalyzed decomposition reaction. It is an endothermic reaction requiring 1.7 $MJ/Nm3H_2$. Good heat exchange must be provided between incoming NH_3 vapor and cracked gases, as the useful reaction temperatures lie in the range 600-800°C. Unreacted ammonia can be limited to 3-5 % and recovered by a simple water washing and then recycled via rectification or sold as fertilizer. As suggested in (3), the cracking furnace is heated by the dilute H_2/N_2 mixture (30 vol % H_2) purge from the fuel cells. Good discussions of the cracking of NH_3 are given in (3) and (5).

The fuel cells are assumed to be of the 1990 state of development, exhibiting a 70 % efficiency when fed with H_2/N_2 mixture (3/1 volumetric ratio) and with oxygen. Oxygen from the electrolysis could be used but this would require liquefying and storing it. An oxygen separation plant is used instead (different from the nitrogen separation plant for the ammonia synthesis, as the requirements are widely different). The capacity of the fuel cells is adjusted to the peak power demand.

Electrical storage efficiency of the reference solar power station

Fig. 3 shows the electricity to electricity storage efficiency as 40 %, based on day-time operation of the ammonia plant, i.e. with the electrical load of the ammonia plant satisfied directly by the solar power plant.

Fig. 4 shows the efficiency dropping to 35 % when based on night-time operation, i.e. with the ammonia plant using electricity from the fuel cells fed by electrolytic hydrogen.

The resulting efficiency is 36.7 % if the average day-time operation is taken as 8 hours/day. A figure of 36 % was used in the subsequent calculations.

As can be seen, there is a large scope for optimizing the technology since the corresponding efficiency of a large plant was found to be 53 %.

Ammonia yield

A similar calculation shows that the average consumption of primary electricity (from the solar power plant) is 40 GJe/t NH_3 produced.

Population served by the reference solar power station

Although a power station using the ammonia storage concept could provide all types of energy, only the two noblest ones, electricity and ammonia for fertilizer use, will be considered here.

If the French yearly 1976 consumptions per capita are used as a basis for evaluation of the average needs of a population, namely :

- 50 kg NH3 for fertilizer use
- 12240 MJ of electricity

the following formula results :

$$P = \frac{29.4\ h}{(1 + 1.778\ f) + 0.163} \tag{1}$$

where :

P = population number
h = utilization of photocells, hours/year
f = fraction of electricity sold made from ammonia

Fig. 5 shows this equation graphically. The values h = 1000 and f = 0.7 could apply to France and yield P = 12,200. The area to cover with photo-cells having an efficiency of 10 % is 82 m2 per person in this case.

In a desert where h = 3000 and f = 0.3, equation (1) yields P = 52,000. In this case, 19 m2 of photocells per person would be adequate, such that the house roofs would offer ample space to install the photocells. This leads to the concept of individual generating capacities connected to the central "storage and back-up" station via two-way electric meters.

Specific investment for the reference solar power station

The specific investment S is defined as the investment per ton of oil equi-valent produced per year by the system, on the basis of :

1 t.o.e. = 14.4 GJ of electricity (= 4000 kWh)
1 t.o.e. = 1.25 t NH3

The installed power (fuel cells) is assumed to be 1 kWe per person (equiva-lent to the French generating capacity), and the investment for ammonia cracker + fuel cells + oxygen plant and ancillary equipment is taken as $ 200/kWe output.

The total investment for the reference station is 70 + 200 P, $ million (see Fig. 1), where P is the population number.

The specific investment is then given by :

$$S = 2,675,000\ \frac{(1 + 1.778\ f + 0.163)}{h} + 235 \tag{2}$$

Fig. 6 shows equation (2) graphically. In French conditions, S turns out to be around 6000 $/(t.o.e./year), whereas in desert conditions the figure would drop to 1700.

If the specific investment were (more correctly) defined as the investment over and above the cost of the equivalent hydrocarbon based capacities ($ 300/kW for fuel oil - based power plant, and $ 230/(t NH_3/year) for hydrocarbon based ammonia synthesis plant), its value would be :
S - 370, $/(t.o.e./year).

Results when only electricity is sold are very similar.

Average cost of electric power from the reference power station

The annual costs for the operation of the power station would amount to approximately $ 13 million. TABLE IV gives the corresponding cost price of electricity as a function of the parameters f and h in the case where no ammonia is sold as fertilizer.

The cost prices range from 18 to 80 $/GJ. Considering that the electricity produced incorporates a fair amount of peak power, these may be acceptable.

If local conditions make it practical to interconnect a total of more than 1000 MW peak solar generating capacity, a central ammonia facility of the modern, efficient type could be installed (ammonia storage and fuel cells would not need to be centralized). TABLE V gives the approximate cost price of electricity in this hypothetical case, using the efficiency and investment figures of large scale operation (as per TABLE II). Resulting prices of electricity are about 1/3 lower than in the reference case.

Such prices appear to be very high, but it must be borne in mind that nuclear power stations would not be much more economical in many regions of the world where large solar power systems are likely to be installed, due to low utilization factors (low base-load).

Another point must be made : the above prices do not include the cost of interconnection, which would necessarily be very high in regions where most of the population live in small, scattered villages or farmsteads. For instance a reference power station serving 30,000 people living in 100 villages would require an investment of about $ 25 million for electrical interconnection.

Since hydrogen and ammonia are both cheaper to transport than electricity, interconnection via H_2, oxygen and ammonia pipe-lines might be the cheapest solution if the terrain permits pipe-line laying. This solution would present other advantages :

- visually non-obstrusive
- large contribution to the storage capacity
- ammonia distribution in the farmlands for fertilizer use.

Need to develop "midget" ammonia plant

As discussed in the preceding paragraph, interconnection between scattered
small consumer units is financially heavy ; in many instances it may be totally impractical. Ideally, each consumer unit should be able to make
ammonia from excess electricity himself. The unit synthesis capacity would
then have to be very small indeed, about 100 times smaller than the reference
plant, while featuring a totally automated operating mode.

Simply scaling down the investment of the reference ammonia synthesis shows
that the resulting total investment per consumer unit would not be larger
than for the reference case including the interconnection cost.

Automated, 1 ton/day ammonia plants could probably be mass produced for
$ 300,000 using existing technology.

But much research is being done to find new catalytic systems to bring down
both temperature and pressure in the ammonia synthesis reactor, down to eventually almost ambient conditions. This would make the construction of
"midget", fully automatic ammonia synthesis units possible, perhaps down to
a capacity corresponding to the needs of individual homes. For increased
safety, ammonia would then be stored in aqueous solution. A system adequate
for an individual home could have the characteristics given in TABLE VI
showing a cost of about $ 9,000 for an annual useful electric production of
36,000 MJ.

DESCRIPTION OF AN EXISTING 75 T/D AMMONIA PLANT

As already mentioned the plant which is going to be described here is by no
means an example of high efficiency, modern technology and it is not optimized for high cost energy conditions. But it is rugged, well proven and it
has the same small capacity as the ammonia plant of the reference solar
power station described in the preceding chapter. As such, it may offer some
interest in the context of this Conference.

Historical background

The ammonia plant is located in the chemical factory of Rhône-Poulenc
Industries at Saint-Auban sur Durance, South-Eastern France. On the same
site was installed the first industrial ammonia synthesis plant using the
Casale process in 1920, to make use of the hydrogen by-product from the
electrolysis of sodium chloride brine. Since 58 years the capacity of
ammonia production on the site was gradually increased from the initial
5 t/d to the present 75 t/day, as the chlorine production itself was increased. Many improvements were made to the original process in the course
of time.

General description

Fig. 7 is a block diagram showing the various sections of the ammonia plant :

- nitrogen separation from air,
- storage of hydrogen and nitrogen,
- mixing of the two gases in the required proportion,
- compression to 650 bar and purification of the gaseous mixture,
- synthesis loop
- storage of ammonia.

Nitrogen separation

Nitrogen is extracted from air using the Claude process. Filtered air is compressed to 15 bar and washed with caustic soda to remove carbon dioxide, dried over potassium hydroxide, and carefully filtered to remove oil traces. After heat exchange with the nitrogen and oxygen coming from the distillation column at respectively - 194 and - 180°C, and partial condensation, uncondensed air is expanded to 5 bar in an engine, generating electricity, before entering the Air Liquide 2 - stage distillation column.

Nitrogen is produced at a fairly high purity (50 vpm oxygen) but no effort is made to obtain pure oxygen which exits as a by-product containing 5 % nitrogen.

The separation plant is oversized relative to the ammonia plant : 43 % of the nitrogen produced is used for other purposes in the factory.

It is important to note that it takes about 20 hours to start up the nitrogen separation section after a shut-down : this is the time required to cool the equipment down to its operating temperature. Although the air compressor is oversized to 1350 HP to accelerate the cooling during start-up, obviously it must be operated on a continuous basis.

The nitrogen separation unit has to be shut down twice a year to defrost the equipment.

During shut-downs, nitrogen is made by burning hydrogen in air in a special combuster which can be started in half an hour and is quite flexible. In order to limit the flame temperature so as not to produce nitrogen oxides, 60 % of the nitrogen is recycled around the combuster using a Roots compressor. The gases are cooled by water washing without recovering the heat of combustion. The nitrogen produced contains about 1 % hydrogen and 0,1 % oxygen which must be removed by catalytic deoxidation. It is interesting to consider that the ammonia plant could be operated without an air separation unit, by burning 17 % of the hydrogen feed : if the heat of combustion were used to power a highly efficient Stirling engine, and if the electricity recovered were recycled to the water electrolysis, the extra-consumption of hydrogen would be reduced to 10 %. Deducting the electricity saved by not operating the air separation plant, this figure would drop to 7.5 %. This shows that the air separation unit is much cheaper than the combustion.

Storage of hydrogen and nitrogen

Buffer storage under slightly higher pressure than atmospheric is provided :
1200 m3 for hydrogen (12 minutes operating time) and 800 m3 for nitrogen.

Mixing of hydrogen and nitrogen

A very precise system is used for controlling the volume ratio of the two
gases within 0.3 % of a value near the stoechiometric value of 3. This is
because there is normally no purge of inert gases from the synthesis loop.
Two in-line analyzers measure the composition of the feed and of the recycle loop and control the flow of nitrogen via a ratio relay.

Compression and purification

The mixed gas stream is compressed by two 6-stage Burckhardt compressors to
a final pressure between 580 and 650 bar (depending on operating conditions).
The heat of compression is removed between compression stages, using water
recycled from a wet cooling tower. Between the second and third stages, the
mixture is purified from oxygen traces (by catalytic deoxidation at 280°C)
and from carbon dioxide traces (by washing with a solution of caustic soda).

Synthesis loop

The synthesis reactor is schematically depicted on Fig. 10. Its outside cylindrical shell is made of a special nickel-chrome-molybdenum steel with the
following dimensions : 9.3 m height, 0.9 m outside diameter and 0.55 m inside diameter. The total weight of the reactor is 38 tons. The catalyst is
iron made by low temperature reduction of Fe_3O_4 containing 3 % alumina.
The volume of catalyst used is 1.1 m3 (2.4 tons).

The electrical resistance in the middle of the reactor is used only for
start-up. In normal operation (75 t/d), the temperature of gases at the
outlet of the reaction zone is 550°C, and the incoming mixture is preheated
to 450°C by heat-exchange, while the gases exiting the reactor are at 250°C.
The minimum operating rate without electrical heating is 11 t/d (heat losses
compensated by the exothermic reaction).

About 1/3 of the hydrogen reacts per pass. Between each pass, the gases are
cooled and all of the ammonia produced is condensed (cooling water temperature is under 18°C).

The uncondensed gases are recycled via an ejector (Fig. 11) powered by the
gaseous mixture from the compressors. The pressure at the inlet to the reactor is 560 to 580 bar (depending on reaction conditions).

Storage of ammonia

The plant is equipped with six tanks containing each 20 tons of ammonia at
ambient temperature (under 10 to 15 bar pressure). The liquid level inside
each tank is monitored with precision using a radioactive source.

Tank-trucks and tank-cars are loaded by displacing the liquid by a vapor compressor.

Ammonia vented from tanks and lines is recovered as aqueous solution.

Economic data

The manpower to run the ammonia plant continuously is 21 shift operators plus two head-operators.

The total electrical consumption including nitrogen separation and cooling water is 7200 MJ/t, 90 % of which is taken by the Burckhardt compressors.

The hydrogen consumption is 2000 Nm3/t. Catalysts and chemicals amount to $ 1.6/t.

Technical level required

Although the plant is not highly sophisticated, the point must be made that a relatively high qualification of the technical environment is necessary for successful operation and maintenance of the high pressure equipment used. But with suitably trained technical personel, most of the major maintenance work can be done locally.

CONCLUSION

It is encouraging to realize that seasonal storage and/or transportation of electricity or hydrogen could be implemented with existing technologies without any dependence on geological or other local constraints.

But the price to be paid is high, especially on small scale.

Future technological developments on water electrolysis and hydrogen fuel cells and the discovery of low pressure, low temperature, practical catalytic systems for ammonia synthesis should make it possible to mass-produce small scale equipment for economic local seasonal storage of hydrogen as ammonia.

Research towards this goal should be encouraged.

REFERENCES

1. R.L. Graves, J.W. Hodgson and J.S. Tennant, Proceedings of the Hydrogen Economy Miami Energy Conference (1974), paper S8-15

2. R.A. Reynolds and W.L. Slager, ibidem, paper S2-1

3. G.L. Dugger, E.J. Francis and W.H. Avery, Solar Energy, 20, 259 (1978)

4. L.J. Nutall, A.P. Fickett and W.A. Titterington, Proceedings of the hydrogen Economy Miami Energy Conference (1974), paper S9-33

5. P.O. Carden, Solar Energy, 19, 365 (1977)

TABLE I

MAJOR CHARACTERISTICS OF VARIOUS
LIQUID ENERGY CARRIERS

Characteristics	Units	Hydrogen	Methane	Propane	Ammonia	Methanol
Specific gravity	kg/l	0.071	0.425	0.58	0.82	0.79
Boiling point	°C	-252.6	-161.4	-42	-33.4	64.7
Vapor pressure at 50°C	bar	$(> t_c)$	$(> t_c)$	20	20	< 1
Heat of combustion of gas (high)	MJ/kg MJ/l	141.9 10.1	55.4 23.6	50.3 29.2	22.4 18.4	23.8 18.8
Heat of vaporization	MJ/kg	0.5	0.5	0.3	1.4	1.1

TABLE II

DATA FOR LARGE SCALE SEASONAL
STORAGE OF ELECTRICITY

Section	Capacity	Specific consumptions	Investment 1975 $ million
Electrolysis (1)	500 MWe	13.3 MJe/$Nm3H_2$	44
Ammonia synthesis (2)	1500 t/d	2 $Nm3H_2$ + 0.56 MJe/kg NH_3	38
Ammonia storage	180,000 t	50 TGe/year (3)	43
Fuel cell (4)	500 MWe	69 kg NH_3/GJe	90
			215

Notes

(1) General Electric "SPE" system
    Includes reverse osmosis for pure water feed

(2) Includes N_2 separation

(3) Assumes an average outside temperature of 24°C ; 1 TJ = 10^6 MJ

(4) Includes O_2 separation and ammonia cracker, heated with the H_2
    purge from the fuel cells.
    Assumes 100 % cracking rate (actually 3 to 4 % uncracked ammonia
    is recovered as a solution sold as fertilizer)

TABLE III

COST PRICE OF ELECTRICITY FROM
LARGE SCALE SEASONAL NH3 STORAGE

Cost Price = $1.38 + 1.89 p + \dfrac{33805}{h}$				
p \ h	2000	3000	4000	5000
1	20.2	14.5	11.7	10.0
2	22.1	16.4	13.6	11.9
3	24.0	18.3	15.5	13.8

Where :

h = utilization of the peak capacity of the ammonia plant, hours/year

p = cost of excess electricity, $/GJ

Basis :

- Chemicals, labor, overhead at $ 20/t NH$_3$
- Fixed charges at 15 % of investment (including maintenance)
- Annual production : 954 h, GJ/year

TABLE IV

COST PRICE OF ELECTRICITY FROM THE
REFERENCE SOLAR POWER STATION

Cost Price in $/GJ			
h f	1000	2000	3000
0.3	54	27	18
0.5	67	34	22
0.7	81	41	27

Where :

h = utilization of the peak power of photocells, hours/year

f = fraction of electricity sold made from ammonia

Basis :

- 30 MWe fuel cells

- Annual costs	$ million
. Operating personel	1.0
. Catalysts, chemicals, overhead	0.4
. Fixed charges at 15 % of investment (including maintenance)	11.4
	12.8

- Annual production (electricity alone, no NH_3 sold) $\dfrac{360\ h}{1 + 1.778f}$, GJ/year

TABLE V

COST PRICE OF ELECTRICITY FROM SOLAR
POWER PLANTS INTERCONNECTED TO A LARGE
CENTRAL AMMONIA SYSTEM

	Cost Price in $/GJ		
f \ h	1000	2000	3000
.3	35	18	12
.5	40	20	13
.7	45	23	15

Where :

h = utilization of the peak power of solar cells, hours/year

f = fraction of electricity sold made from ammonia

Basis :

- 1700 MWe solar cells (investment $ 850 million)
- 1500 t/d ammonia synthesis plant
- 500 MWe fuel cells
- Annual costs .... $ million
  - operating personel 4
  - catalysts, chemicals, overhead 6
  - fixed charges at 15 % of investment
    (including maintenance) 160
    ___
    170

- Annual production (electricity
  alone, no NH_3 sold) .... $\dfrac{6120\ h}{1 + 0.887\ f}$, GJ/y

TABLE VI

POSSIBLE SPECIFICATIONS OF AN INDEPENDENT
ELECTRICAL SUPPLY SYSTEM FOR AN INDIVIDUAL HOME

Item	Capacity	Cost, $
Solar cells	7.5 kW peak	3,750
Water electrolysis	7.5 kW	1,000
H_2 gas storage	10 hours peak = 20 Nm3 (2 cylinders)	250
Ammonia synthesis	8 kg/day	2,200
Ammonia storage and rectification	750 kg = 2.5 m3 as 33 % solution	900
Ammonia cracker and fuel cell	3 kW	900
		9,000

Basis :

- Overall electricity to electricity efficiency = 50 %
- Fraction of electricity consumed made from ammonia, f = 0,5
- Utilization of peak power of solar cells, h = 2000 hours/year
- Annual consumption of electricity 36,000 MJ (10,000 kWh)
- Ammonia storage equivalent to 10,000 MJ
- Rectification of aqueous ammonia uses by-product heat from fuel cell or a heat pump

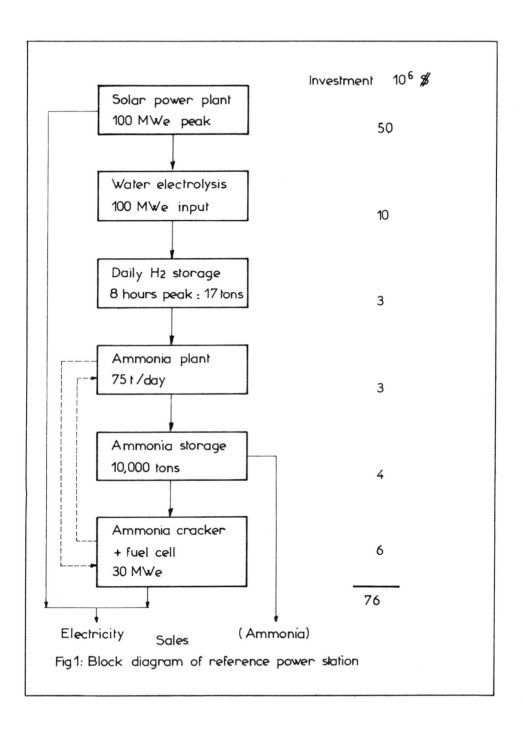

Fig 1: Block diagram of reference power station

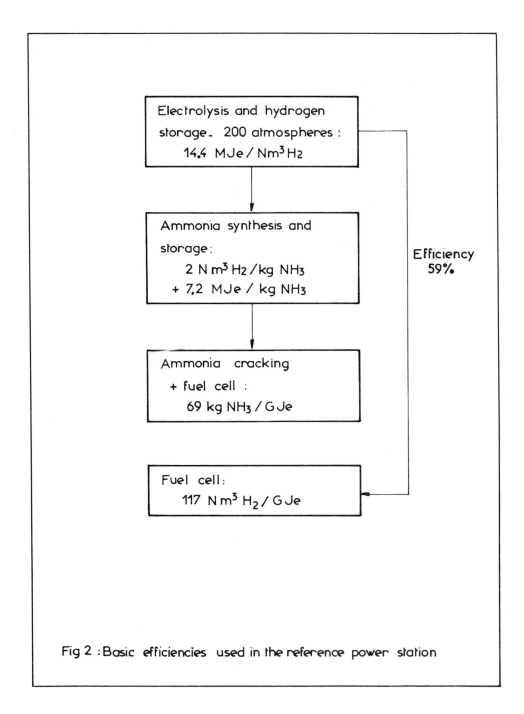

Fig 2 : Basic efficiencies used in the reference power station

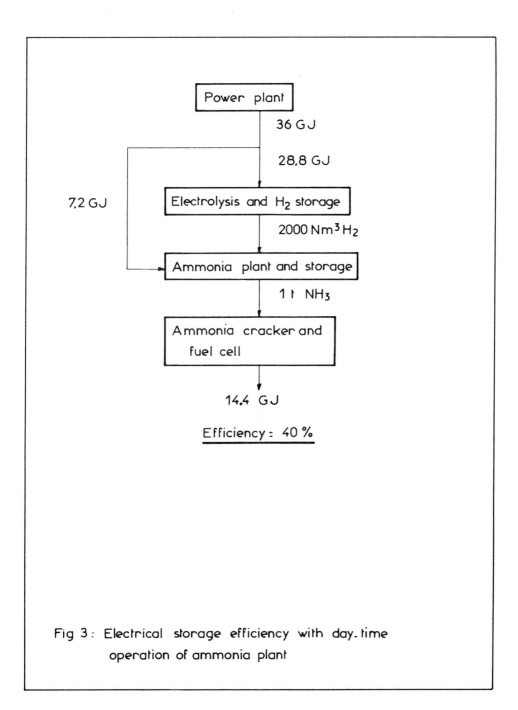

Fig 3 : Electrical storage efficiency with day.time
operation of ammonia plant

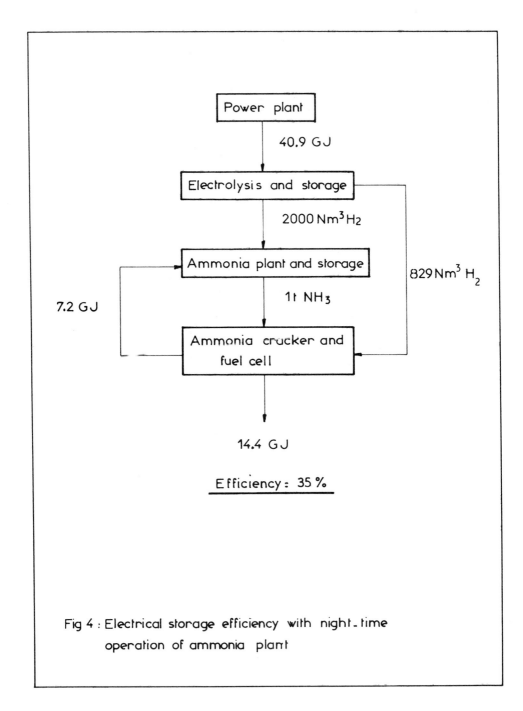

Fig 4 : Electrical storage efficiency with night_time
operation of ammonia plant

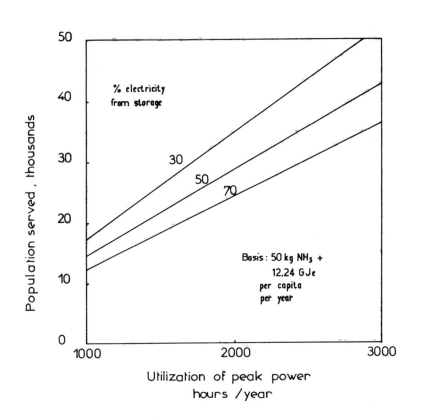

Fig 5 : Population served by a 100 MWe peak solar power
station producing both electricity and ammonia
for fertilizer use

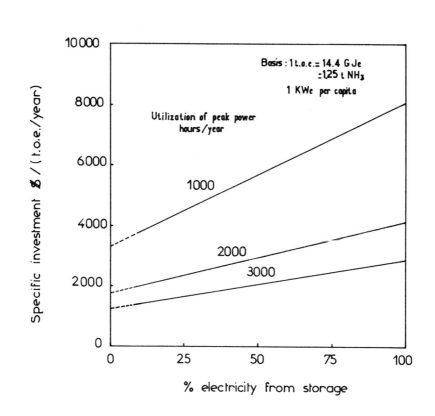

Fig 6 : Specific investment for a 100 MWe peak solar
power station producing both electricity and
ammonia for fertilizer use.

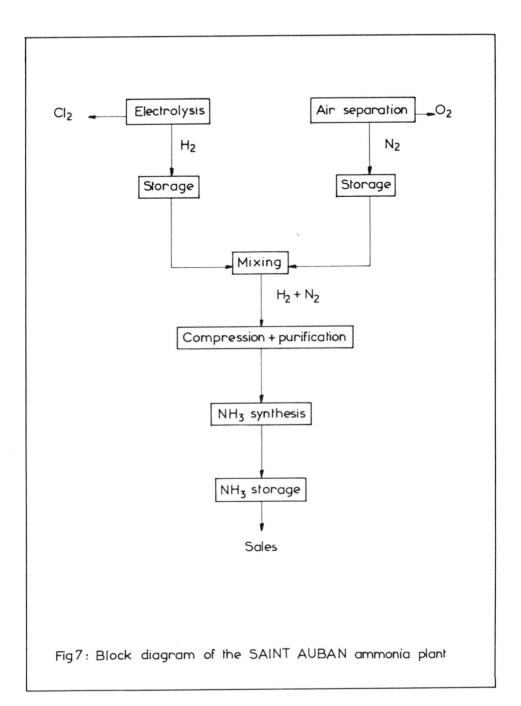

Fig 7 : Block diagram of the SAINT AUBAN ammonia plant

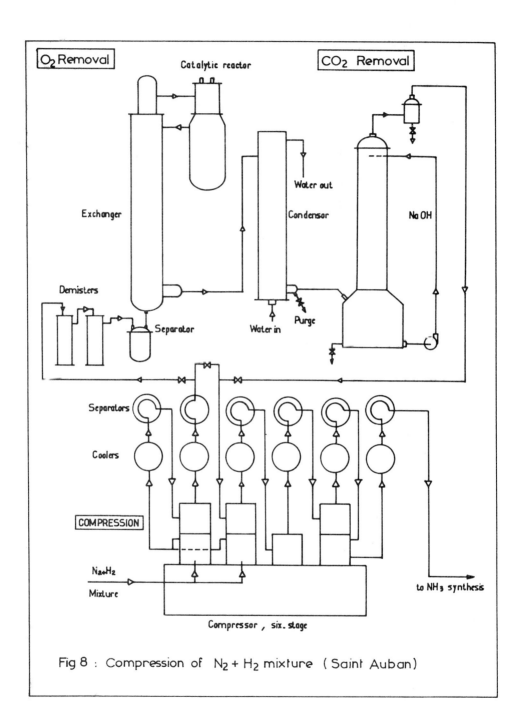

Fig 8 : Compression of $N_2 + H_2$ mixture (Saint Auban)

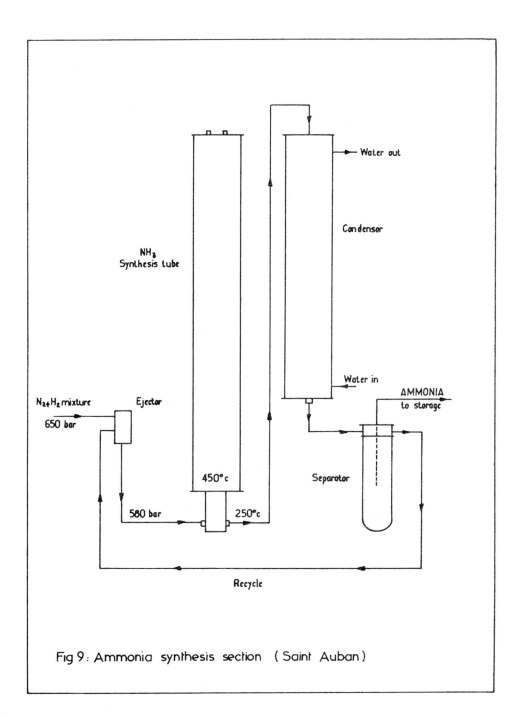

Fig 9: Ammonia synthesis section (Saint Auban)

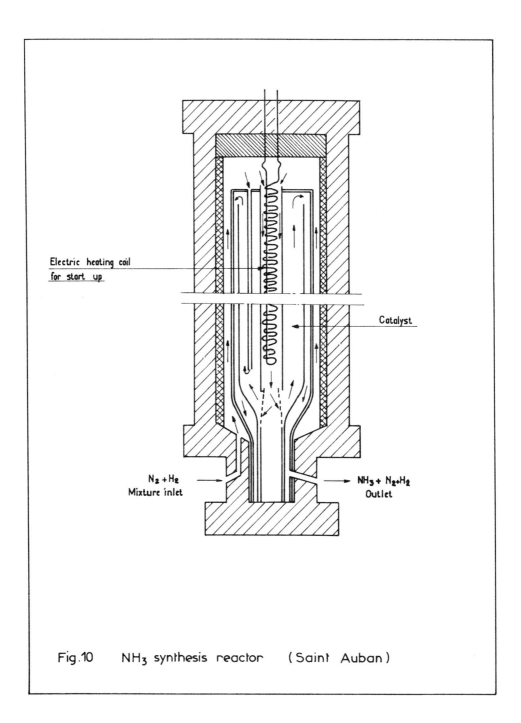

Electric heating coil
for start up

Catalyst

$N_2 + H_2$
Mixture inlet

$NH_3 + N_2 + H_2$
Outlet

Fig.10 NH_3 synthesis reactor   (Saint Auban)

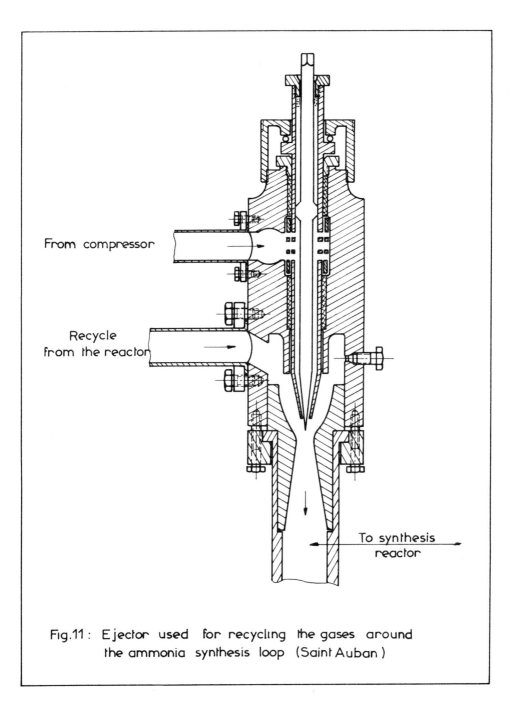

From compressor

Recycle
from the reactor

To synthesis
reactor

Fig.11 : Ejector used for recycling the gases around
the ammonia synthesis loop (Saint Auban)

HYDROGEN STORAGE I

A STUDY ON HYDROGEN STORAGE BY USE OF

CRYOADSORBENTS

C. Carpetis and W. Peschka
DFVLR-Institute for Technical Physics
Stuttgart, F. R. Germany

ABSTRACT

The paper reports investigations on the techniques and economics
of hydrogen storage by means of cryoadsorption. Also a comparison with alternative storage methods is included. The hydrogen
storage capacity of several adsorbents in the temperature range from 65 K to 150 K has been investigated experimentally.
Basing on these data economics and operating conditions for minimum total costs of the system are calculated. Utilization-
factor and capacity-factor parameters are shown to be decisive
for outlining the favourable ranges of application for competitive hydrogen storage methods.

INTRODUCTION

The importance of secondary energy storage is generally recognized. The storability of Hydrogen as a carrier of secondary
energy is widely cited as one of its dominant advantages. Concerning the storage of large amounts of hydrogen for buffering
the daily and seasonal difference between production capability
and demand, it seems to be no doubt that compression of the gas
in underground cavernes is a technically feasible and economically favourable storage method. However, there is little doubt
that the availability of underground cavernes will not cover
the demand for storage capability. Consequently further hydrogen
storage methods, e.g. overground compressed gas storage, liquefaction and storage in metal hydrides has been examined as alternatives. The authors have proposed in /1/ to take into account also the use of cryoadsorbents (i.e. storage of hydrogen
in cryogenically cooled vessels containing adsorbent material)
for large scale hydrogen storage. In this paper preliminary results of the related study are reported. The main objective of
the paper is the examination of techniques and economics of
the cryoadsorption storage. However, the economics of the alternatives mentioned above are also considered. An attempt is made
to get a well founded base of comparison between the different
alternatives. It must be emphasized that the comparison of economics between storage facilities involves more significant parameters, than the comparison between different power plants.
This is mainly due to the fact, that the costs of the storage

1433

facility must be splitted off into a power dependent part and into an energy capacity dependent part. The relation between these two cost factors is a crucial parameter for the comparison of the different storage methods.

THE RESEARCH PROGRAM

The results presented in this paper have been obtained during an one-year research program supported by funds of the European Community Commission. The research program has included the following points:

a) an overview-examination of the state of the art on adsorbbing materials suitable for storage of hydrogen in large quantities at cryogenic temperatures.

b) concept definition of a cryoadsorbent hydrogen storage facility as a part of a distribution system and investigation of the related components or modules.

c) investigation of the economics of this system and comparison with alternative storage systems

d) experimental work to support the above investigations.

The initiation of this work was based on the fact that the adsorption capacity of adsorbents is drastically raised at cryogenic temperatures and moderate pressures. Since the hydrogen gas has to be cooled down only to e.g. 80 K, the installed cooling capacity and energy consumption is substantially lower than in the case of hydrogen liquefaction. Thus, advantages can be expected in comparison with the liquid hydrogen storage. Advantages in comparison with hydride storage can be expected from the fact that the stored hydrogen per unit mass of adsorber is higher than in the case of large scale hydride storage. Further the costs of the adsorber material per unit mass are expected to be lower than for metal alloys suitable for large scale hydride storage. Finally, for compressed gas storage the volume and/or the operating pressure of the storage vessels represent a much more adverse cost factor than for the previous alternatives. To clarify this situation, the adsorption-desorption isothermes of hydrogen should be available for the conditions considered. Thus, experimental work for the evaluation of this characteristic for selected materials was a substantial part of the research program.

CRYOADSORBENT MATERIAL INVESTIGATION

General

For the first period, the selection of suitable materials for cryoadsorption was supported by a literature research. The result was that "classical" adsorbents, for example activated carbon and silicagel variations, as well as variations of the ortho-para-catalyst of the type NiOSiO$_2$ have been included in the experimental programme. For the later material (the well known catalyst "Apachi") a very high hydrogen adsorption capability at cryogenic temperatures has been cited in recent reports [2], [3]. The experimental investigation of the adsorption isotherms of hydrogen at several temperatures in the region 65 to 150 K led to some surprising results:

a) The adsorption capability of the NiOSiO$_2$ catalyst was found to be very low and not much different from the capability of the silicagel group.

b) The silicagel adsorption capability was lower than this of the activated carbon and not adequate for the technical application considered here.

c) The adsorption of many activated carbon specimens was higher than generally reported [4], [5].

Concerning the nickeloxyde-silicate catalyst the measured values were found to be in good accordance with the original measurements [6]. These original measurements have been repeatedly cited, unfortunately with a scale error introduced in newer reports. Literature values for activated carbon ([4], [5], as above mentioned) concerned only the dense carbon of "Pittsburgh" type. In our experiments, high adsorption values were achieved by the very porous type F12 of DEGUSSA. More dense activated carbon types showed lower adsorption capability both on volume (grams H$_2$ per cm^3) or mass (grams H$_2$ per gram of adsorbent) basis. Because of these results, the second series of experiments has been concentrated on the investigation of diverse very porous variations of activated carbon only.

Experimental device and methodics of the investigations

The main objective of the experiments has been the investigation of the adsorption-desorption isothermes of hydrogen on the selected materials in the temperature range of 65 to 150 K. Beside this, also the registration of kinetic phenomena during charging-discharging cycles and the influence of large number of cycles on the behaviour of material should be enabled. This implied that the apparatus should involve possibly large adsorber volume and an automatic device for performance of the process cycle and registration of the experimental results. In the experimental device the adsorber volume consisted of a group of six cylinders of a total volume of 1200 cm^3. The system could be immersed in a dewar filled with liquid nitrogen (Fig. 1). In this case the isotherme temperature was that of liquid nitrogen. Lower temperatures could be achieved by evacuation of the space above the liquid nitrogen and control of the vapour pressure. For higher temperatures a device was used (incorporated in the LN_2-dewar) for controlling a flow of cold nitrogen vapour along the adsorber cylinders. Pressure and temperature sensors were located at several points, both in the adsorber volume and in the feeding system. The hydrogen flow to and from the adsorber volume (during charging resp. discharging-periods) was controlled by a system of electric activated flow-control valves. The hydrogen flow rate was measured by a Hastings-flowmeter and the output signal integrated in time. Therefore, the flow conditions and the total hydrogen amount in the adsorber volume were registered. The process control, the registration and storage of data, and the output of results was performed by a data processing unit incorporating a DEC 11/04 minicomputer. In addition to measuring and controlling, the data processing system performed during the experiment a rather large number of calculations:

a) calibration corrections (e.g. the flowmeter has a pressure dependent calibration characteristic) and conversions to technical units from the sensor voltages

b) calculations of adsorption data (which of course cannot be measured directly) from the measured data, taking into account the dead volumes involved. The later have been carefully measured prior to the experiment series.

Fig. 2 is a diagram of the experimental device. A charging-discharging cycle was performed as follows: Beginning at the initial pressure level, hydrogen at a controlled flow rate was fed into the adsorber volume. When the measured pressure has achieved a preset value the feeding valves were closed. In general, the pressure diminished slightly because of adsorption hysteresis. After steady state was achieved, the corresponding measured data were edited in a protocol form (corresponging to one isotherme point) and stored in the data memory for later evaluation. Then the feeding procedure was continued in a similar way (stepwise) up to the maximum pressure level (typically 42 bar). The discharge procedure took place in a similar manner by a stepwise reduction of the pressure. The data corresponding to these steps were used (after the experiment) to plot the isothermes and for other evaluations. Fig. 3 shows an example of a record of such data (flow rate, total hydrogen flow, pressure, temperature) for a charge-discharge cycle. Adsorbent specimens used for the first time in an experiment series were carefully treated by heating them under vacuum (in the adsorber cylinders), so that reproducible "regenerated" initial states for all specimens were achieved.

Experimental results

The plots of the adsorption isotherms show the amount H_a of isothermally adsorbed hydrogen per unit mass of the adsorbent (or per adsorber unit volume) as a function of pressure. For a hydrogen storage device not only this quantity (that is the amount in the so-called micropores of the adsorbent accessible to adsorption forces) but rather the entire gas quantity H_s in the adsorber vessel (including the gas between the adsorber particles) must be considered. Both quantities, H_a and H_s, are therefore given in the plots. In the technical storage application (including charge-discharge cycling of pressurized adsorbent) only part of the maximally stored amount H_s can be obtained from the storage vessel by discharging isothermally to the lower pressure level, which will be normally higher than one bar. This fraction of H_s, denoted as H_r, is the gas amount reversible stored during an isothermal charge-discharge cycle between the considered pressure levels. A systematic description of the experimental results for all materials and temperatures cannot be given within this paper. A few results are given in the following and a more comprehensive discussion in the next subsection. Table 1 gives a survey of the stored hydrogen mass (in grams per kg of the adsorbent) at 78 K after pressurization at 42 bar for several adsorbents investigated. Fig. 4 and 5 show the pressure dependence of adsorbed (H_a) and stored (H_s) hydrogen amount during charging and discharging at 78 K. The curve W in these diagrams is a plot of the hydrogen amount released during the discharge procedure. The maximum value of W represents the reversible stored hydrogen amount $H_{r,max}$, corresponding to the upper pressure level. Fig. 4 corresponds to the dense activated carbon with lower performance, Fig. 5 to the activated carbon type F12/350 with the best performance. Fig. 6 shows the same curves as Fig. 5 for isothermal conditions at 65 K. The influence of temperature level can be seen from Fig. 7. In this figure the isothermes for the stored hydrogen amount H_a (charging) and of the released amount W (discharging) are given for three temperature levels. A summary of the results for a great number of activated carbon adsorbents can be depicted with the plots given in Fig. 8. The coordinates are grams of hydrogen per kg of material (Y-axis) and grams of hydrogen per liter of the adsorber volume (X-axis). The points shown are corresponding to the stored amount after charging to 42 bar (not filled signs) or to the amount released after discharge from 42 bar to about 2 bar (filled signs). Note that all points for one material are on a line, the slope of which corresponds to the density of the adsorber material.

Discussion of the results

A summary of the experimental results can be given as follows:

1) From all investigated materials activated carbons show the best performance. For technical hydrogen storage the properties of investigated materials other than activated carbon seem to be unadequate.

2) Among the active carbon types, there are substantial differences in performance. The adsorption capability of dense carbon is considerably low. Therefore, the use of these carbons does not yield any advantage, also on a volume basis. Best performance could be achieved with the carbon types of highest effective porosity.

3) The characteristic "saturation form" of the adsorption isotherme (meaning a substantial part of adsorption activity is observed at low pressures) implies that a part of the whole amount of stored hydrogen is still remaining in the adsorber vessel after an isothermal discharge. Even so, the hydrogen storage yield per cycle between 42 bar and 2 bar is about 4.2 percent (at 78 K) or 5.2 percent (at 65 K) related to the adsorbent mass.

4) The storage capability is increasing for lower temperatures. The increase in the region from 150 K down to about 80 K seems to be rather linear, but the rate increases notably below it down to the lowest measured temperature of 64 K.

5) Observed hysteresis and kinetic effects are small and have no importance for the storage cycle.

THE ECONOMICS OF THE STORAGE SYSTEM AND COMPARISON WITH ALTERNATIVE HYDROGEN STORAGE METHODS

General remarks

Before examining the cost situation it is necessary to discuss in detail some important peculiarities involved in the cost calculation of energy storage systems. As mentioned before, this involves more parameters than for cost calculations of power stations. Considering for example an electrical power station, the cost situation is well defined by the cost of produced energy unit, e.g. DM/kWh and the utilization

$$A = N \cdot H \text{ (hours per year)} \qquad (1)$$

H is the production period per day and N the number of production days per year. Costs given without relation to A, or even costs per installed power unit e.g. DM/kW (which per definition does not include A) are generally accepted as a measure of the economics for a power plant. That is because the utilization factor of baseload plants or peak power plants is considered as generally known for the specialized type (e.g. turbine-plant). For energy storage systems, the situation is much different. At first, for all energy storage systems a large application range with H from some hours to 10 and more hours per day is anticipated. At second and more important, the costs of storage facilities are composed from a power-dependent term (like in the power plants) and, in addition, from a term depending on the storage capacity. Consider for example a hydrogen storage facility loaded for H hours per day and N days per year with a power L measured in kW of thermal equivalent of the incoming gas. The storage capacity S measured in kWh (thermal equivalent) must be

$$S > L \cdot H$$

in order to match the special conditions of the energy system. The capacity factor U defined as

$$U = (L \cdot H)/S \qquad (1)$$

governs the cost situation as a measure for the relation between S and L:

$$S/L = A/(N \cdot U) = H/U \qquad (2)$$

Furthermore, for diverse alternative storage methods the proportions of the power unit costs to the storage capacity unit costs are very different. For instance a liquid hydrogen storage facility shows high power depending costs (liquefaction plant) but relatively low storage vessel costs. For compressed gas storage the situation becomes just inversed. With these remarks it is easy to see why a number of "cost calculation results" must be discarded for large scale stationary applications:

(1) Cost on power basis e.g. DM/kW
(2) Cost on basis of storage capacity unit e.g. DM/kWh
    or DM/kg, or $ per 1000 SCF of hydrogen storage capacity.
(3) Cost on stored energy basis e.g. DM/kWh (stored) without
    defining the values of U and A (or U, N and H).

It is a simple matter of analysis to see that costs according to (1) resp. (2) include the factor U/H = L/S as a weight factor for the power resp. capacity dependent part of costs. Unfortunately, the user of such cost assessments (wanting just to know the "storage cost" for one kg of hydrogen) cannot be aware that such broadly varying parameters (like H and U) are "hidden" without declared values in the cost analysis. In the case (3) the utilization N · H is (per definition) taken into account, but similarly the factor L/S is also "hidden" as weight-factor for the capacity depending cost. The way to avoid this confusion is simple. The costs mandatory are to be given per energy unit stored (undergoing a storage cycle) as in above case (3), but only for well defined cases in accordance to the capacity factor U, using N and H as parameters. Therefore, a comparison will be valuable only for defined U, N and H.

Method of cost calculation

In the following some details of the storage cost calculation for four alternative hydrogen storage methods (cryoadsorber, metalhydride, compressed gas and liquid hydrogen) will be given. The case of a diurnal storage facility, also accomodating the week variations, includes U values varying from U = 0.25 to 0.5 (some remarks on seasonal storage will be given later). The loading power has been selected to be L = 200 MW (thermal equi- valent of hydrogen). Values of N are considered to be N = 260 or 360 and H is varied as parameter.

Capital cost calculations are made basically according to the method described for example in /7/. The main modules of the facility are defined (e.g. refrigerators, compressors, storage vessels) and their delivery (equipment) costs are estimated or calculated. To accomodate for installation (material and labour) costs, interconnections cost etc., a module factor has been used. In this work rather moderate module factors (1.6 for machine blocks and vessels, 1.20 for material handling) were considered (site development, buildings and indirect costs did not need to be considered for the purpose of comparison aimed here). The capital costs K are then the sum of module costs (equipment cost times module factor). From the specific costs $C_0 = K/L$ (DM/kW) the fixed charges per year $C_1 = (f + a) \cdot C_0$ (DM/kW, a) can be calculated, where the capital charge factor f has been calculated for a fixed charge rate of 15% and a plant lifetime of 20 years. The operating and maintenance cost factor a was assumed to 5%. From the power demand and utilization of the machine-modules the power cost E were also calculated on annual basis (DM/kW, a). The total costs per kW and year are then

$$C_t = C_1 + E, \quad DM/kW, \, a \qquad (3)$$

and the costs per stored energy unit (storage cost)

$$C = C_t/(N \cdot H), \quad DM/kWh \qquad (4)$$

More details on the cost calculation are included in the final report /8/. In the following only a few detailed informations are given:

1) Power costs have been considered as costs of electric drive for all involved machines. The unit price has been

assumed to be 0.05 DM/kW$_e$.

2) For refrigerator costs the NBS-Surveys [9] have been taken
   into account. Cryogenic vessel costs estimation from [10],
   [11] have been used.

3) For pressure vessels, besides of the possibility to use steel
   vessels, the use of large vessels consisting of cast iron
   segments hold together with prestressed cables (VGB-Behälter)
   must be considered. They offer substantially lower prices.
   Such vessels with liners and inner thermal insulation recently
   came in use [12]. Their use for cryoadsorbers (being now investigated) would permit a further reduction of the vessel costs.
   On the other hand, for pressurized gas storage the necessary
   volume becomes so large that even the use of the biggest of
   such vessels (about 8,000 m^3 each at about 2,000 DM/m^3) is
   connected with unacceptable costs. For the pressurized gas
   storage a more optimistic situation has been considered according to [13]: Use of huge vessels (typically 150,000 m^3)
   at low pressure level (at 11 bar), with extremely low price
   per volume unit.

4) Material prices have been assumed as follows:

   a) Activated carbon: 3 DM/kg (recent market)
   b) Metalalloy (FeTi) for hydride storage: 8 DM/kg [14]

5) For the cryoadsorber storage only the reversible storable
   hydrogen amount H$_r$, corresponging to isothermal cycling
   (42 bar max. pressure), has been considered.

Discussion of the results of cost calculation

Prior to the comparison between alternative storage methods it
would be useful to take a look on the influence of material
and temperature range selection for the cryoadsorber storage
facility. Fig. 9 shows the fixed and total costs per year and
the storage costs in DM/kWh of stored energy, for different
cryoadsorber temperatures, as a function of utilization. Two
values of U (0.5 and 0.25) have been considered for N = 360
days per year. The lowest temperature offers the best results
because the stored amount increases rapidly at lower temperature, which balances the higher cooling power costs. In the
case U = 0.5 there are only small differences in costs between
T = 78 and 65 K. In general, there is no expectation for considerable cost reduction at temperatures below 65 K. The results of Fig. 9 are corresponding to activated carbon F12/350
adsorbent. Fig. 10 shows the difference for a more dense activated carbon, type F12/470, for U = 0.5 and two temperature
levels. In general, the investigation shows that the active carbon F12/350 in the temperature range about 65 K offers the

lower costs for the cryoadsorber case. For the following comparison only this material and temperature level is considered.
In Fig. 11 the annual costs and the resulting storage costs
for the alternative storage methods (KA = cryoadsorber, MH =
hydride storage, DR = compressed gas, LH = liquid storage) are
compared. Results are for N = 290 and U = 0.5. The differences
in the dependence of fixed and total costs from the utilization
is very crucial. For compressed gas the ("initial") costs for
low utilization are very low, but they increase rapidly with
H or A. For liquid storage the situation becomes inverse, hydride and cryoadsorber storage show a mediocre behaviour. The
importance of this situation is well illustrated by changing
the parameters U and N. Fig. 12 shows the results for U = 0.25
and N = 360. In this case the storage facility with highest
power requirements and lower storage vessel costs (the liquid
storage) becomes advantageous in the high utilization region.
Fig. 13 (I to III) depicts the storage costs for some selected cases corresponding to hydrogen storage facilities on diurnal or week
storage basis. More information on the cost structure could
be taken from tables like tables 2 and 3. Only one example for
U = 0.5 is given here. It is valid for N = 360, H = 7, i.e.
it corresponds only to one point of the above diagrams.

As a result, the investigation shows the competition of hydrogen
storage by means of cryoadsorbers to alternative storage methods. The advantages depend mainly upon the utilization case
and the capacity factor. Table 4 gives a crude summary of the
situation. It must be pointed out that for the above results
no allowance has been made for the recovery of the energy used
in the cryogenic processes. For example, it is possible to use
output cold gas to reduce the work required to compress the air
fed for a coupled gas turbine plant, as planed in the case of
liquid hydrogen /9/ or liquid natural gas /14/, /15/. In the
case of cryoadsorber system no heat-exchange and coupling to
another process is necessary, since cold gas compression can
be a part of the storage cycle. Compression of the discharged
gas up to the line pressure level starting from the temperature
level of about 135 K would substantially reduce power requirements and costs. Fig. 14, otherwise identical to Fig. 13 (I)
includes this recovery configuration for the cryoadsorber case.

Some remarks on the efficiency of the storage methods

In the tables 2 and 3 values for energy consumption per kg
of stored hydrogen and the corresponding storage efficiencies
are given. Table 5 summarizes these values. As expected, these
efficiencies alone do not characterize the economic situation, since storage methods with the lowest efficiency can
be economically preferable. This should be obvious even from
an elementary analysis of the economics. Unfortunately the
importance of "energetic efficiency" of storage systems some

times becomes exaggerated as the decisive factor. For cases,
when higher efficiency seems to be achievable by means of
higher capital investment, one should be aware that a substantial part of capital investment corresponds also to energy consumption costs.

Seasonal hydrogen storage

The above analysis can be easily extended to the case of seasonal hydrogen storage. Although the results cannot be presented here, the final result is quite simple: Only storage
facilities with extreme low capacity-depending costs are competitive. Beside the underground caverne storage, only the
liquid hydrogen storage is to be regarded as alternative storage method.

CONCLUSIONS

According to the experimental results the hydrogen storage
capacity of cryogenically cooled adsorbent materials amounts up
to 68 grams per kg of adsorbent at 78 K, increasing to 82 grams
(8.2 % b.wt) at 65 K, both at the 42 bar pressure level.
Isothermal expansion to the 2 bar level yields 42 g/kg resp. 52 g/kg
for the temperature levels mentioned above. With these values (i.e.
effective isothermal capacity of 4% - 5.2% on adsorbent mass
basis) the concept of cryoadsorbent storage system is competitive with the alternative methods. Although storage at higher
temperatures (up to 140 K), non-isothermal discharging etc. have
yet to be further examined, it seems that best results can be
achieved by use of very porous activated carbon in the 65 K
temperature region. For comparison between alternative storage
methods the consideration of the utilization and of the capacity
factor is decisive. For low capacity factors the advantages of
cryoadsorbent storage are expected in high utilization region.
With increasing capacity factor the advantages of the cryoadsorber concept are shifted to the low utilization region, whereas the advantages of liquid hydrogen storage for higher utilization becomes predominant.

REFERENCES

1   CARPETIS, C.   "On the Storage of Hydrogen by
    PESCHKA, W.   Use of Cryoadsorbents"
                          Paper 9C - 45, 1st World Hydrogen
                          Energy Conference, March 1976

2   SINGLETON, A.   "Rate Model for Ortho-Parahydro-
    LAPIN, A.   reaction in a Highly Active Ca-
                          talyst"
                          Advances in Cryogenic Engineering,
                          Vol. 13, p. 409, (1968)

3   BMFT-Studie   "Neuen Kraftstoffen auf der Spur"
                                Bonn, F.R.G., p. 426, (1974)

4   KIDNAY, A.   "High Pressure Adsorption Iso-
    HIZA, M.   therme of Neon, Hydrogen and He-
                                lium at 76 K"
                                Advances in Cryogenic Engineering,
                                Vol. 12, p. 730, (1967)

5   COOK, W. H.   "Physical Adsorption of Gases on
                                Activated Carbon"
                                Ph.Dr. Thesis, University of
                                Ottawa, Canada (1965)

6   CLARK, R.   "Investigation of the Para-Ortho-
    KUCIRKA, J.   Shift of Hydrogen"
    JAMBHEKAR, A.   A.F. Aero Propulsion Laboratory,
    SCHMAUCH, G.   ASD-TDR-62-833, Part II, (Oct.
                                1963)

7   GUTHRIE, K.   "Data and Techniques for Prelimi-
                                nary Cost Estimates"
                                Chem. Eng. 114 (March 1969)

8   CARPETIS, C.   "Untersuchung der Wasserstoff-
    PESCHKA, W.   speicherung mit Kryoadsorbern"
                                Final Report on Project FA 057-
                                76 FHD, Research Program of the
                                Commission of the European Com-
                                munities (to be published)

9   STROBRIDGE, T.   Cryogenic Refrigerators - An
                                Updated Survey"
                                NBS Technical Note 655, Boulder,
                                Col., (1974)

10 HALLET, N.   "Study, Cost and System Analysis
                                of Liquid Hydrogen Production"
                                NASA Report CR-73226 (1968)

11 KELLEY, J. H.   "Hydrogen tomorrow"
    e.a.   Report of the NASA HEST-Study,
                                JPL, Pasadena (Dec. 1975)

12   Information sheets of the Siem-
                                pelkamp Comp., Krefeld, F.R.G.
                                "Energiespeicherung in vorge-
                                spannten Guß-Druckbehältern
                                (VGD)"

13 BMFT-Studie   "Auf dem Wege zu neuen Energie-
                                systemen", Bonn, F.R.G., (1975)

14 SCHMITT, R.   "Technische und wirtschaftliche
                                Aspekte der Wasserstoffspeiche-
                                rung in Metallhydriden"
                                A.I.M.-Congress: "L'hydrogen et
                                ses perspectives", Liège,
                                (Nov. 1976)

TABLE 1

Material	Bulk density g/cm^3	Pore volume cm/g^3	Pore-surface m^2/g	Max.stored hydrogen at 42 bar, 78 K $g(H_2)/kg$ adsorbent
Nickeloxyde-silicate 1200–88–02	0·67	0.64	530	34
Nickeloxyde-silicate G 49	0.85	0.31		23
Silicagel KIESELGEL 40	0.59	0.65	650	31
Siligacel KIESELGEL-PERL	0.72	0.38		31
Activated carbon SCHUMASORB	0.85	0.68		25
Activated carbon NK 12	0.36	1.17		49
" " BS 12	0.39	1.04		52
" " F12/470	0.46	0.80		49
" " F 12	0.38	1.08	1250	66
" " F12/350	0.35	1.25		68

Max. measured value (Activated carbon F12/350 at 65 K and 42 bar):

82 $g(H_2)/kg$ of adsorbent.

TABLE 2   Comparison of hydrogen storage facilities for L = 200 MW eq.th. (6006 kg(H$_2$)/hour), S = 2.8 · 10^6 kWh (84084 kg H$_2$), U = 0.5 H = 7 hours per day / N = 360 days per year. (Specific data are referred to equivalent thermal power or energy of the storage unit)

Storage facility	Cryoadsorber	Metalhydride
Characteristics	Active carbon, 380 kg/m^3 , 3 DM/kg H$_r$ = 51 g(H$_2$)/kg at 65 K, 42 bar	FeTi alloy, 3550 kg/m^3 , 8 DM/kg H$_r$ = 17 g(H$_2$)/kg
Main modules (equipment costs)	Refrigerator (13.6 MW) 14 · 10^6 DM other cryogenics 1.67 · 10^6 DM Compressor 7.37 · 10^6 DM Act.Carbon (1648 ton) 4.95 · 10^6 DM Adsorber vessels 17.3 · 10^6 DM	Compressors and heat exchangers 15.05 · 10^6 DM Metal alloy (4946 ton) 39.57 · 10^6 DM Metalhydride vessels 5.05 · 10^6 DM
Specific storage data	1.7 kWh/kg ; 0.645 kWh/l	0.566 kWh/kg ; ≈ 2 kWh/l
Energy consumption	3.19 kWh/kg(H$_2$)	Electric: 1.74 kWh/kg(H$_2$) ; low temperature: 4.2 kWh/kg(H$_2$)
Efficiency	0.912	0.95 (on electric energy consumption basis only)
Capital costs	71.07 · 10^6 DM	83.64 · 10^6 DM
Spec.capital costs	355 DM/kW	418 DM/kW
Spec.fix cost per year	83.2 DM/kW, a	92 DM/kW, a
Sp.Energy costs per year	12 DM/kW, a	9.7 DM/kW, a
Total cost per year	95.2 DM/kW, a	101.8 DM/kW, a
Storage cost	0.0377 DM/kWh	0.0404 DM/kWh

1447

TABLE 3   Comparison of hydrogen storage facilities for L = 200 MW eq.th.
(6006 kg(H$_2$)/hour), S = 2.8 · 10^6 kWh (84084 kg(H$_2$)), U = 0.5
H = 7 hours per day / N = 360 days per year
(specific data are referred to equivalent thermal power or energy of the storage unit)

Storage facility	Compressed gas	Liquid hydrogen
Characteristics	Pressure level 11 bar Vessel module: about 30 m diameter (at about 51 DM per m^3)	Cryogenic tank at 650 DM/m^3
Main modules (equipment costs)	Compressor (4.9 MW) 4.18 · 10^6 DM Storage vessels (935520 m^3) 48.02 · 10^6 DM	Cryogenic equipment (68.2 MW) 43.45 · 10^6 DM Liquid hydrogen tank (1360 m^3) 0.89 · 10^6 DM
Specific data	0.03 kWh/l	2.36 kWh/l
Energy consumption Efficiency	0.367 kWh/kg(H$_2$) ~0.99	10.4 kWh/kg(H$_2$) 0.76
Capital costs Spec.Cap.costs	83.52 · 10^6 DM 414 DM/kW	70.9 · 10^6 DM 354 DM/kW
Spec. fix costs per year	100.6 DM/kW, a	85.5 DM/kW, a
Spec. Energy costs per year	1.4 DM/kW, a	39.36 DM/kW, a
Total cost per year	102 DM/kW, a	124.8 DM/kW, a
Storage cost	0.0405 DM/kWh	0.0495 DM/kWh

TABLE 4

Least cost storage facility for various cases

Capacity factor	Utilization factor	
	Low	High
U = 0.5	Compressed gas	Cryoadsorber
U = 0.25	Cryoadsorber	Liquid hydrogen

TABLE 5

Storage	Typical values of storage-efficiency defined as $\eta = H_o/(H_o + W)$
Cryoadsorber	~ 0.91 - 0.97 (temperature dependent)
Metallhydride	~ 0.95
Compressed gas	~ 0.98 - 0.99
Liquid hydrogen	~ 0.75

H_o = thermal equivalent of stored secondary energy carrier (= lower heating value of hydrogen), kWh/kg(H_2)

W = secondary energy consumption of storage system (electric power pr.mover) kWh/kg(H_2)

Fig. 1

Adsorber unit, lifted off
from the LN$_2$-dewar

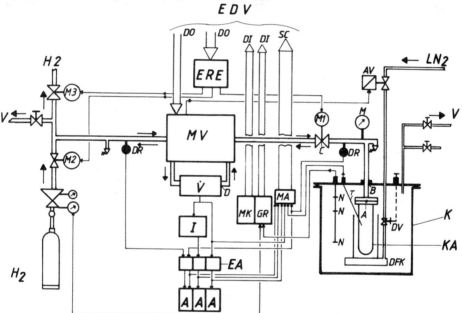

Fig. 2   Schematic diagramm of the experimental device. KA = Adsorber unit;
DFK = Nitrogen-gas-flow cryostat for temperatures > 78 K; DR =
pressure sensors; $\dot{V}$ = flowmeter; I = integrator; MV = flow control valves; ERE = control unit for motor valves. Information
from and to the data acquisition system is denoted by the arrows
(dig. input, output and scanner connection).

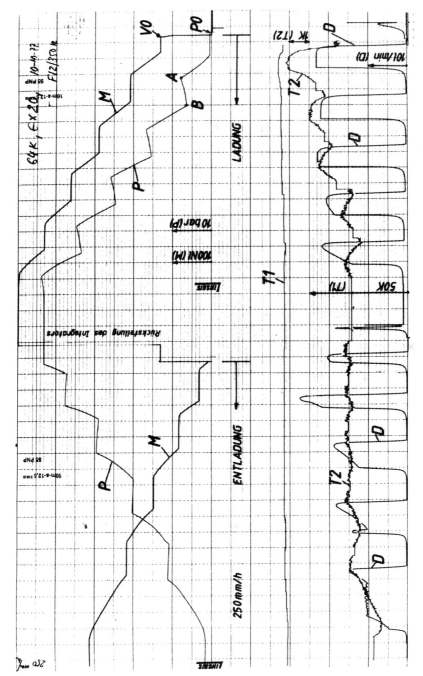

Fig. 3   Strip record of a charge-discharge cycle. P = pressure; D = flow rate; M = stored (right part) or released (left part) hydrogen amount; T1, T2 = temperature T of adsorbent (two scales). Use the vertical arrows for scaling the different curves.

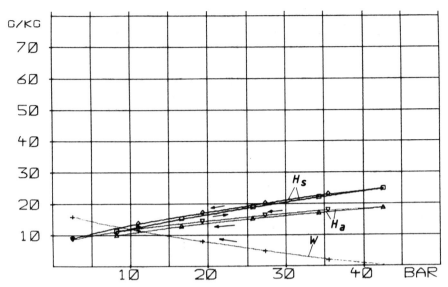

Fig. 4 Isothermes H_a and H_s, discharged hydrogen amount W, for 78 K.
Dense activated carbon (ρ = 0.85 g/cm^3).

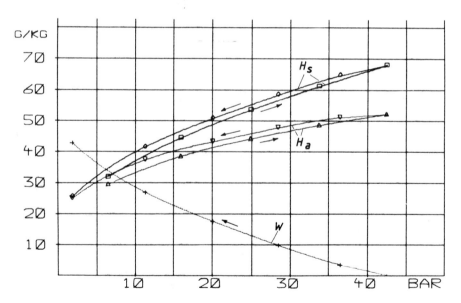

Fig. 5 Isothermes H_a and H_s, discharged hydrogen amount W, for 78 K.
Activated carbon F12/350

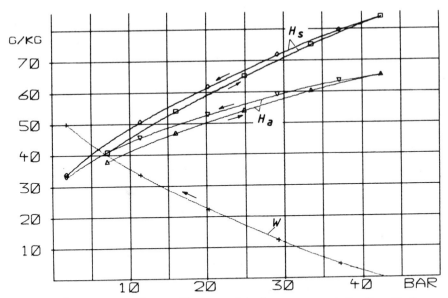

Fig. 6 Isothermes H_a and H_s, discharged hydrogen amount W, for 65 K. Activated carbon F12/350

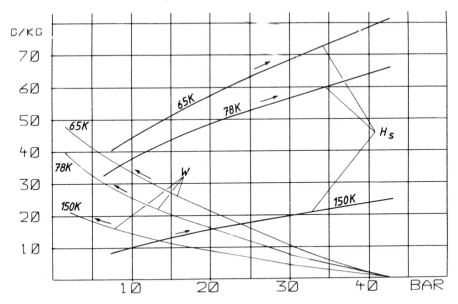

Fig. 7 Isothermes for stored amount H_s and for discharged hydrogen amount W at three temperature levels. Activated carbon F12.

1453

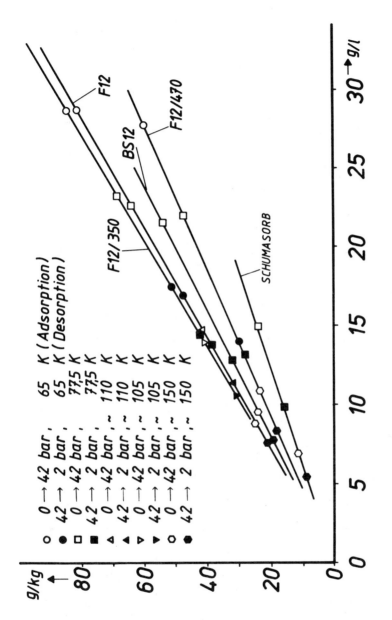

Fig. 8 Adsorbent mass- and adsorber volume-related amount of stored hydrogen (H_s) for different materials and temperature levels. X-Axis: grams of hydrogen stored per liter of adsorber volume. Y-Axis: the same per kg of adsorbent. Non filled signs correspond to maximum values at 42 bar, filled signs are for discharge values.

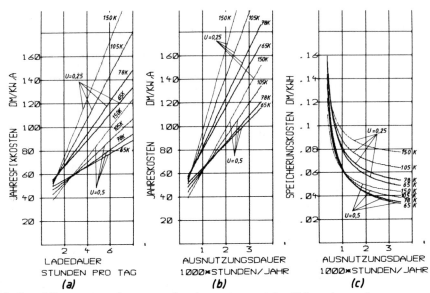

Fig. 9 Specific costs of a cryoadsorber storage facility (L = 200 MW
N = 360) for several storage temperatures and two capacity factors.
(a): fixed costs as a function of charge time per day. (b): annual
costs as a function of utilization in 10^3 hours per year). (c):
storage costs as a function of utilization. Adsorbent: Activated
carbon F12/350.

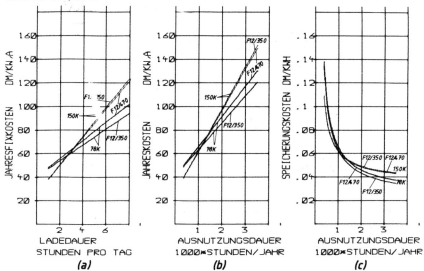

Fig. 10 Specific costs as in Fig. 9. Data are for 78 K and 150 K and
for U = 0.5 only. Compared are two activated carbon variations:
F12/350 and F12/470

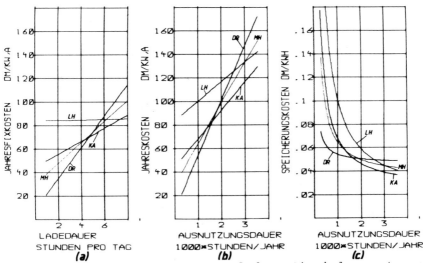

Fig. 11 Comparison of the specific costs of alternative hydrogen storage
facilities: KA = cryoadsorber; MH = metalhydride; DR = compressed
gas; LH = liquid hydrogen. (a): fixed costs as a function of charge
time per day, (b): annual costs as a function of utilization (in
10^3 hours per year), (c): storage costs as a function of utilization. L = 200 MW (th.equ.); U = 0.5; N = 290

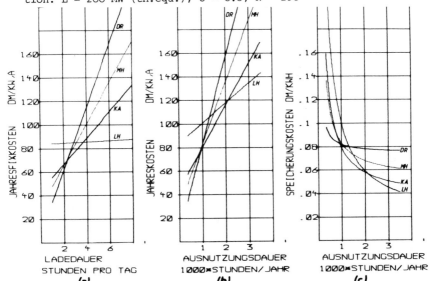

Fig. 12 The same situation as in Fig. 11. Only the parameters U and N
have been altered: U = 0.25; N = 360

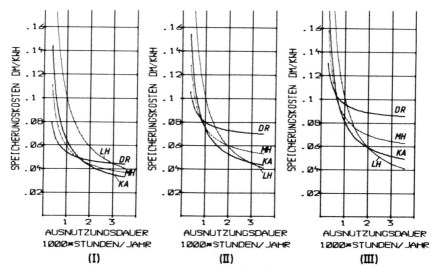

Fig. 13 Comparison of the specific storage costs (as a function of utilization in 10^3 hours per year). KA = cryoadsorber; MH = metalhydride; DR = compressed gas; LH = liquid hydrogen.
(I)   : for N = 360; U = 0.5 (L = 200 MW)
(II) : for N = 360; U = 0.3 (L = 200 MW)
(III) : for N = 290; U = 0.3 (L = 200 MW)

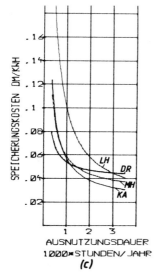

Fig. 14 This figure corresponds to the data of Fig. 13(I), with the exception that for the cryoadsorber case an energy recovery (by use of cold gas compression) has been assumed. L = 200 MW (th.equ.); U = 0.5; N = 360.

THE CRYOGENIC STORAGE OF HYDROGEN

J.J. Thibault
L'Air Liquide   C.E.C.
38360 Sassenage, FRANCE

ABSTRACT

The liquefaction processes will be roughly indicated. The problems with large size liquefiers will be pointed out. Examples on industrial liquefiers will be given.

The storage and distribution equipment will be reviewed : storage tanks (fixed and mobile), transfer lines, cryogenic pumps.. The state of the art will be given and several examples described.

The different factors entering in the choice of a given distribution process will be discussed.

INTRODUCTION

The aim of this paper is to show that the cryogenic storage and distribution of hydrogen have to be taken into account in comparison with classical gaseous distribution : the cryogenic technology has been fully developed through the space projects involving LH2-LO2 propulsion and the liquid form retains several intrinsic advantages over the gaseous form ; i.e. compactness and high purity. These advantages can be determinant in several present or potential applications : hydrogen for electronics, fuel for aircrafts...

1. LIQUEFACTION OF HYDROGEN

11. Technical aspect

For industrial liquefiers, the thermodynamic cycle must include at least one expansion device with external work. Let us recall the main characteristics of a simplified Claude cycle as shown in Fig. 1 : hydrogen gas is compressed by compressor C and then successively cooled down through heat exchanger E 1, heat exchanger E 2 (cooled down by liquid nitrogen), partly through heat exchanger E 3 and partly through the expansion device T, heat exchanger E 4 and Joule-Thomson expansion valve J.T.

In order to improve the efficiency of the cycle for large production liquefiers, the number of expansion devices is increased. A typical example of a 3 600 litre/hour liquefier cycle is given in Fig. 2.

Only the main problems met in such facilities will be mentioned here as they have already been discussed in several precedent publications :

. purification of H2,
. ortho-para conversion,
. choice of the compressor (reciprocating or rotating)
. choice of the expansion device (reciprocating piston or turbine).

The largest facility in the world has been installed in Sacramento, California (USA), a photograph is given in Fig. 3.

The production rate is sixty tons a day, the hydrogen gas being supplied by a natural gas steam reforming plant. It uses two reciprocating compressors for N2 and H2 working at 27 bars and one centrifugal compressor for N2 working at 5.5 bars and two expansion turbines.

A modern average sized liquefier is presently manufactured by L'Air Liquide in Japan for the production of 730 litres/hour^{-1} (or 1.2 tons a day). For the first time, a new cycle will be used : the cold power is produced in a Claude helium cycle with two gas bearing expansion turbines (Fig. 4) and hydrogen is cooled down and liquefied in a separate circuit. This process has already been tested in Japan in a small hydrogen liquefier (40 litres/hour) built by L'Air Liquide.

The main technical advantages are better reliability for the components used in a closed helium cycle (compressor, turbines) and the simplification of safety problems.

12. Economic aspect

The Carnot theoretical efficiency of a hydrogen liquefier is 3.3 kilowatt-hour per kilogram of H2.

For large industrial facilities, the practical efficiency can reach 10 kwh/kg.

The running cost will be the sum of the cost of energy and the cost of maintenance, the latter being more difficult to estimate.

The investment cost can be roughly estimated using the results of the National Bureau of Standards enquiry (1) which gave the

following correlation formula in 1973 :
$$C_{73} = 6 \times 10^3 \ P^{0.7}$$

where

C_{73} = cost in 1973 US dollars
P^{73} = installed input power in kilowatts.

If P is expressed as
$$P = \frac{10}{24} \times M$$

M being the LH2 production rate in kilograms per day, we have :
$$C_{73} = 3.25 \times 10^3 \ M^{0.7}.$$

For example, in 1973, the investment cost of a 10 tons per day LH2 plant would have been :
$$C_{73} = 3.25 \times 10^3 \times (10^4)^{0.7}$$
$$C_{73} = 2 \times 10^6 \ US \ \$.$$

13. Extrapolation for very large production plants

It is clear that if LH2 is considered as an energy vector, larger liquefaction plants will have to be designed and manufactured.

For example, if LH2 is used as an aircraft fuel, the normal consumption for one airport would be in the range of 10^3 to 10^4 tons per day. Let us take the case of a 2.5×10^3 ton per day liquefier.

The compressors will have to compress 5×10^6 standard cubic metres per hour at about 20 bars. It will be necessary to develop huge multistage axial centrifuge compressors. The input power would be 10^6 kw which implies one electricity production plant for one liquefier.

For such capacities, it appears that the best solution would probably be to develop 100 to 500 ton modular liquefaction units. In such an approach, the technology does exist now.

2. DISTRIBUTION OF LIQUID HYDROGEN

21. Available components

211. Fixed storage tanks

Such tanks have been commercially available for several years. They are difficult to construct, due to the brittleness of ma-

terials at cryogenic temperatures and to the low latent vaporization heat of LH2.

For example, the typical thermal losses of a 4 000 litre tank
and a 50 000 litre tank are respectively 1 % and 0.5 % per day.

The biggest tank which has been manufactured to date is set up
in Cape Kennedy, Florida (USA) : it is a spherical 900 000 gal-
lon tank (3.2 x 10^6 litres)(Fig. 5).

It is presently possible to design and manufacture spherical
tanks (perlite insulated and under vacuum) of 225 000 to
650 000 litres capacity which a 0.1 % loss rate per day.

212. Mobile tanks

LH2 can be transported by the following means :

semi-trailers (Fig. 6),
containers for trains, ships and planes (Fig. 7),
waggons (Fig. 8),
canal boats or ships.

Only the three first means have been used to date.

Boats have not been used yet but it is possible to make an ana-
logy with the transport of liquefied natural gas.

Generally, transport by road is preferred. The capacity of the
semi-trailers is officially limited in each country. For example in France, 42 000 litres is a maximum. The daily loss rate
of such a semi-trailer is lower than 1 %.

213. Cryogenic transfer lines

It is theoretically possible to transport subcooled LH2 through
vacuum insulated cryogenic lines provided with cryogenic pumps
and refrigerators when needed to preserve the single-phase flow
(2). Economic study shows that this process is only cost-effective for short distances : some 10 to 10^3 metres.

214. Cryogenic accessories

The cryogenic pumps (Fig. 9) have been adapted to LH2 ; this
caused certain difficulties for the circumferential speed must
be high due to the low density of LH2.

Other accessories like valves, couplings, safety disks control
and measurement equipment have also been developed.

3. SAFETY PROBLEMS

The transfer of LH2 from one component of the distribution link
to another must take into account the particular inflammability
properties of H2. The H2 proportion in air must be maintained
below 4 %.

This feature added to the ability of LH2 to condense O2, makes
it necessary to take special care in the handling.

Before the transfer of LH2, the equipment must be purged with
N2 or He gas. The surroundings must be ventilated, leaks must
be checked for (hydrogen detectors) and the risks of spark or
flame must be avoided.

For the thermal insulation of the equipment plastic foams must
be rejected because of possible concentration of condensed solid oxygen.

When large quantities of LH2 are handled, safety regulations
must be set up and the workers must be carefully taught about
the hazards of H2 and LH2.

4. APPLICATIONS

41. LH2 as a gas source

The main advantages of LH2 in the distribution when H2 is used
as a gas are :

> . compactness which gives out a lower distribution
>   cost for large quantities and long distances,
> . high purity of the product,
> . large storage capacity in the customer's shop in
>   a relatively small area,
> . suppression of handling of gas cylinders in the
>   customer's shop.

A typical example is the supply of H2 to the semi-conductor
industry. L'Air Liquide has been delivering LH2 to the semiconductor industry in France for several years and this type
of distribution trends to be expanded.

LH2 is finally vaporized in the same type of equipment as generally utilized for LN2 or LO2.

42. LH2 as a rocket fuel

LH2 associated to LO2 presents the highest specific impulsion

among possible chemical rocket fuels, the mixture fluorine-hy-
drogen excepted. It has been used in the USA for many years
and the European launching project called Ariane is presently
running with this fuel used in the third stage. The first
flight is scheduled for mid 1979.

LH2 is produced in a facility built by L'Air Liquide in the
north of France (800 litres/hour)(Fig. 9) and transported to
French Guyana in four 35 000 litre containers. These containers
are designed to withstand a one month trip without any loss of
H2. The first two of these have been correctly delivered.

On the launching area, two vertical storage tanks have been
installed, 100 000 litres each. A set of cryogenic transfer
lines, 200metres long, connects the different components.

In France and more recently in Germany, LH2 is currently delivered by L'Air Liquide to different facilities involved in the
space programs.

43. LH2 for cars

The most promising applications of H2 for cars are air-hydrogen
fuel cells.

Cryogenic distribution of H2 from the production plants to the
service stations will probably be preferred. Apart from the general problems of distribution of LH2, it will be necessary to
develop a cryogenic service-station which will either transform
low pressure LH2 into high pressure H2 for use in cars or deliver LH2 directly.

A similar project has been experimentally studied by L'Air Liquide for LNG (3).

44. Distribution of large quantities of H2

A typical example could be the future development of LH2 as a
fuel for aircrafts.

We will compare hereunder the costs of distribution for the
different solutions.

Let us consider the supply of 400 and 4 000 tons per day to
50 and 500 km.

- By road
        It is assumed that 42 000 litre semi-trailers are
used, the loading time at the production plant is 2 hours, the

speed on the road is 50 km/h and the unloading time is 3.5 hours.

The number of semi-trailers is given in the following table :

	400 tons/day	4 000 tons/day
50 km	70	677
500 km	140	1 354

If the cost per transported kilogram to 50 km is c, the cost for 500 km will be 5 c whatever the daily quantity. Our cost estimation includes the capital cost and the running cost.

- By train

It is assumed that the cryogenic wagons have a capacity of 100 000 litres and the loading or unloading time is 5 hours. The cost of the cryogenic substructure of the loading and unloading stations (one cryogenic pump per wagon) is taken into account.

The number of wagons is the following (15 wagons per train) :

	400 tons/day	4 000 tons/day
50 km	60	600
500 km	90	900

The cost for 50 km would be 0.9 c and for 500 km 2.4 c whatever the daily quantity.

- By canal-boat

It is of course assumed that a canal does exist between the production plant and the place of consumption and in that case only the 500 km distance will be considered.

Canal-boats of 4.4. million litres capacity would be used. So, 1 boat is needed for 400 tons/day and 10 boats for 4 000 tons/day, the transportation cost will be 1.4 c.

- By gas pipeline

This is not a cryogenic distribution process but it must be compared to the others. The cost will be the following:

	400 tons/day	4 000 tons/day
50 km	1.20 c	1.10 c
500 km	2.50 c	1.50 c

- The final results of the cost comparison appear on the following table :

C = cost of transportation of the unit-quantity	400 tons/day		4 000 tons/day		
	50 km	500 km	50 km	500 km	
Semi-trailer	c	5 c	c	5 c	⎫ LH2
Train	0.9 c	2.4 c	0.9 c	2.4 c	⎬
Boat	-	1.4 c	-	1.4 c	⎭
Gas pipeline	1.2 c	2.5 c	1.1 c	1.5 c	H2 gas

For 50 km, the road remains relatively cost-effective.

For 500 km, the boat is the cheapest mean but it can only be taken into consideration in particular applications.

REFERENCES

1. Refrigeration for superconducting and cryogenic systems
   by T.R. Strobridge
   Particle Accelerator Conference Washington March 1969
   Actualization in 1973
2. Long distance transfer of liquefied gases
   by R.B. Jacobs
   Advances in Cryogenic Engineering n° 2
3. Liquid natural gas service-station for vehicles
   by P. Pelloux-Gervais
   14th Refrigeration International Conference - Moscow
   September 1975.

FIGURE 1 - CLAUDE CYCLE FOR THE LIQUEFACTION OF H2

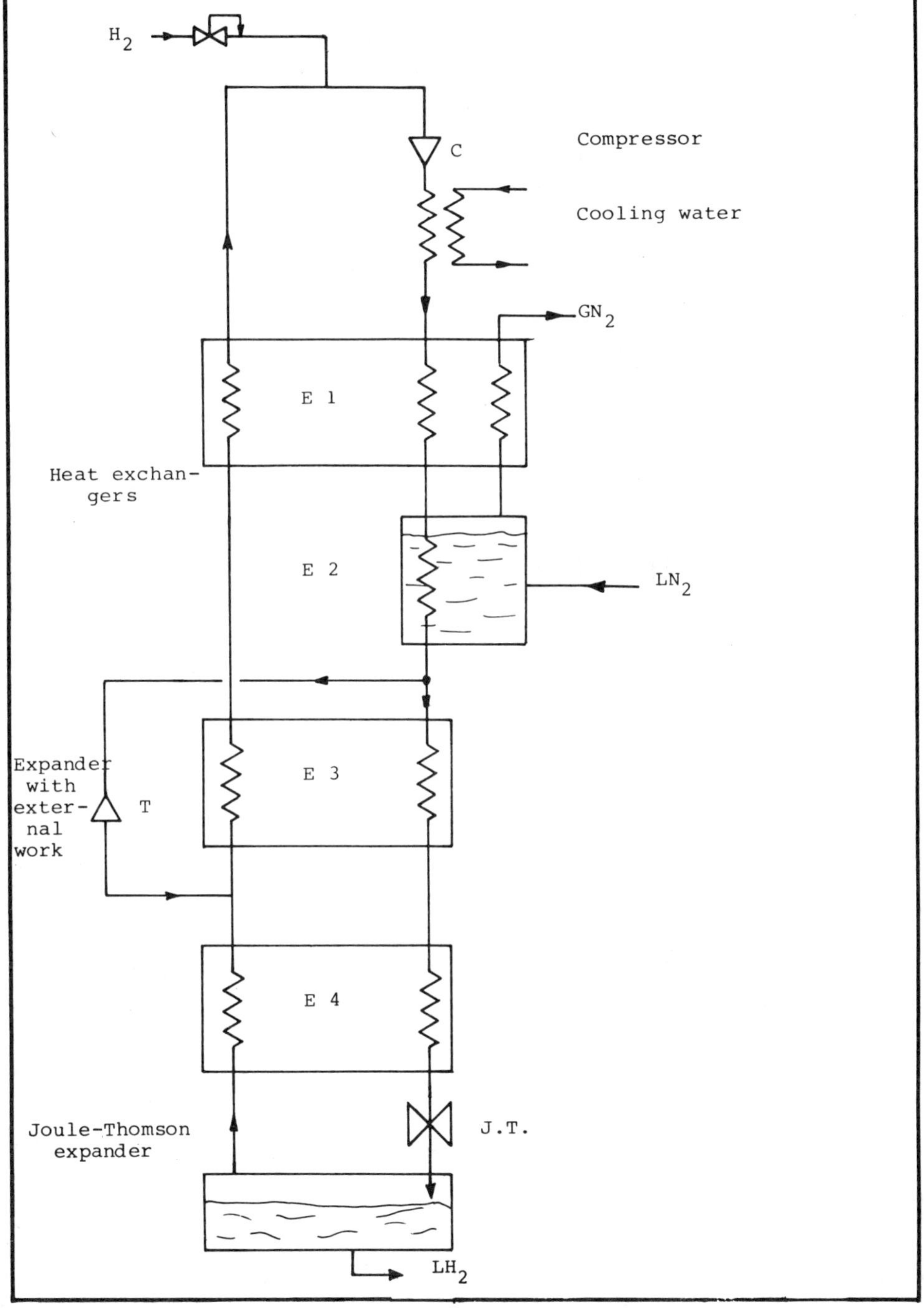

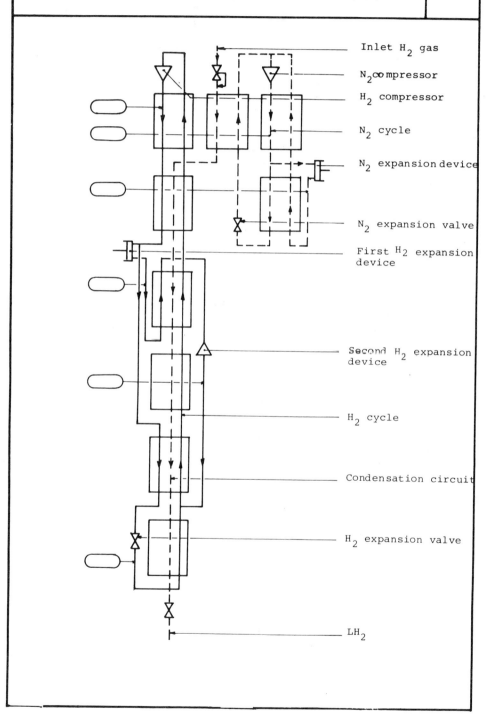

FIG. 2 - 3,600 L/H HYDROGEN LIQUEFIER BUILT BY L'AIR LIQUIDE
IN CANADA AND INSTALLED IN PEDRICKTOWN (USA)

Inlet H_2 gas

N_2 compressor

H_2 compressor

N_2 cycle

N_2 expansion device

N_2 expansion valve

First H_2 expansion device

Second H_2 expansion device

H_2 cycle

Condensation circuit

H_2 expansion valve

LH_2

FIGURE 3 - 60 tons/day H2 liquefaction plant built
by Union Carbide Corporation in
Sacramento (USA)

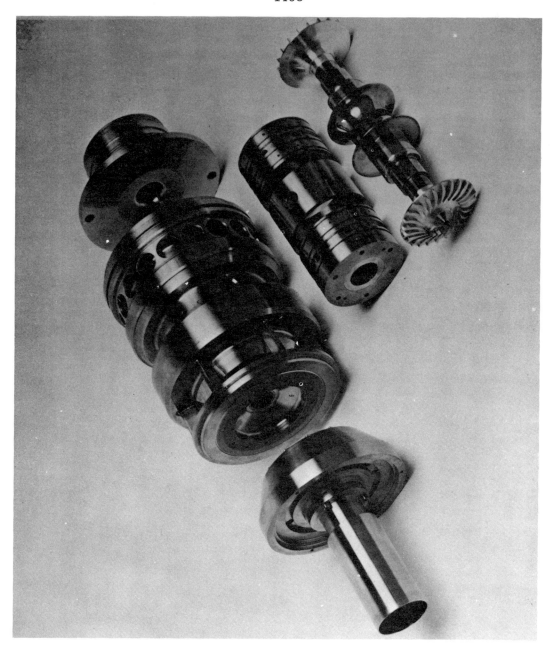

FIGURE 4 – L'Air Liquide gas bearing expansion
turbine for 730 l/h H2 liquefier

FIGURE 5 - 900 000 gallons LH2 storage tank in
Cape Kennedy (USA)

FIGURE 6 - LH2 semi-trailer built by L'Air Liquide

FIGURE 7 – 35 000 litre LH2 container built by
L'Air Liquide for the Ariane launching
area

FIGURE 8 - Cryogenic wagon built by L'Air Liquide

FIGURE 9 – 800 litres/hour H2 liquefier built by
L'Air Liquide in Frais-Marais (France)

THE ROLE OF METAL HYDRIDES IN HYDROGEN STORAGE AND UTILIZATION

F. E. Lynch
Denver Research Institute
University of Denver
Denver, Colorado 80208, USA

E. Snape
Ergenics Division
MPD Technology Corporation
Waldwick, New Jersey 07463, USA

ABSTRACT

The reactions involved in the formation and decomposition of several metallic hydrides are rapid enough to consider using such materials in many types of stationary devices where weight is not an overriding consideration. A wide variety of applications, ranging from simple, low-volume hydrogen storage in cylinders to complex heating and refrigeration systems, has been proposed.

The purpose of this paper is to review some of the more immediate, practical applications of hydrides and discuss some of the technical and economic limitations of hydride systems. Examples of specific hardware which utilize hydrides are described and their cost/performance characteristics compared with more conventional systems.

The economic and technical advantages of using metal hydride storage tanks in conjunction with electrolyzers is illustrated. Compressors and pumps based on metal hydrides are also described.

Some of the methods of characterizing hydrides for specific applications, including safety testing and evaluation of the effects of gaseous contaminants and hydriding response, are also reviewed.

INTRODUCTION

The primary reason why metal hydrides have been proposed as hydrogen or energy storage media is the extremely high volumetric density of hydrogen in these compounds. It is possible to pack more hydrogen in a metal hydride than in the same volume of liquid hydrogen (Table I).

Another fundamental and useful property of metal hydrides is the heat of formation and dissociation. This is the amount of heat released and absorbed when the metal hydride is formed and dissociated, respectively (Table I). The amount of heat can be quite large and is roughly proportional to the stability of the hydride.

In order to serve as a hydrogen or energy storage medium in practical systems, a number of criteria must be satisfied:

· The hydride should be readily formed and decomposed, and effectively act as a chemical analog of a battery, with hydrogen taking the place of electricity.

· The absorption/desorption kinetics must satisfy the charge/discharge requirements of the system.

· The equilibrium pressure corresponding to the hydride decomposition temperature at peak desorption rate should be compatible with safety requirements of the hydride containment system.

· The hydride should be useful over many charge/discharge cycles.

· The hydride itself should be at least as safe as other energy carriers.

In addition to the above technical characteristics, certain economic criteria must be met:

· Cost of the hydride per unit of reversibly stored hydrogen should be as low as possible.

· Storage vessel and ancillary equipment cost and the fabrication/installation costs should be reasonable.

· Operating and maintenance costs and energy requirements (other than waste energy or energy extracted from ambient air) per storage cycle should be low.

Fortunately, recent studies of the formation of ternary hydrides from intermetallic compounds have uncovered several economic materials with hydriding properties suitable for a variety of applications [1]. More importantly, these studies indicate that the properties of metal hydrides can now be manipulated to suit specific applications. This paper describes a number of potential applications of metal hydrides and considers some of the technical and economic constraints.

CHARACTERISTICS OF METAL HYDRIDES

The usefulness of metal hydrides can best be appreciated by considering the behavior of an ideal system (Figure 1). As hydrogen is taken up by the metal at constant temperature, the equilibrium pressure increases and then remains constant.

The upper curve, 1-2-3-4 is the absorption isotherm, while the lower curve 4-5-6-1 is the desorption isotherm. Beginning at point 1, the alloy is essentially free of hydrogen. As the hydrogen pressure is increased, the alloy absorbs hydrogen until the pressure and hydrogen content correspond to point 2. As gas is absorbed, the heat of solution must be removed or the system will become warmer and deviate from Figure 1. The state of the system at point 2 may be described as hydrogen gas in equilibrium with a hydrogen-saturated metal phase.

If more hydrogen enters the system, a new phase is formed according to a relatively pressure-invariant reaction:

$$\text{I.} \quad M + xH_2 \rightarrow MH_{2x}$$

where M represents the metal phase, and x is the number of H_2 molecules required to form a unit of the hydride phase, MH_{2x}. The composition consequently increases along the absorption isotherm from point 2 toward point 3. In this region two solid phases coexist; hydrogen saturated metal and metal hydride. Reaction I is exothermic, so, as before, heat must be removed if the system is to remain at constant temperature. The slope of the curve from 2 to 3 is ideally zero according to Gibb's phase rule, but for real hydriding alloys a positive slope is usually observed, so the pressure at 3 is greater than at 2. This region of the isotherm is commonly referred to as the absorption plateau.

Upon arriving at point 3, the hydrogen-saturated metal phase, M, has been consumed, Further increases in hydrogen pressure will cause a small increase in hydrogen content from point 3 to point 4. This represents the solid solution of hydrogen in the hydride phase and is exothermic, just as the previous absorption processes.

If the metal-hydrogen system has additional hydride phases (e.g., FeTi-H), then point 4 is equivalent to point 1 with respect to the formation of the new phase. If not, then increasing pressure after point 4 is reached will bring diminishing increases in hydrogen content, so point 4 may be regarded as the "charged" condition.

If the hydrogen pressure in the system decreases, the composition will decrease from the charged condition at point 4, but not along the same path by which it arrived, 1-2-3-4. Instead, it will take the path 4-5-6-1, the desorption isotherm. From 4 to 5, hydrogen is desorbed from the solid solution in the hydride phase. From 5 to 6, the hydride phase decomposes into hydrogen-saturated metal and hydrogen gas. This region of the isotherm is called the desorption plateau. Finally, from 6 to 1, hydrogen desorbs from the solid solution in the metal phase and the system is once again essentially free of hydrogen or "discharged". All of the decomposition processes along path 4-5-6-1 are endothermic, so heat must enter the system or its pressure-composition characteristic will deviate from the isothermal conditions of Figure 1.

The region between curves 1-2-3-4 and 4-5-6-1 is the absorption-desorption hysteresis. The width of the hysteresis region can range from nearly zero to more than 100 atm. A metal hydride which has large hysteresis is undesirable, because it requires higher charging pressure, to achieve a given discharge performance, than a hydride which has small hysteresis. This, in turn requires stronger, heavier and more expensive containers and system components and dilutes the argument that hydrides are safer than high-pressure storage vessels. A recent paper [2] reviews historical theories of hysteresis and offers a new theory based on geometric factors and strain in the hydriding process.

Another result of strain in the hydriding process is fracture of the metals into fine powders ranging down to a few microns in size. Filters are therefore required to prevent these powders from being carried out of the container with the hydrogen flow.

Figure 2 is a typical family of desorption isotherms which demonstrate the variation of desorption pressure* with temperature.

APPLICATIONS OF METAL HYDRIDES

Table II lists several applications of metal hydrides which are or may be economically and technically viable. The most obvious application, and one which has received the most attention, is hydrogen storage. This can be in the form of (a) small laboratory-size containers to replace or supplement compressed gas cylinders, or (b) large containers for safe hydrogen transportation in the form of lightweight hydrides and major storage facilities such as those envisioned for load leveling and reserve power generation in utilities. Requirements for these containers differ, depending on such factors

---

*Pressure is usually scaled logarithmically for convenience in thermodynamic analysis of hydride data.

as: pressure and temperature of hydrogen supply, purity of
hydrogen, rate of charging and discharging, volume limitations
and weight restrictions.

a)   Small Portable Containers

Small stationary or portable hydride containers can play a
valuable role in providing a new source of high purity hydrogen for government and industrial laboratories and in a variety of light industries. At least three companies are offer-
ing such containers for research, welding, gas analysis and a
variety of other uses. Some advantages of this type of container over conventional compressed gas cylinders are:
light weight, much smaller volume, high purity hydrogen,
safety and ability to accept hydrogen from relatively low
pressure sources.

Kernforschungsanlage Julich GMBH (KFA) has developed a small,
inexpensive and safe hydrogen storage container [3]. This
container is of stainless steel construction and has a volume
of approximately 1.4 liters (Figure 3). It is filled with
activated iron-titanium alloy and heat removal during hydriding is provided by pressurized air flowing through a spiral
cooling tube running through the hydride bed and accessible
from the outside. Hydriding is accomplished at 60 bar. A
comparison of this container with other hydrogen storage
systems is shown in Table III and some of its operating
characteristics are given in Table IV.

Billings Energy Corporation [4] produces a portable aluminum
container, again filled with an activated iron-titanium alloy.
It is provided with a controlled temperature heating jacket
to facilitate hydrogen withdrawal.

This tank, designated AHT-5, may be used at a pressure of 34
atmospheres (500 psig). The temperature has to be maintained
below 93°C during charging. Hydrogen flow rates of between
1 and 2 liters per minute are attainable, depending on the use
and the environment. Specifications of the AHT-5 tank are
shown in Table V.

The Ergenics Division of MPD Technology Corporation offers
custom - built containers ranging from simple cylinders to
complex modules with internal and external heat exchangers in
addition to an aluminum container of the type described above
(Figures 4 and 5). Such containers can be supplied with a
variety of hydriding (HY-STOR) alloys (Table VI). This includes a family of nickel-mischmetal alloys which can absorb

hydrogen under ambient conditions. Thus, a key feature of
these containers is their ability to be used in conjunction
with low pressure, laboratory electrolyzers. Such a combination can provide considerable savings over purchased compressed gas cylinders. This is because an electrolyzer
coupled to a hydride tank can be operated continuously and can
produce hydrogen during off-peak, non-working hours. Thus,
hydrogen requirements, particularly high purity hydrogen, demands can be satisfied on site and the expense and inconveni-
ence of hydrogen purchase, cylinder rental, etc. can be
avoided. Figure 6 shows a typical hydride container/control
panel/electrolyzer assembly being promoted by the Ergenics
Division of MPD Technology Corporation.

The rate of supply of hydrogen from the above containers is
controlled by heat- and gas-transfer phenomena. A detailed
analysis of the heat transfer into and out of a metal hydride
container is beyond the scope of this paper but, since the
heat transfer problem is common to all hydride applications,
it is worthwhile to consider the 'simple' case of an aluminum
container cooled and heated by natural air convection. In
this case, the sensible heat from the convected air together
with the radiation from the surroundings and the latent heats
of condensation and freezing are conducted through the water
and ice layers, through the wall of the cylinder (which is the
only insignificant thermal resistance), through a layer of
metal powder whose thickness is changing, and into the charged
metal hydride, causing hydrogen to flow into the collector
tube and convectively transfer heat toward the center of the
container (Figure 7).

For the purpose of this analysis, it is convenient to consider the "equilibrium temperature" at a given pressure even
though hydride stability is usually discussed in terms of
"equilibrium pressure" at a given temperature. The equilibrium temperature of a hydride is that temperature at which the
hydride is in equilibrium with hydrogen gas at a specified
pressure. By knowing the equilibrium temperature at one pressure and the reaction enthalpy, ΔH, the equilibrium temperature at any other pressure may be deduced. In general, we
have the relationship:

$$\text{II.} \quad T_2 = \left[\frac{1}{T_1} + \frac{R \ln(P_2/P_1)}{\Delta H} \right]^{-1}$$

In the following discussions of an individual hydride cylinder,
equation I will be used to derive empirical heat transfer coefficients relating the equilibrium temperature of the hydride,
the ambient temperature and the maximum sustainable flow from
each cylinder.

Figure 8 plots cylinder pressure versus time for several rates of withdrawal from a prototype aluminum cylinder with the following specifications:

Internal Volume	- 4.75 liters (283 in^3)
Length	- 654 mm (25-3/4 in)
Outer Diameter	- 111 mm (4-3/8 in)
Minimum Wall Thickness	- 5 mm (0.198 in)
Empty Cylinder Weight	- 3.76 Kg (8.3 lb)
Hydride Charge	- 19 Kg HY-STOR Alloy (42 lb)
Filtration	- 0.953 cm O.D., 0.635 cm I.D., 30.5 cm long, 2µ pore size (0.375" O.D., 0.250" I.D., 12" long)

The particular hydride in this example forms a very unstable hydride whose 25°C desorption plateau pressure is in the neighborhood of 4 MPa (580 psia). The difference between the hydride container and a simple pressure vessel is apparent from the curvature of the plots (a pressure vessel would yield a nearly constant rate of pressure drop).

The most rapid withdrawal (circled data) terminates so quickly (640 sec) that very little heat was supplied from the ambient air. Most of the heat consumed by the hydride was extracted from itself and its container ($C_p\Delta T$).

More gradual withdrawals permit increasing amounts of heat to be transferred from the ambient air into the hydride container. The dashed lines in Figure 8 indicate the occurrence of condensation and freezing of water vapor from the ambient air during the course of the experiments. These are casual observations which are, of course, affected by the ambient humidity at the time of the experiment.

Figure 9 shows that an instantaneous discharge of this particular system would be expected to deliver 800-900 liters of hydrogen. By plotting total hydrogen available versus flow period, a practical description of the system's heat transfer properties is obtained. The straight line provides a fair fit to the data and shows that about 0.23 liters/sec (0.49 CFM) may be used as the empirical heat transfer rate. Since no isotherms were available for this alloy, equation II is of no use. Assuming a decomposition enthalpy of 29 KJ/mole H_2 (12 KBTU/lb mole H_2) the corresponding heat transfer rate is 272 J/sec (928 BTU/hr). Since the ambient temperature was 26°C and the equilibrium temperature was about -40°C for this alloy, the effective heat transfer coefficient for the system is:

$$\frac{272 \text{ J/sec}}{[26-(-40)]\,°C} = 4.1 \text{ J/sec-}°C \quad (7.9 \frac{\text{BTU}}{\text{hr-}°F})$$

Actually the data indicate a curve whose slope is decreasing from left to right in Figure 9 as the thickness of the discharged metal region (see Figure 8) increases. A slightly lower heat-transfer rate than calculated above should be used to estimate the time required to discharge the hydride completely.

Another way to view the data is presented in Figure 10 where sustained flow capability is plotted versus withdrawal rate. At flows somewhat less than 0.2 liters/sec (0.4 CFM) nearly the entire hydrogen content of the cylinder (>3200 liters = 113 cubic ft.) is available, whereas for more rapid withdrawals, 25% or less may be withdrawn.

This information, as well as that presented in Figures 8 and 9, is only strictly applicable to a steady discharge of a fully charged cylinder against the prevailing atmospheric pressure (0.083 MPa, 12.1 psia). Performance for other demand pressures may be determined by drawing a new flow period axis onto Figure 8 at the pressure of interest.

b) Medium/Large Stationary Vessels

The first complete hydrogen energy storage system based on metal hydrides was installed by Public Service Electric and Gas Company in cooperation with Brookhaven National Laboratory. The vessel (Figure 11) weighs 1,240 pounds, contains 880 pounds of iron-titanium alloy and can store as much as 12 pounds of hydrogen. The design flow rate for hydrogen delivery is 1.5 lbs/hr but the vessel has been operated at twice this flow rate for a period of several hours. A complete description of this unit is contained in references [5, 6].

A detailed economic analysis has shown that hydride storage systems compare favorably with other techniques for storage of off-peak power. Also, the volumetric storage density is at least a factor of 2 better than advanced electrochemical storage systems. This, combined with the inherent safety of hydride storage, has encouraged Brookhaven National Laboratory to proceed with the development of this technology. Two Variable-Parameter Test Units (VPTU) are being tested to obtain basic data on hydride expansion and hydrogen charging and discharging rates. The first VPTU is shown in Figure 12. It contains six porous metal tubes for loosening the hydride bed and three other porous metal tubes for normal hydrogen

flow in or out of the vessel. The second VPTU has panels of
porous metal sheet for hydrogen distribution and filtration
(Figure 13). It also contains a hollow center body which is
designed to collapse before the vessel expands beyond the
yield point. Hydrides for these vessels will be supplied by
the Ergenics Division of MPD Technology Corporation. Figure
14 shows a typical shipment of HY-STOR alloys for this purpose.

Later on work will commence on a scaled-up version of the
storage vessel 0.9M (3 ft.) in diameter and 6.16M (20 ft.)
long. This vessel will be used as part of the Hydrogen
Chlorine Energy Storage program sponsored by the U.S. Department of Energy.

MPD Technology Corporation is presently designing a large
stationary hydride unit for use in conjunction with on-board
hydride tanks for fleet vehicles. This unit will contain
approximately 4000 pounds of metal hydride.

Safety Considerations

The following comments on safety apply not only to storage
containers but to any hydride application where the hydride
is contained within a vessel.

Tests on hydride containers of the type discussed above have
shown that severe swelling can occur if the hydriding alloy
migrates to the bottom of the cylinder, say, by repetitive
vibration. This swelling occurs because hydrides undergo a
volume expansion during hydriding and as powder particles
break down with repetitive hydride/dehydride cycles, they
settle and cause further expansion. Figure 15 shows the
swelling which was observed in a container which was repeatedly dropped onto a steel plate between cycles to accentuate
the setting and resulting expansion. Even the first cycle
swelled the container beyond the elastic limit.

Similar, although less severe, swelling problems have been
observed in horizontal tubes. Recognizing this swelling
problem, MPD Technology Corporation, Brookhaven National
Laboratory and others are now incorporating design features
which reduce or eliminate swelling. Some approaches are
described in the section on medium/large hydride vessels. It
should also be pointed out that extensive cycling of containers filled to slightly less than maximum capacity and
maintained in the horizontal position has not resulted in
strains beyond the elastic limit. Also, expansion is a function of alloy composition and for some alloys (FeTi, in par-

ticular), if packing density is kept relatively low, the elastic limit is not exceeded and hence, there is no permanent deformation of the container. Thus, swelling, although a potential problem, can be coped with through careful design and testing by the manufacturer.

The other requirements for safe handling of hydride containers are similar to the familiar procedures for handling flammable, compressed-gas cylinders. Hydrides have, in common with propane and other liquefied gases, a logarithmic relation between pressure and inverse temperature. These relationships have the form,

$$\ln P = \frac{A}{T} + B$$

where P is absolute pressure, T is absolute temperature and A and B are constants. Therefore, hydride cylinders must be kept below certain limits of temperature or the pressure will become excessive.

The worst case of overpressurization occurs when very unstable alloys, which have been fully charged to the rated pressure of the cylinder, become overheated. Permissible temperature limits should therefore be established for each hydride so that excessive pressure will not build up when a completely charged hydride is left in a warm location. As in the case of compressed gas cylinders, cylinder surface temperatures should be maintained below 65°C (150°F) to prevent pressure buildup and, in the case of aluminum containers, overaging of the aluminum alloy. This precaution is especially necessary during rapid recharging.

In addition to the safety aspects of hydride containers, the inherent properties of the alloys and their hydrides are also important. To a chemist who is accustomed to dealing with highly reactive hydride reducing agents such a $LiAlH_4$, the suggestion of dealing with tonnage quantities of hydrides would be cause for raised eyebrows. None of the hydrides currently considered for commercial uses have properties which vaguely resemble $LiAlH_4$. In order to place the safety aspects of storage-type hydrides in a proper perspective with other means for chemical energy storage, representative materials including $LaNi_5$, $FeTi$ and (mischmetal-Ca) Ni_5 have been extensively tested. The battery of tests employed in these evaluations includes DOD explosive sensitivity evaluation procedures, BOM bulk and dust explosibility testing plus additional procedures developed specifically for hydrides. A general summary of these studies indicates:

(1) DOD explosive sensitivity is essentially nil.
    (Worst result is normal combustion.)

(2) BOM explosibility data for several hydrides are
    described in Table VII.

(3) Special hydride tests identified spontaneous ignition
    followed by moderate burning as the worst result of
    hydride spillage. Many finely divided powders exhibit
    pyrophoricity to varying degrees and, provided adequate
    safety precautions are taken and proper handling proce-
    dures are followed, this need not pose a serious problem.
    There are also several hydride materials available which
    do not have this tendency, e.g., FeTi-H.

COMMERCIALIZATION PROSPECTS

Metal-hydride containers are capable of storing hydrogen gas
very compactly at moderate pressure. A compressed-gas cylinder
of equal size would require about 69 MPa (10,000 psi) to achieve
a storage density equivalent to a relatively low pressure metal-
hydride container. The hydrogen density in metal-hydride containers ranges around 0.06 gm H_2/cc of internal cylinder volume.

Most of the hydrides of interest may be charged in a reasonable
period of time with two or three MPa of applied H_2 pressure.
Under most anticipated conditions of use, only a few tenths of
a MPa of hydrogen pressure exists in the container. This considerably reduces the pneumatic rupture hazard in comparison to
that of the common, compressed-gas cylinders pressurized to 14
MPa (2000 psi). In the event of a rupture of a fully charged
container, the quantity of hydrogen which escapes instantly is
roughly 10-25% of the total contents, depending on initial conditions. The remaining gas will be released over a period of
time from a few minutes to many hours, depending on the type of
hydride and the temperature of the surroundings. This further
reduces the pneumatic hazard and greatly diminishes the immediate explosion hazard compared to a compressed-gas cylinder of
hydrogen.

Notwithstanding these improvements in safety compared to compressed gas, hydride cylinders must be handled properly or
serious consequences may result. Hydride cylinders of the
type discussed above should always be maintained in the horizontal position and placed in a nonvibrating location to minimize settling of the hydride powder during cyclic use. Settling
can result in plastic strain and rupture of the container.

Axial filtration is recommended to reduce the tendency for end-to-end powder migration within the cylinder even though much smaller filters are permissible from the standpoint of gas-flow capacity. Powder migration and subsequent packing may otherwise result in container strain and loss of flow capability during cyclic use of the container.

Considerable experience is being gained on large hydride storage systems and data to date indicate that no unsurmountable technical problems exist and economics are favorable.

HYDRIDE COMPRESSORS

Because of the low molecular weight of hydrogen, compression is normally performed in reciprocating compressors which operate at a relatively low speed. These reciprocating machines require a relatively high investment on a unit energy basis as compared to modern centrifugal and axial compression equipment. Reciprocating compressors also have serious sealing problems due to the rapid diffusion of the relatively small hydrogen molecules and the attack by hydrogen on sealing materials in nonlubricated designs.

Centrifugal compressors will not produce the required pressure ratios without extensive multistaging. Regenerative compressors have a much higher head coefficient, but have a lower efficiency.

The use of metal hydrides to compress hydrogen has been proposed by Powell and Salzano [7]. North American Philips Corporation has also demonstrated a compressor module using LaNi$_5$ which operates in the temperature range 25°C to 80°C. Denver Research Institute and MPD Technology Corporation are developing a prototype compressor based on proprietary hydrides with superior performance and lower cost than the LaNi$_5$ alloy. The essential features of this compressor in its simplest form are shown in Figure 15.

Selection of the hydride was a key element in the success of this compressor. Not only does the hydride permit hydrogen absorption at atmospheric pressure and room temperature, but it also exhibits stability at the temperature of hydrogen discharge, whereas LaNi$_5$ partitions into the higher, more stable hydrides at the same temperature.

The sequence of operations is as follows:

1) Hydrogen enters the system at pressure P_{Lo} through check valve #1 and saturates the hydride to an extent determined

by its absorption pressure-composition isotherm at the temperature of the cooling means, T_C.

2) Heat is supplied at temperature T_H warming the hydride. As its temperature increases the hydrogen desorption pressure, P_d, increases in accordance with the desorption van't Hoff relationship which is expressible as

$$P_d = \exp\left(\frac{A}{T} + B\right)$$

where A (actually $\frac{\Delta H}{T}$) and B are specific to the hydride used in the compressor.

3) When the value of P_d exceeds the pressure on the output side of the compressor, P_{Hi} check valve #2 opens and hydrogen exits the compressor at P_{Hi}.

If we neglect the hysteresis effects, i.e., assume that the absorption and desorption pressures are identical at the lower temperature, T_C, the pressure ratio is

$$R_p = \frac{P_d(T_H)}{P_d(T_C)} = \exp A(1/T_H - 1/T_C)$$

Assuming a hydride enthalpy of -7kcal/mol H_2 with T_C = 298°K (76°F) and T_H = 470°K (386°F) we find that

$$A = \frac{-7 \times 10^3 \text{cal/mole}}{1.987 \text{ cal/mole-°K}} = -3523°K$$

and hence

$$R_p = 75.7$$

so if the input pressure is 2 atm, the output will be 151 atm or about 2220 psia.

Since most hydride materials do exhibit some absorption-desorption pressure hysteresis this should be considered. An easy way to deal with hysteresis analytically is to define T_C^1 such that $P_d(T_C^1) = P_a(T_C)$ and redefine the pressure ratio as

$$R_p = \exp A[1/T_H - 1/T_C^1]$$

With the current knowledge of unstable hydride materials it is possible, within reasonable limits, to engineer hydrides to a set of process specifications. If the input pressure in the above example were increased to 10 atm, the hydride material should be altered to achieve near saturation at T_C at 10 atm rather than supersaturating the original material. By doing so, we may either increase the available output pressure holding T_H constant or maintain the previously calculated 151 atm output at a reduced value of T_H.

The design of a full scale unit would proceed from considerations of waste heat availability and quality, and output pressure requirements for a particular application. The hardware design would be largely concerned with the costs of hydriding metals and heat exchanger surface area which would dictate the amount of hydride and the cycle rate needed to give the required hydrogen output rate.

An important practical concern in the design of a hydride compressor is the behavior of the material after thousands of repetitive cycles. Fortunately, extensive tests have shown that metal hydrides can be cycled many times without deterioration in kinetics or hydrogen capacity. The predominant physical change is a reduction in hydride particle size and this can be handled by design.

With the flexibility afforded by different hydride compositions, it is possible to conceive of a variety of compressors consisting of multistage units using two or more different hydrides which would compress hydrogen from one to over 100 atm with low-grade heat as the energy source.

HYDROGEN PURIFICATION

The formation of metal hydrides in the presence of hydrogen and certain hydrogen-gas mixtures suggests that hydrides could be used as purifiers by selectively stripping hydrogen from mixed gases. Unfortunately, all known hydrides are poisoned to varying degrees by common gaseous species including oxygen, carbon monoxide and nitrogen. Once poisoned, the ability of hydrides to accept hydrogen is severely limited and, in some instances, completely destroyed. Since, however, the poisoning effect can be reversed by following the activation procedure for virgin material, it is possible to conceive of a purification system in which the hydride operates for one or more cycles, accepting and rejecting hydrogen, with contaminant gases being periodically removed by reactivating the hydride. For continuous operation, a moving hydride bed or a series of beds could be employed.

Certain gas mixtures may be separated without the need for reactivation. For example, Cholera and Gidaspow [8] have shown the possibility of complete separation of hydrogen from a hydrogen-methane mixture using an Fe-Ti-Ni alloy supplied by the Ergenics Division of MPD Technology Corporation.

Table VIII gives the results of a study by Reilly and Wiswall [9]. They exposed $LaNi_5$ and $LaCu_4Ni$ to gas mixtures corresponding to those which might be encountered in gas produced by steam reforming of a hydrocarbon fuel. Hydrogen pickup was observed in all but one case and the copper containing hydride appeared to be more tolerant to poisoning, suggesting the possibility of developing poisoning resistant compositions.

Current purification systems involve the use of palladium-silver membranes, cryogenic separation or pressure swing adsorption. All involve high capital cost and high operating cost. Depending on the recycle rate of metal hydrides, it is possible to envision a relatively small hydride plant handling large volumes of gases. In such circumstances, lower capital and operating expenses are possible.

Reaction rates in practical metal hydride devices will usually be governed by the heat transfer and gas dynamic aspects of the system rather than by chemical kinetics. Several studies of reaction kinetics [10,11] have, despite considerable lack of agreement, established that these reactions will go to near completion in seconds. Reasonable container dimensions, on the order of a few inches, generally require many minutes and sometimes hours to transfer enough heat to charge or discharge a metal hydride to near completion.

The preceding discussion of heat transfer assumed that the flow of heat to or from the container is limiting the sorption rates. The use of impure hydrogen, however, may seriously impede the hydriding-dehydriding rates so that the assumption of a heat transfer limited process is incorrect. One of the most important hydriding alloys, FeTi, is especially sensitive to common impurities found in commercial hydrogen. Work is currently under way at International Nickel Research and Development Center and DRI to quantify the effects of gaseous impurities on the rate and extent of hydrogen absorption by various alloys.

Contamination tolerance in hydrides for use in commercial hydrogen handling systems is highly desirable. A moderate tolerance, such as that exhibited by $LaNi_5$, permits the use of commercial grade hydrogen ($\sim$99.95%) without serious degradation in capacity or sorption rates. Binary FeTi would quickly deteriorate using such gas unless some means for purification were employed to keep the contaminants out of the

hydride container. The most attractive potential use for contamination tolerant hydriding alloys is the extraction of
hydrogen from mixed gas streams. The synergistic effects of
the major components of hydrogen containing commercial gas
streams is being studied in addition to the effects of individual contaminant gases.

The philosophy of our experimental approach in characterizing
hydrides for particular applications emphasizes practical utility rather than intrinsic chemical kinetics. A common reaction
vessel of about 1 cm I.D. and 3 cm in length, closed at one end
is used. It is recognized that such a container imposes a limit
on the resulution of small losses in sorption rates because its
heat transfer properties will mask them. It seems unlikely,
however, that effects which are below our sensitivity will be of
commercial importance. Also, since the reactor is closed at one
end there is no circulation and, as hydrogen is absorbed from a
gas mixture, the concentration of impurities increases inside
the reactor. This is not easy to treat theoretically but is
preferred over the flowing sample method used by others [12]
because of experimental simplicity and similarity to the most
common hydride systems designs.

Automatically timed cycles of absorption and evacuation are
monitored by a strip chart recorder so that cumulative effects
of exposure to a given gas mixture can be observed and compared
to baseline data obtained with high purity hydrogen. Figure 17
shows the two basic types of deterioration which are observed
in these experiments.

Loss of capacity is usually accompanied by rate degradation so
that combined effects are often observed. Rate reduction can
occur without loss in capacity, however, and this is the crux
of the evaluation. If the rate of absorption of hydrogen from
a contaminated source is rapid enough to serve the application
then the heat transfer design may proceed as usual (although
the assumption of a sharp reaction front is less realistic than
before). If the rate is not sufficient, then other alloys and/
or operating conditions (i.e., pressure and temperature) must
be investigated.

ISOTOPE SEPARATION

Research at Daimler-Benz [13], Brookhaven National Laboratories
[14] and General Electric [15] have shown that certain hydrides
preferentially absorb light hydrogen. Thus, inexpensive H_2/D_2
separation processes are possible with an energy consumption
far below that of conventional methods.

Buchner [13] found that TiNi and Ti_2Ni and their mixtures se-

lectively absorb hydrogen. He envisioned a continuous closed
cycle process in which enrichment might reach a factor of more
than 50,000 in a relatively small number of steps. Since modern heavy-water-moderated power reactors may require up to 200
tons of pure D_2O for cooling and neutron moderation and 40 tons
of D_2 is needed for this purpose, the use of a Ti-Ni separation
process would also yield 3×10^5 tons of high purity hydrogen
as a by-product. This would cover the annual requirements of
one million hydrogen powered vehicles. Thus, producing deuterium with metal hydrides could bring about a considerable
reduction in the cost of producing hydrogen.

VEHICULAR SYSTEMS

As a near-term alternative to the bulky and limited range bat-
tery-driven vehicles now being used and developed, the hydrogen
fueled internal combustion engine appears promising. Metal
hydrides offer an attractive alternative to cryogenic or compressed gas storage. Hoffman et al [16] presented the first
detailed treatment of vehicular metal hydride fuel systems in
1969. Since then, several hydrogen powered vehicles utilizing
metal hydrides have been built. Hydrogen powered fleet vehicles serviced by central garages offer particularly attractive
possibilities. Such vehicles, because of the almost zero pollution characteristics of hydrogen fuel, could be very desirable in congested urban and industrialized areas. Billings
Energy Corporation [17] and Daimler-Benz [18] have extensively
tested hydride systems in buses (Figures 18a and 18b). Denver
Research Institute and MPD Technology Corporation are developing a prototype hydrogen powered forklift truck (Figure 19).

The search for lightweight hydrides is continuing since, for
most surface vehicles, the hydride weight ultimately limits
the range of performance of the vehicle by limiting the quantity of fuel carried. Hybrid systems may facilitate the introduction of hydrides. For example, the concept of hydrogen
injection into gasoline fueled IC engines is being studied by
the Jet Propulsion Laboratory. The use of metal hydrides
would avoid the complexities of an on-board reformer. The
latter approach has been demonstrated by Daimler-Benz A.G.

A detailed discussion of hydrogen vehicle developments is provided in another paper [19]. The pace of development in this
area is accelerating. Thus, while it may be some time before
private automobiles employ hydrides extensively because of the
massive investment required for production and distribution,
several fleet vehicle applications will undoubtedly arise during the next decade.

HYDROGEN STORAGE ELECTRODE

An electrode which can quasi-reversibly store large quantities
of hydrogen is of interest as a construction material. Will
[20] has patented a hermetically sealed secondary battery using
a lanthanum nickel anode in an alkaline electrolyte. Work at
the Centre National de la Recherche Scientifique [21] has led
to a series of $LaNi_5$-based alloys by partially substituting
other elements for La and Ni for use as hydrogen storage electrodes in alkaline batteries. These alloys are characterized
by a high hydrogen storage capacity, can be charged at atmospheric pressure within a wide range of temperatures, and are
resistant to contamination by aqueous alkaline solutions.

Other studies of $LaNi_5$ and similar alloys for hydrogen storage
electrodes are proceeding in several countries. This intense
research activity suggests that high energy density batteries
based on metal hydrides may be close at hand.

In addition to $LaNi_5$ systems, Buchner et al have developed
reversible electrodes based on TiNi alloys [22,23]. These
electrodes have many advantages including: high storage capacity, 70 to 80 Wh/kg with NiOOH electrodes (10-hour discharge
rate), high current density at 0°C, long life (between 300 and
500 discharge cycles), insensitivity to oxygen and water below
100°C, high charging efficiency, low maintenance requirements,
high chargeability at low temperatures and long storage and
service life without maintenance. They appear to be useful
both as electrochemical storage units for off-peak power storage and low temperature automotive batteries.

HEATING, HEAT PUMPS, AIR-CONDITIONING,
REFRIGERATION AND POWER GENERATION

The heat effects of hydrogen-metal interactions offer the possibility of using hydrides for heat storage, generation, pumping, air-conditioning, refrigeration and power generation. An
ideal working fluid for refrigeration cycles is one having good
thermal stability, high cycle efficiency, lower corrosiveness,
favorable critical properties and low toxicity. Hydrogen fits
this requirement very well.

An integrated system for thermal storage, space-conditioning,
power generation and refrigeration (HYCSOS) has been developed
by Argonne National Laboratory [24]. This system consists of
four stainless steel tanks (Figure 20). Two hydrides, $LaNi_5$
and $CaNi_5$ supplied by the Ergenics Division of MPD Technology
Corporation are used. A summary of the data obtained in three
cycles in the storage mode (the high temperature decomposition

of CaNi$_5$ hydride and formation of LaNi$_5$ hydride) and three
cycles in the recovery mode (the low temperature decomposition
of LaNi$_5$ hydride and formation of CaNi$_5$ hydride) are shown in
Tables IX and X, respectively. These data represent the first
15 minutes of a 45-minute cycle when 75% to 90% of the hydrogen
was transferred. Considering the fact that this system was
operated without insulation, these data clearly demonstrate
that a two-hydride concept for storage and recovery of thermal
energy for heating, cooling and energy conversion in an integrated system is viable. An economic study by TRW [25] indicates that this system will be cost-competitive with conventional absorption refrigeration systems.

A 100-ton baseline system was estimated to have a coefficient
of performance of 0.46 at a total cost of $17,292. This com-
pares to a cost of $20,000 for a 100-ton lithium bromide ab-
sorption heat pump which exhibits a COP of 0.68.

A particularly attractive feature of this system is its ability to use solar thermal input when little heating or cooling
is required. This means that the cost of solar collectors can
be amortized over the whole year.

HYCSOS is undoubtedly the first of a whole new class of heat-
ing/refrigeration/power systems based on metal hydrides. New
generations of alloys based on the LaNi$_{5-x}$Al$_x$ or MnNi$_{5-x}$Al$_x$
are being developed for this application.

In the HYCSOS system, power is generated by expanding hot
hydrogen isothermally by adding heat to the expander during
expansion. This additional heat is much less than the desorption energy (about 1100 Btu/lb H$_2$ compared to 6400 Btu/lb H$_2$
desorbed). The theoretical efficiency is 0.15, comparable to
other power cycles working between 250°F and 95°F. Figure 21
shows the theoretical power cycle operation.

A sophisticated power generation/refrigeration system has also
been proposed by Terry and Schoeppel [26]. This comprises four
or more reactors, a means for periodically supplying hydrogen
gas and alternately heat, in out-of-phase staggered cycles to
each reactor, a means for removing heat and continuously converting the energy of pressurized hydrogen to shaft power.
Figure 22 is a schematic flow diagram illustrating the use of
this system as a bottoming unit for a nuclear plant.

Spent stream from the nuclear plant goes to the hydride reactors. These reactors are operated in staggered or out-of-
phase sequence during a cycle, with each reactor undergoing

a hydriding, pressurizing, dehydriding and depressurizing
phase. At the outset, the alloy in reactor 24 is charged with
hydrogen from the turbine (from a previous cycle) at a typical
pressure of 6.8 atmospheres and a temperature of about 210°K
(-81°F). The exothermic heat of reaction and cold water circulation maintains the hydride bed temperature at or below
100°F, the equilibrium temperature for this particular hydride.

Residual steam now circulates through this reactor and the
pressure increases to about 55 atmospheres. Complete pressurization can be achieved by increasing the temperature to 386°K
(235°F). The reactor is now dehydrided by expanding through
a turbine to produce shaft power. Dehydriding continues while
the bed temperature is maintained relatively constant until
all the hydrogen is discharged. Hydrogen from the turbine
returns to the reactor and the cycle is repeated.

By properly synchronizing the out-of-phase operation of the
four reactors, hydrogen gas under high pressure is continuously
supplied to the turbine and continuous power generation is realized. In the sequential mode of operation of the four reactors, one reactor is undergoing dehydriding, a second reactor
is undergoing pressurizing, a third is undergoing hydriding
and the fourth, depressurizing. In order to more fully explain and clarify the synchronization and sequential operation
of the phases of this process, Table XI shows the status of
the various valves used in the heat exchange medium delivery
system as the several reactors undergo phase changes.

APPLIANCES

Recently, the world's first hydrogen powered home was built in
Provo, Utah [27]. This contains a hydrogen fueled range, oven
and bargeque grill. Burner conversion to hydrogen was achieved
by eliminating primary air and adding flame colorants. Burner
construction comprises a spoke shaped spreader nozzle to dis-
tribute hydrogen within a wire mesh catalyst. Stainless steel
shavings serve as the catalytic element and the mesh inhibits
mixing of hydrogen and air. Combustion begins at air concentrations exceeding 25% by volume.

Thus, the practicality and simplicity of hydrogen conversion
has been demonstrated and one can envision the coupling of
hydride storage containers to a variety of heating appliances.
Hydrides are also being considered as a source of ignition for
gas appliances, thereby eliminating the need for pilot lights
and conserving energy [28]. An important advantage of such
catalytic burners is that the operating temperature is under

1750°K, below the temperature of formation of significant amounts of nitrogen oxides. Also, an inoperative ignition source or burner accidentally left on in a natural gas appliance poses a serious problem. A hydrogen catalytic heater, on the other hand, would only produce unwanted heat if accidentally left on.

It appears that hydride catalytic devices could be used for water heating, cooking, process heating and space heating. The only likely technical obstacles to development of practical systems are the availability of catalytic material and an economical source of hydrogen.

SUMMARY AND CONCLUSIONS

In the few years since the first discovery of moderate temperature reversible metal hydrides, much progress has been made in understanding the factors controlling hydride formation and hydride performance. This has permitted the development of a wide variety of systems based on metal hydrides.

Many problems such as hydride packing, swelling, disproportionation and poisoning remain to be resolved before these systems can be considered commercially viable. However, results to date are encouraging and no insurmountable technical problems have been identified. As more companies become aware of the potential of metal hydrides, new applications will undoubtedly emerge.

Before the end of this century, it is predicted that metal hydrides will play an important role in industry. As the cost of hydrogen is reduced and more cost effective, lighter weight hydrides are developed, metal hydrides could have the same impact on society in the next century as gasoline has had in the 20th century.

REFERENCES

1. Sandrock, G.D., "A New Family of Hydrogen Storage Alloys Based on the System Nickel-Mischmetal-Calcium," Proc. 12th IECEC, Vol. 1, 95, 1977, Washington, D.C.

2. Lundin, C.E. and Lynch, F.E., "A New Rationale for the Hysteresis Effects Observed in Metal-Hydrogen Systems." Proc. International Symposium on Hydrides for Energy Storage. Geilo, Norway. August 1977.

3. "Hydrogen in Metals, Hydrogen Storage, Hydrogen Purifica-

tion." Technical Information No. 6, Kernforschungsanlage Jülich GMBH, 5170 Jülich, W. Deutschland.

4. Hydrogen Progress, 2nd Quarter 1977, published by Billings Energy Corporation.

5. Strickland, G. and Reilly, J.J., "Operating Manual for the PSE&G Reservoir Containing Iron Titanium Hydride." BNL-19725, February 1974.

6. Strickland, G., Reilly, J.J. and Wiswall, R.H. Jr., "An Engineering-Scale Energy Storage Reservoir of Iron Titanium Hydride," The Hydrogen Economy Miami Energy Conference, Miami, Florida, March 18-20, 1974.

7. Powell, J.R. and Salzano, F.J., "Hydride Compressor," U.S. Patent Serial No. 646,703, January 5, 1976.

8. Cholera, V. and Gidaspow, D., "Hydrogen Separation and Production From Coal-Derived Gases Using Fe_xTiNi_{1-x}," Proceedings of the 12th Intersociety Energy Conversion Engineering Conference.

9. Wiswall, R.H. Jr. and Reilly, J.J., "Metal Hydrides for Energy Storage," Proceedings of the 7th Intersociety Energy Conversion Engineering Conference, September 25-29, 1972, San Diego, California.

10. Lundin, C.E. and Lynch, F.E., "A Detailed Analysis of the Hydriding Characteristics of $LaNi_5$," Proceedings of the 10th Intersociety Energy Conversion Engineering Conference, August 1975, p. 1380.

11. Boser, O. and Lehrfeld, D., "The Rate Limiting Processes for Sorption of Hydrogen in $LaNi_5$," Proceedings of the 10th Intersociety Energy Conversion Engineering Conference, August 1975, p. 1363.

12. Reilly, J.J. and Wiswall, R.H. Jr., "Hydrogen Storage and Purification Systems," Brookhaven National Laboratories Publication 17136, August 1, 1972.

13. Buchner, H., "Method for Preparation of Deuterium by Isotope Separation," U.S. Patent 3,940,912, March 1976.

14. Wiswall, R.H. Jr. and Reilly, J.J., "Inverse Hydrogen Isotope Effects in Some Metal Hydride Systems."

15. Römpps Chemie Lexikon, Bd. 2, p. 803, 7, Auflage 1973.

16. Hoffman, K.C., Wische, W.E., Wiswall, R.H., Reilly, J.J., Sheehan, T.V. and Waide, C.H., "Metal Hydrides as a Source of Fuel for Vehicular Propulsion," SAE paper 690232 presented at the International Automotive Engineering Conference, Jan. 13-17, 1969, Detroit, U.S.A.

17. Wooley, R.L., Simons, H.M. and Billings, E.R.C., "Hydrogen Storage in Vehicles - An Operational Comparison of Alternative Prototypes," Fuels and Lubricants Meeting, SAE Paper 760570, June 1976, St. Louis, U.S.A.

18. Buchner, H. and Saüfferer, H., "Results of Hydride Research and the Consequences for the Development of Hydride Vehicles," 4th Symposium on Low Pollution Power Systems Development, CCMS, NATO, April 1977, Washington, U.S.A.

19. Lynch, F.E. and Snape, E., "Technical and Economical Aspects of In-Plant Hydrogen Fueled Fleet Vehicles."

20. Will, F.G., "Hermetically Sealed Secondary Battery with Lanthanum Nickel Anode," U.S. Patent 3,874,928, April 1, 1975.

21. Percheron, A., Achard, J-C, Loriers, J., Bonnemay, M., Bronoel, B., Sarradin, J. and Schalapbach, L., French Patent Specification 75-16160, May 21, 1975.

22. Gutjahr, M.A., Buchner, H., Beccu, K.D. and Saüfferer, H., "A New Type of Reversible Electrode for Alkaline Storage Batteries Based on Metal Alloy Hydrides," 8th International Power Sources Conference, Sept. 1972, Brighton.

23. Buchner, H., Gutjahr, M.A., Beccu, K.D. and Saüfferer, H., "Wasserstoff in Intermetallischer Phasen am Beispiel des Systems Titan-Nickel-Wasserstoff," Z. Metallkunde 63, p. 417, 1972.

24. Gruen, D.M., McBeth, R.L., Mendelsohn, M., Nixon, J.M., Schreiner, F. and Sheft, I., "HYCSOS: A Solar Heating, Cooling and Energy Conversion System Based on Metal Hydrides," Presented at IECEC Conference, Sept. 1976, Lake Tahoe, Nevada, U.S.A.

25. Gorman, R. and Akridge, W.L., "Hydride Heat Pump System for Building Air-Conditioning Using High Temperature Solar Input," TRW Energy Systems Group, Work Done Under Auspices National Laboratory Purchase Order No. 898403 and SN No. 97164, 1977, McLean, Virginia, U.S.A.

26. Terry, L.E. and Schoeppel, R.J., U.S. Patent 3,943,719, March 16, 1976.

27. "Hydrogen Homestead," Hydrogen Progress, Fall Quarter 1977, published by Billings Energy Corporation.

28. Gregory, D.P., "A Hydrogen-Energy System," American Gas Association, August 1972.

TABLE I

Comparison of Hydrogen Storage Media

Medium	Hydrogen Content Wt %	H Storage Capacity g/ml of vol	Energy Density Heat of Combustion** (higher) cal/g	cal/ml of vol
MgH_2*	7.0	0.101	2373	3423
Mg_2NiH_4	3.16	0.081	1071	2745
VH_2	2.07		701	
$FeTiH_{1.95}$	1.75	0.096	593	3245
$TiFe_{.7}Mn_{.2}H_{1.9}$	1.72	~0.09	583	~3050
$LaNi_5H_{7.0}$	1.37	0.089	464	3051
$R.E.Ni_5H_{6.5}$	1.35	~0.09	458	~3050
Liquid H_2	100	0.07	33900	2373
Gaseous H_2 (100 atm pressure)	100	0.007	33900	244
N-Octane			11400	8020

*Starting alloy 94%Mg-6%Ni
**Refers to H only in metal hydrides

TABLE II

IMMEDIATE/NEAR-TERM APPLICATIONS OF METAL HYDRIDES

- HYDROGEN STORAGE
- HYDROGEN COMPRESSION
- HYDROGEN PURIFICATION
- ISOTOPE SEPARATION
- VEHICULAR SYSTEMS
- HYDROGEN STORAGE ELECTRODES
- HEATING
- HEAT STORAGE
- HEAT PUMP
- AIR-CONDITIONING
- REFRIGERATION
- POWER GENERATION
- WASTE HEAT RECOVERY
- APPLIANCES

TABLE III

COMPARISON OF SEVERAL
HYDROGEN STORAGE SYSTEMS

Storage Volume 1 Nm^2 = 90g Hydrogen	Pressure-container	Fe-Ti-container	Cryostat (liquid)
Volume (l)	10	1.0	1.3
Total mass (kg)	17	6.5	4.0
Working pressure (bar)	0-100	5-30	1.0
Safety	-	+	-

TABLE IV

OPERATING CHARACTERISTICS OF
KFA HYDRIDE CONTAINER

Max. stored hydrogen	1 Nm^3 = 90g
Container volume	1,41
Working pressure at room temperature	5-30 bar
Storage metal	Fe-Ti-alloy 50A%Fe/50A%Ti as grains, size 0.5-1mmϕ
Mass of grains	4,670 kg
Total mass of container	7kg
Minimal H$_2$-withdrawal and loading time	90% of the stored H$_2$ in approx. 15 minutes

TABLE V

SPECIFICATIONS OF BILLINGS AHT-5* HYDRIDE TANK

Cylinder Capacity:	230g of Hydrogen
Volume:	4.75 liters
Length:	65.4 cm
Diameter:	11.1 cm
Weight:	18.6 kg
Material:	Aluminum
Hydride:	Iron-Titanium

*AHT-5 is a proprietary term
of Billings Energy Corporation

TABLE VI

SOME HYDRIDING "HY-STOR" ALLOYS CURRENTLY BEING SOLD BY MPD TECHNOLOGY CORPORATION

Trade Name	Alloy
HY-STOR 100 Series (Fe-Base)	
HY-STOR 101	FeTi
102	$(Fe_{.9}Mn_{.1})Ti$
103	$(Fe_{.8}Ni_{.2})Ti$
HY-STOR 200 Series (Ni-Base)	
HY-STOR 201	$CaNi_5$
202	$(Ca_{.7}M_{.3})Ni_5$*
203	$(Ca_{.2}M_{.8})Ni_5$
204	MNi_5
205	$LaNi_5$
206	$(CFM)Ni_5$**
207	$LaNi_{4.7}Al_{.3}$
HY-STOR 300 Series (Mg-Base)	
HY-STOR 301	Mg_2Ni
302	Mg_2Cu

TABLE VII

RELATIVE EXPLOSIBILITY OF COAL DUST AND METAL HYDRIDES

Material	Bureau of Mines Explosibility Index
Pittsburgh Coal Dust	1
FeTi Hydride	<.1
CaNi$_5$ Hydride	.1
LaNi$_5$ Hydride	<.1
Uranium Hydride	10

TABLE VIII

Hydriding of Various Alloys in Hydrogen-Containing Gas Mixtures

Alloy	Composition of gas	Temp. °C	Pressure, psia	Flow rate liters/min.	Total flow liters	Product composition
LaNi$_5$	74 H$_2$ 26 CO$_2$	25	400		0.6	LaNi$_5$H$_{6.1}$
LaNi$_5$	99 H$_2$ 1 CO	25	400-250	0.4	116	(no H pickup)
LaNi$_5$	99.95 H$_2$ 0.05 CO	25	487-200		8.0	LaNi$_5$H$_{5.7}$
LaNi$_5$	97.0 H$_2$ 3.0 air	25	475-375	0.1	1.0	LaNi$_5$H$_{5.1}$
LaNi$_5$	72 H$_2$ 28 CO$_2$ sat. H$_2$O	25	286-195		6.0	LaNi$_5$H$_{4.1}$
LaNi$_5$	79.3 H$_2$ 20.3 CO$_2$ 0.3 CH$_4$ 700 ppm N$_2$ 20 ppm CO sat. H$_2$O	25	175	0.17	4.00	LaNi$_5$H$_{6.4}$
LaCu$_4$Ni	100 H$_2$	22	575	static	-	LaCu$_4$NiH$_{4.9}$
LaCu$_4$Ni	72 H$_2$ 24 CO$_2$ 4 CO sat. H$_2$O	124	430-460	static	-	LaCu$_4$NiH$_{2.7}$

TABLE IX

STORAGE MODE HYDROGEN TRANSFER:
CaNi$_5$ HYDRIDE TO LaNi$_5$

Run	H$_2$ Moles	CaNi$_5$ HYDRIDE			LaNi$_5$ HYDRIDE		
		ΔH_B kcal	Heat added kcal	Heat Add. ΔH_B %	ΔH_B kcal	Heat Rec. kcal	Heat Rec. ΔH_B %
6	20.65	154.88	177.20	115	148.68	135.03	91
8	22.59	169.42	203.36	120	162.65	122.66	75
10	21.09	158.18	164.60	104	151.85	124.22	82

TABLE X

RECOVERY MODE HYDROGEN TRANSFER:
LaNi$_5$ HYDRIDE TO CaNi$_5$

Run	H$_2$ Moles	CaNi$_5$ HYDRIDE			LaNi$_5$ HYDRIDE		
		ΔH_B kcal	Heat Rec. kcal	Heat Rec. ΔH_B %	ΔH_B kcal	Heat added kcal	Heat Add. ΔH_B %
7	23.36	175.20	149.75	85	168.19	119.09	71
9	18.10	135.75	137.22	101	130.32	97.04	74
11	21.91	164.32	164.30	100	157.75	157.75	76

TABLE XI

	Reactor 24	Reactor 26	Reactor 28	Reactor 30
Hydriding Phase				
Valves Open	91,68,76,150	152,98,78,70	100,72,80,154	156,102,74,82
Valves Closed	104,42,126,166,120	168,106,120,122,44	108,46,170,124,122	172,110,126,124,48
Pressurizing Phase				
Valves Open	126,68,104	120,70,106	122,72,108	124,74,110
Valves Closed	91,166,120,42,76,150	122,98,44,78,152,168	80,124,154,170,46,100	126,48,156,172,102,82
Dehydriding Phase				
Valves Open	42,120,166	44,122,168	170,46,124	48,126,172
Valves Closed	150,91,104,68,76,126	78,152,70,98,106,120	154,72,122,100,108,80	82,74,156,102,110,124
Depressurizing Phase				
Valves Open	91,68,76	98,78,70	100,72,80	102,74,82
Valves Closed	150,166,42,120,126,104	152,168,106,44,122,120	154,170,46,122,124,108	156,172,126,124,48,110

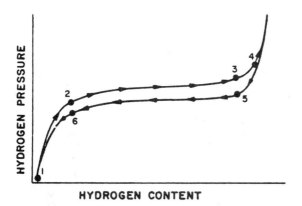

Figure 1. Typical absorption (upper curve) and desorption (lower curve) isotherms for a metal-hydrogen system. The separation between the curves is the hysteresis.

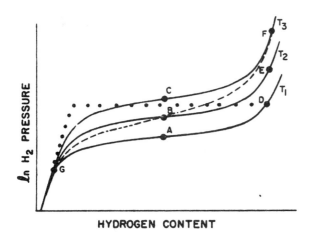

Figure 2. Set of isotherms for temperatures T_1, T_2 and T_3. Dashed line depicts non-isothermal discharge. Dotted line shows constant input pressure charging.

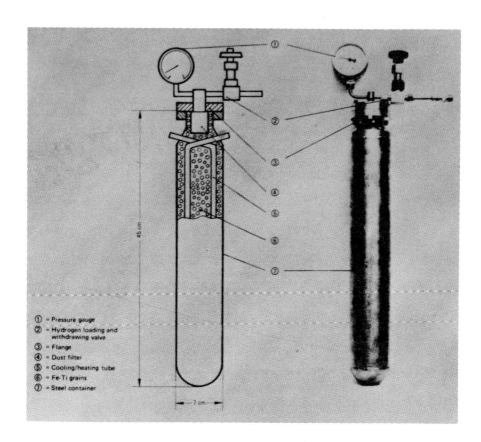

① = Pressure gauge
② = Hydrogen loading and
     withdrawing valve
③ = Flange
④ = Dust filter
⑤ = Cooling/heating tube
⑥ = Fe-Ti grains
⑦ = Steel container

Figure 3.   KFA Laboratory Hydrogen Storage Container

Figure 4. Comparison of Ergenics Division (MPD Technology
Corporation) HY-STOR Container With Conventional
Compressed Gas Cylinder

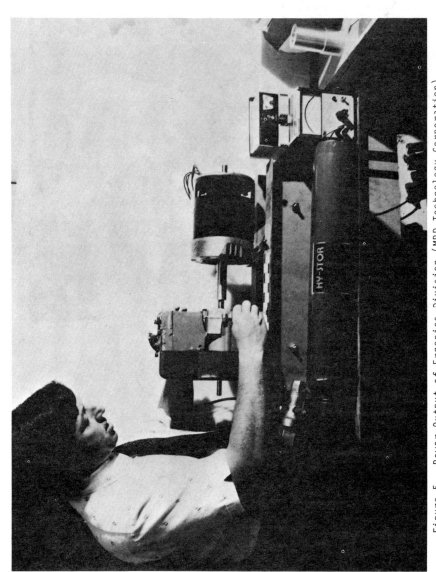

Figure 5. Power Output of Ergenics Division (MPD Technology Corporation) HY-STOR Container

Figure 6.   Ergenics Division (MPD Technology
             Corporation) HY-STOR Container and
             Control Panel Coupled to a General
             Electric SPE Electrolyzer

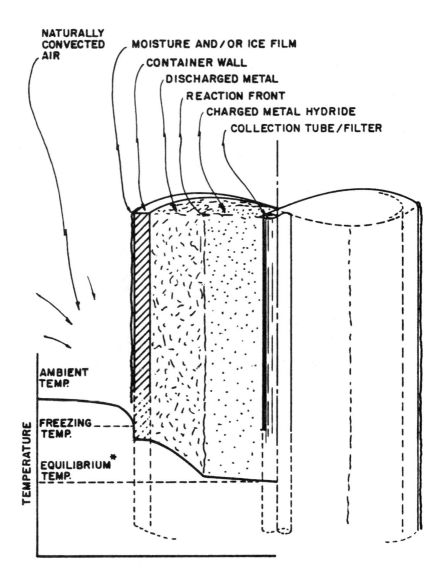

NATURALLY CONVECTED AIR

MOISTURE AND / OR ICE FILM

CONTAINER WALL

DISCHARGED METAL

REACTION FRONT

CHARGED METAL HYDRIDE

COLLECTION TUBE / FILTER

AMBIENT TEMP.

FREEZING TEMP.

EQUILIBRIUM* TEMP.

TEMPERATURE

Figure 7. TEMPERATURE PROFILE FOR A HEAT TRANSFER LIMITED HYDRIDE CYLINDER DURING DISCHARGE.
*EQUILIBRIUM TEMP. IS THE TEMP. AT WHICH THE DESORPTION PLATEAU PRESSURE IS EQUAL TO THE DEMAND PRESSURE.

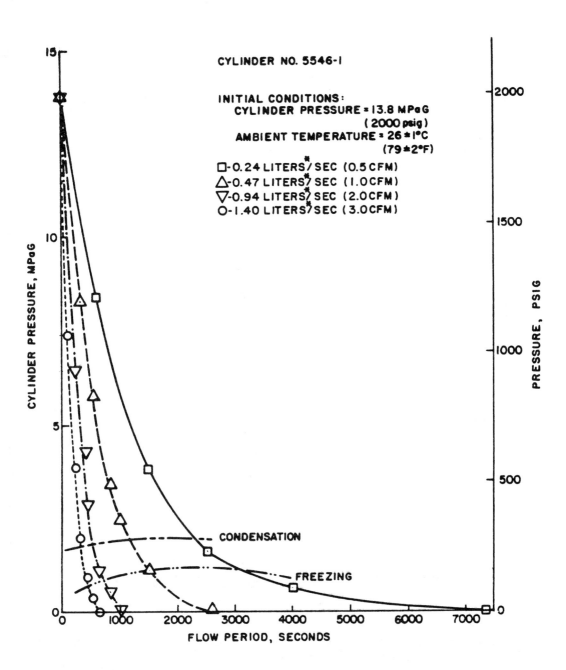

Figure 8.   CYLINDER PRESSURE vs FLOW PERIOD FOR SEVERAL WITHDRAWAL RATES
*21°C, 0.101 MPa (70°F, 14.7 psia)

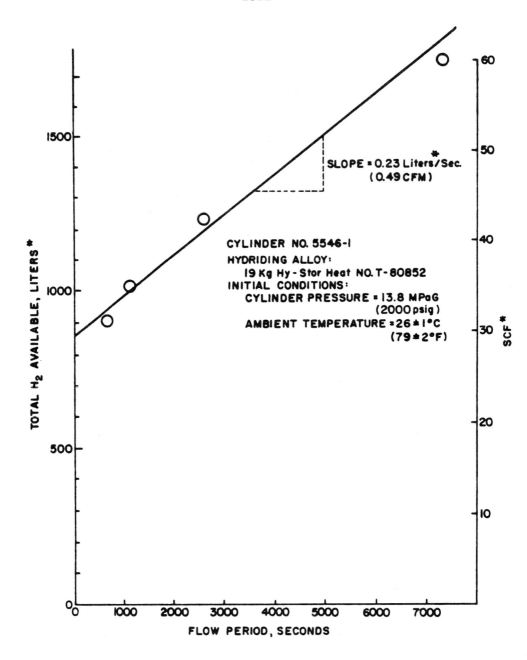

Figure 9.   TOTAL HYDROGEN AVAILABLE vs TIME, SLOPE IS PROPORTIONAL TO
EFFECTIVE HEAT TRANSFER RATE OF THE SYSTEM. * 21°C, 0.101 MPa
(70°F, 14.7 psia)

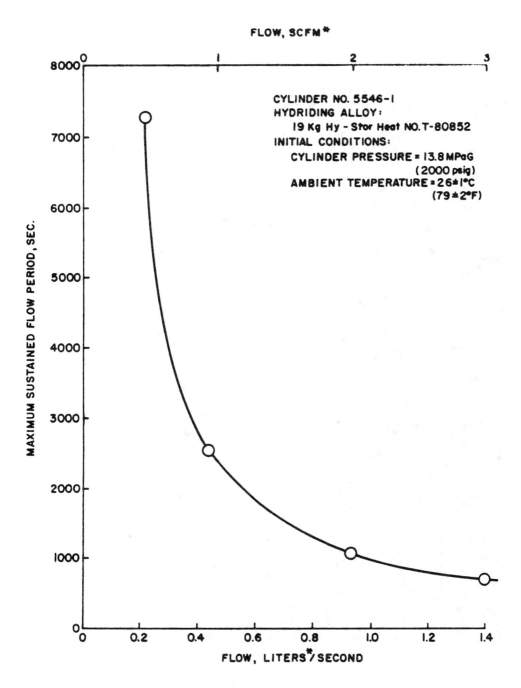

Figure 10. MAXIMUM SUSTAINED FLOW PERIOD vs FLOW
* 21°C, 0.101 MPa (70°F, 14.7 psia)

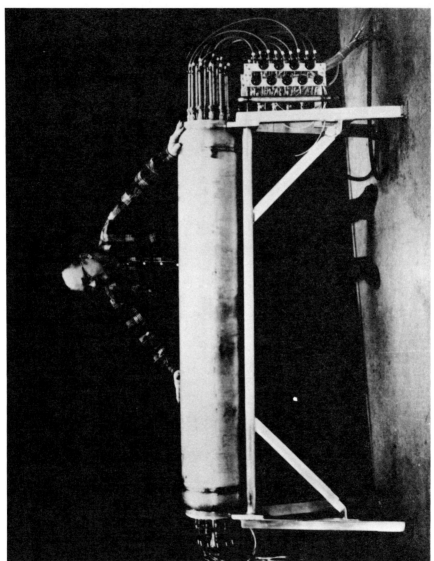

Figure 11. =Metal Hydride Storage System for Public Service Electric and Gas Company of New Jersey Peak Shaving Demonstration

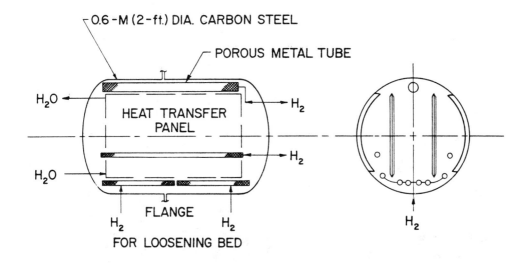

Figure 12.   Sectional View of BNL Variable-Parameter
Test Unit Showing Major Components

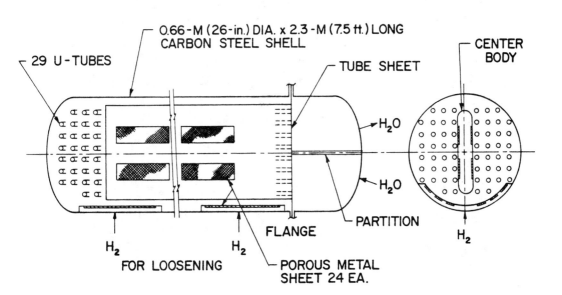

Figure 13.   Sectional Views of FWEC Variable-Parameter
Test Unit Showing Major Components

Figure 14. HY-STOR Alloys Supplied to BNL by the Ergenics Division of MPD Technology Corporation

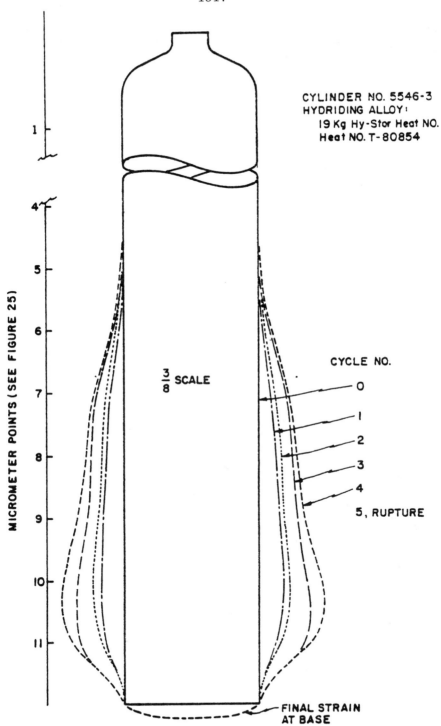

CYLINDER NO. 5546-3
HYDRIDING ALLOY:
19 Kg Hy-Stor Heat NO.
Heat NO. T-80854

CYCLE NO.

0
1
2
3
4
5, RUPTURE

MICROMETER POINTS (SEE FIGURE 25)

$\frac{3}{8}$ SCALE

FINAL STRAIN
AT BASE

Figure 15.   PROGRESSIVE SWELLING AND FAILURE OF A VERTICAL
HYDRIDE CYLINDER DUE TO SETTLING.   DEFLECTIONS
ARE DRAWN AT 27x FOR CLARITY.

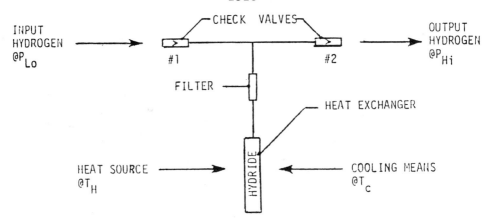

Figure 16. Schematic of Prototype Compressor Under
            Development by the Ergenics Division of
            MPD Technology Corporation

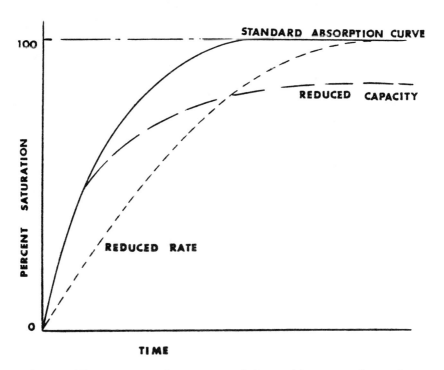

Figure 17. Two Basic Types of Capacity Deterioration
            Observed in Poisoning Experiments

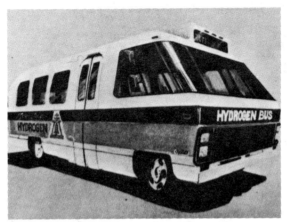

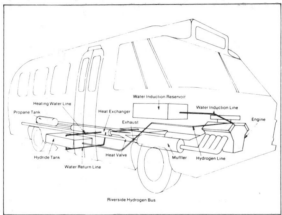

The Argosy Airstream bus designed specifically for Riverside, California has several improvements over the first hydrogen-powered bus. Aluminum hydride storage vessels reduced the weight and cost and a new hydrogen flow system helps to utilized more of the stored hydrogen. Also included in the Riverside bus is a backup fuel system.

Figure 18a. Demonstration Billings Hydrogen Bus
Designed for Riverside, California

Figure 18b. Daimler-Benz Hydrogen-Powered Bus

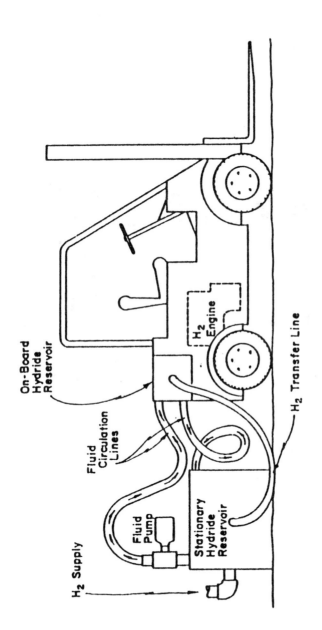

Figure 19. Schematic of Hydrogen-Powered Forklift Truck Under Development by Denver Research Institute and MPD Technology Corporation

Figure 20. HYCSOS Demonstration Test Facility

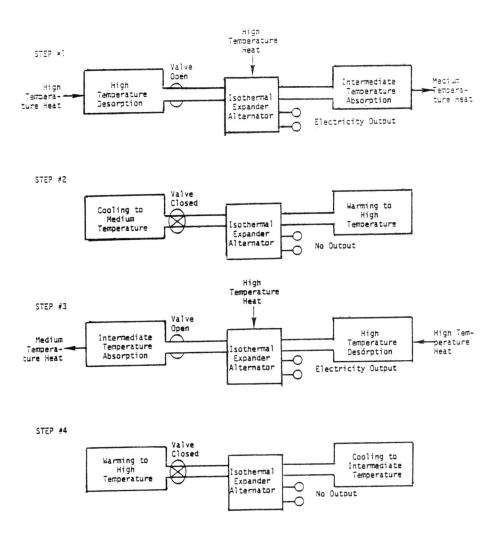

Figure 21. Theoretical Hydride Power Cycle Operation

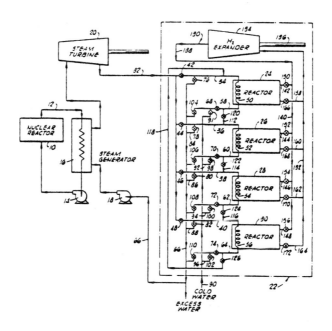

Figure 22. Use of Terry-Schoeppel Hydride
System as Bottoming Unit in a
Nuclear Plant

USE OF BINARY TITANIUM ALLOYS
FOR HYDROGEN STORAGE

O. de Pous and H.M. Lutz
Battelle Institute
Geneva, Switzerland

ABSTRACT

Owing to the strong affinity shown for hydrogen by titanium, several binary alloys of this element can be used for the storage of hydrogen in the form of solid metal hydrides. Nevertheless, only a few alloys are suitable for practical application. For example, alloys with aluminium and the transition metals are of particular interest as storage materials for use between room temperature and $300°C$.

The present study concerns the systematic examination of intermetallic binary compounds such as Ti_3M, Ti_2M, TiM and TiM_2. The reversibility of hydrogen exchange depends on the relative thermal stabilities of intermetallic compounds, defined as the enthalpy of formation. This parameter also represents the heat required for hydrogen release from the hydrogen storage system. The permanence of the intermetallic composition during hydrogen absorption and desorption operations is, for each system, related to the range of variation of the operating temperature and pressure.

This study highlights those special features of the iron-titanium system which explain the extensive developments that are presently underway with this material.

COMPLEX METAL HYDRIDE FORMATION

Considering the reaction between a binary alloy $Ti_\varepsilon M_{1-\varepsilon}$ and hydrogen, the enthalpy of reaction may have one of three values:

$\Delta H < 0$:   there is formation of a complex hydride

$\Delta H = 0$:   there is a high hydrogen solubility in the alloy but no complex hydride formation (absence of metal-hydrogen bonds)

$\Delta H > 0$:   there is a low hydrogen solubility in the alloy which increases with the temperature.

Suitable materials for chemical hydrogen storage correspond to the first case, for which a typical composition-pressure isotherm diagram is represented in Figure I. At least one plateau is observed corresponding to the coexistence in equilibrium of two distinct phases; the intermetallic

1525

compound $Ti_\varepsilon M_{1-\varepsilon}$ and the complex metal hydride $Ti_\varepsilon M_{1-\varepsilon} H_z$, as described by equation (1):

(1) $\qquad Ti_\varepsilon M_{1-\varepsilon} + \dfrac{Z}{2} H_2 \xrightarrow{\Delta H(1)} Ti_\varepsilon M_{1-\varepsilon} H_z$

This reaction is reversible and the relation between equilibrium pressure and temperature is given by the Vant'off equation:

(2) $\qquad - RT \ln P_{eq.} = \Delta H(1) - T\Delta S(1)$

$\Delta H(1)$, the enthalpy of reaction, can, to a first approximation, be considered as independent of the hydrogen concentration in the solid phase. The flat nature of the composition-pressure isotherm diagram strengthens the argument in favour of this assumption.

$\Delta S(1)$, the entropy of the reaction, results mainly from the presence of hydrogen in a gaseous form. Thus, to a first approximation, $\Delta S(1)$ can be considered as being equal to $S^\circ_{H_2}$ and consequently nearly independent of the temperature (31.2 cal/mole H_2 at $300^\circ K$ and 36 cal/mole at $600^\circ K$ [1]).

The equilibrium conditions (hydrogen pressure 1 atmosphere) being defined as:

(3) $\qquad \Delta G(1) = \Delta H(1) - T_{eq}\Delta S(1) = 0$

it follows that:

(4) $\qquad \Delta H(1) = - T_{eq.} \Delta S(1) = -T_{eq.} S^\circ_{H_2}.$

Based on the previous approximations, the equilibrium temperature is plotted as a function of the reaction enthalpy in Figure 2. The dependance between equilibrium temperature and the enthalpy of formation of elementary hydrides argues in favour of the assumption on the entropy.

In conclusion, suitable titanium alloys for chemical hydrogen storage are those for which the enthalpy of the reaction with hydrogen lies in the range of -9.3 to -11.6 Kcal/mole H_2 if the expected working temperatures are ambient and $100^\circ C$ respectively.

PARAMETERS INFLUENCING THE ENTHALPY OF REACTION OF TITANIUM BINARY ALLOYS WITH HYDROGEN

The value of $\Delta H(1)$ is influenced by several factors. Reaction (1) can be broken down as follows:

$$(5) \quad Ti_\varepsilon M_{1-\varepsilon} \xrightarrow{\ -\Delta H_{fTi_\varepsilon M_{1-\varepsilon}}\ } \varepsilon Ti + (1-\varepsilon) M$$

$$(6) \quad \varepsilon Ti + \frac{X}{2} H_2 \xrightarrow{\ \Delta H_{fTi_\varepsilon H_x}\ } Ti_\varepsilon H_x$$

$$(7) \quad (1-\varepsilon) M + \frac{Y}{2} H_2 \xrightarrow{\ \Delta H_{fM_{1-\varepsilon} H_y}\ } M_{1-\varepsilon} H_y$$

$$(8) \quad Ti_\varepsilon H_x + M_{1-\varepsilon} H_y \xrightarrow{\ \Delta H(8)\ } Ti_\varepsilon M_{1-\varepsilon} H_z$$

with $X + Y = Z$.

Then:

$$(9) \quad \Delta H(1) = -\Delta H_{fTi_\varepsilon M_{1-\varepsilon}} + \Delta H_{fTi_\varepsilon H_x} + \Delta H_{fB_{1-\varepsilon} H_y} + \Delta H(8)$$

where:

$\Delta H_{fTi_\varepsilon B_{1-\varepsilon}}$ represents the enthalpy of formation of the binary alloy,

$\Delta H_{fTi_\varepsilon H_x}$ and   represent the enthalpy of formation of each elementary hydride,

$\Delta H_{fM_{1-\varepsilon} H_y}$

$\Delta H(8)$ represents the enthalpy of reaction (8). This term is characteristic of the difference between the metal-hydrogen bond in the complex metal hydride and in each elementary hydride. The value of $\Delta H(8)$ must be negative in order to avoid the decomposition of $Ti_\varepsilon M_{1-\varepsilon} H_z$ into $Ti_\varepsilon H_x$ and $M_{1-\varepsilon} H_y$.

In a series of isostructural materials, the value of $\Delta H(8)$ can be considered as constant, which explains the rule of "reversed stability" proposed by Miedema [2] for the evaluation of the enthalpy of formation of complex metal hydrides.

The values of $\Delta H_{fTi_\varepsilon M_{1-\varepsilon}}$, $\Delta H_{fTi_\varepsilon H_y}$ and $\Delta H_{fM_{1-\varepsilon} H_y}$ are known or can be evaluated, whereas the value of $\Delta H(8)$ represents the principal unknown parameter.

The formation of a titanium-based complex metal hydride requires that $\Delta H(1)$ be more negative than $-Z/2\ TS^o_{H_2}$. Then the following relationship must be satisfied in order that stable metal hydrides be obtained:

$$(10) \quad \boxed{\Delta H_{fTi_\varepsilon M_{1-\varepsilon}} > \Delta H_{fTi_\varepsilon H_x} + \Delta H_{fM_{1-\varepsilon} H_y} + \Delta H(8) + \frac{Z}{2} TS^o_{H_2}}$$

The second requirement concerns the relative stability of the complex and elementary hydrides. At the pressure and temperature defined as equilibrium conditions for reaction (1), the free energy variation of the following reactions must be positive i.e. no reaction must take place.

$$(11) \quad Ti_\varepsilon M_{1-\varepsilon} H_z \longrightarrow Ti_\varepsilon H_x + \frac{Z-X}{2} H_2 + (1-\varepsilon)\ M$$

$$(12) \quad Ti_\varepsilon M_{1-\varepsilon} H_z \longrightarrow M_{1-\varepsilon} H_y + \frac{Z-Y}{2} H_2 + \varepsilon Ti$$

The equilibrium conditions for reaction (1) are defined as:

$$(13) \quad \Delta G(1) = \frac{Z}{2} RT \ln P_{H_2}$$

and the conditions to be satisfied in order to avoid the decomposition of the complex hydride into the elementary hydrides are, respectively:

$$(14) \quad \Delta G(11) = \Delta G_{fTi_\varepsilon H_x} - \Delta G_{fTi_\varepsilon M_{1-\varepsilon}} + \Delta G(1) > 0$$

$$(15) \quad \Delta G(12) = \Delta G_{fM_{1-\varepsilon} H_y} - \Delta G_{fTi_\varepsilon M_{1-\varepsilon}} + \Delta G(1) > 0$$

Taking into account the relation between free energy enthalpy and entropy changes ($\Delta G \sim \Delta H - TS^o_{H_2}$), the previous conditions can be written as

$$(16) \quad \Delta G(11) \sim \Delta H_{fTi_\varepsilon H_x} + \frac{X}{2} TS^o_{H_2} - \frac{Z}{2} RT \ln P_{H_2} - \Delta H_{fTi_\varepsilon M_{1-\varepsilon}} > 0$$

(17) $\quad \Delta G(12) \sim \Delta H_{fM_{1-\epsilon}H_y} + \dfrac{Y}{2} TS^O_{H_2} - \dfrac{Z}{2} RT \ln P_{H_2} - \Delta H_{fTi_\epsilon M_{1-\epsilon}} \quad > 0$

In conclusion, the second series of inequalities which must be satisfied in order to obtain stable complex hydrides are:

(18) $\quad \boxed{\Delta H_{fTi_\epsilon B_{1-\epsilon}} \quad < \Delta H_{fTi_\epsilon H_x} + \dfrac{X}{2} TS^O_{H_2} - \dfrac{Z}{2} RT \ln P_{H_2}}$

(19) $\quad \boxed{\Delta H_{fTi_\epsilon B_{1-\epsilon}} \quad < \Delta H_{fM_{1-\epsilon}H_y} + \dfrac{Y}{2} TS^O_{H_2} - \dfrac{Z}{2} RT \ln P_{H_2}}$

The selection of suitable titanium-based alloys for hydrogen storage can be made on the basis of their enthalpies of formation which must simultaneously satisfy the inequalities (10), (18) and (19).

Too low a value of the heat of formation of $Ti_\epsilon M_{1-\epsilon}$ will, in the presence of hydrogen, lead to decomposition with, for example, the formation of titanium hydride.

From too high a value of the heat of formation, on the other hand, will result the impossibility of formation of complex metal hydrides.

The selection of titanium binary alloys on the basis of their enthalpy of formation is schematically illustrated in Figure 3. The value of $\Delta H(8)$ which has to be taken into account for a quantitative evaluation of $\Delta H(1)$ remains an unknown and can be evaluated only on the basis of experimental results.

This development concerns only complex metal hydrides for which the composition-pressure isotherm exhibits a distinct plateau and for which consequently the enthalpy of the reaction with hydrogen can be considered as independent of the hydrogen concentration in the material.

APPLICATION TO IRON-TITANIUM COMPLEX HYDRIDE FORMATION

In the titanium-hydrogen system [3], two hydrides have been observed: $Ti_\epsilon H_x$ (β) with $X < \epsilon$ and an enthalpy of formation of -25 Kcal/mole H_2 [4] and, $Ti_\epsilon H_x$ (γ) with $X < 2\epsilon$ and an enthalpy of formation which has been found to vary with the hydrogen concentration in the solid [5].

Composition	Enthalpy of formation [5]	
	Kcal/mole H_2	Kcal/at.
$TiH_{1.6}$	-34	-27
$TiH_{1.7}$	-33	-28
$TiH_{1.85}$	-31	-28.7
$TiH_{1.97}$	-30	-29.5

Formation of TiFeH

The hydrogenation reaction of TiFe can be schematically considered as:

$$TiFe + 1/2\ H_2 \xrightarrow{\Delta H(1)} TiFeH$$

In order to calculate the corresponding enthalpy of hydrogenation $\Delta H(1)$, this reaction can be split into:

$$TiFe \longrightarrow Ti + Fe$$

$$Ti + 1/2\ H_2 \longrightarrow TiH$$

$$TiH + Fe \xrightarrow{\Delta H(8)} TiHFe$$

so that:

$$\Delta H(1) = -\Delta H_{fTiFe} + \Delta H_{fTiH} + \Delta H(8).$$

Hultgren [6] gives ΔH_{fTiFe} = -9.7 Kcal/mole.

From the equilibrium decomposition pressure-temperature relationship measured by Reilly [7], the enthalpy of hydrogenation has been found to be -3.36 Kcal/mole TiFeH.

Thus, $\Delta H(8) = -9.7 + 12.5 - 3.36 = -0.6$ Kcal/mole TiFeH.

Considering the second inequality (18) that has to be satisfied in order to obtain a stable complex hydride,

$$\Delta H_{fTiFe} < \Delta H_{fTiH} + 1/2\ TS^{o}_{H_2} - 1/2\ RT\ \ln P_{H_2}$$

which, at 300K corresponds to:

$-9.7 < -12.4 + 4.5 - 0.3 \ln P \, H_2$.

This means that, for a hydrogen pressure in excess of 1000 atmospheres, the followng reaction would be observed:

$TiFe + 1/2 \, H_2 \longrightarrow TiH + Fe$.

Formation of $TiFeH_2$

The hydrogenation reaction of TiFe can be schematically considered as:

$TiFe + H_2 \xrightarrow{\Delta H(1)} TiFeH_2$

In order to calculate the value of $\Delta H(1)$, two models can be proposed which differ with respect to the hydrogen distribution in the lattice. The first model considers a possible segregation of hydrogen in the vicinity of titanium atoms (TiH_2Fe) resulting from the greater affinity of this metal for hydrogen compared to that between iron and hydrogen.

The hydrogenation reaction can then be split into:

$TiFe \longrightarrow Ti + Fe$

$Ti + H_2 \longrightarrow TiH_2$

$TiH_2 + Fe \xrightarrow{\Delta H(8)} TiH_2Fe$

with: $\Delta H(1) = -\Delta H_{fTiFe} + \Delta H_{fTiH_2} + \Delta H(8)$. The value of $\Delta H(1)$ can be calculated as -7.66 Kcal/mole TiH_2Fe from the experimental variation of enthalpy measured by Reilly [7] for the following reaction:

$TiFeH + 1/2 \, H_2 \xrightarrow{\;-\;4.3\ Kcal\;} TiFeH_2$.

The value of $\Delta H(8)$ is consequently:

$\Delta H(8) = -7.66 + 9.7 + 30 = 32$ Kcal/mole TiH_2Fe.

This positive value is incompatible with the existence of the stable TiH_2Fe compound.

The second model considers a statistical distribution of hydrogen in the lattice (TiHFeH).

In order to calculate the corresponding enthalpy of hydrogenation $\Delta H(1)$, the reaction can be split into:

$$TiFe \longrightarrow Ti + Fe$$

$$Ti + 1/2\ H_2 \longrightarrow TiH$$

$$Fe + 1/2\ H_2 \longrightarrow FeH$$

$$TiH + FeH \xrightarrow{\ \Delta H(8)\ } TiHFeH$$

and

$$\Delta H(1) = -\Delta H_{fTiFe} + \Delta H_{fTiH} + \Delta H_{fFeH} + \Delta H(8).$$

Due to the fact that iron hydride does not exist, ΔH_{fFeH} must be considered as positive, and a value of $+ 4$ Kcal/mole FeH can be calculated using the Miedema model [8].

The previous equation becomes:

$$- 7.66 = + 9.7 - 12.5 + 4 + \Delta H(8)$$

and

$$\Delta H(8) \sim - 8.8 \text{ Kcal/mole TiHFeH.}$$

This negative value is compatible with the existence of a stable TiHFeH complex hydride.

In conclusion, the second model is preferred and a statistical distribution of hydrogen in TiFe without segregation can be considered as more probable. Thus, only one plateau is observed in the composition pressure isotherm for annealed TiFe samples. In unannealed samples [9], the composition-pressure isotherm is characterised by the absence of a marked plateau due to the inhomogeneous distribution of titanium and iron and to the coexistence in the structure of interstitial hydrogen atoms surrounded by different combinations of metal atoms i.e. 4Fe, 3Fe 1Ti, 2Fe 2Ti, 1Fe 3Ti, 4 Ti.

In some cases, hydrogen has been observed in the vicinity of that element for which the lower affinity would be expected. For example, in $ZrCr_2$, crystallographic positions in the vicinity of chromium have been found by Beck [10] to be preferentially occupied by hydrogen atoms.

APPLICATION TO TITANIUM-BASED COMPLEX HYDRIDE FORMATION

For a complete investigation of the formation of titanium-based complex hydrides, the following thermodynamic characteristics must be known:

- the enthalpy of formation of titanium binary alloys,

- the enthalpy of hydrogenation of the elementary metals,

- the value of ΔH(8), which cannot be neglected in the calculation but which could be considered as nearly constant for a series of isostructural alloys.

Unfortunately, experimental data on the enthalpy of formation of binary alloys and the enthalpy of hydrogenation of the elements is extremely limited. An evaluation has been proposed using the Miedema model [11]. Since the calculation of the enthalpy of hydrogenation of binary alloys involves the addition of positive and negative numbers, the value of each parameter (experimental or calculated) is required to a high degree of accuracy. This is particularly true for complex hydrides of low thermal stability since the enthalpy of formation lies between 0 and -9 Kcal/mole H_2.

In any case, considering the calculated enthalpy of formation of complex hydrides from intermetallic compounds composed of Ti, Zr, Y and a transition metal [11], the following tendencies can be defined in Table I:

- a decrease of the thermal stability along the transition element series due to the increase in the external electron density [12],

- an increase of the complex metal hydride thermal stability on substituting titanium by zirconium or yttrium, which corresponds to an increase of the interatomic distance [12].

The difference between the calculated value, for example -15 Kcal/mole $TiFeH_2$, and the experimental value of -7.7 Kcal/mole $TiFeH_2$ results from the fact that the Miedema model does not take into account the value of ΔH(8), which is of the order of -8.8 Kcal/mole $TiFeH_2$.

EXPERIMENTAL RESULTS

The enthalpies of formation of TiMH and $TiMH_2$ have been determined from the enthalpies of decomposition (Tables 2 and 3) calculated from measured equilibrium decomposition pressure-temperature relationships. Study of these values shows that the effect of an alloying element depends on its nature as follows:

Low thermal stability	Fe-Co-Mn-V-Cu ————————————>	high thermal stability

With chromium, a strong tendency to form $TiCr_2$ is observed, while with copper an increase of the hydrogen pressure leads to the formation of TiH_2 and Cu.

The difference between the predicted and experimental results is due to the approximate value used in the theoretical evaluation of the enthalpy of formation of CrH, MnH, FeH, CoH, NiH, ...

CONCLUSION

The shape of the composition isotherm diagram, the heat required for hydrogen release, as well as the storage capacity can be varied as a function of the alloy composition. However, the prediction of the character-istics of hydrogen storage of a metal hydride is limited by the lack of experimental thermodynamic data concerning both the stability of binary and ternary titanium alloys and the reactions of the elements with hydro-gen. Moreover, the hydrogen localisation in the lattice, at least for some typical structures (A2, B2, C14, C15 ...), should be known in order to evaluate the importance of the parameters influencing the value of the enthalpy of formation of such hydrides.

The enthalpy of formation is an important parameter not only in order to evaluate the stability of the metal hydride but also in order to determine the heat required for hydrogen release from the storage system. In any case, a large number of titanium ternary alloys (Table 4) can be consid-ered as potentiel candidates for hydrogen storage. Considering the cost of the material, iron-titanium looks to be particularly attractive for an industrial development.

BIBLIOGRAPHY

[1]   D.R. STULL and H. PROPHET
       JANAF Thermochemical Tables NSRDS-NBS 37 (1971)

[2]   H.H. VAN MAL
       Philips Res. Report Suppl. 1976, No 1

[3]   A.D. McQUILLAN
       Proc. Roy. Soc. (London), Ser. A 204, 309 (1950)

[4]   R.M. HAAG and F.J. SHIPKO
       J. Amer. Chem. Soc. 78, 5l55 (1956)

[5]   B. STALINSKI and Z. BIEGANSKI
       Bull. Acad. Sci. Poc. Sci. Chim. 8, 243 (1960)
       Bull. Acad. Sci. Poc. Sci. Chim. 10, 247 (1962)

[6]   J.R. HULTGREN
       "Selected values of the thermodynamic properties of binary alloys"
       Am. Soc. of Metals (1973)

[7]   J.J. REILLY and R.H. WISWALL
       Inorg. Chem. 13, 2l8 (1974)

[8]   A.R. MIEDEMA, R. BOOM and F.R. DE BOER
       J. Less. Common Metals 41, 283 (1975)

[9]   J.J. REILLY and J.JOHNSON
       1st World Hydrogen Energy Conf., Miami Coral Yables, Fla.
       (1976)

[10]   J.R.L. BECK
       Res. Report LAR-55 (1961), Denver Univ.

[11]   K.H.J. BUSCHOW and A.R. MIEDEMA
       Int. Symp. on hydrides for energy storage, Geilo/Norway,
       (1977)

[12]   O. de POUS and H. LUTZ
       "Effect of the interstitial hole size and electron concentration on
       complex metal hydride formation"

TABLE 1 - CALCULATED ENTHALPIES OF FORMATION OF HYDRIDES FORMED
FROM INTERMETALLIC COMPOUNDS OF Zr, Ti AND Y WITH
TRANSITION METALS, FOR AB_3, AB_2 AND AB COMPOUNDS
RESPECTIVELY [11]

Elements	Enthalpy of hydride formation Kcal/mole H_2		
	Zr	Ti	Y
V	-25 -27 -32		
Cr	-16 -20 -28	-13 -16 -21	
Mn	-16 -21 -29	-15 -17 -21	-27 -32 -45
Fe	- 5 -10 -21	- 3 - 7 -15	-19 -26 -41
Co	+ 1 - 5 -18	+ 1 - 5 -12	-13 -21 -37
Ni	+ 2 - 6 -16	0 - 5 -12	-12 -20 -37

TABLE 2 - ENTHALPY OF $TiMH_4 \rightarrow TiMH_2$ REACTION AND CALCULATED
ENTHALPIES OF FORMATION OF $TiMH_4$ AND $TiMH_2$

Hydride atomic composition	Enthalpy of decomposition Kcal/mole H_2	Calculated enthalpy of formation	
		$\Delta H_{fTi_\varepsilon M_{1-\varepsilon} H_1}$	$\Delta H_{fTi_\varepsilon M_{1-\varepsilon} H_2}$
100Ti	44	- 15	- 29
75Ti 25V	21	- 7	- 14
53Ti 47V	33	- 11	- 22
50Ti 50V	21	- 7	- 14
35Ti 65V	16	- 5.3	- 10.6
90Ti 10Cr	39.3	- 13.1	- 26.2
80Ti 20Cr	34	- 11.3	- 22.6
90Ti 10Mn	40.8	- 13.6	- 27.2
85Ti 15Al	38.3	- 12.7	- 25.5

TABLE 3 - ENTHALPIES OF TiMH$_2$ → TiMH AND TiMH → TiM REACTIONS, AND CALCULATED ENTHALPIES OF FORMATION OF TiMH$_2$ AND TiMH

Binary alloy	Enthalpy of decomposition Kcal/mole H$_2$		Calculated enthalpy of formation Kcal/mole	
	TiMH	TiMH$_2$	TiMH	TiMH$_2$
50Ti 50CV		+14		−14
50Ti 50Cr	−	−	−	−
50Ti 50Mn	+ 9	+13	− 4.5	−11.0
50Ti 50Fe	6.3	7.5	− 3.15	− 6.4
50Ti 50Co	10.6	11	− 5.3	−10.8
50Ti 50Ni	−	−	−	−
50Ti 50Cu	13.6	15.5	− 6.8	−14.55

TABLE 4 - TOTAL HYDROGEN ABSORBED BY TITANIUM ALLOYS

ALLOY COMPOSITION atom %		ABSORBED HYDROGEN weight %
100Ti		4
93Ti	7Mn	3.89
66Ti	33V	3.88
89Ti	11Nb	3.82
82Ti	18Mo	3.81
50Ti	50V	3.63
81Ti	19Cr	3.42
89Ti	11Cr	3.32
84Ti	26Nb	3.32
66Ti	33Co	3.29
25Ti	75V	2.98
50Ti	50Zr	2.90
66Ti	33Mn	2.52
75Ti	25Al	2.34
33Ti	66Cr	2.26
66Ti	33Ni	2.13
66Ti	33Th	1.83
50Ti	50Fe	1.70
50Ti	50Sb	1.40
50Ti	50Co	1.35
50Ti	50Cu	1.23
50Ti	50Ag	1.00
75Ti	25Bi	0.91
66Ti	33Mn	0.80

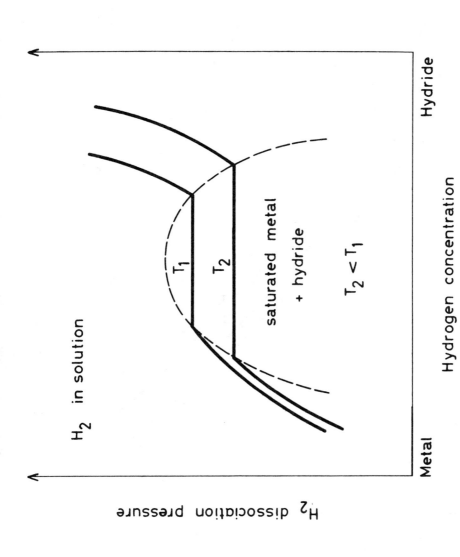

FIGURE 1 - SCHEMATIC CPI DIAGRAM FOR A METAL-HYDROGEN SYSTEM FORMING AN HYDRIDE

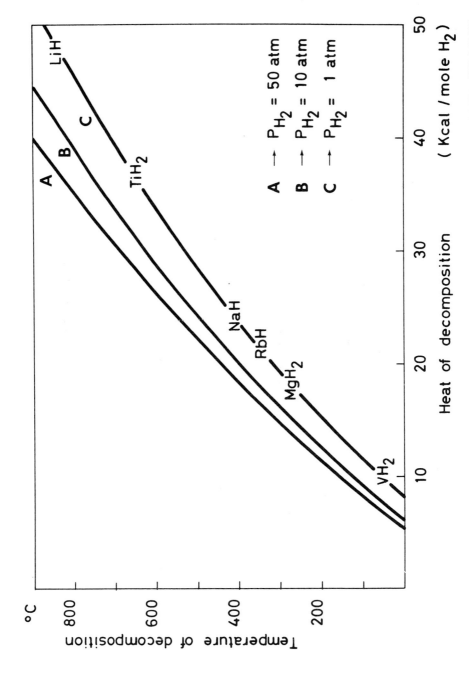

FIGURE 2 – RELATION BETWEEN TEMPERATURE AND HEAT OF DECOMPOSITION FOR SELECTED HYDRIDES

1541

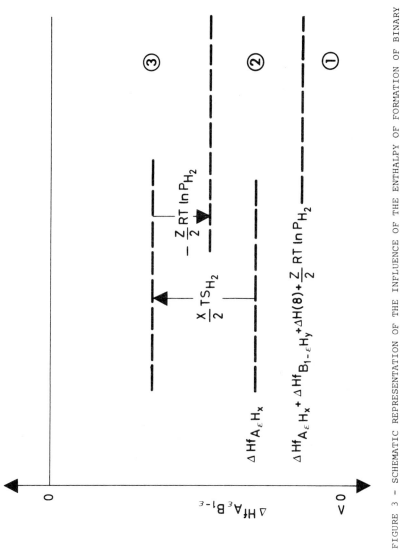

FIGURE 3 - SCHEMATIC REPRESENTATION OF THE INFLUENCE OF THE ENTHALPY OF FORMATION OF BINARY
ALLOYS ON THE EXISTENCE OF COMPLEX HYDRIDES

(1) NO COMPLEX HYDRIDE FORMATION

(2) COMPLEX HYDRIDE FORMATION

(3) NO COMPLEX HYDRIDE FORMATION DUE TO THE DECOMPOSITION INTO ELEMENTARY HYDRIDES

CUBIC METAL-ALLOYS FOR HYDROGEN STORAGE

H. Buchner, M. Stohrer, O. Bernaur
Daimler Benz AG, Dept. Technical Physics
D-7000 Stuttgart 60, Germany

ABSTRACT

Technical application of hydrogen storage in metal alloys, especially in vehicle transportation systems, claims low tank weight, respectively low molecular weight of the components of the alloy, and cheapness, respectively large abundance. Therefore about 12 elements of really practical interest have been investigated.

In the limits of vehicle application ($p < 50$ bar, $-20 \le T$ $500°C$) the pressure and temperature dependence of the hydrogen to metal ratio of the alloys was determined and the hydride capacity characterized. In addition the magnetic susceptibility of the phase was measured; from the paramagnetic susceptibility the band structure at the Fermi level and hydrogenation effects were analyzed.

Only alloys of the hydride forming elements Mg and Ti form hydrides. The data indicate correlations between the paramagnetic susceptibility and the hydrogenation behaviour, the magnetic properties, however, are of Mg_2NiH_x and $TiNi_x$ independent from the hydrogen: metal ratio. Based on three results a hypothesis for further hydride development is proposed.

INTRODUCTION

A survey of binary metal alloys shows [1] , that in general one hydrogen atom per metal atom seems to be an upper limit of hydrogen storage; therefore, for the technical application of mobile hydrogen storage in vehicle transportation only the first 30 elements of the periodic system, i.e. the lightest atoms, are of interest in getting hydrogen densities of at least 2 percent per kg hydride, which yields a favourable energy density of more than 600 Wh per kg storage material. On cost reasons only the 12 most abundant, i.e. cheapest elements are of really practical interest; they have been compiled in Fig. 1, the hydride forming elements are hatched. Alloys of these components are of great promise for the requirements of the Hydrogen Hydride Energy Concept [2] as well as for the functions of the automotive tank system [3] .

Because of the favourable experiments with some outstanding alloys [6 - 9] (Mg$_2$Ni, TiFe, TiNi) investigations of the cubic binary phases are of special interest. In the limits of vehicle application, i.e. pressure 1...50 bar and temperature -20...+400°C, the literature rules of hydride formation have been tested; especially the effects of hydrogenation on the electronic structure of the metal hydride. Usually the electronic band factor is neglected: thus the model of Lundin[4] et. al. correlates the thermodynamic properties of hydrides of intermetallic compounds with the interstitial hole sizes, i.e. the geometrical factor of hydride formation; whereas the concept of Miedema [5] et. al. for hydride formation is based on nearest neighbour effects and the local geometrical bonding of hydrogen. The Miedema rule, indeed, fails in case of the hydride predictions of the binary intermetallic compounds Ti-Fe, Co, Ni.

The magnetic susceptibility coheres with the electronic structure of the alloys [10 - 11] ; the paramagnetic susceptibility is essentially determined by the density of states at the Fermi energy. Therefore, measurement of the magnetic susceptibility should enable the analysation of changes in the conduction band of metal alloys [12, 13] (density, curvature) and resolve correlations between electronic band structure and hydride formation.

EXPERIMENTAL

Sample preparation:

the alloys have been prepared by hot pressing and sintering of powder samples. The origin substances were either pure elements or in case of instable components, e.g. Ti, hydrides, which were decomposed before reaction. The aluminium and magnesium compounds with highly different melting points must be melt in an induction furnance.

Quality control:

the structure and homogenity of the prepared alloys were determined by powder x-ray diffraction and analyzed by visual peak height comparison; repeated annealing increased the homogenity of the phases to more than 90 percent. As examples the x-ray diffraction patterns of the cubic (CsCl structure) intermetallic compound series TiFe, TiCo, TiNi are shown in Figs. 2 - 4. A second quality control is the determination

of the ferromagnetic part of the magnetic susceptibility [13].
All samples have only ferromagnetic phases of less than 1 percent.

Hydrogenation:

all specimens have been activated under vacuum conditions at
about $400°C$ and by several cycles of hydrogenation and decomposition. The characteristics of the metal hydride were
determined by isothermal measurements of the hydrogen pressure dependence from the concentration of hydrogen: metal in
the hydride at different temperatures in the limits of vehicle
application, i.e. between $-20°C$ to $300°C$. We use a conventional cpi-apparatus [6 - 7] with a volumetric determination
of the absorbed and decomposed hydrogen by means of a calibrated volume. The investigations have been restricted on decomposition isotherm, because this situation comes up to vehicle operation.

Magnetic susceptibility:

the magnetic properties are measured by the Faraday method,
i.e. a highly sensitive balance determines the magnetic force
of a sample in an inhomogenious magnetic field. In the case
of small ferromagnetism the x component of the force F_x on a
specimen of volume V is

$$F_x = \frac{1}{2} x \cdot V \frac{d}{dx} H^2 \qquad (1)$$

provided the derivative of the magnetic field H is constant
over the volume [10]. From the magnetic field dependence of
the magnetic susceptibility x (H) one can determine [13] the
paramagnetic x^{para} and ferromagnetic susceptibility d^{ferro}.
H^{-1} :

$$x = x^{para} + d^{ferro} \cdot \frac{1}{H} \qquad (2)$$

In first approximation [11 - 12] the paramagnetic susceptibility is related to the density of states N (E) at the Fermi
energy E_F by

$$x^{para} = \frac{4}{3} \cdot \mu_B^2 N (E_F) \qquad (3)$$

where μ_B is the Bohr magneton. Thus changes in the paramagnetic susceptibility indicate effects on the electronic band
structure at the Fermi level.

RESULTS AND DISCUSSION

Fairly all binary cubic alloys of the elements of Fig. 1 have been prepared and examined in order to compare their hydrogenation and magnetic properties under the same conditions of specimen purity, preparation, and measuring methods. The results are compiled in Tab. 1. Partly we have reexamined published systems of binary intermetallic compounds [1, 6 - 9], respectively extended them to the pressure and temperature regimes of vehicle application.

Our results are in principal agreement with these examinations, deviations may be due to annealing and homogenisation effects. A survey over the hydrogenation results shows the following items:

- Alloys from components, which form no hydrides themselves, indicate only very small hydrogen solubilities (H/Me $<$ 0,05), they do not form stable hydrides. The measurements support Reilly's rule [15] : at least one component of hydride forming alloys must be a hydride forming element.

- Aluminium based binary alloys do not form hydrides from the gas phase. The irreversible preparation of AlH_3 dominates in the Al-alloys likewise.

- According to material costs and Reilly's rule the hydride forming alloys of interest for vehicle application consist of the elements magnesium or titanium, respectively.

- If there exist a monohydride as well as a dihydride phase of the metal hydride, the alloy shows the same behaviour as the hydride forming component itsself:
  Ti-based alloys show two pressure plateaus with different hydrogen pressure niveaus, as titanium hydride TiH_2 does, too. In contrast Mg-based alloys have always a single flat plateau, exactly like MgH_2.

In Tab. 1 the values of the magnetic susceptibility of the prepared alloys are also summerized and opposed to the hydrogenation results of the cpi-measurements. The paramagnetic part of the susceptibility x_g is compiled as susceptibility x_g^{para} per gramm of the specimen, which is directly measured value, and also as converted molar susceptibility per atom x_{g-A}^{para} , which allows to compare the different alloys. In addition the prefactor d^{ferro} of the inverse magnetic field dependence of the magnetic susceptibility is listed, which is a measure of the ferromagnetism of the specimen; for example for pure iron powder this prefactor is a constant of about 0,1 in the magnetic field range of the presented measurements. All susceptibility values were determined at room temperature. There are the following trends:

- All the binary hydride forming alloys are paramagnetic. The ferromagnetic portion is very small; if one relates the ferromagnetic prefactor α^{ferro} of the alloys to the value of pure iron one can see, that these ferromagnetic phases are less than some percents and can be attributed to incomplete reactions or contaminations of the starting elements.

- There is no common trend in all the alloys; especially not between the stable Mg-based high-temperature hydrides with high formation enthalpies and the fairly unstable Ti-based low-temperature hydrides with small enthalpies of hydride formation.

- But within the same crystal structure there are comparable tendencies of the hydrogenation and the paramagnetic susceptibility, or the density of energy states at the Fermi level E_F respectively.

For the hydride systems with the favourable hydrides Mg_2Ni and TiFe for automotive hydrogen storage different metal hydride properties are summerized in the Figs. 8 and 9. They show schematically the trends of the heat of formation of the alloy and the hydride, of the free cell volume, i.e. the difference between the cell volume and the atomic volumes, of the maximum hydrogen capacity, of the number of hydrogen pressure plateaus as well as of the paramagnetic susceptibility. One may generally notice that the density of states $N(E_F)$ is strongly influenced by the increasing number of atomic d-electrons of the alloyed components of Ti or Mg. In both, the Mg- and Ti-based hydrides, the trends of the hydride formation enthalpy are in parallel with the trends of the free cell volume and the paramagnetic susceptibility, as well. On the other hand an increased paramagnetic susceptibility correlates with a decreased hydrogen storage capacity and is, in the case of the Ti-based hydrides, accompanied with a reduced number of hydride phases, respectively number of cpi-plateaus. An increased value of $N(E_F)$ seems to stabilize the hydride formation, but on costs of the entire hydrogen uptake.

These correlations between the electronic band structure and the hydrogenation properties gave rise to the question whether these are changes of the magnetic susceptibility by hydrogenation and whether these are differences between the Mg- and Ti-based hydrides. Because of their good stability at room temperature and of the same alloyed component Ni we measured the dependence of the magnetic susceptibility of Mg_2Ni and TiNi from the hydrogen to metal ratio of the hydrides. The experimental results are shown in Figs. 10 and 11. The magnetic susceptibility, respectively the band structure at E_F, is independent of the hydrogen concentration of the hydrides within experimental errors. A pronounced maximum in the x-curves, as resolved in the dihydride TiH_{2-x} [12], is not found in the monohydrides Mg_2Ni and TiNi; the effect of hydrogenation on the electronic band structure does not affect the range of E_F. The results can be explained if one transfers the hydride theories of the pure hydrides Mg_2H_2 (covalent bonding) and TiH_2 (metallic s-d band hybridisation) to their Ni-based alloys, although there are large differences in the strength of the hydride bonding: hydrogen bonding of the hydride forming element predominates also in its (mono)hydride forming alloy, the localized character of the hydride forming elements (even d-electrons of Ti) determine the hydride formation of the alloys.

CONCLUSION

By means of the $TiFe/Mg-Mg_2Ni$ hydrogen combination tank system one gets favourable values of hydrogen storage for automotive propulsion and a sufficient range [2]. For special purposes (climatisation, auxiliary heating etc.), however, the further hydride development is required. For this we draw the following hypothesis from our results:

- Ternary alloys from non-hydride forming elements are no hydrogen absorbing systems in principal, because of lack of the necessary localized hydride formation

- Low densities of state $N(E_F)$ at the Fermi level E_F, e.g. reduced numbers of d-electrons, reduce the hydride formation enthalpy of the alloys; $N(E_F)$, however, is not related with the low- or high-temperature behaviour of the Mg- or Ti-based alloys.

- The hydrogen bonding affects the band structure of the alloys below the band edge and is connected with fairly localized states; ESCA-investigations should confirm this conclusion for hydrogenation.

- All of the binary Ti-based alloys should show a dihydrid phase because of local hydrogen bonding properties of Ti; low temperature as well as low pressure cpi-investigations should elucidate that statement.

- A possible crucial exception of the above conclusions indicates the hydride formation of $TiCr_2H_{3.5}$, where the number of hydrogen atoms is higher than the expected 2 hydrogens of the Ti-component; ternary alloys based on $TiCr_2$ should help to clarify the hydride properties.

The hydride research work is sponsored by the German Federal Ministry of Research and Technology.

REFERENCES

1   H. W. Newkirk; Hydrogen Storage by Binary and Ternary Intermetallics for Energy Applications.
    (Report UCRL - 52 110. 1976)

2   H. Buchner; Application of Low- and High-Temperature Hydrides: The Hydride/Hydrogen Energy Concept.
    (2nd World Hydrogen Energy Conference, Zürich. 1978. Session 10)

3   H. Buchner, H. Säufferer; Entwicklungstendenzen von Wasserstoff-getriebenen Fahrzeugen mit Hydridspeicher.
    (ATZ 79, 1977, p. 1)

4   C. E. Lundin, F. E. Lynch, C. B. Magee;
    A correlation between the Interstitial Hole Sizes in Intermetallic Compounds and the Thermodynamic Properties of the Hydrides Formed from those Compounds.
    (J. Less-Common Metals. 1978; to be published)

5   A. R. Miedema, K. H. J. Buschow, H. H. van Mal;
    Which Intermetallic Compounds of Transition Metals form stable Hydrides?
    (J. Less-Common Metals 49, 1976, p. 463)

6   J. J. Reilly, R. H. Wiswall; Formation and Properties of Iron Titanium Hydride
    (Inorg. Chem. 13, 1974, p. 218)

7   J. J. Reilly, R. H. Wiswall; The Reaction of Hydrogen
     with Alloys of Magnesium and Nickel and the Formation
     of Mg_2NiH_4.
     (Inorg. Chem. $\underline{7}$, 1968, p. 2254)

8   H. Buchner, M. A. Gutjahr, K. D. Beccu, H. Säufferer;
     Wasserstoff in intermetallischen Phasen am Beispiel des
     Systems Titan-Nickel-Wasserstoff.
     (Z. Metallkunde $\underline{63}$, 1972, p. 497)

9   K. Yamanaka, H. Saito, M. Someno;
     Hydride Formation of Intermetallic Compounds of TiFe,
     TiCo, TiNi and TiCu.
     (J. Chem. Soc. Japan $\underline{8}$, 1975, p. 1267)

10   C. Kittel; Introduction to Solid State Physics.
     (J. Wiley & Son, New York. 1966. p. 428)

11   A. Sommerfeld, H. Bethe; Elektronentheorie der Metalle.
     (Springer-Verlag, Berlin. 1967)

12   H. Nagel, H. Goretzki; Magnetic Properties of Titanium-
     Vanadium-Hydrides.
     (J. Phys. Chem. Solids $\underline{36}$, 1975, p. 431)

13   R. Hempelmann. E. Wicke; Irreversible Change of the Ma-
     gnetic Properties of TiFe by Hydrogenation.
     (Ber. Bunsenges. phys. Chemie $\underline{81}$, 1977, p. 425)

14   E. Källne; X-ray Study of Line Shifts and Band Spectra
     in the Alloys TiNi, TiCo and TiFe.
     (J. Phys. F. Metal Phys. $\underline{4}$, 1974, p. 167)

15   J. J. Reilly; Synthesis and Properties of Useful Metal
     Hydrides.
     (Int. Symp. on Hydrides for Energy Storage, Geilo. 1977)

16   A. C. Switendick; Electronic Band Structures of Metal
     Hydrides.
     (Solid State Comm. $\underline{8}$, 1970, p. 1463)

oy	Molecular-weight	Crystal structure	Concentration-Pressure-Isotherms			Magnetic Susceptibility		
			max.w%	Hydr.Phases	$\Delta H/kJ.mol^{-1}$	$\chi_g/cm^3.g^{-1}$	$\chi_m/cm^3.g^{-1}$-A	$/cm^3.G.g^{-1}$
o	85,9	B2	<0,05	nih.		$7,5.10^{-5}$	$3,2.10^{-3}$	$8,4.10^{-4}$
r_2	131,0	$C11_b$	<0,1	nih.		$1,1.10^{-5}$	$4,9.10^{-4}$	$1,02.10^{-4}$
Cu_9	679,8	$D8_3$	<0,12	nih.		$2,4.10^{-4}$	$1,2.10^{-2}$	$2,2.10^{-3}$
Mg_{17}	737,0	A12	1,6	1	65	$2,94.10^{-4}$	$7,5.10^{-3}$	$2,6.10^{-3}$
n	81,9	B2	<0,05	nih.		diamagn.		
i	85,7	B2	<0,07	nih.		$7,4.10^{-5}$	$3,1.10^{-3}$	$7,4.10^{-4}$
i	74,9	$L1_0$	<0,05	nih.		$9,5.10^{-5}$	$3,5.10^{-3}$	$8,6.10^{-4}$
Ti	128,8	DO_{22}	0,66	solution		$2,5.10^{-4}$	$8,0.10^{-3}$	$2,3.10^{-3}$
$_5$	470,6	$D8_2$	<0,05	nih.				
i	87,0	B20	<0,05	nih.				
i_2	115,1	C1	0,06	solution		$4,6.10^{-5}$	$1,7.10^{-3}$	$4,8.10^{-4}$
i	106,8	B2	1,56	1	50	$3,0.10^{-5}$	$1,6.10^{-3}$	$2,2.10^{-4}$
i_2	154,7	fcc	1,79	solution				
Si	184,1	A15	<0,05	nih.		$2,7.10^{-5}$	$1,2.10^{-3}$	$2,2.10^{-4}$
Ti	151,9	C15	1,1	3	18,5(Pl.II)			
$_2$	112,2	fcc	2,6	1	62	$3,7.10^{-5}$	$1,3.10^{-3}$	$3,8.10^{-4}$
Si_4	1065,0	$D8_6$	<0,05	nih.		$1,6.10^{-4}$	$9,3.10^{-3}$	$1,5.10^{-3}$
Si	218,7	D8	<0,05	nih.		$1,9.10^{-4}$	$1,1.10^{-2}$	$1,6.10^{-4}$
i_2	159,3	fcc	0,14	solution		$2,7.10^{-4}$	$1,4.10^{-2}$	$2,4.10^{-3}$
n	128,9	A3	<0,05	nih.		$2,2.10^{-5}$	$1,4.10^{-3}$	$2,4.10^{-4}$
i	83,9	B20	<0,05	nih.				
Si	195,6	DO_3	<0,05	nih.		ferromagn.		
i	103,7	A2	1,87	2	30(I),42(II)	$2,0.10^{-5}$	$1,0.10^{-3}$	$8,2.10^{-5}$
Ni	107,4	hex.	3,6	1	60	$5,4.10^{-5}$	$1,8.10^{-3}$	$9,0.10^{-4}$
i_3	231,1	$L1_2$	<0,05	nih.		$4,6.10^{-4}$	$2,5.10^{-2}$	$2,0.10^{-3}$
Si	192,9	A2	<0,05	nih.		$8,0.10^{-4}$	$4,0.10^{-2}$	$7,6.10^{-3}$
i	83,0	B20	<0,05	nih.		$5,3.10^{-5}$	$2,2.10^{-3}$	$1,1.10^{-4}$
Si	204,2	$L1_2$	<0,11	nih.				
i_2	154,5	fcc	2,2	3		$1,9.10^{-5}$	$9,6.10^{-4}$	$1,54.10^{-4}$
i	106,6	A2	1,36	solution		$5,5.10^{-4}$	$2,9.10^{-2}$	$3,0.10^{-3}$
Zn_{21}	1667,0	D8	<0,05	nih.		$1,0.10^{-4}$	$1,0.10^{-4}$	$1,0.10^{-3}$

Tab. 1   Binary alloys. CPI-results and magnetic susceptibilities.

IA	IIA	IIIB	IVB	VB	VIB	VIIB	VIII			IB	IIB	IIIA	IVA
1 H 1.0													
	12 Mg 24.3											(13 Al 26.9)	14 Si 28.0
	20 Ca 40.0		22 Ti 47.9	(23 V 50.9)	Cr 52.0	25 Mn 54.9	26 Fe 55.8	27 Co 58.9	28 Ni 58.7	29 Cu 63.5	30 Zn 65.3		

Fig. 1   Favourable elements for mobile hydrogen storage in metal hydrides

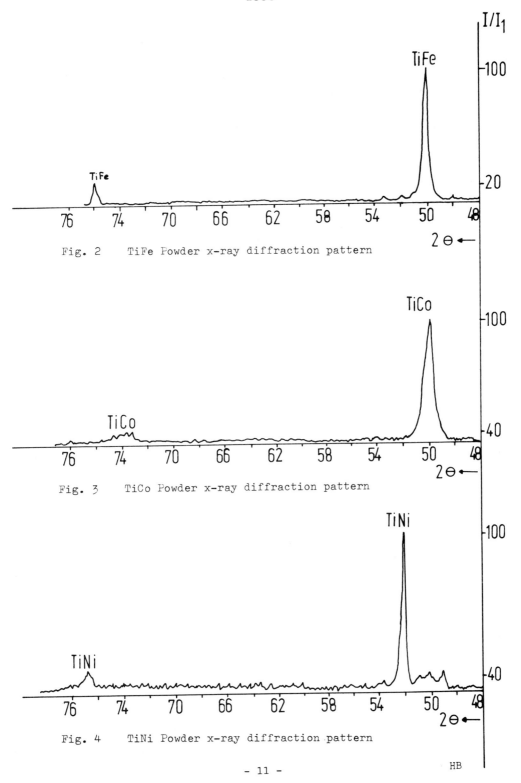

Fig. 2   TiFe Powder x-ray diffraction pattern

Fig. 3   TiCo Powder x-ray diffraction pattern

Fig. 4   TiNi Powder x-ray diffraction pattern

HB

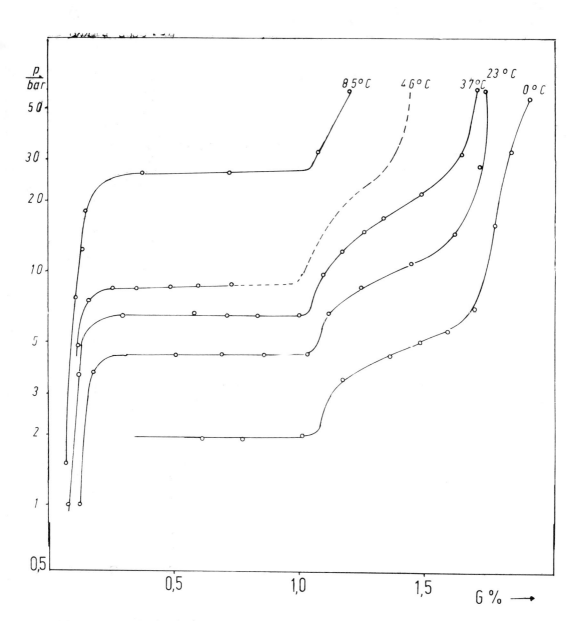

Fig. 5   TiFe Concentration-pressure-isotherm

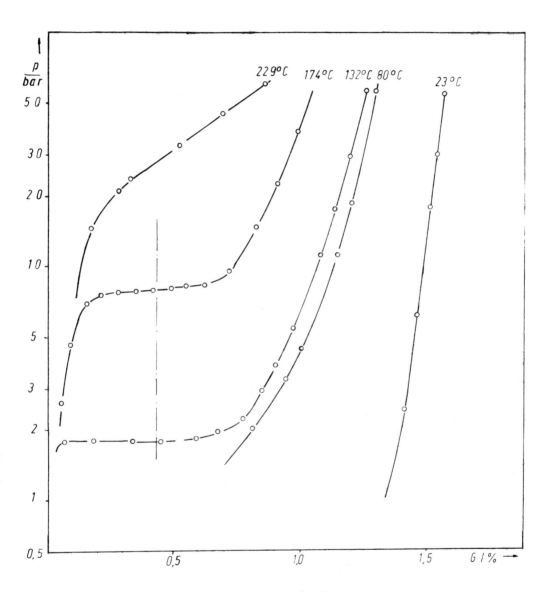

Fig. 6   TiCo Concentration-pressure-isotherm

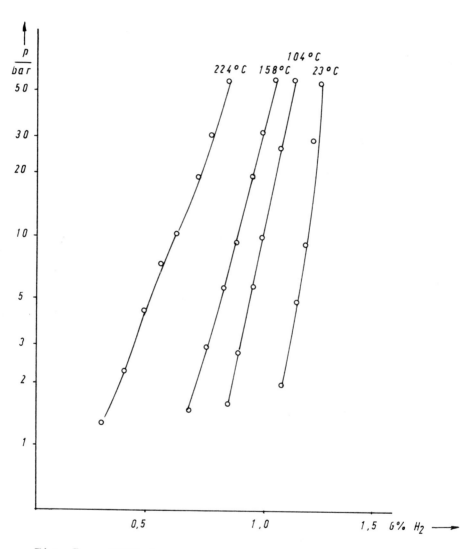

Fig. 7   TiNi Concentration-pressure-isotherm

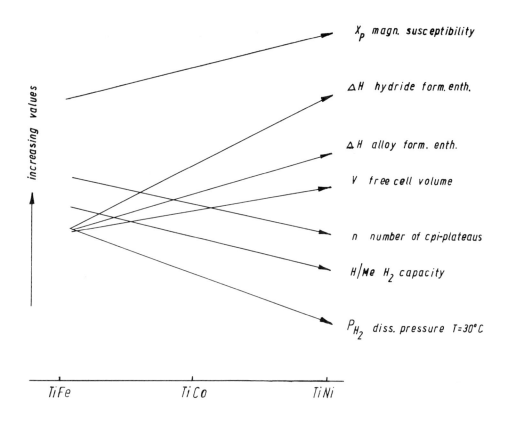

Fig. 8   Binary intermetallic compounds Ti(Fe, Co, Ni)
          Trends in hydride formation

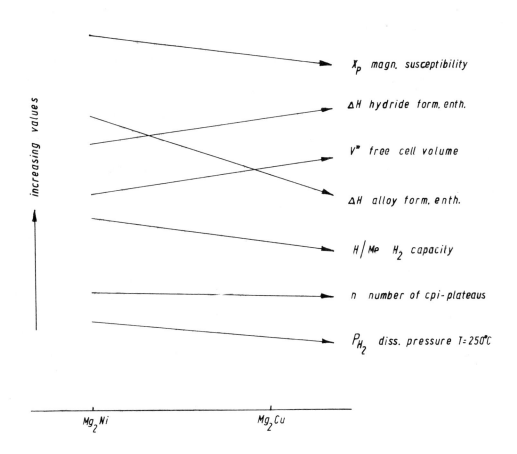

Fig. 9   Binary intermetallic compounds Mg_2(Ni, Cu)
Trends in hydride formation

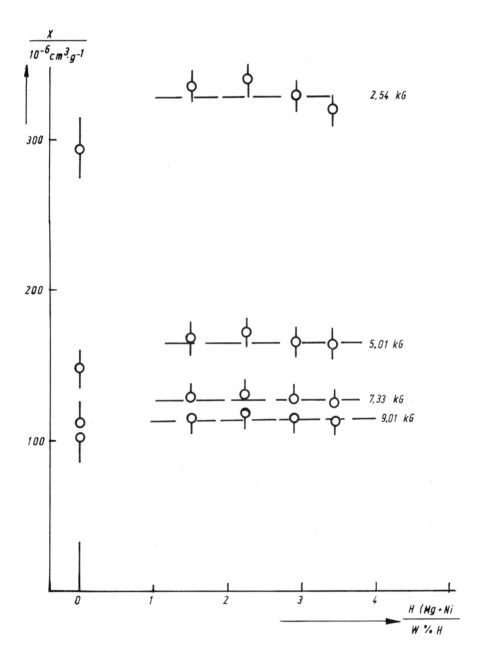

Fig. 10   Mg$_2$Ni Dependence of the magnetic susceptibility from hydrogen: metal ratio

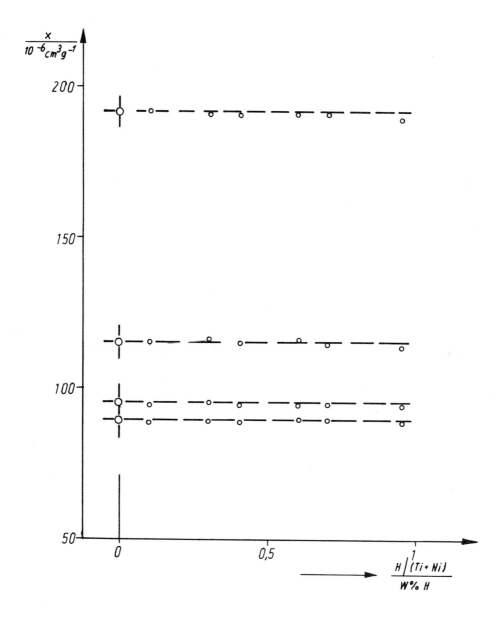

Fig. 11   TiNi Dependence of the magnetic susceptibility from hydrogen: metal ratio

MIXING EFFECTS OF DIFFERENT TYPES OF HYDRIDES

S. Suda and M. Uchida
Kogakuin University
Hachioji-shi, Tokyo, Japan 192

ABSTRACT

A method to synthsize a mixture which has new hydriding and dehydriding properties is presented and a series of hydriding al-
loy mixtures composed of $LaNi_5$ and $Ti_{.8}Zr_{.2}Cr_{.8}Mn_{1.2}$ were selected
to demonstrate the validity of the proposed method.
Experimental results were reported for the equilibrium behaviors
of the $[xLaNi_5+(1-x)Ti_{.8}Zr_{.2}Cr_{.8}Mn_{1.2}]H_y$ system, where x is the weight
fraction of $LaNi_5$ and y is the numbers of hydrogen atoms absorb-
ed in mole of alloy mixture and is ranging from 3.0 to 6.7.
The rate studies were also undertaken briefly under various temperature and composition conditions.
From the results, the mixed hydrides were expected to find various uses in thermodynamic cycles and many types of energy systems, and on the more practical sence, in the safety device developments and uses of hydriding alloy mixtures from economical
standpoints.

INTRODUCTION

Research and development works on metal hydrides and the related technological developments in the field of hydrogen storage
and thermodynamical applications have been advanced greatly in
recent years.
Among various types of hydriding metals and alloys which have
been developed in the past, those expected to have practical
uses would be limited to several materials which are mainly clas-
sified into such groups as magnesium-, titanium-, and rare earth
metal alloys.
Uneven distribution of those hydriding materials to the two re-
sricted regions is observed in the equilibrium pressure and temperature relations plotted as log P vs.1/T shown Fig. 1.
Metal hydrides formed by rare earth- and titanium-alloys are distributed in the relatively high pressure and moderate temperature
regions. On the contrary, magnesium-hydrides are located in the
low pressure and high temperature regions. There has not yet been
investigated such a metal hydride that covers those two separated
regions.
It is difficult to find a single metal hydride which has suitable
properties for a special technical application such as energy
storage and transfer media. At this respect, it is of great significance to develop new hydriding alloys which cover the fore-

1561

mentioned regions and synthesis or improvement of the properties
of known metal hydrides is also meaningful for obtaining material of proper hydriding and dehydriding properties.
Ti-alloys which has a form of $Ti_xZr_{1-x}Cr_yMn_{2-y}$ developed by Machida,
Yamadaya and Asanuma(1,2) and series of elaborated works on 4-
and 5-components titanium alloys by Gamo et al(3,4) are good examples of the most advanced investigations which are aimed to
control and cover the wide ranges of hydriding and dehydriding
properties with keeping high hydrogen retentive properties.
The authors' proposal(5) is to get metal hydride mixtures which
have the improved equilibrium behaviors and kinetic properties
different from those of the original hydriding materials.
It was shown by the previous reports(6,7) for the $LaNi_5$ + TiFe
and the $LaNi_5$ + $Ti_{.8}Zr_{.2}Cr_{.8}Mn_{1.2}$ mixtures that simple mixing of
different kinds of metal hydrides is helpful for practical engineering applications, especially for thermodynamic cycles such
as heat pumps and industrial heat exchanging systems.
The proposed method will provide a simple but a helpful means
for designing new hydriding material which has an "in-between"
properties of original constituents.
Experimental results for the $LaNi_5$ + $Ti_{.8}Zr_{.2}Cr_{.8}Mn_{1.2}$ hydriding
mixtures are presented for the equilibrium behaviors and kinetic
properties.

EXPERIMENTAL

Details of the experimental apparatus and procedures were reported elsewhere(5,6), and the experimental results for the $LaNi_5$ +
TiFe mixtures also published in the references(6,7).
Experimental samples previously mixed with $Ti_{.8}Zr_{.2}Cr_{.8}Mn_{1.2}$ and
$LaNi_5$ were used without any activating procedures. The former
alloy which generally has a form of $Ti_xZr_{1-x}Cr_yMn_{2-y}$ has remarkably high activating properties and several excellent characteristics as a hydriding material with other 4-component alloys
recently investigated by Gamo(8). After several times of hydriding concentration relations were volumetrically measured under
several isothermal conditions for a series of constant composition mixtures.
Samples of about 10 grams were taken for the equilibrium measurement and prepared as the following five combinations; $LaNi_5$,
29.99 wt.% $LaNi_5$, 50.02, 66.34, and $Ti_{.8}Zr_{.2}Cr_{.8}Mn_{1.2}$, respectively.
Rate of dissociation of hydrogen gas also measured volumetrical-
ly as percent loss of hydrogen gas from hydriding mixtures as
a function of time(min) for four samples; $Ti_{.8}Zr_{.2}Cr_{.8}Mn_{1.2}$, 29.99
wt.% $LaNi_5$, 50.02, and $LaNi_5$, respectively. Each sample was adjusted to have total moles of 0.058(9.5-25grams) for rate studies.
Concentration of hydrogen absorbed in hydriding mixtures was expressed by the number of hydrogen atoms per mole of alloy mixture
(=H/mol-M).

EXPERIMENTAL RESULTS

Composition dependence of the equilibrium pressure and hydrogen
concentration relations for the LaNi5 + Ti.8Zr.2Cr.8Mn1.2 system
was examined at isothermal condition of 40°C and illustrated in
Fig. 2. Shapes of the P-C curves for three mixtures were affected by LaNi5 at hydrogen lean conditions and were also influenced
by Ti.8Zr.2Cr.8Mn1.2 at higher hydrogen concentration regions.
Fig. 3 shows the temperature dependence of the dissociation curves
for the 50.02 wt.% LaNi5 mixture in the LaNi5+ Ti.8Zr.2Cr.8Mn1.2
binary system.
Reproducibility of the dissociation curve of the 50.02 wt.% LaNi5
mixture was certified after 100 times of hydriding and dehydriding cycles, no remarkable fluctuation in the pressure-composition
relation was observed where errors in volumetric measurements
were estimated to be within 0.3 percent against the total hydrogen gas dissociated.
Fig. 5 shows the dissociation speeds expressed as the percent
loss of hydrogen from the fully hydrogen-charged hydride mixtures. Significant improvement of the dissociation speed of LaNi5
hydride by adding Ti.8Zr.2Cr.8Mn1.2 was clearly shown in the figure.
Change in the discharging speeds of the 50.02 wt.% LaNi5 was also
given in Fig. 6 expressing as the depression rates of the system
pressure under four isothermal conditions, where sample had a total moles of 0.058(9.5-25grams) for each run. Isothermal percent
loss of hydrogen from the fully charged hydriding alloy mixture
of 50.02 wt.% LaNi5 was also observed and resulted data are graphically given in Fig. 7 where the rate of dissociation of the hydride mixtures is shown to be affected remarkably by temperature.

DISCUSSION

As shown in Fig. 4, the mixed hydrides have reproducible isothermal equilibrium P-C relations. From the P-C curves for various alloy compositions under several isothermal conditions, it is possible to draw straight lines in connection with the dissociation
pressure versus temperature relations as given in Fig. 1 and the
heat of dissociation(ΔH) were calculated from the slopes of log P
vs.1/T plots as tabulated in Table 1 where linear changes of ΔH
with alloy composition are evident(Fig. 8).
Maximum hydrogen absorbed by unit mass of alloy mixture(cm^3/g) is
also a linear function of compositions of the hydriding alloys
(Fig. 9) and the hydride was determined to have a form of [xLaNi5
+(1-x)Ti.8Zr.2Cr.8Mn1.2]Hy, where x is the weight fraction of LaNi5
and y has a value ranging from 3.0 to 6.7.
From rate studies on the mixed metal hydride, addition of
Ti.8Zr.2Cr.8Mn1.2 to LaNi5 serves considerably to improve the dissociation speed of hydrogen from hydride phase(Fig. 5).
A general trends that the rate of dissociation is enhanced by increasing temperature for the 50.02 wt.% LaNi5 hydride mixture
(Fig. 7).

Equilibrium relations of the hydrogen-hydride system are quite
analogous to those of the saturated vapor pressure of organic
compound and high temperature causes considerable increase in
pressure, and therefore, it is important to keep hydride pressure within a given tolerable pressure range when hydride reservoir and system are designed.

Practically, the mixing method will offer some safety devices,
for example, in the case where high equilibrium pressure should
be avoided under ambient conditions or a sudden increase in temperature be anticipated. When the surrounding temperature shifts
greatly, it is also useful to mix a small amount of metal hydride
as a pressure absorbant which exerts relatively low pressure at
higher temperature conditions into a metal hydride that has a
higher P-T relation.

Mixing of inexpensive alloy with relatively high cost alloys is
also of interesting from economical standpoints, if no signif-
cant degradation would be observed in the hydriding and dehydriding properties.

Gas liquilification plant and heat pumps reported by van Mal(9)
and by Sheft et al(10) employ two different kinds of metal hydrides, LaNi$_5$ and CaNi$_5$, and they made good use of the difference
in the P-T relations of each hydride.

The mixed metal hydride is also expected to be useful and adapt-
able to such technological applications that no simple set of
metal hydrides can be reached, and in those case, improved P-T
relations and kinetic properties will serve for the optimization
of the system design.

Recent advancements in the field of thermodynamical application
of metal hydride aiming at the waste heat recovery, solar energy
utilization, and heat pumping of the low grade energies, require
to develop new hydriding alloys and to improve the properties of
known hydrides suitable for those energy applications as well as
to search new hydride of high hydrogen contents.

From above observations, mixing of two metal hydrides is known
to provide an alternative method for improving or modifying the
equilibrium P-T-C relations and kinetic properties.

CONCLUSIONS

Through experimental observations for the LaNi$_5$ + Ti$_{.8}$Zr$_{.2}$Cr$_{.8}$Mn$_{1.2}$
hydriding alloy mixtures, the P-T-C and kinetic properties were
known to be improved considerably according to their compositions.
The P-T relations of the mixed hydrides were known to be control-
lable between those of the two hydriding constituents.

Rates of sorption and desorption of hydrogen gas were also found
to be improved remarkably by mixing of those two hydriding materials.

Metal hydride mixtures are of practical importance for designing
thermodynamic cycles and several types of energy systems.

REFERENCES

(1) Y. Machida, T. Yamadaya and M. Asanuma, Proceeding of the 3rd Meeting of Hydrogen Energy System Society(Japan), p.43, November 25, 1976.
(2) Y. Machida, T. Yamadaya and M. Asanuma, International Symposium on Hydrides for Energy Storage, Geilo, Norway, August 14-19, 1977.
(3) T. Gamo, Y. Moriwaki, T. Yamashita and M. Fukuda, The Titanium and Zirconium, Japan Titanium Society, 25(4), 159(1977).
(4) T. Gamo et al, J. Japan Inst. Metals 41, 148(1977).
(5) S. Suda and M. Uchida, International Symposium on Hydrides for Energy Storage, Geilo, Norway, August 14-19, 1977.
(6) M. Uchida and S. Suda, Proceeding of the 4th Meeting of Hydrogen Energy System Society(Japan), p.48, November 25, 1977.
(7) M. Uchida and S. Suda, Proceeding of the 42nd Annual Meeting of the Society of Chemical Engineers, Japan, p.575, April 1-4, (Hiroshima), 1977.
(8) T. Gamo, private communication.
(9) H.H. van Mal, "Stability of Ternary Hydrides and Some Applications"(1976).
(10) I. Sheft et al, International Symposium on Hydrides for Energy Storage, Geilo, Norway, August 14-19, 1977.

TABLE 1. Heat of Dissociation of the $[LaNi_5 + Ti_{.8}Zr_{.2}Cr_{.8}Mn_{1.2}]$ system.[#]

Alloy	Heat of Dissociation (Kcal/mol-H$_2$)
LaNi$_5$	6.81
29.99 wt.% LaNi$_5$	6.58
50.02 wt.% LaNi$_5$	6.81
66.34 wt.% LaNi$_5$	6.71
Ti$_{.8}$Zr$_{.2}$Cr$_{.8}$Mn$_{1.2}$	6.65

[#] Heat of Dissociation was calculated from the P-T relations wherepressure were determined at the middle point of the Isothermal P-C curves.

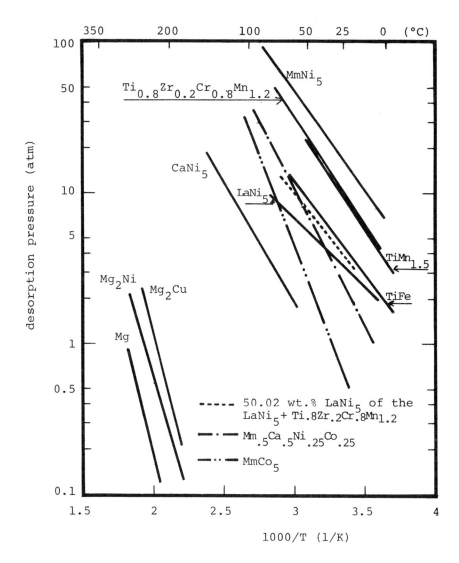

Fig. 1 Dissociation Pressure and Temperature relations of
Rare earth-, Titanium-, and Magnesium Hydrides.

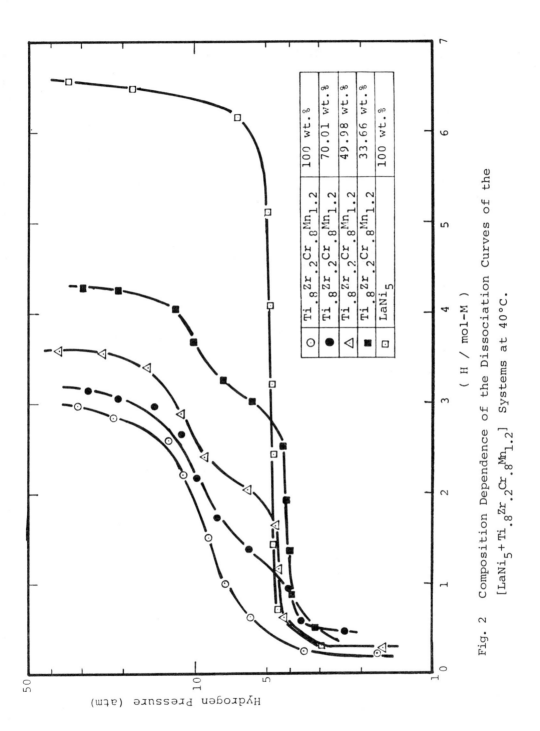

Fig. 2   Composition Dependence of the Dissociation Curves of the
[LaNi$_5$ + Ti$_{.8}$Zr$_{.2}$Cr$_{.8}$Mn$_{1.2}$] Systems at 40°C.

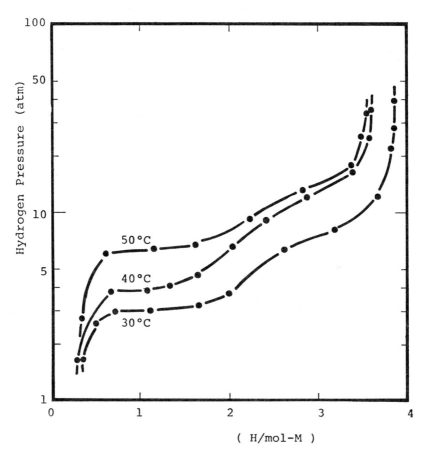

Fig. 3 Temperature Dependence the Dissociation curves for the 50.02 wt.% LaNi$_5$ hydriding mixture of the [LaNi$_5$ + Ti$_{.8}$Zr$_{.2}$Cr$_{.8}$Mn$_{1.2}$] System.

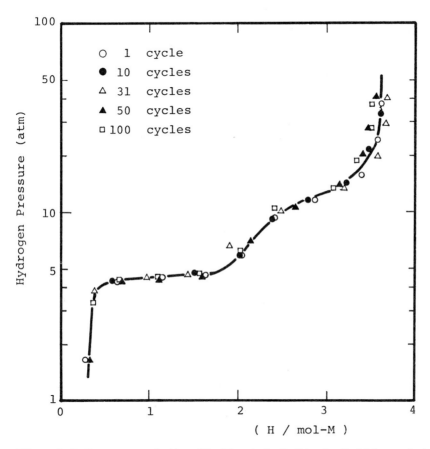

Fig. 4 P-C curve of the 50.02 wt.% LaNi$_5$ hydriding mixture at 40°C after the 100 times of hydriding-dehydriding cycles.

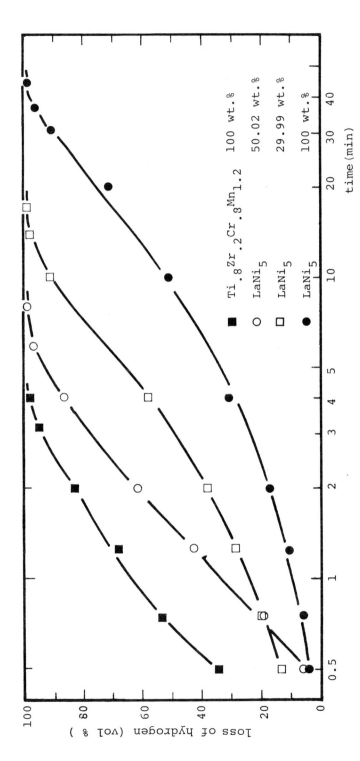

Fig. 5 Composition effects on the dissociation speeds under a constant temperature of 30°C.

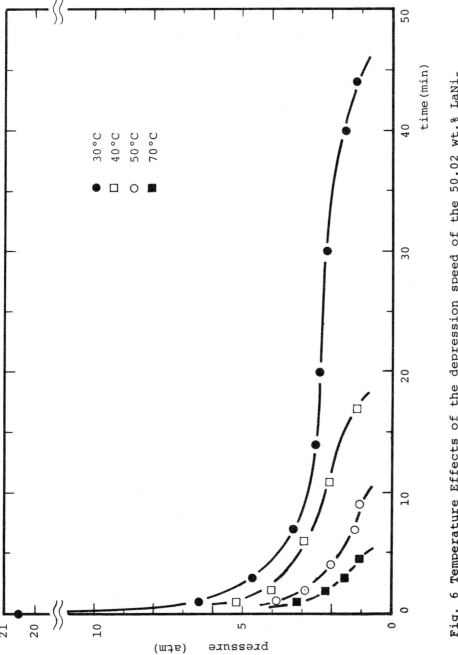

Fig. 6 Temperature Effects of the depression speed of the 50.02 wt.% LaNi$_5$ Hydriding Mixture.

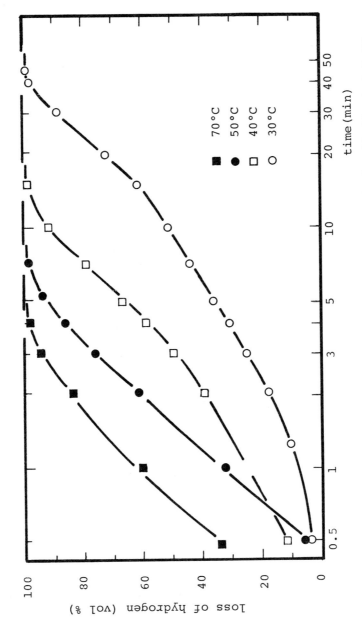

Fig. 7 Temperature Effects on the Dissociation Speeds of the 50.02 wt.% LaNi$_5$ hydriding mixture.

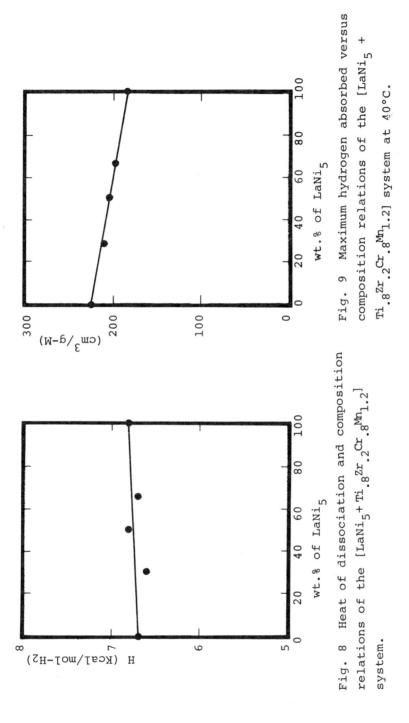

Fig. 8 Heat of dissociation and composition relations of the [LaNi$_5$+Ti$_{.8}$Zr$_{.2}$Cr$_{.8}$Mn$_{1.2}$] system.

Fig. 9 Maximum hydrogen absorbed versus composition relations of the [LaNi$_5$ + Ti$_{.8}$Zr$_{.2}$Cr$_{.8}$Mn$_{1.2}$] system at 40°C.

HYDROGEN (DEUTERIUM) IN VANADIUM

T. Schober and A. Carl
Inst. für Festkörperforschung, KFA Jülich
517 Jülich, Germany

ABSTRACT

The systems VH and VD were investigated by differential thermal analysis and metallography. The techniques for specimen
preparation and handling are described in great detail. Accurate VH and VD diagrams are presented. Isotopic differences are
already present in the solvus. For the VH (VD) solvus we obtained ΔH = 0.1413 eV (0.135 eV) and c_0 = 510 % (700 %). The
phase diagrams exhibit also large isotope effects. The disordering temperatures of the phases β_H and β_D are 40° apart. A
new low-temperature phase was found in the VD diagram. The VD
system could be extended to about 200 %. The VD transformation
enthalpies were measured and compared with similar VH and VD
data of Asano et al. The transformation enthalpies again show
considerable isotopic differences. The VH, VD transition en-
tropies are also included. A discussion of the reasons for the
large isotopic differences in V is presented.

INTRODUCTION

Vanadium is an important alloying element for steels and high
melting point alloys. It is also used as a canning and structural material in reactors, as well as in superconductors
(V_3Si, V_3Ga). One of the most serious factors limiting its use
is the affinity of V to light interstitial elements as H, O, N,
C etc. Contrary to the other group Vb metals niobium and tantalum, V is very difficult to purify as far as oxides, nitrides
and carbides are concerned. As to hydrogen and its isotopes
deuterium and tritium, the group Vb metals Nb, Ta and V have
been prototype substances for the last decades for the study
of the interactions of hydrogen with metals. The background of
these studies was not purely academic. Possible applications of
metal hydrides in hydrogen storage and heat pumps, as well as
in fusion reactor technology were clearly seen. Compared with
Nb and Ta, V had the advantage that V hydrides and deuterides
have a relatively high dissociation pressure which makes these
hydrides suitable materials for the construction of small-scale
laboratory hydrogen storage and heat pump units. Also, V-dihy-
drides are readily formed (in the case of Ta, only H/Ta values
of 80 percent may be achieved). Furthermore, it was found that
there are drastic isotope effects between the systems VH and

VD |1-3|. Finally, V hydrides are well known for the occurrence of tetrahedral and octahedral occupancy of H and transitions (possibly with mixed occupancy) between these phases |4,5|. For the above reasons, the VH-D-T systems are very interesting to study since they contain many complexities such as the above isotopic effects and different crystallographic sites for the hydrogen, but are nevertheless not too difficult to deal with experimentally.

We report here on a more fundamental study of V-hydrides and deuterides. Differential thermal analysis (DTA) was extensively used to determine the VH, VD phase diagrams, and to determine the involved transformation enthalpies. This information is vital for the construction of laboratory-scale heat pumps or heat storage devices. Considerable progress was achieved in the preparation of homogeneous bulk specimens with very high hydrogen concentrations. Partly, this was due to the construction of a high-pressure charging apparatus capable of operating up to $6 \cdot 10^7$ Pascal. Structural and metallographic aspects of V hydrides are also discussed.

EXPERIMENTAL

The material used in the beginning of this study was electro-refined V prepared by the U.S. Bureau of Mines, Boulder City, Nev. The interstitial impurity levels were approximately C < 86, N < 10, O < 70 weight ppm. In the later stages, ultra-high purity V (prepared in the Ames Lab., Iowa) was employed |6|. With this material, the impurity levels were considerably lower: C < 2, O < 15, N < 1 weight ppm. The resistivity ratio was determined to be about 600 |6|. We used in all cases hydrided bulk samples. Powder samples were deliberately avoided since they lead to inhomogeneities in concentration and to a pick-up of O and N during charging.

Gas phase charging of V was accomplished either in a standard low-pressure apparatus, or in a special high-pressure apparatus operating up to $6 \cdot 10^7$ Pa (see Fig. 1). In the latter equipment, V could be charged with ultra high purity H or D obtained through the thermal decomposition of technical V-hydrides or deuterides. Intermediate concentrations were obtained by controlled vacuum hot extraction of hydrides with higher concentrations. In this way, a given sample was measured after each of several dehydriding steps. More specifically, charging was accomplished around 400 °C. Concentrations up to 60 % were obtained with the low pressure equipment when the samples were cooled to room temperature at rates > 1 °/min. Slow cooling ($\leqslant$ 1 °/min) from 400 °C to room temperature resulted even at moderate pressures in 80 to 85 %-samples. Concentrations above 85 % were either achieved with a novel technique described

below, or else by allowing the sample to crack in high-purity
hydrogen at temperatures below 150 $^{\circ}$C. Cracks were often
formed under rapid cooling conditions. Electrolytic charging
was also used. Here, dilute solutions of H_2SO_4 in H_2O or
D_2SO_4 in D_2O were used. The electrolytic method often leads
however to inhomogeneities in the samples. The technique was
mainly used to boost the concentration of precharged V to
values which would be difficult to obtain conventionally. For
example, 70 % specimens could easily be turned into 90 - 100 %
samples with electrolytic charging. - The highest concentrations obtained in this work were 180 %. In line with previous
studies, we observed that ultra-high purity material is more
difficult to charge than more impure stock.

A new palladium coating technique was developed in the course
of this work |7|. Conventionally, Pd is evaporated onto oxide-
free metal surfaces to allow for a free exchange of hydrogen
between the gas-phase and the metal |8|. The present technique
(see Fig. 2) consisted of spotwelding a thin foil (10 - 25 μm
thickness) of Pd onto the sample to be charged with hydrogen.
Provided a certain number of welds was performed (approximately
50/cm^2) the sample could be hydrogenated in the gas phase at
temperatures of 100 to 150 $^{\circ}$C. A typical charging time is 1 hr.
The advantages of this new Pd coating process over the previous
one |8| are: (1) the basic simplicity; no UHV apparatus and
degassing equipment is required, (2) any material may be
coated. No oxide layers have to removed prior to coating (3)
the hydrogen used for charging may be of technical purity only,
(4) no oxidation of the samples may occur during charging. (It
is well known that the formation of suboxides and oxides occur
in V, Ta and Nb already at temperatures around 250 $^{\circ}$C). Finally, it may be anticipated that even cold-welded contact areas
between palladium and an oxide-free metal surface are "hy-
drogen-transparent" around 150 $^{\circ}$C. - Removing the Pd-layer
after hydrogenation may be done chemically with Hg |9| or me-
chanically with fine SiC-paper. This effectively seals the V
sample again.

The differential thermal analysis (DTA) equipment used was a
Mettler TA 2000 (B) apparatus. It has a very high thermal
sensitivity due to the use of evaporated thermocouples. DTA-
runs were made in two different modes. Firstly, low heating-
rate scans (0.1 - 1° min^{-1}) were used to exactly determine the
transition temperatures of hydride phase transformations. Secondly, standard heating rates (approximately 5° min^{-1}) were
employed to measure the reaction enthalpies. For purposes of
calibration, indium was used. Details of the DTA techniques
as applicable to hydrides may be found elsewhere |10,11|. As
to the accuracy of the measurements, transition temperatures
could be determined with a precision of ± 0.2 - 0.5°. The enthalpies, however, may be in error by as much as ± 15 - 20 %.

Optical metallography was also used extensively in the present
study. Earlier work in this laboratory |12-14| had shown that
metallographic techniques such as polarized light microscopy
and Nomarski interference contrast are very helpful in the
study of ordered hydrides, their domain configurations and
morphologies. Also, dynamic experiments with heating and
cooling stages are instrumental in establishing transition
temperatures between various hydride phases |15|.

The main pitfall in the metallographic preparation of V-deute-
rides is the undesirable hydrogen uptake of the sample during
wet grinding and polishing |10|. Since this uptake may easily
amount to 5 to 10 %, the isotopic ratio would be changed during
the preparation. This problem was eliminated in this study by
following the sequence listed below: (1) V crystals are given
a perfect metallographic polish |14| and are freed subsequently
from the embedding resin. (2) These samples are annealed short-
ly at 800 $^{\circ}$C in a vacuum of 10^{-7} Pa and are deuterium charged
from the gas phase. (3) The samples are investigated metallographically without any further polishing or etching. - Surfaces prepared in this manner were suitable for polarized
light studies. Also, they displayed nicely in Nomarski interference contrast the morphology of the hydride plates, as well
as the internal domain structure. (The above technique may be
essential in future metallographic studies of metal tritides).

RESULTS

The Phase Diagrams

Focussing our attention on the VH and VD solvus, we note that
DTA allows the transition temperatures to be determined with
high precision ($\pm 0.5 - 1^{\circ}$). All the data were collected during
heating runs to avoid any effects of supercooling. There is
ample evidence in the literature for hysteresis effects in the
transition temperatures. For a sharp definition of these temperatures, we used large pieces (≈ 250 mg) of the ultra-high
purity V; as well as low scanning rates ($0.1^{\circ} - 0.5^{\circ}$ min^{-1}).
The results in Fig. 3 are plotted in the usual logarithmic
scale. A clockwise rotation of 90° of Fig. 1 will bring it into
the more familiar T-c presentation. Included are the results
(open circles) of V charged electrolytically with an isotope
mixture of D/(H + D) = 0.27. We definitely see a drastic isotope effect in the solvus for the 3 cases. At a given temperature, the equilibrium concentration may be different by a factor of 2 for H and D. We thus may confirm the less accurate
results of the VH (D) solvus found elsewhere |1|. A short
review of other VH (D) solvus studies was recently given in
|10, 16|. Expressing the solvus at low concentrations in the

usual form:

$$c = c_o \cdot \exp(-\Delta H / kT) \tag{1}$$

where c is the atom ratio in % and ΔH the enthalpy we obtain for c_o and ΔH (in the approximation of very low concentrations) the values listed below.

	c_o (%)	ΔH (eV)
pure H	510 ± 50	0.1413 ± 0.005
pure D	700 ± 70	0.135 ± 0.004
D/(H + D) = 0.27	600	0.1413

The VH-phase diagram resulting from our DTA-work is seen in Fig. 4a. Fig. 4b shows an enlarged section of the high-temperature part of Fig. 4a. We list below the most important information on the various VH-phases |1-5; 16|.

α_H-phase: bcc, random distribution of H (D), tetrahedral occupancy
β_H-phase: monoclinic, ordered solution, octahedral occupancy
ε_H-phase: bct, partly ordered, octahedral occupancy
ζ_H-phase: similar to ε, higher concentration
γ_H-phase: fcc, composition: VH_2, CaF_2 structure
η_H-phase: similar to β
δ_H-phase: monoclinic, composition: V_3H_2, octahedral sites

The following transition temperatures were observed: eutectoid reaction $\varepsilon_H \rightarrow \alpha_H + \beta_H$: (164 ± 1) °C, decomposition of β_H: (173 ± 1)°C, decomposition of ε_H at (197 ± 1) °C and peritectoid reaction $\delta_H \rightarrow \eta_H + \varepsilon_H$ (-51 ± 1)°C. A detailed comparison of the diagram in Fig. 4a with previous work |17, 18| was given elsewhere |10|. The VH system is dominated by the high-temperature transition of β_H into the partially disordered phase ε_H. For structural details see Asano et al. |17|. The transition at -20 °C into the proposed phase η_H is accompanied by only a very small enthalpy change. A confirmation of the existence of this phase η should come through independent methods. At present, we cannot rule out that effects have caused the -20 °C peak which are not associated with a VH-phase transition. The main open point, however, in the VH diagram is the question of further ordered phases at very low temperatures.

An excellent review of the VH system was recently given by Fukai and Kazama |19| focussing on the questions of site occupancies, the $\beta \rightarrow \varepsilon$ transformation and the energy levels of various sites in phases α_H, β_H, ε_H, δ_H. Westlake [20] concluded that the VH-diagram in Fig. 4a, b was in full agreement (except for phase η_H) with his resistometric measurements. The V-D diagram obtained in this DTA-study is shown in Fig. 5a. In brief, the various phases may be characterized as follows:

α_D-phase: bcc, random solution of D in V, tetrahedral sites

β_D-phase: monoclinic ordered solution, octahedral occupancy, O_{z1}-sites filled

γ_D-phase: orthorhombic, fully ordered solution, composition V_4D_3, tetrahedral occupancy

δ_D-phase: orthorhombic, ordered solution, composition 75 %, tetrahedral occupancy

ε_D-phase: fcc, composition VD_2, CaF_2-structure

ζ_D-phase: ordered solution, composition $\approx V_4D_3$

We note the most important transition temperatures: order → disorder transformation ($\beta \rightarrow \alpha$) at c = 50 %: 133.5 $^{\circ}$C, $\gamma_D \rightarrow \beta_D + \delta_D$ at -122 $^{\circ}$C, $\alpha_D \rightarrow \beta_D + \zeta_D$ at -56 $^{\circ}$C, $\zeta_D \rightarrow \beta_D + \delta_D$ at -62 $^{\circ}$C, $\zeta_D \rightarrow \alpha_D + \delta_D$ at -39 $^{\circ}$C. The uncertainty in the above temperatures is ± 0.5 $^{\circ}$C. The differences from Asano's diagram [2] are mainly the introduction of phase ζ, a clear description of the range of existence of γ and δ, and the extension to much higher concentrations. Furthermore, the transition temperatures were defined much more precisely in the present work. There is already strong evidence in the work of Asano et al. [2] and Arons et al. [3] of the existence of phase ζ. In order to exclude the possibility of ζ being a metastable phase, a few samples between 55 and 75 % were cooled to -100 $^{\circ}$C to form phase δ. Subsequently, annealing experiments just below -62 $^{\circ}$C demonstrated that the $\delta \rightarrow \zeta$ transformation could not be suppressed. Likewise, cooling runs at different cooling rates reproducibly led to the double peaks at -56 $^{\circ}$C and -62 $^{\circ}$C. We therefore could only conclude that ζ is thermodynamically a stable phase. Electron diffraction experiments are under way to study the structure of phase ζ.

As to the question of further ordered low temperature phases, we note the work of Entin et al. [21] on the VD system who reported appreciably different transition temperatures for phases γ and δ. No evidence of phase ζ_D is presented in their work. They report, however, on a new phase transformation at 80 K for c = 80 % resulting in a gradual disappearance in γ - reflections. The same authors [22] presented recently an interesting neutron diffraction study of the domains occurring in ordered VD alloys.

A schematic version of our VD diagram without data points is shown in Fig. 5b. It is extended to 200 %. The dashed lines indicate hypothetical phase boundaries. The peritectoid transition $\delta_D + \varepsilon_D \rightarrow \alpha_D + \varepsilon_D$ at -32 $^\circ$C was established for a number of concentrations between 100 and 200 % and thus seems to be safely established. VD samples above 90 - 100 % were usually badly deformed and contained cracks, which is not suprising in view of the bcc $\rightarrow$ fcc transformation. $VD_{1.8}$ -chips could even be ignited in air when heated up with a lighter. They burned intensively, very similarly to magnesium. Using the same experimental conditions, chips of $VD_{0.9}$ could not be ignited. Thus, larger quantities of VD_2-powder could be hazardous.

A typical DTA-curve is shown in Fig. 6a for the case of a 60.7 % alloy. Comparison with Fig. 5a shows that the transitions at -122 $^\circ$C, -62 $^\circ$C, -56 and 106 $^\circ$C are observed. Fig. 6b illustrates that the extremely low heating rates of 0.1° min^{-1} are very instrumental in fixing the exact transition temperatures. The reasons for the faint peak at -15 $^\circ$C are not understood. Conceivably, deuteride phase transitions were observed in the suboxides or oxides in the outer layer of the specimen. A slight pickup of O (or N) can never be totally avoided. Similarly, small but definite peaks were observed in the high-temperature parts of the DTA-curves. These peaks occurred very reproducibly at 79.5°, 98°, 105°, 110 and 116 $^\circ$C. Again, we might have observed here transitions in the V-O-D or V-N-D systems.

Metallography; Calorimetry

An example of the metallographic appearance of V-hydrides is presented in Fig. 7. Here, the α_H-β_H-morphology is shown. The thin plates are original β-phase plates arising from gas phase charging. The broad hydride patches (see arrows) were formed during wet grinding and polishing of the samples. Recently, the phenomenological theory of martensitic transformations was successfully applied to the above case of β_H precipitation in α_H |23|.

A number of VD alloys between 35 and 55 % were investigated with a heating stage on an optical microscope. The samples were slowly heated into the single phase α-field. The β_D-domain structure was observed in polarized light to check for possible high-temperature phase transitions of β_D (In the VH-system, such observations clearly resolve the $\beta_H \rightarrow \varepsilon_H$ transition). The outcome for VD-alloys was negative; the β_D-domain structure persisted without changes up to the dissolution of β_D-phase plates. Thus, there is strong metallographic evidence against a high-temperature modification of β_D.

As to the calorimetric results, we first note the detailed results by Asano et al. on the VH-system |17| and on the VD-system |2|. In their work, experimental values of the various transition enthalpies and entropies were given. Also, they included calculations for the entropy changes of the transformations. We report below on our own calorimetric measurements of the VD system and shall give a comparison between the two sets of results. Fig. 8a shows the results of our measurements of the $\beta_D \rightarrow \alpha_D$, $\delta_D \rightarrow \alpha_D$ and $\gamma_D \rightarrow \delta_D$ transition enthalpies (full lines). The dashed lines denote Asano's work.

The agreement for the $\beta_D \rightarrow \alpha_D$ transition enthalpies is (in view of the different equipment used) remarkably good. The same is true for the $\delta_D \rightarrow \alpha_D$ transformation. Note that the two peaks of the $\delta_D \rightarrow \zeta_D$ and $\zeta_D \rightarrow \alpha_D$ were integrated together in this work. The enthalpies of the 2 reactions at -62 and -56 $^\circ$C had an averaged ratio of about:

$$\frac{\Delta H_{-62^\circ}^{(\delta \rightarrow \zeta)}}{\Delta H_{-56^\circ}^{(\zeta \rightarrow \alpha)}} \approx 3$$

A rather serious discrepancy is seen in the case of the $\gamma \rightarrow \delta$ transformation at -122 $^\circ$C. Here, Asano's values are roughly too low by the factor 1/2. Conceivably, this may have been caused by a missing low-temperature calibration of their adiabatic calorimeter. - For each c, we have added the 3 transition enthalpies to obtain a measure for the total heat absorbed on heating (marked ∇ in Fig. 8a). It is seen that the $\delta_D \rightarrow \alpha_D$ transition is energetically the dominant process above 70 %. The above results may be contrasted with the VH results |17| by Asano et al. shown in Fig. 8b. Plotted in Fig. 8b are the ΔH-values for $\beta_H \rightarrow \epsilon_H$ and $\epsilon_H \rightarrow \alpha_H$. The sum of the two enthalpies was also included. Comparison with Fig. 8a shows that the $\beta_H \rightarrow \epsilon_H$ and $\beta_D \rightarrow \alpha_D$ transitions have roughly the same enthalpies. However, if the total sums (ΔH_t^H) are taken into consideration at c = 50 %, we note a drastic isotope effect: $\Delta H_t^H \approx 2700$ Ws/(mole V) versus $\Delta H_t^D \approx 2000$ Ws/(mole V).

Let us consider now changes in entropy. They are related to the above enthalpies and the transition temperatures by:

$$\Delta S = \int_{T_1}^{T_2} c_p/T \cdot dT \approx \frac{\Delta H}{T} \qquad (2)$$

and are therefore readily available. For calculations of ΔS, the separation into vibrational and configurational terms is used:

$$\Delta S = \Delta S^{vib.} + \Delta S^{conf.} \qquad (3)$$

Here, $\Delta S^{vib} \approx 0$ since the system is thermally not excited. $\Delta S^{conf.}$ is usually calculated from principles of statistical mechanics; for examples, see |2, 17, 19, 24|. The VD entropy changes obtained in this work are shown in Fig. 9a (solid lines) for $\beta_D \rightarrow \alpha_D$, $\delta_D \rightarrow \alpha_D$ and $\gamma_D \rightarrow \delta_D$. The dashed lines denote the previous work by Asano et al. |2|. Good agreement (except for $\gamma_D \rightarrow \delta_D$) is obtained between the two sets. It is interesting to note that the largest entropy difference occurs here for the $\delta_D \rightarrow \alpha_D$ transition near the composition 100 %. Here the almost fully ordered VD_{1-x} alloy transforms to the disordered α_D-phase. For the purpose of comparison, the VH entropy changes are depicted in Fig. 9b (after |17|). It is seen that the $\beta_H \rightarrow \varepsilon_H$ and $\beta_D \rightarrow \alpha_D$ phase changes have almost the same ΔS values.

DISCUSSION

Considering the drastic isotopic difference between the VH and VD solvus where $\alpha \rightarrow \alpha + \beta$, we can write the usual relation $c = c_0 \cdot exp\ (-\Delta H/kT)$ in a similar form as given previously for quasiregular substitutional solutions |25|:

$$c = exp\ ^{\Delta S}/k \cdot exp\ ^{-\Delta H}/kT \qquad (4)$$

Here, ΔS denotes an entropy change from a reference state. Within experimental limits, the slopes of the 3 solvus curves in Fig. 3 (the ΔH-values) are equal. Thus, the same energy is required to transfer a H or D atom from phase β to α. The 3 c_0-values, however, reflect different entropy changes in the transition.

As to the VH and VD phase diagrams, the isotope effect most often discussed is the difference of the disordering temperatures for β_H and β_D at 50 %

$$\left.\begin{array}{l} \beta_H \rightarrow \varepsilon_H \quad at \quad 173\ ^{\circ}C \\ but \quad \beta_D \rightarrow \alpha_D \quad at \quad 133.5\ ^{\circ}C \end{array}\right] \Delta T = 40^{\circ}$$

Entin et al. |24| presented recently a calculation of the difference in the following transition temperatures (c = 50 %):

$$\left.\begin{array}{l} \varepsilon_H \rightarrow \alpha_H \quad at \quad 197\ ^{\circ}C \\ but \quad \beta_D \rightarrow \alpha_D \quad at \quad 133.5\ ^{\circ}C \end{array}\right] \Delta T = 63^{\circ}$$

Their theoretical estimate of the above ΔT was 90° |24|. An ordering transition is usually accompanied by a change in the high-frequency part of the vibrational spectrum. Consider the energies of the local vibrations for octahedral and tetrahedral sites as obtained from inelastic neutron scattering experiments:

	octahedral sites	tetrahedral sites
H	$h\nu$ = 0.05 eV	0.120 and 0.17 eV in α-phase |26|
D	0.05 eV	0.08 and 0.12 eV in α-phase |27|

We see immediately that the zero point energies $1/2$ $h\nu$ in the octahedral and tetrahedral sites are far apart. It is this energy gap which is mainly responsible for the large isotope effect in V \|24\|. Consider disorder $\rightarrow$ order transitions between tetrahedral sites. Here, the energy levels are virtually identical and the shift in the disordering temperature will be small when switching from H to D.

Entin et al. \|24\| write the ordering temperature T_c in the standard form:

$$T_c = (\Delta H^{conf} + \Delta H^{vib}) / (\Delta S^{conf} + \Delta S^{vib}) \tag{5}$$

If the difference between two ordering temperatures $T_c^H - T_c^D$ is taken, the above authors obtained for the change in the transition temperature:

$$\Delta T_c \approx (\Delta H_H^{vib} - \Delta H_D^{vib}) / \Delta S^{conf} \tag{6}$$

Here ΔS^{vib} is neglected. As to the ΔH-terms; Entin et al. take $\Delta H_H^{vib}/\Delta H_D^{vib} \approx \sqrt{2}$ and use the relation

$$\Delta H_H^{vib} = \frac{3}{2} (h\nu^T - h\nu^o) \tag{7}$$

The configurational entropy change was obtained from statistical mechanics. As noted above, a ΔT-value of $90°$ was obtained, which is in reasonable agreement with the experiment.

Considering the calorimetric data in Fig. 8a, b we can also show that these data are consistent with the above ΔT_{exp}. At c = 50 % the total enthalpy changes upon disordering are:

hydrogen: $\Delta H_H^{tot} = \Delta H^{\beta\varepsilon} + \Delta H^{\varepsilon\alpha}$

deuterium: $H_D^{tot} = \Delta H^{\beta\alpha}$

The difference between these enthalpies is about 400 W sec/

mole. Upon dividing this value by the entropy difference between α and β of 5 Ws/K (which is roughly the same for H and D)
we obtain a ΔT of 80° which is in line with the experiment. -
Finally, in TABLE I a resumé of the main isotopic differences
between the systems VH and VD is given.

CONCLUDING REMARKS

The systems VH and VD are rewarding systems for the study of
isotope effects. Their importance lies in the simultaneous
occurrence of O- and T-sites within the same parent crystal
structure. This behaviour, however, is quite exceptional for
M-H(D) systems. Similar isotope effects are to be expected
for the VT-system. Conceivably, isotope separation or enrichment could be achieved with the above isotopic differences.
A detailed theoretical understanding of the isotope effects
is still missing. We finally note the increased attention V
is getting in comparison with the group Vb members Nb and Ta.
This trend is expected to continue during the next years.

ACKNOWLEDGEMENT

The authors are thankful to Prof. H. Wenzl for helpful discussions. - The ultrahigh purity vanadium used in this work
was obtained through the kind cooperation of Professors O.N.
Carlson and D.T. Peterson, Ames Iowa. Many thanks are due to
W.W. Stephens, Boulder City and Dr. G. Hörz, Stuttgart for
providing the high-purity electro-refined vanadium used in
the early stages of this work. Dr. G. Kamm helped with the
English text.

REFERENCES

1. D.G. Westlake, S.T. Ockers, Met. Trans. 4, 1355 (1973)
2. H. Asano, M. Hirabayashi, phys. stat. sol. (a) 15, 267
   (1973)
3. R.R. Arons, H.G. Bohn, H. Lütgemeier, Phys. Chem. Solids
   35, 207 (1974)
4. V.A. Somenkov, I.R. Entin, A.Yu. Chervyakov, S.Sh. Shil-
   shtein, A.A. Chertkov, Sov. Phys.-Solid State 13, 2178
   (1972)
5. A.Yu. Chervyakov, I.R. Entin, V.A. Somenkov, S.Sh. Shil-
   shtein, A.A. Chertkov, Sov. Phys.-Solid State 13, 2172
   (1972)
6. O.N. Carlson, D.T. Peterson (private communication)
7. T. Schober, A. Carl to be published
8. N. Boes, H. Züchner, Z. Naturf. 31a, 754, 31a, 760 (1976)

9. B. Makenas, H.K. Birnbaum, Scripta Met. 11, 699 (1977)
10. T. Schober, A. Carl, phys. stat. sol. (a) 43, 443 (1977)
11. T. Schober, A. Carl, Scripta Met. 11, 397 (1977)
12. T. Schober, U. Linke, J. Less-Common Met. 44, 63 (1976)
13. T. Schober, U. Linke, J. Less-Common Met. 44, 77 (1976)
14. T. Schober, U. Linke, Metallography 9, 309 (1976)
15. T. Schober, H. Wenzl, phys. stat. sol. (a) 33, 673 (1976)
16. T. Schober, H. Wenzl in, "Hydrogen in Metals", G. Alefeld
    J. Völkl Editors, Springer Verlag Berlin 1978
17. H. Asano, Y. Abe, M. Hirabayashi, Acta Met. 24, 95 (1976)
18. Y. Fukai, S. Kazama, Scripta Met. 9, 1073 (1975)
19. Y. Fukai, S. Kazama, Acta Met. 25, 59 (1977)
20. D.G. Westlake, Scripta Met. 11, 887 (1977)
21. I.R. Entin, V.A. Somenkov, S.Sh. Shil'shtein, Sov. Phys.
    Solid State 18, 1729 (1976)
22. I.R. Entin, V.A. Somenkov, S.Sh. Shil'shtein, Sov. Phys.
    Solid State 19, 59 (1977)
23. J.S. Bowles, B.C. Muddle, C. Wayman, Acta Met. 25, 513
    (1977)
24. I.R. Entin, V.A. Somenkov, S.Sh. Shil'shtein, Sov. Phys.
    Solid State 16, 1569 (1975)
25. R.B. McLellan, Mat. Sc. Engr. 9 121 (1972)
26. J.J. Rush, H.E. Flotow, J. Chem. Phys. 48, 3795 (1968)
27. J.M. Rowe, Solid State Comm. 11 1299 (1972)

TABLE 1: THE MAIN ISOTOPIC DIFFERENCES BETWEEN THE SYSTEMS VH AND VD

VH	VD
solvus: ΔH = 0.1413 eV, c_o = 510 %	ΔH = 0.135 eV, c_o = 700 %
	- no high-temperature modification of β_D
	- no low-temperature phase with c = 66 %
- no low-temperature superstructures observed with <u>tetrahedral</u> occupancy	- no low-temperature superstructures observed with <u>octahedral</u> occupancy
- highest order → disorder temperature 197 °C (c = 50 %)	- highest order → disorder temperature 133.5 °C (50 %)
- at c = 80 % and room temperature: partially ordered phase present (ε)	- at c = 80 % and room temperature: disordered α-phase present
- total enthalpy in order → disorder transition at c = 50 % : 2750 Ws	- total enthalpy in order → disorder transition at c = 50 % : 2000 Ws
- above 65 %: - main energy is consumed in ε → α transition (~180 °C)	- above 65 % - main energy is consumed in δ → α transition (~40 °C)
- from 100 - 200 %: partially ordered phase (octahedral sites) is in coexistence with f.c.c. VH_2	- from 100 - 200 %: cubic α-phase (tetrahedral sites) in coexistence with f.c.c. VD_2

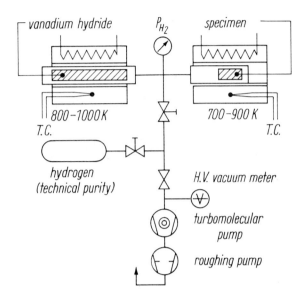

Fig. 1: High pressure hydrogen (deuterium) charging apparatus
(p ≦ 6 · 10⁷ Pa)

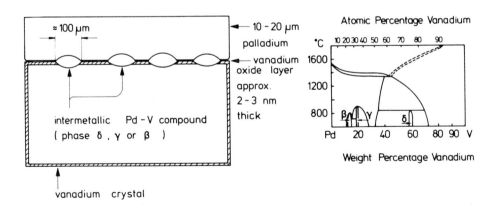

Fig. 2: New technique to make surfaces hydrogen transparent.
20 μm Pd is spotwelded onto the metal

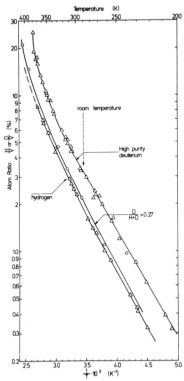

Fig. 3: Solvus for H and D and an isotope mixture

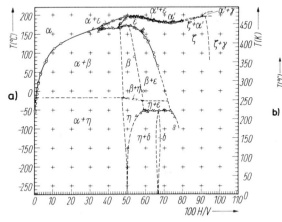

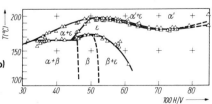

Fig. 4 a) VH phase diagram
b) Enlarged section of a)

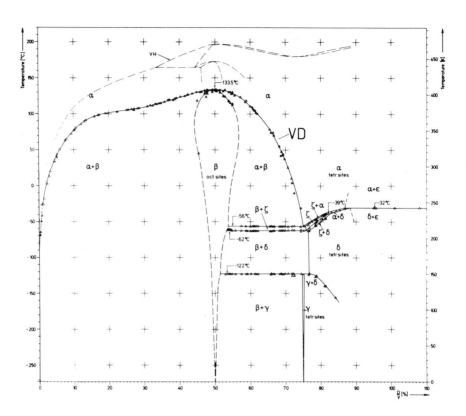

Fig. 5a: VD-diagram (DTA-results)

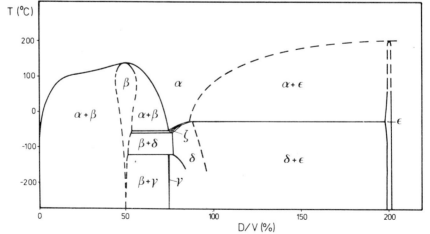

Fig. 5b: Schematic VD-diagram extended to 200 % (D/V)

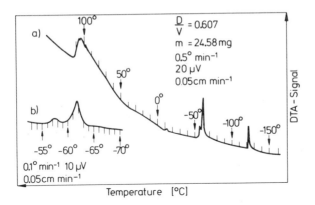

Fig. 6: DTA curve of a 60.7 % VD alloy

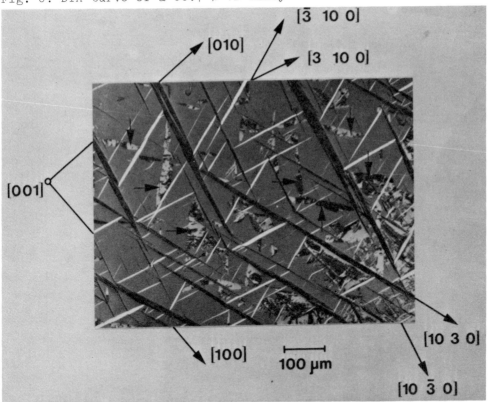

Fig. 7: Polarized light micrograph of Vanadium hydride plates
in V (β in α)

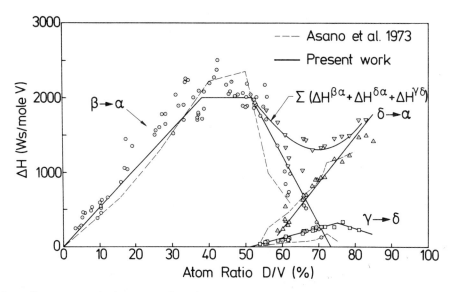

Fig. 8a: Transition enthalpies of VD alloys

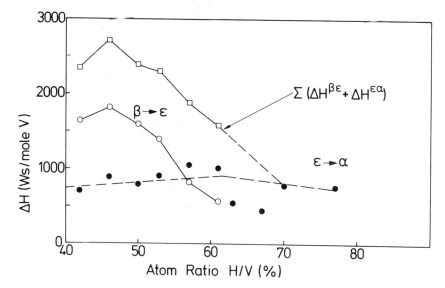

Fig. 8b: Transition enthalpies of VH alloys after Asano
         et al. |17|

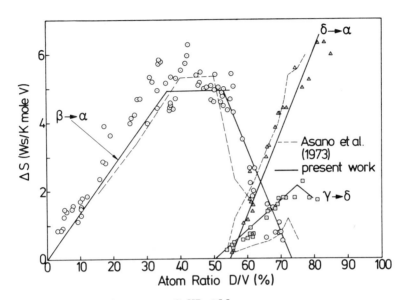

Fig. 9a: Entropy changes of VD alloys

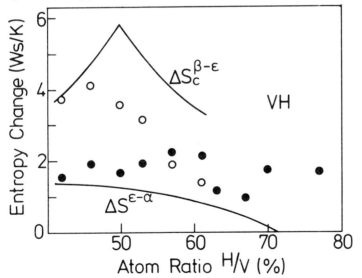

Fig. 9b: Entropy changes of VH alloys after Asano et al. |17|

TECHNICAL SESSION 7B

HYDROGEN STORAGE II

EFFECT OF THE INTERSTITIAL
HOLE SIZE AND ELECTRON CONCENTRATION
ON COMPLEX METAL HYDRIDE
FORMATION

O. de Pous and H.M. Lutz
Battelle Institute
Geneva Switzerland

ABSTRACT

The influence of the interstitial hole size and the electron density of intermetallic compounds is discussed from the view point of utilising these materials for hydrogen storage. Intermetallic compounds with electron to metal atom ratios ranging from four to seven have been found to form complex metal hydrides with temperatures of decomposition above ambient for a hydrogen pressure of one atmosphere. Higher electron concentrations lead to a lowering of the hydrogen content and a consequent decrease in the storage capacity.

In a series of isoelectronic materials, the thermal stability of complex metal hydrides has been found to increase with the metal-metal distance characteristic of the parent metal lattice.

Experimental results shows the extent to which the selection of a suitable material for hydrogen storage is limited.

INTRODUCTION

The utilisation of metal hydrides for hydrogen storage provides certain advantages over the currently used storage techniques [1] [2]. In particular, metal hydrides satisfy the stringent safety requirements for hydrogen utilisation [3] [4]. Effectively, vanadium hydride [5], titanium-iron hydride [6] and lanthanum-nickel hydride [7] can be used at room temperature with a hydrogen storage capacity of around 1.5 wt %. Mg_2Ni can be used at higher temperature ($250^\circ C$) with a hydrogen storage capacity in the range of 3.5 wt % [8].

The definition of a metal hydride working at room temperature with a hydrogen storage capacity greater than 3 wt % is of considerable interest. Several approaches [9] [10] have been proposed for a systematic examination of intermetallic compounds which could possibly represent raw materials for chemical hydrogen storage.

MODELS FOR COMPLEX METAL HYDRIDE FORMATION

The first approach, based on thermodynamics, was proposed by Miedema [9]. In a series of isostructural materials, the thermal stability of the complex metal hydride has been observed to increase as the enthalpy of formation of the parent intermetallic decreases. From this observation, a rule of "reversed stability" has been proposed and defined as:

$$(1) \quad \Delta H_{rA_\varepsilon B_{1-\varepsilon} H_z} = \Delta H_{fA_\varepsilon H_x} + \Delta H_{fB_{1-\varepsilon} H_y} - \Delta H_{fA_\varepsilon B_{1-\varepsilon}}$$

where $\Delta H_{rA_\varepsilon B_{1-\varepsilon} H_z}$ corresponds to the enthalpy of the reaction

$$(2) \quad A_\varepsilon B_{1-\varepsilon} + \frac{z}{2} H_2 \quad ---> \quad A_\varepsilon B_{1-\varepsilon} H_z$$

This empirical rule holds in cases where the electronic structure of the alloy is only slightly affected by the presence of hydrogen in interstitial positions which corresponds to complex metal hydride unstable at high temperatures.

The previous model is also limited by the fact that the total hydrogen content (Z) differs in many cases from the theoretical hydrogen content calculated on the basis of the hydrogen contents corresponding to each of the elementary metal hydrides (X and Y respectively). Then the distribution of hydrogen atoms in the metal lattice (respective coordinance of A and B atoms) has to be known in order to evaluate the respective contribution of

$$\Delta H_{A_\varepsilon H_x} \quad \text{and} \quad \Delta H_{B_{1-\varepsilon} H_y}$$

to the value of the enthalpy of hydrogenation of $A_\varepsilon B_{1-\varepsilon}$.

The hydrogen distribution can be inferred from neutron scattering results on the equivalent deuteride. Nevertheless, the high diffusivity of hydrogen in some metals is such that its residence time in an interstitial hole can be of the same order of magnitude as the time required for a jump from one position to another (e.g. vanadium-hydrogen system). Then the evaluation of hydrogen distribution in the metal lattice is extremely hazardous.

The stabilisation of hydrogen in particular positions would require a strong perturbation of the electronic structure of the binary alloy. Unfortunately, in such cases, the quantitative evaluation of the hydriding enthalpy of the binary alloy $A_\varepsilon B_{1-\varepsilon}$ cannot be carried out using equation (1).

In conclusion, the Miedema [9] model can be used for the prediction of complex metal hydride formation only for well-defined compounds with a hydrogen concentration close to theoretical.

A second model has been proposed by Lawson [10] for the determination of materials for which a large hydrogen solubility can be expected. This development is based on the consideration of the solubility parameter, δ defined as

$$(3) \quad \delta^2 = \frac{\Delta H_s - RT}{V}$$

where ΔH_s is the enthalpy of sublimation of the metal and V the atomic volume. In order to obtain a large hydrogen solubility at room temperature, the value of δ must lie between 100 and 118 $(cal/cm^3)^{1/2}$. Prediction of a large hydrogen solubility in an intermetallic compound can be made on the basis of Hildebrand's mixing rules [10].

$$(4) \quad \delta = \gamma_A \delta_A + \gamma_B \delta_B$$

where γ_A and γ_B are the volumic fractions of A and B, respectively, defined as:

$$\gamma_A = \frac{V_A}{V_A + V_B} \quad and \quad \gamma_B = \frac{V_B}{V_A + V_B}$$

The applicability of such relations is nevertheless limited to two components with nearly the same atomic volume, the volumic fraction being directly deduced from the atomic composition of the material. Table I presents the decomposition temperature corresponding to a hydrogen pressure of 1 atm. and the solubility parameter δ for several metallic materials.

Unfortunately for many intermetallic compounds, the effective volumic fraction differs greatly from the value calculated from the atomic composition, due to a significant electron transfer from one element to the other. Then the atomic volume of each metal varies independently, and the volumic fraction can only be calculated on the basis of a precise knowledge of the crystallographic structure.

In conclusion, the Lawson [10] model can only be applied to binary alloys for which the atomic volumes of each component and their electronegativity are nearly equivalent. The prediction of complex metal hydride formation using the Lawson [10] concept is more convenient for metal solid solutions than for intermetallic compounds.

EVALUATION OF PARAMETERS INFLUENCING COMPLEX METAL HYDRIDE FORMATION

In the present work, in order to select possible candidates for hydrogen storage, the parameters which determine the existence and the thermal stability of complex metal hydrides have been considered [11]. The systematic examination [12] of metallic solid solution and intermetallic phase formation has shown that the relative atomic size is one of the main factors controlling the existence and the structure of alloys.

During the hydrogenation of an intermetallic compound or solid solution as described by equation (2), several structural modifications are observed, such as:

- variation of the metallic lattice parameter

- distortion of the metallic lattice (bcc → bct, fcc → fct)

- a transformation of the metallic lattice structure (hcp → bcc → fcc).

These changes are due to the progressive establishment of metal-hydrogen bond (existence of a definite metal hydride) together with a progressive weakening of the metal-metal bond as the hydrogen content increases. These modifications can be explained in detail by the perturbation of the electronic band structure resulting from the hydrogen absorption. Unfortunately, the interpretation of the influence of hydrogen on the electronic band structure is only completely valid for pure metals, and a more empirical approach has to be used for intermetallics as well as for solid solutions. Parameters to be considered are:

- the interatomic distance in the parent alloy

- the local atomic coordination,

- the electron to atom concentration.

GEOMETRICAL PARAMETERS

Hydrogen is absorbed by metallic materials in a dissociated form and is located principally in the tetrahedral (regular or irregular) holes of the metal lattice. The possibility for hydrogen to be present in such positions depends on the hole-size defined by the distances between the four metal atoms which constitute the tetrahedron. For metals and intermetallic compounds, the interatomic distance D (M - M) depends on:

- the single bond interatomic distance D_o (M - M) characteristic of each metal,

- the coordination number, CN, characteristic of the lattice structure, and

- the number of electrons, ne, participating in the metal-metal bond.

The intermetallic distance can be calculated as follows:

$$(5) \quad D\ (M-M)\ =\ D_o\ (M-M)\ -\ 0.6\ \log\ \frac{ne}{CN}$$

This general equation, proposed by Pauling [13], can be used successfully for the calculation of intermetallic distances in elementary hydrides [11]. For complex hydride formation the average interatomic distance characteristic of each intermetallic compound must be considered.

Considering the effect of transition elements present in a binary alloy the relation between the geometrical parameter (metal-metal distance) and the complex metal hydride stability can be defined as follows:

- in the same series of element a decrease of the complex hydride thermal stability is observed as D (M-M) decreases.

- in the same group of elements an increase of the complex hydride thermal stability is observed as D (M-M) increases.

For example, in a series of isoelectronic materials such as MCo_5, MNi_5, $LaCo_{5-5x}Ni_{5x}$, the linear dependence between the pressure of decomposition and the interatomic distance has been clearly established by Buschow [15]. As shown in Figure 1, the stability increases as the cell volume, and consequently the interatomic distance decreases.

In any case, a minimum thermal stability of the binary alloy is required in order to avoid a possible decomposition of the binary alloy such as:

$$(6) \quad A_\varepsilon B_{1-\varepsilon}\ +\ \frac{X}{2}\ H_2\ \longrightarrow\ A_\varepsilon H_x\ +\ B_{1-\varepsilon}$$

For each crystallographic structure, an optimum intermetallic distance can be found which corresponds to the possibility of formation of a complex metal hydride stable at room temperature. In table 2, the critical metal-metal distance is reported, determined experimentally from the screening tests of Beck [14] for several structures.

The value of the critical intermetallic distance is observed to increase as the metal coordination number increases.

In conclusion, a suitable intermetallic material for hydrogen storage is represented by a metal lattice containing tetrahedral holes and an average metal-metal distance of 2.7 to 3.10 Å. The number of hydrogen atoms absorbed will depend on the electron concentration.

ELECTRON CONCENTRATION

The empirical equation (5) requires that the intermetallic distance decrease as the metallic valency increases [16]. In the first transition element series, the metal-metal bond distance decreases and the cohesive energy increases up to chromium (group VI) and remains approximately constant for the subsequent groups. This observation would indicate that the metallic valency increases to 6 for chromium, corresponding with the group number, and that it remains constant for the subsequent elements (ne ~ 6 electrons/ atom). However, isostructural materials such as TiMn, TiCo, TiFe and TiNi have been found experimentally to be markedly different with respect to hydrogen affinity.

Thus, in the present evaluation, the electron concentration of the elements has been taken as the sum of the external s and d electrons. For the binary alloy, the electron concentration is taken as the average of the electron concentration of each of the components.

- If ne is lower than 2 electrons/atom, the metallic character of the material disappears rapidly as the hydrogen content increases. This results in a decrease of the metal lattice energy and the formation of an ionic hydride, the stability of which depends only on the metal-hydrogen bond strength.

- If ne lies between 2 and 4 electrons/atom metal-metal and metal-hydrogen bonds can coexist, resulting in the formation of thermally stable complex metal hydrides.

- If ne lies between 4 and 7 electrons/atom, the metallic character of the material is maintained In spite of the presence of hydrogen: a complex metal hydride is formed with an equilibrium temperature which depends on the interatomic distance. An increase of the electron density lowers the stability of the complex metal hydride and consequently, in many instances, the storage capacity.

- If ne is greater than 7 electrons/atom, no complex metal hydride formation is observed.

DISCUSSION

In Figure 2, the electron concentration is plotted versus the average value of the metal-metal distance. Materials for which the formation of complex metal hydrides has been observed are clearly indicated.

Any intermetallic or solid solution with an electron concentration ranging from 4 to 7 electrons/atom looks attractive for hydrogen storage application. As shown in Figure 3, the hydrogen concentration seems to be maximum for an electron concentration of 4 electrons/atom. An increase of the electron concentration leads to the destabilisation of the complex metal hydride, from which results a lowering of the hydrogen storage capacity.

In a series of isoelectronic alloys the thermal stability of the complex metal hydride will depend on the average intermetallic distance characteristic of the alloy. In order to get complex metal hydrides stable at room temperature, the metal-metal distance must be greater than 2.65 $\mathring{A}$' 2.75 $\mathring{A}$ and 2.80 $\mathring{A}$ respectively for A2, C14 and C15 structures. These limits are due to the minimum hole size required for the localisation of hydrogen in interstitial tetrahedral positions. Using ternary alloys as raw materials for hydrogen storage, the chemical composition can be adjusted in order to simultaneously obtain high hydrogen storage capacity and suitable thermal stability. For example, substituing a transition element present in the alloy by another element of the same series characterised by a larger intermetallic distance, an increase of the complex hydride thermal stability will be obtained. A decrease of the complex hydride thermal stability can be obtained on substituing one element by another element of the same group having a lower intermetallic distance. New materials with an AB_2 composition and their derived ternary alloys such as:

- $Cr_2(Zr,Ti)$

- $(Cr,Mn)_2(Zr,Ti)$

- $(Cr,V)_2Zr$

- $(Cr,Mn)_2Ti$

- $(Cr,Mn)_2Zr$

- $(Cr,Mo)_2Ti$

- $(Cr,Mo)_2Zr$

could be of particular interest. The influence of elements such as Cu, Zn, Al has to be considered in more detail. For such elements, the presence of p electrons, the possible s-p hybridisation, and the electron transfer are such that no simple model can be used in order to evaluate systematically the effect of such elements on the thermal stability of complex metal hydrides. In some cases the localisation of electrons resulting from the presence of these elements can lower the apparent electronic density, from which could result an increase of the hydrogen concentration. Systematic experiments have to be undertaken in order to evaluate the potential represented by ternary alloys containing Zn, Cu or Al as candidates for new hydrogen storage materials.

BIBLIOGRAPHY

[1]   P. JONVILLE, H. STOHR, R. FUNK, M. KORNMANN
        Hydrogen Economy Miami Energy (THEME) Conference,
        March 1974, S8-25

[2]   K.D. BECCU, H. LUTZ and O. de POUS
        Chem. Ing. Tech. $\underline{48}$ No 2, 161, 1976

[3]   C.E. LUNDIN and F.E. LYNCH
        Proc. 10th Intersociety Energy Consumption and Engineering
        Conference, New York, 1975

[4]   C.E. LUNDIN and R.W. SULLIVAN
        Hydrogen Economy Miami Energy (THEME) Conference,
        March 1974

[5]   J.J. REILLY and R.H. WISWALL
        Inorg. Chem. $\underline{9}$, l678 (1970)

[6]   J.J. REILLY and R.H. WISWALL
        Inorg. Chem. $\underline{13}$, 2l8 (1974)

[7]   J.H. van VUCHT, F.A. KUIJPERS and H.C. BRUNING
        Philips Res. Report $\underline{25}$, 133 (1970)

[8]   J.J. REILLY and R.H. WISWALL
        Inorg. Chem. $\underline{7}$, 2254 (1968)

[9]   W.R. MIEDEMA
        J. of Less Comm. Metals $\underline{32}$, 117 (1973)

[10]   D.D. LAWSON, C.G. MILLER and R.F. LANDEL
        Hydrogen Economy Energy (THEME) Conference, Miami,
        March 1974, 9B-3

[11]   O. de POUS and H.M. LUTZ
        Sec. Int. Cong. on Hydrogen in Metals, Paris (1977)

[12]   A.F. WELLS
        Structural Inorganic Chemistry (1975), Clarendon Press

[13]   L. PAULING
        Phys. Rev. $\underline{54}$, 899 (1938)

[14]   R.L. BECK
        Investigation of hydriding characteristics of intermetallic
        compounds, LAR-55-Denver Research Institute (196l)

[15]   K.H.J. BUSCHOW and A.R. MIEDEMA
        "Hydrogen absorption into rare earth intermetallic compounds)
        Int. Symp. Hydrides for Hydrogen Storage, Geilo/Norway
        August 1977

[16]   R.E. RUNDLE
        Theories of bonding in metals and alloys, Intermetallic
        Compounds, J. Wiley Ed. (1967)

TABLE 1 - SOLUBILITY PARAMETER OF MATERIAL USABLE FOR HYDROGEN
STORAGE [10]

Materials	Solubility parameter	Temperature ($^{\circ}$C) for 1 atmosphere H_2
V	119	12
VNb	117	45
ZrNi	107	
TiNi	106	
TiFe	103	- 8 - 20
$TiFe_{0.8}Cr_{0.2}$	102.5	20
$TiFe_{0.7}Mn_{0.3}$	100.6	10
TiCu	99.2	
TiMn	94.4	
Ti	94	630
$MgNi_2$	75.9	no hydride
Mg_2Ni	64.0	253
Mg_2Cu	56.3	237
Mg	50	280
$LaCuNi_4$	117	25
$LaNi_5$	99.9	10
$SmCo_5$	96.1	-20

TABLE 2 - CRITICAL LATTICE PARAMETER AND METAL-METAL DISTANCE
DETERMINED BY BECK [14] FOR THE FORMATION OF COMPLEX
METAL HYDRIDE

Structure	Critical lattice parameter (Å)	Critical metal-metal distance (Å)	Average coordination number
A2	a = 2.98	2.58	8
B2	a = 3.16	2.60	7
A15	a = 4.93	2.82	9
C15	a = 7.15	2.87	9
L12	a = 4.30	3.04	10
C14	a = 5 b = 8.2	3.10	12

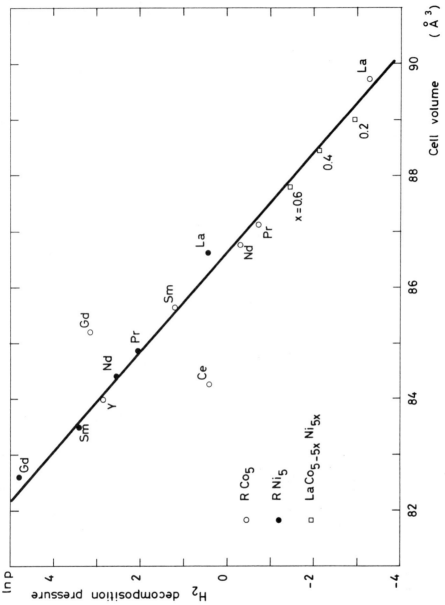

FIGURE 1 – DEPENDENCE BETWEEN THE PRESSURE DECOMPOSITION AND THE CELL VOLUME OBSERVED BY BUSCHOW [15]

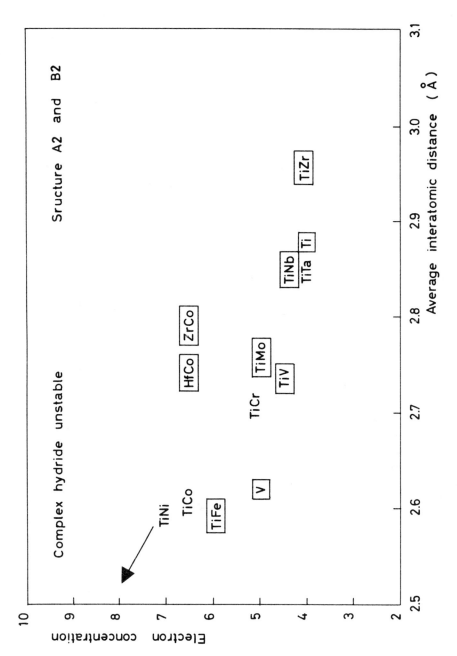

FIGURE 2a – DEPENDENCE BETWEEN THE ELECTRON CONCENTRATION AND THE AVERAGE INTERATOMIC DISTANCE
FOR MATERIAL WITH A2 AND B2 STRUCTURES

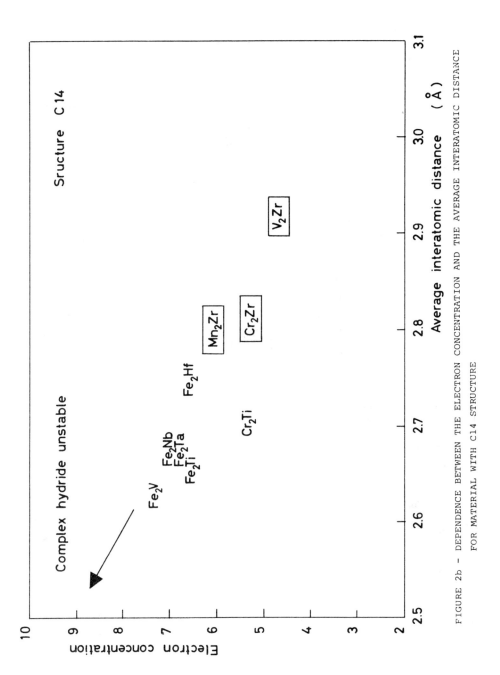

FIGURE 2b – DEPENDENCE BETWEEN THE ELECTRON CONCENTRATION AND THE AVERAGE INTERATOMIC DISTANCE FOR MATERIAL WITH C14 STRUCTURE

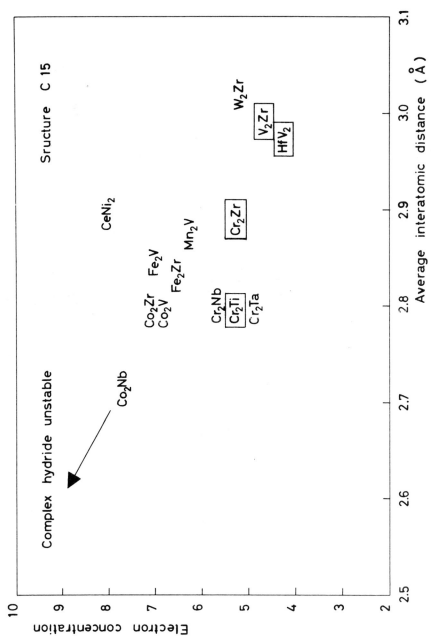

FIGURE 2c – DEPENDENCE BETWEEN THE ELECTRCN CONCENTRATION AND THE AVERAGE INTERATOMIC DISTANCE FOR MATERIAL WITH C15 STRUCTURE

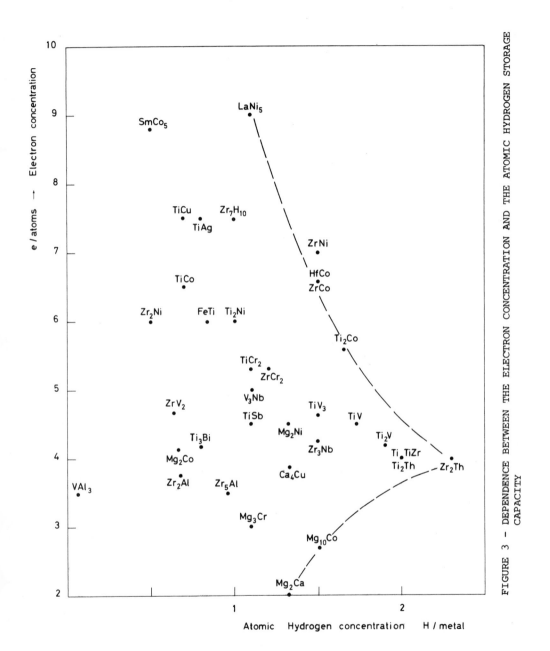

FIGURE 3 – DEPENDENCE BETWEEN THE ELECTRON CONCENTRATION AND THE ATOMIC HYDROGEN STORAGE CAPACITY

EFFECT OF NI,CE AND CO ON HYDROGEN ABSORPTION BY LA-NI ALLOYS

Y.C.Huang,M.Tada,T.Watanabe and K.Fujita
School of Engineering,Tokai University
Hiratsuka, Kanagawa, Japan

ABSTRACT

An assessment on the influence of Ni, Ce, and Co to the solubility and equilibrium pressure of hydrogen, in the systems
for La-Ni, and Misch metal-Ni alloys have been made by means
of measuring Pressure-Temperature-Composition relationships.
The hydrogen pressure at plateau, which indicates two-phase
coexistence, was shown almost at the same value, above Ni/REM
ratio is more than 2 to 5, and the plateau region of isotherms
was increased with increasing of Ni content. Ce would increase
the plateau pressure of hydrogen in La-Ni alloys, and the experiment was carried out up to the $La_{0.4}Ce_{0.6}Ni_5$ alloy.However,
an addition of Co would decrease the plateau of hydrogen
pressure remarkably for La-Ce-Ni$_5$ alloy.

INTRODUCTION

Among various AB$_5$-H systems, where A is a rare earth and B is
Co or Ni, the LaNi$_5$-H system is of the best known system, and
of important prospective compound for the hydrogen storage
madia, in stationary applications, or where weight is not a
limiting consideration.
LaNi$_5$-H system was found by van Vucht et al.[1~5], and a more
extended work has been done by J.J.Reilly et al.[6~12], at
Brookhaven. Since, this system has a unique feature of readily
absorbing and desorbing a large quantity of hydrogen at modest
pressures and room temperatures. Also, a considerable research
from various aspect has been done on this system which could be
seen in literatures elsewhere.[13~16]
The aim of the present paper is focussed upon the assessment
of the influence of La/Ni atomic ratio to the absorption limit
and the equilibrium pressure of hydrogen at the plateau,
through the measurement of La-Ni$_x$, where x stands for 0, 1.0,
1.4, 2.0, 3.0, 4.0, and 5.0 atoms of Ni, and Misch metal-Ni$_5$
alloys. The Pressure-Composition isotherms at different
temperatures for these alloys were measured at the hydrogen
pressure ranging from 10^{-5} to 40 kg/cm^2, and at that temperature, ranges mostly from 20° to 90°C, and 550° to 750°C for
some cases. The influence of Ce, which is commonly found in
Misch metal, and Co, on the equilibria in La-Ni-H alloys were

also discussed through the results of P-C-T measurements.

THERMODYNAMICS

There are many thermodynamic properties to be considered for
Metal-Hydrogen Systems, and they can be derived from the equilibrium pressure of hydrogen to the composition, at a different
temperature of samples. Procedure for deriving those properties
can be seen from many literatures eleswhere,[17~19] and only
those properties calculated out from P-C isotherms will be
proposed in this paper.

EXPERIMENTAL

The samples were prepared by arc-melting under purified argon
from starting materials of 3N purity of La, Ce, Ni, and 2N of
Co. The chemical composition of Misch metal produced in Japan
was shown in Table 1. To improve homogeneity, the button
samples were repeatedly remelted for several times and no
further annealing treatment was performed. Samples of about
5 gr. of 1 mm in diameter were taken from the crusched particles of button ingot, for the high pressure measurement. A 0.5
mm thick sample was sliced from the button, and was used for
the low pressure measurement. Hydrogen gas used for high
pressure measurement was 7N purity of Nippon Oxygen Gas Co.,
and that for low pressure measurement was obtained from desorption of TiH_2, and purified by Pd tubing permeation method.
The experimental apparatus was separated into two parts; one
for a high pressure reactor and the other for a low pressure
reactor. However, the principle of measurement was fundamentally not different at all for both.
A diagram of manifold and reactor assembly is shown in Fig.1.
The high pressure tubing and reactor were made of stainless
steel tubing to a bellow-sealed valves, which are capable of
sustaining up 200 kg/cm^2 of hydrogen. The pressure was measured
by the transducer or pressure gauges. On the other hand, the
low pressure part was made of pyrex glass and quartz tubing
for the reactor. Pressure was measured by mercurry manometers
or the McLeod gauge, which is capable of reading up to 1×10^{-6}
Torr of hydrogen. Both reactors were heated by means of electrical resistance furnaces.
The volume of the entire system and each individual subsections
were calibrated to a known amount of hydrogen. The operation
of the equipment is fundamentally the same as that of Siverts'
apparatus.
After introducing the sample into the reactor, the sample was
outgassed at 700° to 750°C for several days to reach the
leakage rate of 8×10^{-10} 1.Torr/sec, at 1×10^{-6} Torr level for
low pressure measurement. For the high pressure, the sample
was outgassed at 1×10^{-3} Torr, and then introducing 100 kg/cm^2
of hydrogen into the reactor for activation of samples.

Activation of the sample was easily achieved by absorption-desorption cycles for more than 10 times. Avoiding the loss of sample in the reactor, the stainless steel filter inserted in the reactor was essential.

Pressure-Composition isotherms were obtained by the addition of a known amount of hydrogen to the reaction system, and by withdrawing of a known amount of hydrogen from the reaction system. The change of composition of hydride was calculated from PVT measurements. The time for equilibration was much depended on their compositions, temperature and samples, and is varied from a few minutes to several hours. Blank runs showed the loss of hydrogen by diffusion through the wall of the reactor and tubing, is negligible at the experimental temperatures and pressures.

RESULTS AND DISCUSSION

[1] Effect of Activation.

The sample was evacuated at 1×10^{-6} Torr level until the leakage rate could reach 8×10^{-10} l.Torr/sec, at 700°C for the low pressure measurements. However, high vacuum evacuation technique was not the effective method of measuring P-C isotherms for La-Ni-H alloys. The better process for these alloys is to expose them to hydrogen at a pressure of 50 to 150 kg/cm^2, at ambient temperatures. In general, it takes only a few minutes for the alloy to absorb hydrogen even for the first time. This activation process is always accompanied by the sample breaking into small particles. Thus forming a highly active powder with a large uncontaminated surface. Fig. 2 shonws the effect of the activation cycle to the equilibration even at temperature of 0°C. The pressure of the introduced hydrogen was 56.2kg/cm^2, and the exposing time was 1 to 60 min. It shows that equilibria could be achieved in less than 10 cycles, and the absorption time can be shortened by raising the temperature to 20° to 50°C. The absorption capacity of this alloy for hydrogen was not much changed after 200 cycles. The sample was broken to 20 to 50 μ at the first exposition to hydrogen, and it was broken up to 1 to 5 μ. No further effect on rate of absorption or breaking down was seen after 10 cycles of activation.

[2] La-H, and Ce-H systems.

The phase diagrams of La-H, and Ce-H systems were established by measuring P-C-T relationships. Fig.3 is the P-C-T diagram of La-H system, and in the case of Ce-H system, it is almost as the same as that of La-H. Some of their thermodynamical functions were shown in Table 2.

[3] Effect of Ni on La-H alloys.

An addition of Ni to La-H alloy will effect the absorption of hydrogen and the equilibrium pressure, and therefore, the energy of formation of hydries will differ with their Ni content. The P-C isotherms were shown in Fig.4. Alloys poor in Ni($La_{1.4}$ to La_3Ni)were compared at 450°C, and alloys rich in Ni($LaNi_2$ to $LaNi_5$) were compared at 20°C. The equilibrium pressure of hydrogen was about 2 kg/cm^2 at 20°C for the later alloys, and among them $LaNi_5$ alloys absorbs the largest quantity of hydrogen. The energy of formation of hydride changes with Ni content as shown in Table 2. An increase in Ni makes a decrease of ΔH_f of hydrides in La-Ni-H alloys. No further comment can be made on this result at present.

[4] Misch metal-Ni_5-H system.

Hydride of $LaNi_5$ is the most prospetive compound up to date. However, La is an expensive metal. Misch metal containing La and Ce, as could be seen in Table 1, is a relatively inexpensive mixture. The P-C isotherms of $LaNi_5$ has been plotted in, for comparing with their plateau of hydrogen equilibrium pressure. The equilibrium pressures of hydrogen for these Misch metal-Ni alloys were higher than that of $LaNi_5$, but the quantity of absorbed hydrogen was the same as that of $LaNi_5$ alloy.

[5] $La_{1-x}Ce_x.Ni_5$-H system.

Among the constituent of Misch metal, La and Ce are the majority. Theretore, a number of alloys in $La_{1-x}Ce_xNi_5$-H system were measured on P-C isotherms as a function x, where x stands 0.1, 0.2, 0.3, 0.4, 0.5, 0.6. A partial relacement of La by Ce increases the absorption capacity somehow, for example, the alloy of $La_{0.7}Ce_{0.3}Ni_5$ can absorb nearly 7 atoms of hydrogen per formula unit. The increament of Ce in $LaNi_5$ alloy will affect the lattice parameter of the compound as was investigated by K.H.J.Buschow et al.[3] The change in the slop of the curve near, x=0.5, may be attributed to a change in valency of Ce atoms. Since, sample of x is more than 0.5, activation of the samples became difficult and alloys of up to x=0.6 can be measured by P-C isotherms, in this experiment. The effect of Ce on P_{H_2} in $La_{1-x}Ce_x$-Ni_5 hydrides was shown in Fig.8, and the plateau pressure were 50 kg/cm^2 for absorption and 15kg/cm^2 for desorption. Comparison has been made on P_{H_2} in various La-Ce-Ni_5 alloys as shown in Table 3. Alloy with x beyond 0.8 was difficult to prepare and difficult to determin the P-C isotherms.

[6] Effects of Co on P-C isotherms of La-Ce-Ni$_5$ alloy.

An addition of Co will effect the P-C isotherms of La-Ce-Ni$_5$ alloys. For the instance, the P_{H_2} of plateau for La$_{0.4}$Ce$_{0.6}$Ni$_5$ was 15 kg/cm^2 at 20°C, and it would be decreased to 0.5 kg/cm^2 by replacing 3 atoms of Co in Ni. Thus the composition would be written to La$_{0.4}$-Ce$_{0.6}$-Ni$_2$Co$_3$, and it will reduce the energy of formation of hydride slightly. It did not change hydrogen absorption capacity. Therefore, aranging the composition of AB$_5$ compound, where A stands for La or Ce and B stands for Ni or Co, it was possible to obtain some kinds of alloys which can show a suitable compounds with hydrogen for proper service and cost.

CONCLUSION

Present paper dealt with the Pressure-Composition-Temperature measurement for alloys of LaNi$_x$, Misch metal-Ni$_5$, and La$_{1-x}$-Ce$_x$-Ni$_5$, and the effect of Ni on La-H, Ce and Co on LaNi$_5$-H were confirmed.
Ni will increase the capacity of absorption of hydrogen, and an atominc ratio at Ni/La=5, is the largest capacity to take up 6.0 to 6.5 atoms of hydrogen. However, the equilibration pressure of hydrogen at plateau will increase remarkably, and meantime, the ΔH_f, energy of formation of hydride is decreasing from -51.2 kca/mol for LaH$_2$ to -5.94 kcal/mol for LaNi$_5$H$_6$.
In Misch metal-Ni$_5$-H$_{6.7}$, the ΔH_f is -6.4 kcal/mol for absorption, and -6.8 kcal/mol for desorption.
In the studies of hydrogen energy system, among many hydrides, LaNi$_5$ alloy is best known for its unique feature and the most convenient for further studies. For example, studying on an application of desorbed hydrogen from hydride.
The expensive La metal can be replaced by an inexpensive Misch metal.

REFERENCES

1. J.H.van Vucht, F.A.Kuijpers and H.C.A.M.Bruning,
   Philips Res.Repts.25(1970)133.
2. F.A.Kuijpers and H.H.van Mal, J. less-common Metals.
   23(1971)395.
3. K.H.J.Buschow and H.H.van Mal, J. less-common Metals.
   29(1972)203.
4. F.A.Kuijpers, Philips Res. Repts. Supplements. 1973,No.2.
5. H.H.van Mal, Philips Res. Repts. Supplements. 1976,No.1.
6. J.J.Reilly and R.H.Wiswall, Jr., Inorg. Chem.9(1970)1678.
7.   "   "   13(1974)218.
8.   "   Proc. 7th Intersociety
   Energy Conversion Engineering Conference, San Diego,25-29
   September 1972, p.1342.
9. G.Strickland and J.J.Reilly, BNL 50421(1974)
10. J.J.Reilly and R.H.Wiswall,Jr., BNL 17136(1972)
11.   "   BNL 19436(1974)
]2.   "   BNL 21322(1976)
13. J.L.Anderson,T.C.Wallce,A.L.Bowman,C.L.Radosevich,and
    M.L.Courtney, LA-5320-MS,Los Almos Sci. Lab., U.of C.
    July (1973)
14. Masahiro Kitada, Japan Inst. of Metal. 41(1977)412.
15.   "   "   41(1977)420.
16. C.Kato, Master thesis (1977), Tokai University.
17. W.Mueller,J.P.Blackledge and G.G.Libowitz; "Metal hydrides"
    Academic Press, New York(1968)
18. B.Baranowski," Thermodynamics of Meta-Hydrogen Systems
    at High Pressures"(1972)
19. D.T.Hurd,"An Introduction to the Chemistry of the Hydrides"
    John Wiley and Sons Inc., New York,(1952)

TABLE I Chemical Composition of Misch Metal

Element	wt%		
	A	B	C
La	74.38	62.61	49.45
Ce	14.62	25.00	15.24
Nd	7.00	8.39	31.87
Pr	2.73	3.96	8.00
Sm	0.67	0.5	0.5
Others		0.95	0.95
Fe	1.03	1.18	1.00
T.R.E.	98.0	97.87	98.05

T.R.E. : Total Rare Earth

TABLE 2 Comparison of P_{H_2} and ΔH_f for $LaNi_x$-H system

	P_{H_2} (kg/cm^2) Absorption	ΔH_f (-kcal/mol)
La H$_2$	3.5×10^{-6} }at 550°C	51.2 ± 1.40
Ce H$_2$	8.0×10^{-6}	52.0 ± 1.60
La$_3$Ni H$_{1.6}$	1.2×10^{-4}	35.0 ± 1.05
La Ni H$_{1.4}$	2.8×10^{-4} at 450°C	21.8 ± 0.65
La Ni$_{1.4}$H$_{0.7}$	1.2×10^{-4}	18.6 ± 0.56
La Ni$_2$H$_{1.1}$	2.0	7.0 ± 0.21
La Ni$_3$H$_{1.5}$	1.9	6.4 ± 0.19
La$_2$Ni$_7$H$_{2.7}$	1.65 at 20°C	6.2 ± 0.19
La Ni$_4$H$_4$	1.95	5.7 ± 0.17
La Ni$_5$H$_6$	2.3	5.94 ± 0.46

TABLE 3 Comparison of P_{H_2} and ΔH_f for $La_{1-x}Ce_x$-Ni_5 alloys

	P_{H_2} (kg/cm^2 at 20°C)		ΛH_f
	Absorption	Desorption	(-kcal/mol)
$La\ Ni_5H_6$	2.45	2.0	5.94 ± 0.46
$La_{0.9}Ce_{0.1}Ni_5H_{5.8}$	3.10	2.25	6.20 ± 0.15
$La_{0.8}Ce_{0.2}Ni_5H_{6.5}$	8.20	4.33	6.05 ± 0.23
$La_{0.7}Ce_{0.3}Ni_5H_{6.7}$	12.40	5.0	5.80 ± 0.4
$La_{0.6}Ce_{0.4}Ni_5H_{6.3}$	17.50	7.33	5.52 ± 0.15
$La_{0.5}Ce_{0.5}Ni_5H_{6.2}$	25.50	8.20	5.28 ± 0.51
$La_{0.4}Ce_{0.6}Ni_5H_6$	46.50	13.0	5.06 ± 0.07

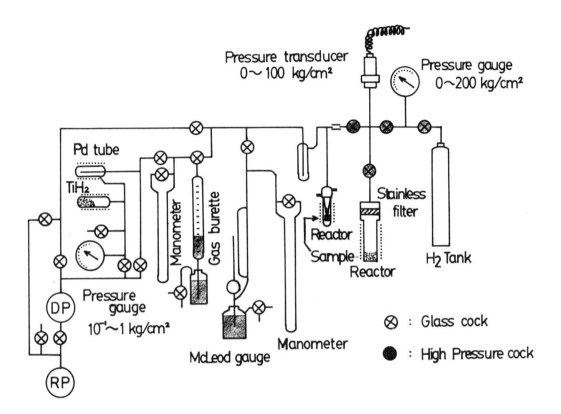

Fig.1 Apparatus for Pressure-Temperature-Composition measurement

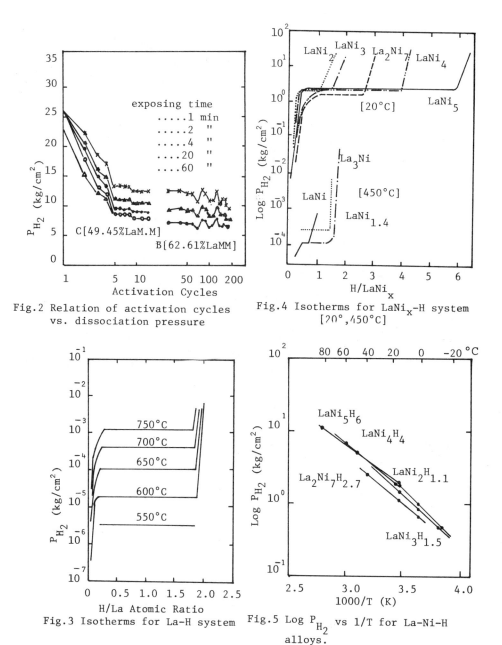

Fig.2 Relation of activation cycles
vs. dissociation pressure

Fig.4 Isotherms for LaNi$_x$-H system
[20°,450°C]

Fig.3 Isotherms for La-H system

Fig.5 Log P$_{H_2}$ vs 1/T for La-Ni-H
alloys.

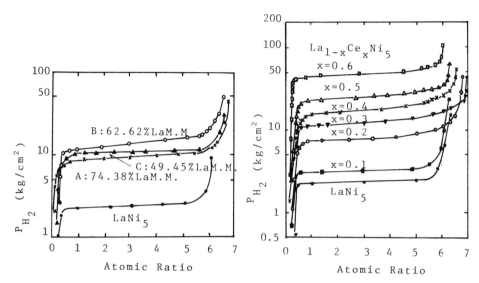

Fig.6 Absorption isotherm for Misch
metal-Ni$_5$ alloys at 25°C.

Fig.8 Desorption isotherms for
La$_{1-x}$Ce$_x$Ni$_5$-H system at 20°C.

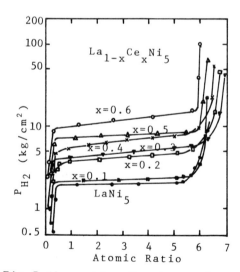

Fig.7 Absorption Isotherms for
La$_{1-x}$Ce$_x$Ni$_5$-H system at 20°C.

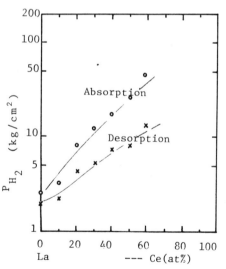

Fig.9 La$_{1-x}$Ce$_x$Ni$_5$: P$_{H_2}$ as a function
of x at 20°C.

DEVELOPMENT OF LOW COST NICKEL-RARE EARTH
HYDRIDES FOR HYDROGEN STORAGE

G. D. Sandrock
The International Nickel Co., Inc.
Inco Research & Development Center
Sterling Forest
Suffern, NY 10901, U.S.A.

ABSTRACT

The intermetallic compound $LaNi_5$ has excellent room-temperature
hydrogen storage properties. However, one major disadvantage
is its high cost. A lower cost AB_5 alloy can be made by substituting the unrefined rare earth mixture mischmetal (M) for
the more expensive refined La. MNi_5, however, forms a very
unstable hydride with impractically high hysteresis. A systematic survey was made to develop ternary substituted MNi_5
alloys with reasonable plateau pressures. Substitutions
included Ca for M and Cu, Fe, Mn, or Al for Ni. All were
successful in lowering the room-temperature plateau pressure
of MNi_5 to practical levels. Properties surveyed include
plateau pressure, hysteresis, H-storage capacity, density, raw
materials cost per unit of hydrogen storage capacity, and crystallographic parameters.

INTRODUCTION

There is rapidly growing interest in the use of rechargeable
metal hydrides for hydrogen storage, compression, and purification, as well as for heat storage and pumping. Thus far, the
main interest in the area of ambient-temperature hydride for-
mers has centered around two intermetallic compounds: the
AB_5 compound $LaNi_5$[1-7] and the AB compound FeTi[8].

For many applications, $LaNi_5$ seems to offer three technical
advantages over FeTi: (a) it is easier to activate, (b) it
exhibits lower hysteresis, and (c) it seems to be somewhat less
sensitive to degradation and deactivation by impurities present
in the H_2 used. In addition, somewhat greater metallurgical
care must be taken in the preparation of FeTi and related
alloys[9,10]. On the other hand, the greatest disadvantage of
$LaNi_5$ is its high cost relative to FeTi, more than five times
on a present raw materials cost basis. For this reason, most
large scale vehicular and stationary hydride storage demonstration projects to date have used FeTi[11-13].

The purpose of this paper is to summarize the results of a
systematic effort to lower the cost of hydrogen storage in the
AB_5 compounds. The experimental research philosophy is shown

schematically in Figure 1. First, it is recognized that a major contributor to the high cost of LaNi$_5$ is the presence of expensive La (about \$44.00/kg at present[14]). It is well known that mischmetal (abbreviated herein as M) can be substituted for La[15]. Mischmetal is an unrefined alloy of a number of rare earth elements (see Table I) that is much lower in cost than La. Presently the price of mischmetal is \$8.93/kg[14]. Unfortunately, MNi$_5$ forms a relatively high pressure hydride with high hysteresis, and is not practical for most room-temperature storage purposes.

It was the approach of this work to at least partially substitute other elements X and Y for the M and Ni atoms in a series of attempts to lower the absorption pressure and reduce hysteresis. One of the criteria used was that the elements substituted should be lower in cost than M or Ni, respectively, and should be readily available. As shown in Figure 1 and Table I, the A-substitution (X-element) tried was Ca and the B-substitutions (Y-elements) tried were Cu, Fe, Mn, and Al.

Ca was selected on the basis of the observation that MNi$_5$ has the so-called CaCu$_5$ prototype crystal structure, so that it might be reasonably expected that Ca would substitute for M atoms in the A sublattice. Cu and Fe were selected because of previous observations that at least partial substitution of these elements could be made into the B-sublattice of LaNi$_5$, resulting in reductions in plateau pressure[4]. Al was selected because of similar observations of partial substitution for Ni in LaNi$_5$ made recently[18,19]. In the case of Mn, Lundin and Lynch have recently shown that it will at least partially substitute for Ni in either LaNi$_5$ or MNi$_5$ and is thereby very effective in lowering plateau pressure[20]. Some of that Mn-substituted MNi$_5$ data will be included in this paper for comparison purposes.

There are a number of important properties of rechargeable metal hydrides that should be quantified for utilization purposes[10]. In a broad survey such as this, where a large number of alloys are tested, it becomes impractical to quantify every important property for each alloy. I have concentrated on the following properties: (a) desorption plateau pressure, (b) hysteresis, (c) H-storage capacity, (d) density, (e) effective raw materials cost per unit of H-storage capacity, and (f) crystallographic correlations.

EXPERIMENTAL PROCEDURE

A number of compositions were made according to general formulas $X_xM_{1-x}Ni_5$ and $MNi_{5-y}Y_y$ where X = Ca and Y = Cu, Fe, Mn, or Al. Alloys were made by induction melting in lot sizes of 8-10 kg and using materials of the following minimum purities: Ni-99.9%, Cu-99.9%, Fe-99.9%, Mn-99.5%, Al-99.8%, M-99%, and Ca-99%. Ca-free alloys were made by vacuum induction melting. Because of the high vapor pressure of Ca, Ca-containing alloys could not be vacuum melted. Low to medium Ca alloys (x <0.5) were melted in a vacuum induction furnace that had been backfilled with argon. Air melting was found to be most effective for high Ca alloys (x >0.5). All alloys prepared were chemically analyzed and examined metallographically. Most were close to the aim stoichiometry, within a few hundredths of an atom for each element in the AB_5 formula. All compositions were brittle and readily crushable.

Equilibrium hydrogen desorption (and in selected cases absorption) isotherms were determined in a high pressure (68 atmosphere capacity) apparatus of carefully calibrated volume. Samples for hydriding (typically about 8 grams) were crushed in air to about 10 mesh prior to testing. With the exception of MNi_5, all compositions could be activated at room temperature within the 68 atmosphere limit of the apparatus. Because of its high absorption pressure, MNi_5 was cooled to -30°C for activation. Most compositions could be activated at room temperature and H_2 pressures considerably below 68 atm. For example, $CaNi_5$ could be activated at H_2 pressures as low as 1 atmosphere. For the sake of comparison, all samples were charged to 68 atmospheres H_2 pressure prior to obtaining a desorption curve. Ultra-high purity H_2 (99.999[+]%) was used. Temperature was controlled to ±0.1°C with a circulating water bath around the specimen reactor. To obtain the low pressure offset on the desorption curves, samples were heated to 100°C and outgassed to a pressure of 0.005 atm or less.

Step by step isotherms were obtained by removing or adding small aliquots of hydrogen and waiting for the pressure to stabilize as the specimens achieved thermal equilibrium. Six to eight hours were taken to run a complete isotherm to assure thermal equilibrium at each point. Non-ideal gas corrections were applied by computer and the data reduced to the form of plots of hydrogen pressure versus atomic hydrogen/metal ratio. To convert to the subscript z in the general formula AB_5H_z, the hydrogen/metal ratio can be simply multiplied by 6.

X-ray diffraction was performed on ground samples in the as-cast condition. A Siemens diffractometer with monochromated Cu radiation was used. XRD was used both to survey for possible second phases and to obtain precision lattice parameter data of the AB_5 phase as a function of composition. Precision lattice parameters were obtained using a linear, least-squares $\cos^2\theta$ extrapolation of the diffractometer data. No XRD analyses were attempted on hydrided samples.

RESULTS AND DISCUSSION

Effect of Ca-Substitution

The overall effect of Ca-substitution for mischmetal in the general formula $Ca_xM_{1-x}Ni_5$ is shown by the series of 25°C desorption isotherms in Figure 2. Ca dramatically lowers the basic curve. By varying x from 0 (MNi_5) to 1 ($CaNi_5$), the 25°C dissociation plateau pressure (at hydrogen/metal = 0.5) can be reduced from about 28 atmospheres to about 0.5 atmosphere. In addition, the substitution of Ca greatly reduces hysteresis as shown in Figure 3.

In addition to the main plateaus, higher Ca alloys have an additional small plateau in the 23-33 atm range. Multiple plateaus have been seen before in AB_5 hydrides and probably represent the occupation of specific interstitial sites by H atoms[21]. At 68 atm applied pressure all $Ca_xM_{1-x}Ni_5$ compounds hydrided to hydrogen contents comparable to $LaNi_5$ on an atomic basis (i.e., to a stoichiometry of about $Ca_xM_{5-x}Ni_5H_6$).

Although it is evident that Ca substitutes for RE atoms in the $RENi_5$ lattice, this substitution is not uniform and some segregation of Ca and RE occurs on solidification. This can be seen on etched metallographic specimens. An example is shown in Figure 4, which is a photomicrograph of etched $Ca_{0.7}M_{0.3}Ni_5$. Although there was a small amount of porosity (black spots in Figure 4) and a trace of a second phase, x-ray diffraction indicated this alloy was for all practical purposes single phase ($CaCu_5$ structure). Dendritic etching effects can be seen that imply composition fluctuations within the phase. They were metallographically found in all the alloys except MNi_5 and $CaNi_5$ and represent some segregation of Ca and M.

The practical consequence of this segregation effect is that the hydriding or dehydriding plateaus of ternary $(Ca,M)Ni_5$ alloys are not flat (Figure 2). In many applications a flat plateau may not be required. In fact a sloped plateau allows direct determination of the amount of hydrogen left in a hydride tank by simply reading the pressure and temperature of the bed. This procedure, of course, cannot be used for a hydride with a flat plateau. On the other hand, a fairly flat plateau may be desirable for many applications, for example, those requiring fast discharge under heat exchange limited conditions.

The lower plateau of $(Ca,M)Ni_5$ alloys can be significantly flattened by homogenization anneals. But this has its limits because of the Ca evaporation that can occur during solid state annealing. The effect of annealing on the 25°C desorption isotherm of $Ca_{0.7}M_{0.3}Ni_5$ is shown in Figure 5. Only selected curves are presented to avoid complicating the figure. Maximum flattening was observed with 1000°C annealing. The curve is not completely flat and some faint segregation of Ca and M could still be seen metallographically. At temperatures higher than 1000°C, Ca evaporation becomes excessive. For example, at 1100°C a thick Ca depleted zone forms on the surface of samples and, in addition, Ca evaporates up preexisting cracks and grain boundaries deep within the samples, leading to extensive precipitation of free Ni[10]. Even using the "core" material of 1100°C annealed material, a substantial loss in H-storage capacity is noted, as shown in Figure 5. With the Ca evaporation problem kept in perspective, it is possible to substantially flatten the plateaus for practical purposes. It should, of course, be noted that annealing should be done in large section size (to minimize surface area) and in inert atmospheres (to prevent oxidation).

In summary, Ca-substitution serves as a useful means of lowering the high pressure and high hysteresis of MNi_5. Thermodynamic and cost parameters will be discussed later. More details of the $Ca_xM_{1-x}Ni_5$ system, with an emphasis on the properties of $CaNi_5$, are presented in [22].

Effect of Cu-Substitution

Cu was found to substitute for Ni in the MNi_5 structure without the formation of second phases. The effect of this substitution, in the general formula $MNi_{5-y}Cu_y$ on the 25°C isotherm is shown in Figure 6. Plateau pressure is somewhat lowered, showing that the effect observed with Cu substitution experiments in $LaNi_5$[4] also holds for MNi_5. Unlike the case observed with Ca-substitution, the as-cast plateaus remain reasonably flat with Cu-substitution, at least up to y = 1.5 atoms. Unfortunately, however, Cu-substitution reduces the reversible storage capacity. The maximum hydrogen/metal ratio is reduced and the low pressure offset is increased, in effect narrowing the useable plateau width. All of the compositions shown in Figure 6 were virtually free of second phases, so that the effect of Cu in lowering reversible capacities is an inherent AB_5 lattice substitution effect.

The plateau pressure cannot be lowered to the levels of $LaNi_5$ without substantive loss of storage capacity. Simultaneous substitution of Ca and Cu was found to provide no overall synergistic effects[23]. Copper substitution lowers hydrogen storage capacity faster than it lowers alloy cost. Finally, as shown by the $MNi_{3.5}Cu_{1.5}$ (y = 1.5) absorption curve in Figure 6, the hysteresis remains undesirably high with Cu-substitution alone.

Effect of Fe-Substitution

The effect of Fe-substitution in the general formula MNi_{5-y}
Fe_y on the 25°C isotherm is shown in Figure 7. Fe does lower
the plateau pressure in a manner similar to Cu. As-cast
plateau slopes remain reasonably flat. Some loss of capacity
occurs, but it is not serious up to y = 1.0. This is contrary to the substantial loss of capacity reported for
$LaNi_{4.0}Fe_{1.0}$[4]. Above y = 1, $MNi_{5-y}Fe_y$ shows a very marked
loss of capacity, as shown by the y = 1.5 curve in Figure
7. As shown by the photomicrographs of Figure 8, the
marked loss of capacity between 1.0 and 1.5 atoms of Fe-
substitution is a result of substantial quantities of second
phases. These second phases have been identified by electron
microprobe analysis. Referring to Figure 8(b), the grey
phase is the $M(Ni,Fe)_5$ phase of storage interest, the smaller
white phase is an Fe-rich Fe-Ni solid solution, and the
darker etching, highly twinned phase is a high rare-earth
phase of approximate stoichiometry $M_3(Ni,Fe)_{10}$. Fe-Ni and
$M_3(Ni,Fe)_{10}$ apparently have little or no useful room temperature hydrogen storage properties. They will not allow us to
lower the plateau pressure of $MNi_{5-y}Fe_y$ to the level of
$LaNi_5$ without loss of most of the capacity.

A surprising benefit of Fe substitution is extremely low
hysteresis. The $MNi_{4.0}Fe_{1.0}$ absorption and desorption curves
shown in Figure 7 exhibit on the order of 0.5-1 atm hysteresis
at 6-9 atm plateau pressure.

Effect of Mn-Substitution

Mn-substitution for Ni has been shown to be a strong stabi-
lizer of AB_5 hydrides (i.e., Mn reduces the plateau pressure
markedly)[20]. In this section, I rely mainly on the results
of Lundin and Lynch, who prepared and evaluated a number of
Mn-substituted MNi_5 samples[20]. A few melts were made in
the present study which confirmed the Lundin and Lynch results.

The potent effect of partial Mn-substitution in lowering the
25°C plateau pressure of $MNi_{5-y}Mn_y$ is shown in Figure 9. No
loss in capacity was observed. Even though the plateau pressure is reduced, two potentially undesirable pressure-
composition properties remain: (a) relatively high hysteresis
and (b) sloping plateaus. The plateau of $MNi_{5-y}Mn_y$ can be
flattened well by a homogenization anneal after casting[20].
In addition, $MNi_{5-y}Mn_y$ alloys seem to be rather prone to
spontaneous ignition when exposed suddenly to air in the
activated state[20], at least relative to the other AB_5
compounds covered in this paper. This may present safety
problems in potential accidental tank rupture situations.

Effect of Al-Substitution

The effect of Al-substitution, in the general formula
$MNi_{5-y}Al_y$, on the 25°C desorption isotherm is shown in Figure
10. Al-substitution is very effective in lowering the
plateau pressure. There is some loss of capacity with increasing Al-content, as well as tilting of the plateau.
The loss of capacity is the result of the formation of a
second phase when Al is substituted for Ni on a one-for-one
atomic basis. The phase is NiAl and is shown metallographically in Figure 11. The amount of this deleterious
phase increases with increasing Al-content. The loss of
capacity resulting for a reasonable room temperature alloy,
say $MNi_{4.5}Al_{0.5}$, is not serious (Figure 10).

The sloping plateau of as-cast $MNi_{5-y}Al_y$ is a result of
segregation within the AB_5 phase. As shown in Figure 12,
the plateau can be effectively flattened by anneals in the
vicinity of 1125°C. Because we do not have the evaporation
effect during annealing noted with $Ca_xM_{1-x}Ni_5$, it is possible
to achieve much flatter plateaus with annealed $MNi_{5-y}Al_y$
than with annealed $Ca_xM_{1-x}Ni_5$ alloys (Figure 5).

The effect of temperature on the desorption isotherms of
annealed $MNi_{4.5}Al_{0.5}$ is shown in Figure 13. A 25°C absorption isotherm is also included which shows surprisingly low
hysteresis. This is, of course, an extremely desirable and
fortunate side effect of Al substitution. The hysteresis
shown by $MNi_{4.5}Al_{0.5}$ at 25°C is less than 0.5 atm at about 4
atm basic plateau pressure.

Heats of Reaction

The heats of reaction for a number of substituted MNi_5
alloys were obtained by running desorption isotherms at a
number of temperatures and plotting on a Van't Hoff plot.
The compositions selected were those that might serve as
reasonable alternatives for $LaNi_5$ for room temperature
hydrogen storage, i.e., exhibit a few atmospheres dissociation pressure at 25°C. Included were $Ca_{0.7}M_{0.3}Ni_5$,
$MNi_{3.5}Cu_{1.5}$, $MNi_{4.0}Fe_{1.0}$, and $MNi_{4.5}Al_{0.5}$. The resultant
Van't Hoff plots and the heats of reaction calculated from
the slopes are shown in Figure 14. Values of ΔH fell in the
6.3-6.7 kcal/mol H_2 range, slightly lower than the 7.6 kcal/
mol value reported for $LaNi_5$ [7]. Of course, changes in the
levels of the substitutions would change the heat of
reaction somewhat.

Comparisons of Raw Materials Cost

An important hydride property is cost, which for most large
scale applications should be as low as possible. Indeed,
the overall objective of this program was to lower the basic
raw materials costs of hydrogen storage in AB_5 compounds.
In this section, I briefly examine the relative raw materials
costs for various substituted $X_xM_{1-x}Ni_5-yY_y$ alloys. It
should be emphasized that I will present only raw materials
costs, that is composition-weighted averages of the elemental
prices shown in Table I. These are not final projected
selling prices, which must include melting and crushing
costs, recovery considerations, profits, etc.

The principal cost reduction from $LaNi_5$ arises from the
basic substitution of low cost mischmetal for expensive La.
The raw materials cost for MNi_5 is $6.06/kg vs. $17.26 for
$LaNi_5$, using present prices for M and La (Table I). Partial
substitution of Ca for M and Cu, Fe, Mn, or Al for Ni further
lowers alloy cost. However, because some of these substitutions result in a loss of hydrogen storage capacity, it is best
to make comparisons on a basis of raw materials cost per unit
of hydrogen storage capacity. This is done for representative
room temperature hydrides in Table II. The criterion used for
capacity was the amount of hydrogen desorbed during a 10 atm to
1 atm excursion at 25°C. For the purpose of the calculations,
hysteresis was ignored, i.e., it was assumed that the sample
was charged well beyond the plateau before desorption was
started. In fact, the 10 atm to 1 atm portions of curves shown
earlier for 68 atm to about 0.2 atm desorption were used. The
overall results will vary slightly depending upon the application of interest (i.e., maximum charge pressure available,
minimum discharge pressure required, temperature, level of
substitution, etc.).

The last column of Table II gives the resultant raw materials
cost on a per gram hydrogen storage capacity basis. The
lowest cost alloy was the Ca-substituted composition, followed
closely by Mn- and Al-substituted alloys. The Cu- and Fe-
substituted alloys are somewhat more expensive because of
larger losses of capacity relative to alloy raw materials
cost reductions. All of the substituted MNi_5 compositions
were substantially lower in capacity-normalized raw materials
cost than $LaNi_5$. They are even substantially lower in cost
than the price for $LaNi_5$ calculated on the basis of a much
lower La price ($19.78/kg[14]) projected to develop if sub-
stantial new markets for La arise in the future.

Comparison is also made in Table II with FeTi. The FeTi
results are based on data taken from a high quality heat
of FeTi made in this laboratory by the rare earth deoxidation

technique[10] and a sponge Ti price of $6.57/kg[18]. Although the comparisons are a bit misleading because of the relatively high hysteresis in FeTi, it is evident that the raw materials costs (per unit hydrogen storage capacity) of the substituted MNi₅ compositions are substantially higher than for FeTi. There may be a number of applications where the technical advantages of the AB₅ compounds (e.g., low hysteresis, relative resistance to poisoning, ease of activation, etc.) may justify their added cost over FeTi.

Lattice Parameter Effects

Since the first paper on the (La,Ce)Ni₅ system[1], it has been widely recognized that there is usually a correlation between the lattice dimensions of the hexagonal AB₅ inter-metallic compound and the hydride dissociation pressure[18,24]. This correlation can be with interstitial hole size, unit cell volume, the basal plane parameter a_0, or the axial parameter c_0. In general, the larger the interstitital hole size, the unit cell volume (roughly related to hole size), or the a_0 parameter, the lower is the plateau pressure. In a qualitative sense, the larger the lattice, the more easily the hydrogen atom fits in and the more stable is the resul-tant hydride. In this section, I briefly examine the lattice effects observed with the substituted MNi₅ alloys studied in this program and their correlation with plateau pressure. All correlations shown are between as-cast parameters and 25°C plateau pressures (taken at hydrogen/metal ratio = 0.5).

The correlations between unit cell volume and plateau pressure are shown in Figure 15 for Ca-, Cu-, Fe-, and Al-substituted alloys. Typical correlations of decreasing plateau pressure with increasing unit cell volume was observed with Cu-, Fe-, and Al-substituted alloys. Al alloys fell along a steeper line than Fe- and Cu-substituted alloys. Ca-substitution showed virtually no correlation between plateau pressure and unit cell volume. In effect, Ca-substitution over the entire 0-1 range of x in $Ca_xM_{1-x}Ni_5$ resulted in no significant changes in unit cell volume but pronounced changes in the plateau pressure.

There is a better correlation between the individual lattice parameters of $Ca_xM_{1-x}Ni_5$ and plateau pressure, as shown in Figure 16. Correlations between the lattice parameters of the Cu-, Fe-, and Al-substituted alloys and plateau pressure also hold.

With the possible exception of the Cu- and Fe-substituted alloys, the data for different systems do not fall along the same curve of plateau pressure vs. a given crystallographic parameter. This indicates that simple geometric factors, such as interstitial hole size, do not completely describe the stability of an AB$_5$ hydride. There must be other chemical or entropy effects that play a significant role. X-ray diffraction screening of lattice parameters, however, does stand as a useful qualitative screening tool for AB$_5$ hydride stability.

CONCLUSIONS AND GENERAL COMMENTS

It has been shown that the partial substitution of Ca, Cu, Fe, Mn, or Al into MNi$_5$ results in a wide variety of useful low-cost AB$_5$ compounds for hydrogen storage and related purposes. All of the above substitutions lower the plateau pressure and hysteresis of MNi$_5$ to manageable levels. This allows the use of low cost mischmetal and results in alloy raw materials costs that are 35-45% that of present LaNi$_5$ on a per unit hydrogen storage basis. In addition, ternary substituted MNi$_5$ alloys offer a wide range of thermodynamic (pressure-temperature-composition) properties that can be tuned to specific application requirements. This extra-ordinary versatility is one of the most important advantages of the modified MNi$_5$ alloys presented in this paper.

For a given application, a number of considerations must be made in choosing the best hydride. Ca$_x$M$_{1-x}$Ni$_5$ seems to be the lowest in cost of the modified MNi$_5$ alloys on a hydrogen storage basis. But Ca$_x$M$_{1-x}$Ni$_5$ is somewhat more difficult to anneal if very flat plateaus are required. In addition, Ca$_x$M$_{1-x}$Ni$_5$ alloys seem to be more susceptible to disproportionation limiting their use for applications where high temperatures might be involved (e.g., in thermal compressors for hydrogen). MNi$_{5-y}$Mn$_y$ is reasonably cheap but has rather high hysteresis and is relatively pyrophoric in the fine activated state. MNi$_{5-y}$Al$_y$ is an especially versatile system that can be tuned to give a wide variety of plateau pressures by adjusting the value of y. MNi$_{5-y}$Al$_y$ can be readily annealed to give reasonably flat plateaus and has the advantage of showing very low hysteresis. MNi$_{5-y}$Fe$_y$ also has the advantage of very low hysteresis and fairly flat as-cast plateaus, but it is more expensive than the others and the range of plateau pressures achievable is limited because of low solubility of Fe in the AB$_5$ phase. MNi$_{5-y}$Cu$_y$ alloys have properties similar to MNi$_{5-y}$Fe$_y$, except that they show higher hysteresis.

In addition to the above considerations, there are other properties that ideally should be known. Resistance to poisoning by potential impurities in the H$_2$ used is very important. We will soon be generating such information on selected representative ternary substituted MNi$_5$ to further develop alloy selection criteria for various applications. Decrepitation (particle size breakdown) effects should be better quantified. Properties such as pyrophoricity, volume change of the hydriding reaction, specific heat, and thermal conductivity may have a bearing on the appropriate hydride chosen for a given application. Much of this detailed data does not yet exist. The main purpose of this program was to provide a brief survey of versatile new families of AB_5 alloys. The $Ca_xM_{1-x}Ni_5$, $MNi_{5-y}Cu_y$, $MNi_{5-y}Fe_y$, and $MNi_{5-y}Al_y$ families of alloys are covered by U.S. patents or patent applications.

ACKNOWLEDGEMENTS

I wish to thank S. L. Keresztes for his extensive and skilled experimental assistance. I wish to also acknowledge the helpful advice of E. Snape, E. L. Huston and J. J. deBarbadillo.

REFERENCES

1. van Vucht, J.H.N., Kuijpers, F.A. and Bruning, H.C.A.M., "Reversible Room Temperature Absorption of Large Quantities of Hydrogen by Intermetallic Compounds", Philips Res. Report, 25, 133 (1970).

2. Kuijpers, F.A. and van Mal, H.H., "Sorption Hysteresis in the $LaNi_5$- and $SmCo_5$-H Systems", J. Less-Common Metals, 23, 395 (1971).

3. Buschow, K.H.J. and van Mal, H.H., "Phase Relations and Hydrogen Absorption in the Lanthanum-Nickel System", J. Less-Common Metals, 29, 203 (1972).

4. van Mal, H.H., Buschow, K.H.J. and Miedema, A.R., "Hydrogen Absorption in $LaNi_5$ and Related Compounds: Experimental Observations and Their Explanation", J. Less-Common Metals, 35, 65 (1974).

5. Boser, O., "Hydrogen Sorption in $LaNi_5$", J. Less-Common Metals, 46, 91 (1976).

6. Boser, O. and Lehrfeld, D., "The Rate Limiting Processes for the Sorption of Hydrogen in $LaNi_5$", Proc. 10th IECEC, 1370, Newark, DE, 1975.

7. Lundin, C.E. and Lynch, F.E., "A Detailed Analysis of the Hydriding Characteristics of $LaNi_5$", Proc. 10th IECEC, 1380, Newark, DE, 1975.

8. Reilly, J.J. and Wiswall, R.H., "Formation and Properties of Iron Titanium Hydride", Inorganic Chem., $\underline{13}$, 218 (1974).

9. Sandrock, G.D., Reilly, J.J. and Johnson, J.R., "Metallurgical Considerations in the Production and Use of FeTi Alloys for Hydrogen Storage", Proc. 11th IECEC, Vol. I, 965, Stateline, NV, 1976.

10. Sandrock, G.D., "The Metallurgy and Production of Rechargeable Hydrides", Proc. International Symposium on Hydrides for Energy Storage, Geilo, Norway, August 1977, in press.

11. Billings, R.E., "A Hydrogen-Powered Mass Transit System", Proc. 1st World Hydrogen Energy Conference, Vol. III, p. 7C-27, Miami Beach, FL, 1976.

12. Buchner, H. and Säufferer, H., "Results of Hydride Research and the Consequences for the Development of Hydride Vehicles", Proc. Fourth International Symposium on Automotive Propulsion Systems, Vol. II, Arlington, VA, 1977.

13. Burger, J.M., Lewis, P.A., Isler, R.J. and Salzano, F.J., "Energy Storage for Utilities Via Hydrogen Systems", Proc. 9th IECEC, 428, San Francisco, CA, 1974.

14. Cannon, J., Molycorp, White Plains, NY, private communication.

15. Reilly, J.J. and Wiswall, R.H., "Hydrogen Storage and Purification Systems", BNL Report 17136, Brookhaven National Lab, Upton, NY, August 1, 1972.

16. American Metal Market, Vol. 86, No. 34, February 17, 1978.

17. American Metal Market, Vol. 86, No. 35, February 20, 1978.

18. Achard, J.C., Percheron-Guegan, A., Diaz, H., Briancourt, F., Demany, F., "Rare Earth Ternary Hydrides-Hydrogen Storage Applications", Proc. Second International Congress on Hydrogen in Metals, Paper 1E12, Paris, 1977.

19. Mendelsohn, M.H., Gruen, D.M., and Dwight, A.E., "LaNi$_{5-x}$ Al$_x$ is a Versatile Alloy System for Metal Hydride Applications", Nature, $\underline{269}$, 45 (1977).

20. Lundin, C.E. and Lynch, F.E., "Modification of Hydrid-
    ing Properties of AB₅ Type Hexagonal Alloys Through
    Manganese Substitution", Proc. International Conf. on
    Alternative Energy Sources, Miami Beach, FL, December
    1977, in press.

21. Stewart, S.A., Lakner, J.F. and Uribe, F., "Storage of
    Hydrogen Isotopes in Intermetallic Compounds", American
    Chemical Society Symposium on Solid State Chemistry of
    Energy Conversion and Storage, April 4-9, 1976, NY,
    Lawrence Livermore Laboratory Preprint UCRL-77455,
    April 6, 1976.

22. Sandrock, G.D., "A New Family of Hydrogen Storage Alloys
    Based on the System Nickel-Mischmetal-Calcium", Proc.
    12th IECEC, Vol. I, 951, 1977, Washington, DC.

23. Sandrock, G.D., "A Survey of the Hydrogen Storage
    Properties of Nickel-Copper-Mischmetal-Calcium Alloys",
    Proc. International Conf. on Alternative Energy Sources,
    Miami Beach, FL, December 1977, in press.

24. Lundin, C.E., Lynch, F.E., and Magee, C.B., "A Cor-
    relation Between the Interstitial Hole Sizes in
    Intermetallic Compounds and the Thermodynamic
    Properties of the Hydrides Formed from Those Compounds",
    J. Less-Common Metals, 56, 19 (1977).

TABLE I

CURRENT LARGE LOT RAW MATERIALS PRICES
FOR VARIOUS ELEMENTS SUBSTITUTED IN
AB_5 COMPOUNDS

Element	Type	Price, $/kg	Reference
La	A	40.09	14
M*	A	8.93	14
Ca	A	3.44	16
Ni	B	4.56	17
Cu	B	1.37	17
Fe	B	0.09	17
Mn	B	1.26	16
Al	B	1.21	17

* Standard Mischmetal: 48-50% Ce, 32-34% La, 13-14% Nd, 4-5%, Pr, and about 1.5% other rare earths.

TABLE II

PROPERTIES OF SELECTED ROOM TEMPERATURE HYDRIDES

(Based on 10 Atm-1 Atm Desorption at 25°C, As-Cast Condition)

Alloy	Density, g/cm^3	Dissociation P at H/Metal = 0.5, Atm	$\Delta\dfrac{\text{Hydrogen}}{\text{Metal}}$	ΔH, Wt.%	Raw Materials Cost $/g	
					$/kg Alloy	H-Storage Capacity
Ca$_{0.7}$M$_{0.3}$Ni$_5$	7.2	3.8	0.67	1.11	4.98	0.45
MNi$_{3.5}$Cu$_{1.5}$	8.5	8.0	0.69	0.95	5.26	0.55
MNi$_{4.0}$Fe$_{1.0}$	8.3	7.6	0.68	0.95	5.40	0.57
MNi$_{4.5}$Mn$_{0.5}$		4.3	0.82	1.15	5.77	0.50
MNi$_{4.5}$Al$_{0.5}$	8.1	3.7	0.78	1.13	5.92	0.52
LaNi$_5$ (Present)	8.3	1.7	0.97	1.36	17.26	1.27
LaNi$_5$ (Future)	8.3	1.7	0.97	1.36	9.45	0.69
FeTi	6.5	4.2	0.64	1.24	3.08	0.25

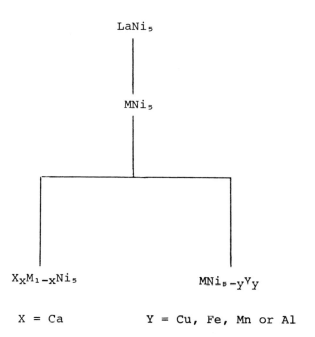

FIGURE 1 - Overall approach to low-cost nickel-rare
earth hydrogen storage alloy development.

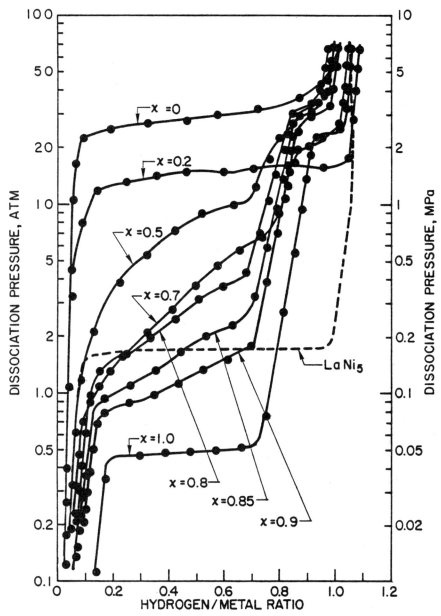

FIGURE 2-EFFECT OF Ca-CONTENT X ON THE 25°C
DESORPTION ISOTHERM OF Ca$_x$M$_{1-x}$Ni$_5$ ALLOYS.

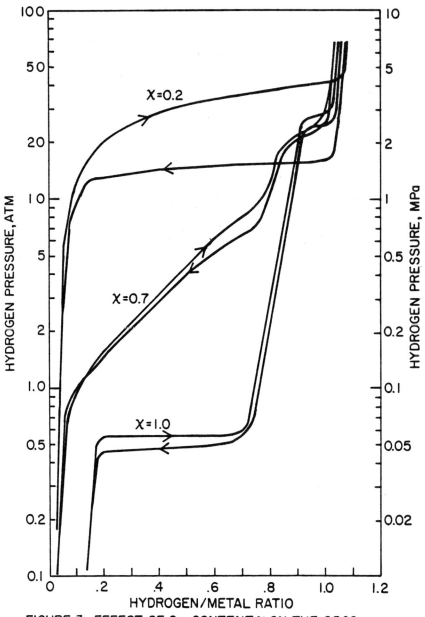

FIGURE 3 - EFFECT OF Ca-CONTENT X ON THE 25°C
HYSTERESIS OF $Ca_x M_{1-x} Ni_5$ ALLOYS.

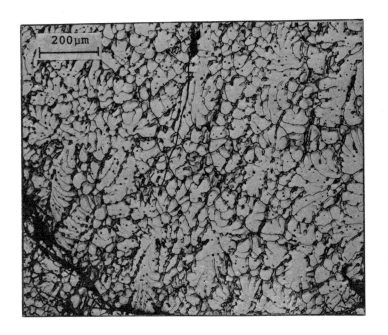

FIGURE 4

AS-CAST MICROSTRUCTURE OF $Ca_{0.7}M_{0.3}Ni_5$

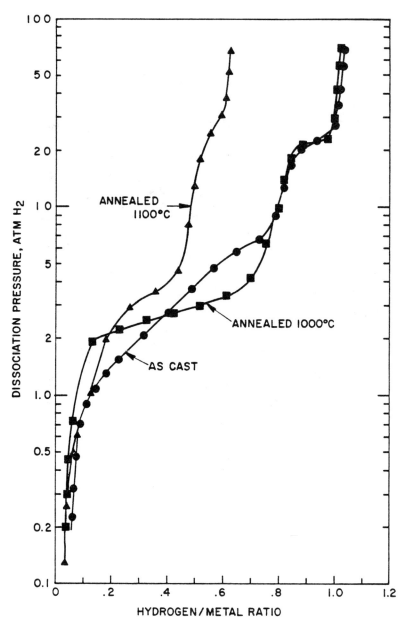

FIGURE 5 - EFFECT OF ANNEALING (24 HOURS) ON THE 25°C
DESORPTION ISOTHERM OF $Ca_{0.7}M_{0.3}Ni_5$.

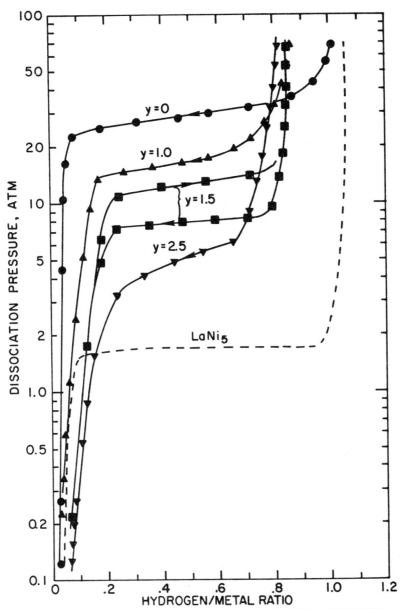

FIGURE 6-EFFECT OF Cu-CONTENT y ON THE 25°C ISOTHERM
OF MNi$_{5-y}$Cu$_y$ ALLOYS.

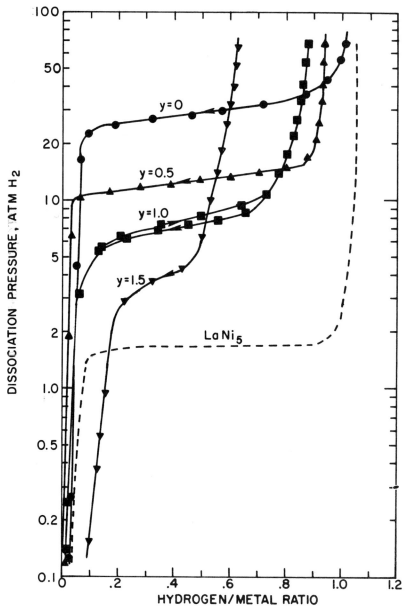

FIGURE 7 - EFFECT OF Fe-CONTENT y ON THE 25°C ISOTHERM OF AS-CAST MNi$_{5-y}$Fe$_y$ ALLOYS.

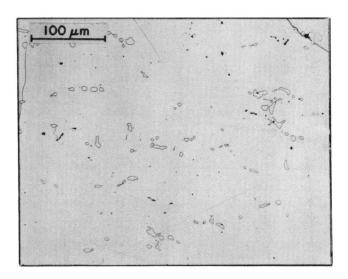

(a) M Ni$_{4.0}$Fe$_{1.0}$

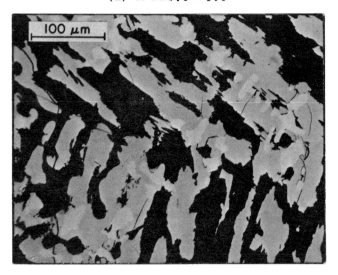

(b) M Ni$_{3.5}$Fe$_{1.5}$

FIGURE 8

AS-CAST MICROSTRUCTURES OF M Ni$_{5-y}$Fe$_y$
ALLOYS.

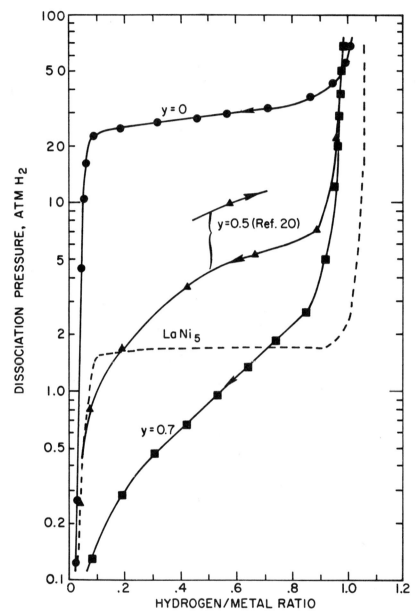

FIGURE 9-EFFECT OF Mn-CONTENT y ON THE 25°C ISOTHERM
OF AS-CAST $MNi_{5-y}Mn_y$.

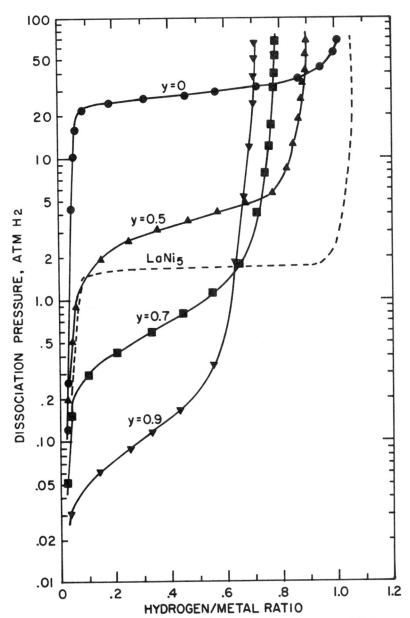

FIGURE 10-EFFECT OF AI-CONTENT y ON THE 25°C DESORPTION
ISOTHERM OF AS-CAST MNi$_{5-y}$Al$_y$ ALLOYS.

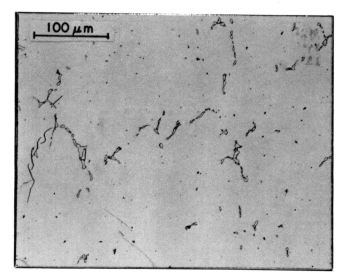

(a) M Ni$_{4.5}$Al$_{0.5}$

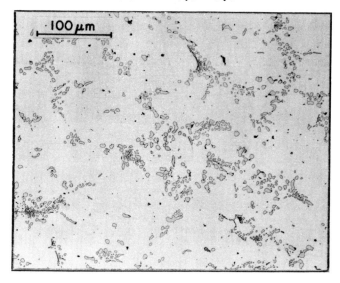

(b) M Ni$_{4.1}$Al$_{0.9}$

FIGURE 11

AS-CAST MICROSTRUCTURES OF M Ni$_{5-y}$Al$_y$
ALLOYS.

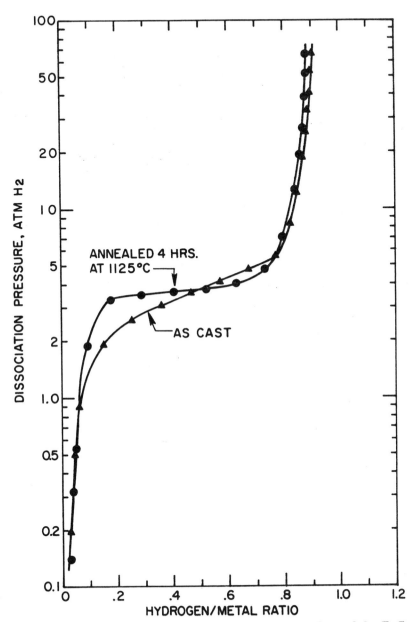

FIGURE 12 - EFFECT OF 1125°C HOMOGINIZATION ANNEALING ON THE 25°C DESORPTION ISOTHERM OF $MNi_{4.5} Al_{0.5}$.

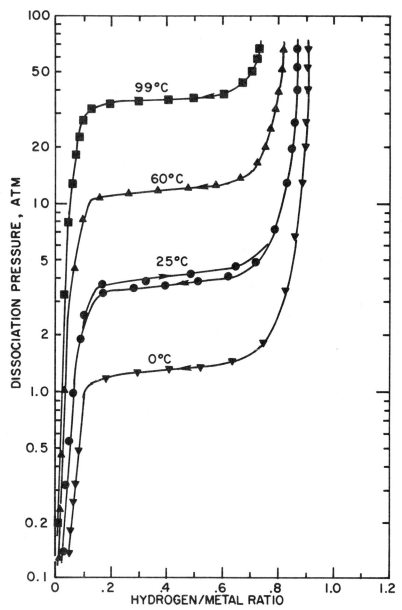

FIGURE 13-VARIOUS ISOTHERMS FOR MNi$_{4.5}$Al$_{0.5}$ANNEALED
4 HOURS AT 1125°C.

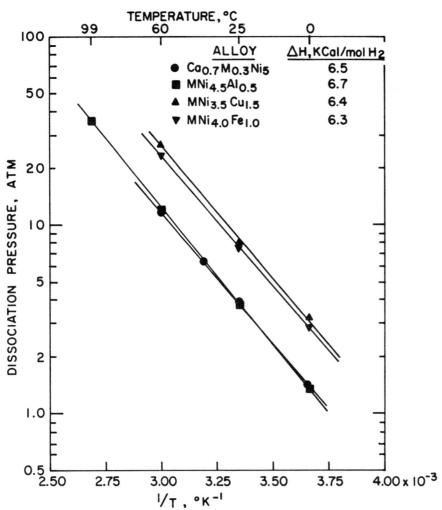

FIGURE 14 - VAN'T HOFF PLOTS FOR SELECTED ROOM TEMP.
HYDROGEN STORAGE ALLOYS.

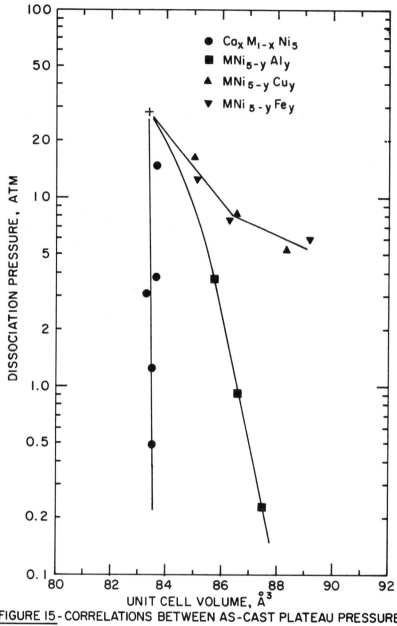

FIGURE 15-CORRELATIONS BETWEEN AS-CAST PLATEAU PRESSURE
(HYDROGEN/METAL=0.5) AND UNIT CELL VOLUME.

1655

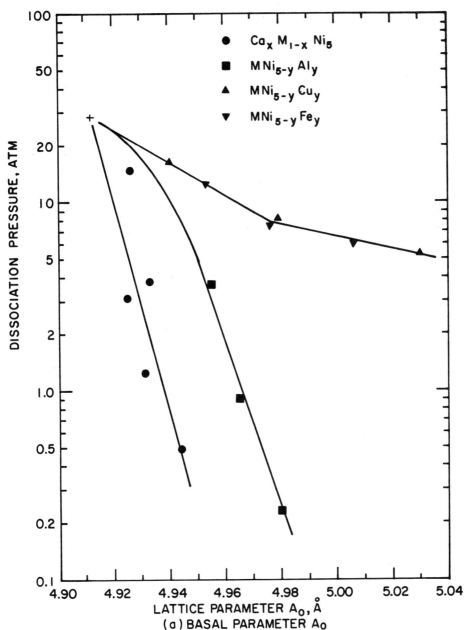

FIGURE 16-CORRELATIONS BETWEEN AS-CAST PLATEAU PRESSURE
(HYDROGEN/METAL=0.5) AND LATTICE PARAMETERS.

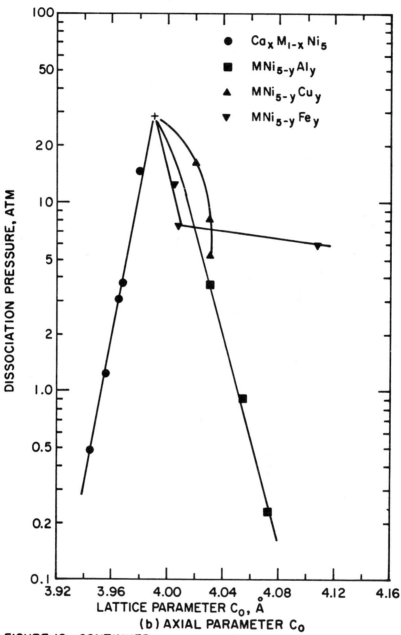

(b) AXIAL PARAMETER C_0

FIGURE 16 - CONTINUED

TECHNOLOGICAL ASPECTS AND CHARACTERISTICS OF INDUSTRIAL HYDRIDES RESERVOIRS

Ph. Guinet, P. Perroud and J. Rebière
Centre d'Etudes Nucléaires de Grenoble
Laboratoire A.S.P., Grenoble, France

ABSTRACT

Several industrial FeTi and Mg alloys have been tested : hydrogen sorption capacity, reaction rate and loss of capacity vs time and hydrogen purity. Two hydride reservoirs for hydrogen storage have been built and tested, in order to emphasize technical problems involved with mechanical design and production of hydrides alloys at industrial scale, and also to specify the in or out flow-rate capability.
The amount of hydrogen stored in the FeTi reservoir is about 110 m^3 STP (10 kg), the maximum out-flow-rate up to 10 m^3 STP/h at nearly constant pressure, and the maximum working pressure is 30 bars. Several heat exchange systems have been tested on a scale model (15 kg FeTi). Improvement of heat transfer capability allows to increase the rate of discharge and the maximum hydrogen content.
The Mg_2Cu reservoir is capable of 22 m^3 STP (2 kg) of hydrogen capacity (75 kg of Mg_2Cu). The maximum working pressure is 30 bars and the maximum temperature 400 C. The internal heat exchanger is a heat-pipe. The in or out flow-rate capability has been measured on a scale model (10 kg Mg_2Cu).

INTRODUCTION

The subject of the present work consists of the study, realisation and testing of industrial scale hydrogen reservoirs, in order to highlight technological problems encountered in the french industrial context : large quantity alloy preparation, design and building of the reservoir. Complementary basic studies have been made with several iron-titanium and magnesium alloys, concerning their behaviour as industrial filling materials.
These products have then been tested in reduced scale models of reservoirs. The working tests analysis allowed us to specify the main features of two reservoir projects : 100 m^3 STP hydrogen capacity with iron-titanium bed and 20 m^3 STP capacity with magnesium-copper bed, and to predict their storage and discharge characteristics.

CHARACTERISTICS OF THE ALLOYS EMPLOYED

Tests have been performed with alloys from various origins, prepared by means of various laboratory and industrial processes.

Iron-titanium alloys

Supplier Baudier Poudmet (Senecourt Liancourt)

- . Elaboration process : double arc melting
- . Starting metals : iron 99.99 %
  titanium sponge
- . Quantity : 5 kg
- . Shape : grains (about 1 mm)
- . Second phase : very weak quantities of Fe_2Ti (X ray analysis)

Supplier SOFREM (Société Française d'Electrométallurgie) - CHEDDE

- . Elaboration process : induction melting in refractory crucible
- . Quantities : 30 and 50 kg
- . Particle size : 13.3 kg   140 μm to 1 mm
  3.2 kg   above 1 mm
  1.8 kg   under 1 mm
  50   kg   (see further)

Supplier Creusot Loire Imphy

- . Elaboration process : induction melting in refractory crucible, followed
  by argon stream "atomization" (industrial process
  PULVIMPHY)
- . Quantity : 114 kg
- . Particle size : 20 to 500 μm, grains
- . Analysis Wt % : C = 0.05 ; Mn = 0.3 ; Si = 0.3 ; O_2 = 0.95

Magnesium-copper alloy (definite compound Mg_2Cu)

Supplier Laboratoire de l'Ecole Nationale d'Electrochimie et Electrométallurgi de Grenoble

- . Quantity : 200 to 300 g ingots

Supplier SOFREM - CHEDDE

- . Starting materials : electrolytic copper and sublimation refined magnesium
- . Elaboration process : induction melting followed by solidification in
  thin moulds
- . Quantities : 2   kg   0   to 0.5 mm
  27.5 kg   0.5 to 10   mm

BASIC STUDIES

Iron-titanium

Baudier alloy

Activation, storage capacity and reaction speed measurements have been made

as reference tests with a high purity alloy supplied by Baudier.
After a 200 C outgassing and drying of the powder under vacuum, the sorption
begins at the same temperature. The following sorptions take place at room
temperature.
The maximum absorption capacity is very close to the maximum theoretical value, and agrees with the formula $FeTiH_{1.48}$ after 2 hours, and $FeTiH'_{1.85}$ after
5 hours.
The alloy keeps its initial absorption capacities after beeing exposed to
air at room temperature for several hours.

Sofrem alloys

The alloy was tested in the model described further by using 19.15 kg powder
(8 kg :0.6 to 1 mm and 11.15 kg :1 to 3 mm grains).

Activation :

It was planned that the "activation" of the load be made "in situ". Therefore, the minimum activation temperature of the alloy is a very important
point. A step by step growing temperatures method was employed : each step
consisting of 150 C outgassing under vacuum, followed by heating at selected
temperatures (100, 200, 300 and 350 C) under hydrogen (30 to 40 bars).
The hydriding reaction began after the 300 C treatment.

Storage capacity :

The maximum storage capacity is not obtained after the first cycle. In order
to achieve it, two methods are possible :
 . High pressurisation for several days,
 . Repeated sorption-desorption cycles under moderate pressure.
The first method is less expensive in the case of an industrial size reservoir. The second may be faster but requires about ten cycles.
The storage capacity increases because of the growth of the active surface,
which is a result of the fragmentation. The maximum value agrees with the
formula $FeTiH_{1.2}$ after 7 or 8 cycles. Above 30 bars hydrogen pressure, the
maximum value does not depend on the actual pressure:see sorption isotherm
curve (Fig. 1) which shows an important slope for high sorption values.
For a high number of cycles, up to 30 cycles, the storage capacity does not
change. We have not noticed any sensitivity to the impurities of the hydrogen used. For a higher number of cycles the verification is yet to be done.

Air-exposure effect :

The first test of air-exposure effect was made after the 13th sorption-desorption cycle. The desorption being achieved at ambiant pressure and the reservoir being opened for one month, only a sample for granulometric test was
taken off. After this opening break in the cycles,the previously obtained
storage properties of the alloy were restored by heating at 150 C under
vacuum.
A second test was made after the 15th cycle, the powder was poured out and
stirred up. The same process used as the first activation was necessary to
regain good storage properties. During the handling in air, a thin oxide or
adsorbed gas layer prevents further hydrogen sorption.

Fragmentation :

This effect is rapid. As early as the 14th cycle, it was noticed that the grain size between 100 to 200 μm was prevalent. The initial size was larger than 600 μm. The results are summarized in table 1.

GRAIN SIZE	Weight [%]		
[μm]	Initial	After 14 cycles	After 20 cycles
< 200	0	52	63
200 à 300	0	22	17
300 à 600	0	17	12
600 à 1000	42	7	6
1000 à 5000	58	2	2

Temperature pressure equilibrium :

Results of equilibrium measurements at 20 C ± 1 C were reported Fig. 1. Each experimental value was measured after several hours in isothermal conditions. The plateau pressure is not well marked. During the desorption from $FeTiH_{1.1}$ to $FeTiH_{0.1}$, the pressure decreases from 12 b to 5.5 b. A maximum working pressure of 30 b allows an useful range of $FeTiH_{1.2}$ - $FeTiH_{0.1}$.

Imphy FeTi alloy

A test sample of 114 kg of FeTi was prepared by the "Pulvimphy" process. This industrial process directly gives small spherical grains whose surface is unoxidized. This alloy contains an important amount of oxygen coming from the hot oxyde crucible (0.95 Wt %); silicium (0.3 Wt %) and manganese (0.3 Wt %) come from base metals. A radiocrystallographic analysis showed that the sample chiefly contained the FeTi phase. So, the fast cooling due to the "atomization" process is not completely connected with the formation of FeTi, but perhaps a little to lattice regularity. Further experimentation is to be performed in order to show the influence of this effect on the hydrogen sorption capacity.

Mg_2Cu (Sofrem)

Alloy composition

The alloy consists of Mg_2Cu phase, and very small quantities of eutectics (Mg/Mg_2Cu or $Mg_2Cu/MgCu_2$) at the grain boundaries.

Activation

Four sorption-desorption cycles were performed:300 C   100 b H_2 99.95 % and 300 C under vacuum alternatively. The hydride sample was exposed to air for 2 weeks at room temperature, then fully de hydrided under vacuum for 3 hr at 400 C and 12 hr at 350 C. The first hydriding reaction was gradual, but

the four following ones began very quickly.
Exposed to air after sorption-desorption cycles, Mg_2Cu did not need reactivation treatment.

Absorption kinetics

Fig. 2 shows the reaction progress λ as a function of time at 300 C, and that of pure Mg at 324 and 347 C. The curves show by how much Mg_2Cu hydrides than Mg : λ reaches 0.8 in 7 hr for Mg_2Cu, in 20 hr at 347 C and in 50 hr at 324 C for Mg.

Absorption capacity

The maximum absorption capacity of Mg_2Cu is :

$$H \% = \text{weight of } H_2 \Big/ \text{weight of hydride}$$

Fig. 3 shows Mg_2Cu at 300 C has a higher capacity than Mg at 350 C (under 10 hr), and Mg at 370 C (under 3 hr).

Loss of storage capacity v.s. time

Previous tests showed the loss of sorption capacity of Mg probably due (Fig. 4) to hydrogen impurities (chiefly oxygen).
In order to determine this effect, two test series were conducted, one with high purity hydrogen (99.9995 %), the other with hydrogen containing 0.1 % O_2.

H_2 99.9995 %

A series of 25 cycles (300 C, 10 hr each) showed that the sorption capacity keeps a constant value of 2.45 % from the 6th cycle to the 25th (Fig. 4). Along similar cycles, the capacity of pure Mg decreases quickly at the beginning, and is lower than that of Mg_2Cu from the 5th cycle.

Hydrogen containing 0.1 % O_2

Eight preliminary cycles were made with H_2 (99.95 %), in order to activate the alloy (Fig. 5). Twelve cycles with H_2 + 0.1 % O_2 (300 C, total duration under hydrogen pressure of 92 hr) led to a limited loss of sorption capacity : 1.27 to 1.04 % in 1 hr, 1.43 to 1.33 % in 3 hr, 1.95 to 1.64 in 16 hr.

Other basic studies

Several other binary magnesium alloys have been studied. Some of them, like Mg_3Cd, Mg_2Sn and MgAl exhibit only a very little absorption.
Mg - 5 at % Cu was found to have a good sorption capacity (4 Wt % in 1 hr at 350 C) and good hydriding kinetics, as previously shown [1], but no test of loss of capacity v.s. time was performed. Mg_2Ni shows a similar behavior.
Mg - 1 at % Si has a capacity of 3 Wt % in 10 hr at 395 C and 4 Wt % at 395 C, which seems to be stabilized at a value of 3 Wt % after 4 sorption-desorption cycles of 10 hr each at 395 C.
Mg_2Si reaches a capacity of 2 Wt % after 10 hr at 300 C.

DESIGN, CONSTRUCTION AND TESTS OF RESERVOIRS

Iron-titanium model

The model contains 19.15 kg of FeTi alloy (grains size larger than 0.6 mm).
The volume of the cylindrical container was 8.4 ℓ (O.D = 140 mm, length = 740mm,
working pressure = 100 b). An internal water-cooled heat-exchanger was made of
a coiled or finned tube. The hydrogen gas flows through a sintered porous
metal axial tube. The model was fitted with 8 thermocouples for bed temperature measurements and 2 thermocouples for inlet and outlet water temperature measurements. The outer wall was equiped with electric heaters in order
to have adiabatic limit conditions. The hydrogen gas used was a commercial
type (99.95 %). Stored volume is 2.5 m^3 STP. The maximum discharge flow-rate
was about 1 m^3 STP/hr. The refuelling time was about 2 hr.

Working conditions

   Charging step (Fig. 6) :

Pressure : Charging time was short when the pressure was higher than 20 b :
80 % of the load was filled in one hour. The charging rate was limited by
the heat exchanger cooling power. The working pressure planned for the
100 m^3 STP hydrogen reservoir was 30 b.

Temperature : The bed was cooled down to ambient temperature during the hydriding step. An optimal value of temperature may be calculated in order to
have a minimum duration of the constant-pressure sorption. This value has
not been clearly noted during experiments.

   Discharging step :

The working pressure is imposed by the FeTi bed temperature. The hydride
storage reservoir is theoretically capable of supplying its content hydrogen
capacity at a constant flow-rate. This peculiar feature has been specially
studied.
A plateau-pressure was noted for small flow-rates and for low temperatures.
Equilibrium pressure values at ambient temperature were shown on the same
plot (Fig. 7) corresponding to each reaction step. The gap between these
values and the actual pressure depends upon the flow-rate and upon the state
of the hydriding progress. At half-discharge, the pressure difference was
2.4 b for 1 m^3 STP/hr flow-rate and 1.8 b for 0.5 m^3 STP/hr. At complete
discharge it was 2.6 b and 2 b respectively.
With the 100 m^3 STP hydrogen reservoir considered as the coupling of 24 cells
having the same characteristics of heat-exchange and gas diffusion as the
model, the expected discharge flow-rate should be about 25 m^3 STP/hr.
Increasing the bed temperature during the discharge may compensate for the
pressure decrease. Fig. 8 shows a constant pressure level during 50 % of the
discharge time for a 1 m^3 STP/hr flow-rate. The final temperature reaches
about 80 C.

This final high temperature value is an unfavourable condition for the subsequent charging step. So, the heating must be stopped some minutes before the end of the discharge to allow the powder to cool.

100 m³ STP hydrogen reservoir design

Main characteristics

FeTi load	= 900 kg
Volume	= 310 ℓ
Weight of the container	= 209 kg
Maximum working pressure	= 30 b
Maximum working temperature	= 100 C
Heat exchanger	= 18 coils of copper tube, water cooled
	total area : 5.3 m²
Total weigh/hydrogen volume =	11 kg/m³

Reservoir technology

Particular attention has been devoted to the aspect of hydrogen compatibility ; other desirable qualities are good welding and good mechanical properties. Nelson's curves give data for the upper limits of pressure and temperature for a number of steels for use with hot and pressurized hydrogen. Carbon steel is reliable below 220 C and under normal industrial pressures. Following specialists advice [2] low sulphur % age boiler quality A42 C1 steel was selected. The reservoir was sand-blasted before welding and annealed up at 620 C under a reducing atmosphere after welding.
The 30 bars upper level of the working pressure was the result of a choice. It must be low if one wants keep the main advantage of hydrides which is low pressure storage. On the other hand, a high pressure is a favourable factor for a rapid hydrogen absorption. The choice is a compromise between the weight of the envelope and the short duration of the sorption step.
We have chosen to "activate" the alloy after the filling of the reservoir ; metallic gaskets tolerate a temperature up to 400 C and removable electric heaters allow the heating of the wall up to the "activation" temperature (350 C).

Magnesium copper model

Main characteristics

The model contains 10 kg of Mg_2Cu alloy. The planned hydrogen volume storage is about 3 m³ STP. The capacity is 8.6 ℓ (O.D. = 140 mm, length = 860 mm). The working pressure is 30 bars and the maximum temperature is 400 C.
A mercury heat-pipe fitted with transversal fins is located in the powder bed and the outer end is heated or cooled by an air-flow (eventually provided by exhaust gases). Five auxiliary electric heaters are used during tests. Temperature of the heat-pipe wall, of the hydride bed and the outer wall of the shell are measured by means of thermocouples.

Test results

The Mg$_2$Cu alloy is heated up to the working temperature (400 C) under vacuum
for two days. The maximum hydrogen volume stored increases during the first
8 cycles. The rate of the reaction rapidly improves but always remains slow :
50 % of the hydrogen capacity is reached in 12 hr. The optimum temperature
for a rapid sorption under 30 bars is about 300 C. The maximum desorption
flow-rate at 380 C is 0.8 m^3/hr at 20 bars and 1.3 m^3/hr at 10 bars.
The actual hydrogen volume stored is about 2.5 m^3 (2.2 Wt %).

20 m^3 STP hydrogen reservoir design

Main characteristics :

Mg$_2$Cu load	= 80 kg
External diameter	= 400 mm
Length	= 500 mm
Heat exchanger	= 5 finned heat-pipe capable of 20 kW - Heating or cooling by air flow.
Max. working pressure	= 30 b
Temperature during charging step	= 300 C
" " decharging step	= 380 C

Materials :

It is necessary to use a chromium steel. Nelson's curves show that 3 % chromium is sufficient to keep clear hydrogen embrittlement up to 500 C. The
choice of the general shape of the reservoir is not yet confirmed. Whatever the choice, the heat-pipe system will be used because it is a suitable
and safe mean to supply or extract the heat of reaction. For a 4 hours charging or discharging time, 1500 W heat power is to be exchanged (0.1 W/cm^3
of bed). The maximum temperature difference in the powder is expected to be
about 15 C. The temperature difference between the fins and the air would
be 50 C or less.

CONCLUSION

The experimental scale models allowed us to point out the functional characteristics of industrial storage reservoirs, whose construction is now in
progress.
The obtained capacity, charging time and flow-rate performances of the latter are hoped to be satisfying, in spite of chosen low pressure and moderate
temperature working conditions. These performances may also be improved,
because the employed alloys were not the best of each type, and laboratory
research is carried on in order to find better ones.
The model first filled with Mg$_2$Cu can be operated with other types of Mg
alloys.
It must be noticed that supplying good quality and low price industrial
alloys seems to be one of the main problems of the process.

Complementary investigations about industrial fabrication methods and cha-racterisation of the alloys are still necessary.
This study showed the advantage of the hydride process from the security and ease of employment points of view.

ACKNOWLEDGEMENTS

This work was first initiated with a financial aid from Institut de Recher-ches sur les Transports. Then, the basic studies on magnesium alloys were partially supported by a contract with Commission des Communautés Européen-nes, and the technology of hydrides reservoirs is being performed under a contract with Délégation Générale à la Recherche Scientifique et Technique. Thanks are due to MM. G. Bertrand, D. Halotier, and A. Saudin for valuable technical assistance.

REFERENCES

1   J.J. REILLY, R.H. WISWALL
     The reaction of hydrogen with alloys of magnesium and copper.
     Inog. Chem. 6, 2220 (1967).

2   FIDELLE (C.E.A.)
     Private communication.

3   H. BUCHNER
     The hydrogen energy concept.
     Int. Symposium on hydrides for storage. Geilo (Norway) 14-19 Aug.1977.

4   J.J. REILLY
     Applications of metal hydrides.
     Int. Symposium on hydrides for storage. Geilo (Norway) 14-19 Aug.1977.

5   G.D. SANDROCK
     The metallurgy and production of rechargeable hydrides.
     Int. Symposium on hydrides for storage. Geilo (Norway) 14-19 Aug.1977.

6   C.H. WAIDE, J.J. REILLY, R.H. WISWALL
     The application of metal hydrides to ground transport.
     BNL Upton New-York, Hydrogen Energy, Part B, Veziroglu, Miami 1974.

7   G.D. SANDROCK
     Metallurgical considerations in the production and use of FeTi alloys
     for hydrogen storage.
     11th I.C.E.C., 1976.

8   J.J. REILLY
     Metal hydrides as hydrogen storage media and their applications.
     Dept of Applied Science, Brookhaven Nat. Lab.
     BNL 21648, 1976.

9   H.H. JOHNSON and A.J. KUMNICK
     Hydrogen and the integrity of structural alloys.
     Cornell Univ. Ithaca N.Y., Hydrogen Energy, Part B, Veziroglu,
     Miami 1974.

10   G. STRICKLAND, J.J. REILLY, R.H. WISWALL
     An engineering scale energy storage reservoir of iron-titanium hydride.
     Proceedings theme Conf. Miami, 1974.

11   C. GALES et P. PERROUD
     Stockage de l'hydrogène sous forme d'hydrures métalliques.
     Application à des moteurs thermiques.
     Journées Internationales d'Etude : L'hydrogène et ses perspectives.
     Congrès A.I.M., Liège (Belgique), 15-16 Nov. 1976.

12   A. BEAUFRERE, F.J. SALZANO, R. ISLER, and W. YU
      Hydrogen storage via FeTi for a 26 MW peaking electric plant, in
      Proc. 1st World Hydrogen Energy Conf., Veziroglu, T.N., Ed.,
      University of Miami, Coral Gables, Fla., March 1976.

13   K.C. HOFFMAN, J.J. REILLY, F.J. SALZANO, C.H. WAIDE, R.H. WISWALL,
      and W.E. WINSCHE
      Metal Hydride Storage for Mobile and Stationary Applications.
      Paper n° 760569, Fuels and Lubricants Meet., Society of Automotive
      Engineers, Warrendale, Pa., 1976.

14   D.L. DOUGLASS
      The Storage and Release of Hydrogen from Magnesium alloy Hydrides
      for Vehicular Applications.
      International Symposium on Hydrides for Energy Storage,
      Geilo Norway (1977).

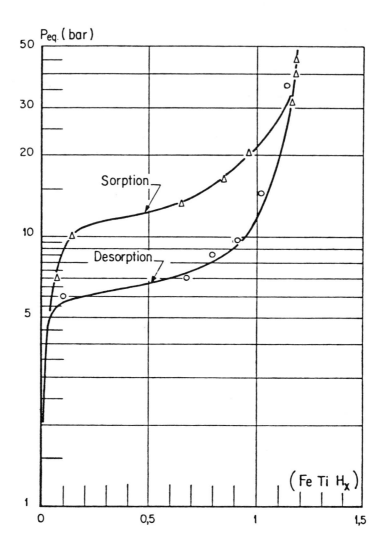

Fig: 1 _ Sorption and desorption isotherm
for Fe Ti SOFREM at 20 °C.

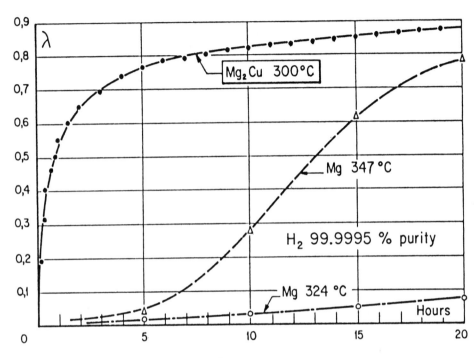

Fig: 2 _ Reaction progress λ of Mg₂ Cu at 300 °C and of Mg at 347 and 324 °C _

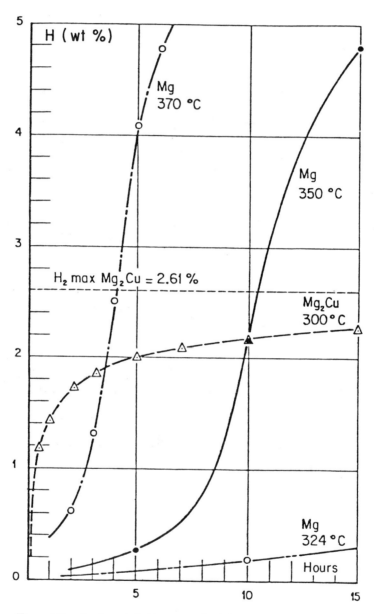

Fig: 3 _ First hydriding of Mg and Mg$_2$Cu.

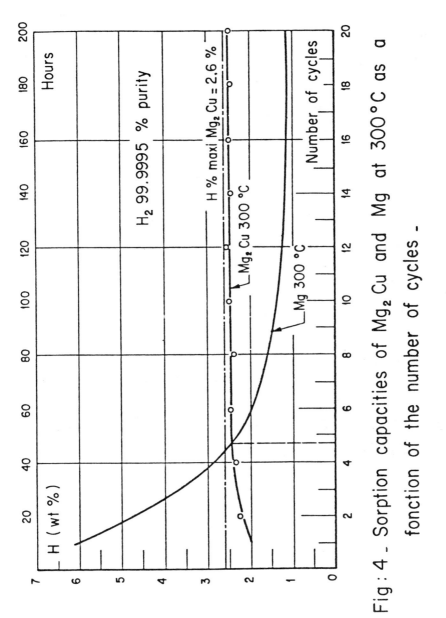

Fig: 4 – Sorption capacities of Mg$_2$Cu and Mg at 300°C as a fonction of the number of cycles –

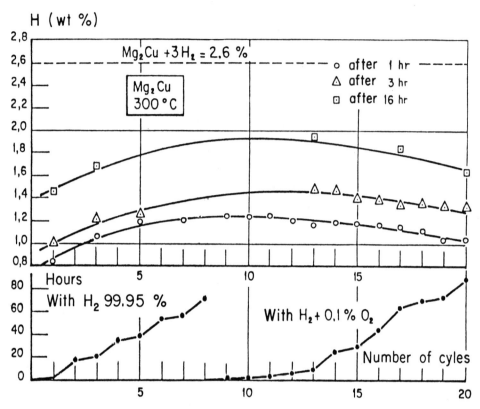

Fig: 5 _ Influence of the number of cycles on the sorption
capacity of Mg₂ Cu _

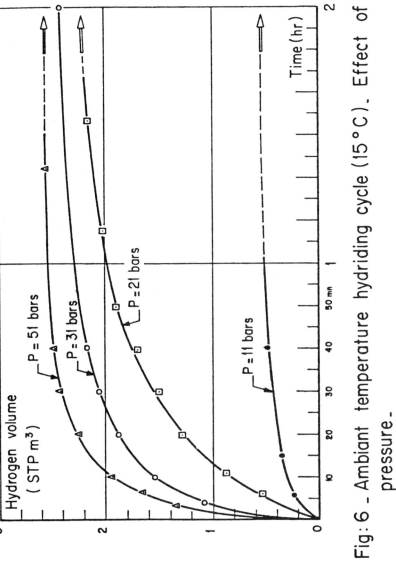

Fig : 6 _ Ambiant temperature hydriding cycle (15 °C). Effect of pressure.

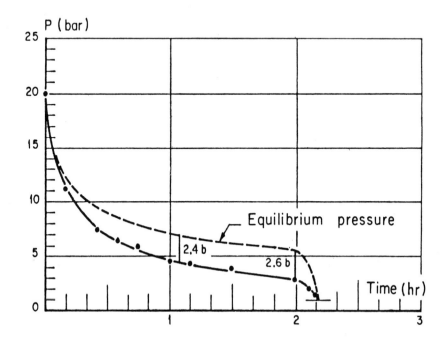

Fig: 7_ Constant flow-rate desorption (1 STP m³/hr)

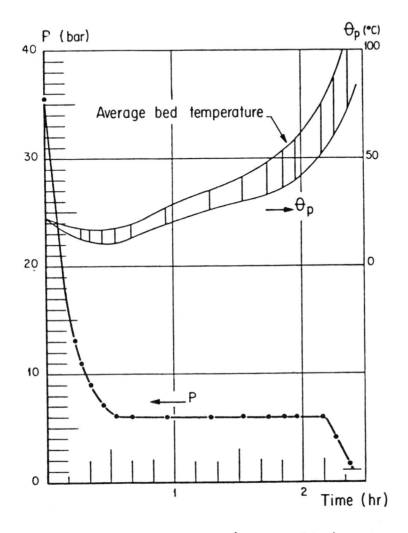

Fig: 8 _ Constant flow-rate (1 STPm³/hr) and constant
pressure (6 bars) desorption.
Bed temperature v.s. time_

DEVELOPMENT OF HIGH-TEMPERATURE HYDRIDES
FOR VEHICULAR APPLICATIONS

H. Buchner, O. Bernauer, W. Strauß
Daimler-Benz AG, Dept. Technical Physics
D-7000 Stuttgart 60, Germany

ABSTRACT

Dilute solutions of magnesium compounds in magnesium improve
the hydrogen exchange kinetics without loosing too much of the
high magnesium-storage capacity.
We investigated the systems $Mg-Mg_2Ni$ and $Mg-Mg_{17}Y_{13}$ to optimize storage capacity, absorption and desorption kinetics and
dissociation enthalpy with respect to vehicular application.

INTRODUCTION

The results gained with the hydrogen-powered vehicles [1 - 3]
and hydride development at Daimler-Benz have shown that a
slightly modified engine and a combination of high- and low-
temperature metal hydrides can be used for automotive propulsion and hydrogen storage (a metal is called a high-temperature hydride, if the equilibrium pressure of the hydride phase
is greater than 1 bar only at temperatures above $100^{\circ}C$, otherwise we call it low-temperature hydride).
In vehicular application a hydride storage tank can be used in
many ways, for

- fuel supply

- air conditioning

- water condensation

- storing and regaining waste heat from the
  engine

- auxiliary heating

The supplementary functions besides fuel supply do not waste
any primary energy. Because of the high hydrogen storage capacity of high-temperature hydrides the total hydrogen capacity
of the storage system can be increased significantly. A main
disadvantage of some high-temperature hydrides are very slow
kinetics during desorption of hydrogen (Mg and some magnesium

alloys). Earlier studies [4, 5, 6] showed the effect of Ag, Bi, Ca, Co, Mn, Sb, Si, Sn, Ni and Y dotations on the hydrogen desorption kinetics of magnesium alloys.

In our research program the magnesium/magnesium-nickel system and the magnesium/magnesium-yttrium system were tested for absorption and desorption kinetics and the pressure-concentration-isotherms were measured.

Energetic Constraints for the Storage Tank Dimensions

In a combustion engine a total energy of 245 kJ/mol H_2 is freed. From this energy 81 kJ/mol H_2 are available in the heat of the exhaust gas at a maximal temperature level between $400°$ and $650°C$.
The dissociation pressure of 1 bar - the pressure the engine needs for working - is reached with Mg_2Ni hydride at $260°C$ and with Mg-hydride at $300°C$. Therefore, only half of the energy in the exhaust gas can be used for the dissociation of about 1/2 mol hydrogen from the high-temperature hydride (the heat of formation of Mg_2Ni-hydride is 63 kJ/mol H_2, and of Mg-hydride 80 kJ/mol H_2). The consequence is, that a combustion engine cannot work with a high-temperature hydride alone, one has to use a combination of high- and low-temperature hydrides. Then the energy of the exhaust gas in the temperature range from $300°C$ to $50°C$ can be used too. To free the missing ~1/2 mol H_2 from the low-temperature hydride another ~15 kJ/mol H_2 are necessary, leaving a rest energy of approximately 25 kJ/mol H_2 in the exhaust gas (the heat of formation of low-temperature hydrides amounts only 20 - 30 kJ/mol H_2).
A short calculation shows, that with a low-temperature hydride of 2 percent storage capacity and a high-temperature hydride of 4 percent storage capacity the weight ratio of high-temperature to low-temperature hydride should be 1:1.

Kinetics of the Different Hydrides

Besides the energy balance one has to consider the kinetics of the hydrides. From tests in laboratory and in cars it can be shown, that independent of the driving situation the kinetics of TiFe and Mg_2Ni are sufficient to produce enough hydrogen for the combustion. There tank desorption characteristics are limited by the heat exchange from the exhaust gas or cooling water to the hydride storage tanks. New Hydrides should have sufficient kinetics too.

Experimental Procedures

The samples were vacuum-melted in molybdenum crucibles for
a period of several hours to build a homogeneous alloy. To
minimize gravity segregation of the two phase constituents the melting was repeated and the samples were cooled
down rapidly. Only samples with a high homogenity were chosen for hydriding purpose.

The hydriding procedure was done as follows. For several
hours the samples were evacuated in a reactor at temperatures
between $400^{o}C$ and $450^{o}C$. Then the samples were hydrided under a pressure of ~10 bar, cooled down to $300^{o}C$ and again desorbed at $450^{o}C$. This procedure was repeated for several times
until the alloys showed maximum hydrogen storage capacity.

The pressure-concentration-isotherms were measured by volumetric method in apparatus schematically drawn in Fig. 1. After
hydriding the valve between the sample container and the calibrated standard volume was closed and the standard volume was
evacuated. Then a defined hydrogen quantity was let into the
standard volume. After the dissociation pressure in the sample
container reached equilibrium the desorption process was repeated as long as the equilibrium pressure was above 1 bar. At
lower pressures the remaining hydrogen content was measured
by baking out the system at $450^{o}C$. The hydrogen content was
calculated using van der Waals equation, the heat of formation
from the temperature dependence of the plateau pressure. To simulate
the driving situation the kinetic measurements were performed
in the same apparatus by steady desorption of hydrogen out of
the reactor into the standard volume where the hydrogen pressure was held constantly at 1 or 2 bar. The temperature of the
reactor amounted in all cases $320^{o}C$.

Results

The system $(Mg-Mg_2Ni)$:

Fig. 2 and 3 show that the hydrogen content of $(Mg-Mg_2Ni)$-
systems increases significantly in comparison to pure Mg_2Ni
We remind that Mg_2Ni stores 3.7 weight percent hydrogen at
about $260^{o}C$ with a dissociation pressure of 1 bar. The pressure-concentration-isotherms of pure Mg_2Ni show a single plateau which is stable up to temperatures of $400^{o}C$. The pressure-concentration-isotherms of an (Mg, Mg_2Ni)-system have
two plateaus (Fig. 2 and 3) which can be attributed to MgH_x

$(0 \leq x \leq 2)$ and to Mg_2NiH_y $(0 \leq y \leq 4)$. Independent of the magnesium to magnesium-nickel ratio the heat of formation of magnesium hydride amounts 79 kJ/mol H_2 while magnesium-2-nickel hydride has a heat of formation of 63 kJ/mol H_2. The maximal hydrogen content in 22% Mg, 78% Mg_2Ni is 4,8w% while 67% Mg, 33% Mg_2Ni is able to store more than 6 w% hydrogen. As in pure Mg_2Ni-systems the dissociation pressure of the Mg_2Ni-hydride in the mixture of $Mg-Mg_2Ni$ amounts 1 bar at 260°C.

Pure magnesium neither can be hydrided nor dehydrided in acceptable times and therefore no pressure-concentration-isotherms were measured.

In contrast to this the magnesium hydride in the system $Mg-Mg_2Ni$ shows sufficient absorption and desorption kinetics to measure the pressure-concentration-isotherms.

As can be seen in Fig. 4 the desorption characteristics of $Mg-Mg_2Ni$ hydride are independent of the magnesium content and comparable to the Mg_2Ni-hydride, which has sufficient kinetics to use it as a fuel storage in a car.
Similar results were obtained in the Mg-MgY-system (Fig. 5). In contrast to $Mg-Mg_2Ni$-systems the pressure-concentration-isotherms only have one plateau, the heat of formation is comparable to the heat of formation of magnesium hydride. The kinetics of hydriding and dehydriding are again much faster than of magnesium hydride.

Discussion

The results in the last chapter demonstrate, that with a combination of high- and low-temperature hydride it is possible to increase the storage capacity in comparison to low-temperature hydride storage system like TiFe-hydride by a factor of 2 to 3 (Table 1).

At the same time such a system makes it possible to regain waste heat in the temperature range from o°C to 650°C. This is mostly due to the fact that in pure Mg-dotations of Mg_2Ni and MgY increase the desorption characteristics enough to make it usable for automotive storage.
The high kinetics of the $Mg-Mg_2Ni$-systems can be explained with a model schematically shown in Fig. 6.

In the mixture of $Mg-Mg_2Ni$ there exists three metal-hydrogen reactions with three different time constants. The first is the reaction of Mg_2Ni with hydrogen to form Mg_2NiH_4. Kinetics of this reaction at temperatures of 300°C are very fast.

$$Mg_2NiH_4 \xrightleftharpoons{K_1} Mg_2Ni + 2H_2 + \Delta H_{Mg_2NiH_{42}}$$

The second is the reaction of the magnesium phase with gaseous hydrogen. This reaction is very slow at temperatures of $300°C$.

$$MgH_2 \xrightleftharpoons{k_2} Mg + H_2 + \Delta H_{MgH_2}$$

In practise this means that it is not possible to desorb the hydrogen from the hydride in a reasonable time.

In a phase mixture of MgH and Mg_2Ni the hydrogen atoms can diffuse out of the Mg phase into the Mg_2Ni to form Mg_2Ni-hydride.

$$MgH_2 + Mg_2Ni \xrightleftharpoons{K_3} Mg_2NiH_2 + Mg + \Delta H_x$$

Out of the Mg_2Ni-hydride the hydrogen can effuse into the gas-phase.

The total reaction can be written:

$$MgH_2 \xrightleftharpoons{K_1 + K_3} Mg + H_2 + \Delta H_{MgH_2}$$

Kinetics of this reaction are ruled by $K_1 + K_3$, this time constant is comparable to the time constant of Mg_2NiH_4 decomposition. All macroscopic features like dissociation pressure and heat of formation are determined by the initial state MgH_2 and the final state, the gaseous H_2.

As the results demonstrate kinetics of Mg-Mg_2Ni-hydrides are comparable to the kinetics of pure Mg_2Ni-hydrides up to a Mg/Mg_2Ni ratio of 2. Slower kinetics are expected when in comparison with the MgH_2-phase the Mg_2Ni-phase becomes too small to ensure a sufficient hydrogen exchange at the surface. The maximal hydrogen storage capacity usable in vehicular applications will be about 7 - 7,5 weight percent.

Like the Mg-Mg_2Ni -system the Mg-$Mg_{17}Y_{13}$-system shows good hydriding and dehydriding character. As the results of other authors show [7] this effect will be inhanced in a Mg- $MgYNi$-system.

REFERENCES

1   J. J. Reilly, K. C. Hoffmann, G. Strickland, R. H. Wis-
     wall - Iron Titanium Hydride as a Source of Hydrogen
     Fuel for Stationary and Automotive Applications
     (26 th Aun. Proc. Power Sources Conference, May 1974)

2   H. Buchner - Application of Low- and High-Temperature
     Hydrides: The Hydride/Hydrogen Energy Concept.
     (2nd World Hydrogen Energy Conference, Zürich 1978)

3   H. Buchner, H. Säufferer - Entwicklungstendenzen von
     Wasserstoff-getriebenen Fahrzeugen mit Hydridspeicher.
     (ATZ 79, 1977, p. 1)

4   J. J. Reilly, R. H. Wiswall, K. C. Hoffmann, C. H. Waide-
     Metal Hydrides as Hydrogen Storage Media
     (7 th Alternative Automotive Power Systems Division
     Contractors Coordination Meeting, Michigan, May 1974)

5   J. J. Reilly, R. H. Wiswall, C. H. Waide - Motor Vehicle
     Storage of Hydrogen using Metal Hydrides.
     (Report EPA Grant - R 802 579)

6   J. J. Reilly, R. H. Wiswall - The Reaction of Hydrogen
     with Alloys of Magnesium and Nickel and the Formation
     of Mg_2NiH_4.
     (Inorg. Chem. 7, 1968, p. 2254)

7   D. L. Douglass - Metalurgical Trans. A., 6 A, 1975,
     p. 2179

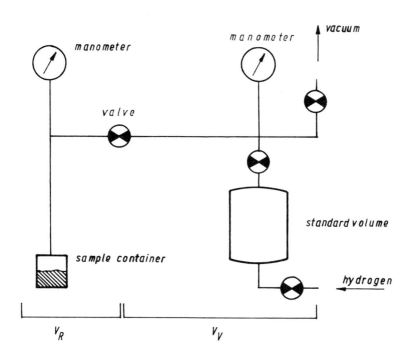

Fig. 1   Pressure-concentration-
          isotherm apparatus

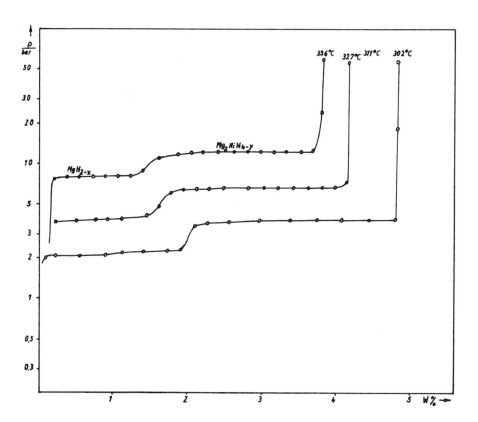

Fig. 2   Concentration-pressure-isotherms
for the system 22% Mg - 78% Mg$_2$Ni

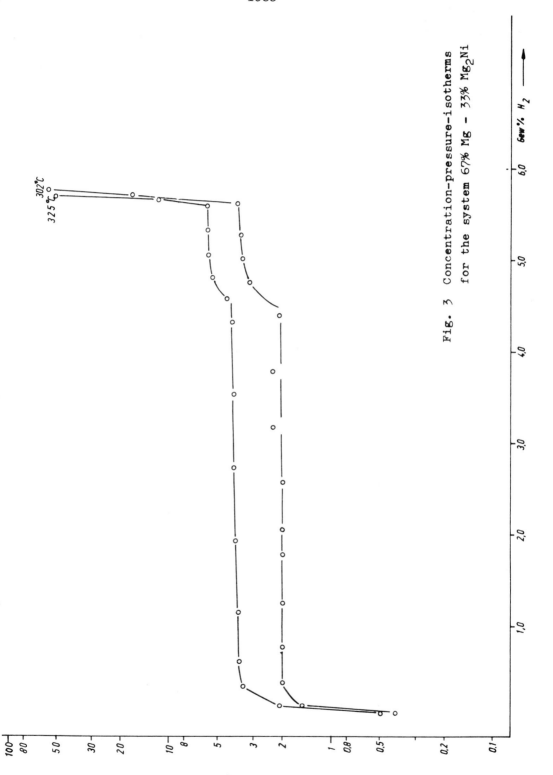

Fig. 3 Concentration-pressure-isotherms for the system 67% Mg – 33% Mg$_2$Ni

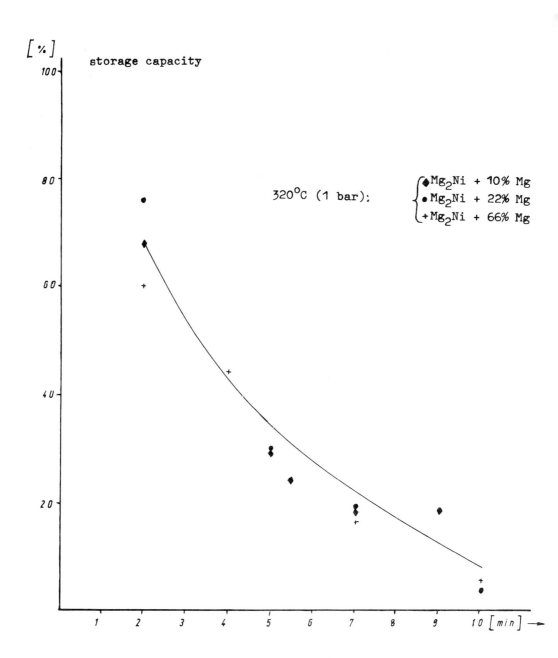

Fig. 4   Desorption characteristics of
different Mg-Mg$_2$Ni - systems

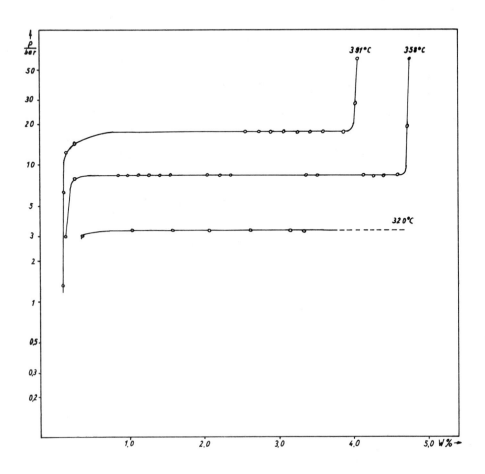

Fig. 5   Concentration-pressure-isotherms
for Mg-Y alloys

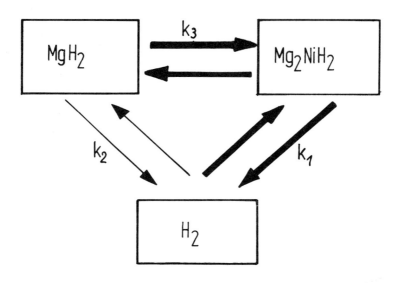

1) $MgH_2 + Mg_2Ni \xrightarrow{k_3} Mg_2NiH_2 + Mg$

2) $Mg_2NiH_2 \xrightleftharpoons[k_1]{k_1} Mg_2Ni + H_2$

3) $MgH_2 \xrightleftharpoons[k_2]{k_2} Mg + H_2$

Fig. 6   Model for the kinetics of the
          Mg-Mg$_2$Ni system

THE INFLUENCE OF Al ON THE HYDROGEN SORPTION PROPERTIES OF INTERMETALLIC COMPOUNDS*

I. Jacob and D. Shaltiel
Racah Institute of Physics, Hebrew University of Jerusalem
Jerusalem, Israel

ABSTRACT

The hydrogen sorption properties of Al containing Laves phase inter-metallics were studied. The absorption capacity of $Zr(Al_xV_{1-x})_2$ (x=0.2, 0.5), $Zr(Al_xCr_{1-x})$ (x=0.3) and $Gd(Al_xCo_{1-x})_2$ (x=0.3, 0.5) was measured at hydrogen pressure of 20 atm and room temperature. Desorption characteristics of the obtained hydrides were studied at various temperatures.

Of particular interest is the behaviour of the systems $Zr(Al_xB_{1-x})_2$ (B=Fe, Co; $0 \leq x \leq 1$) whose hydrogen capacity was measured at $\sim$ 70 atm and room temperature, and at $\sim$ 40 atm and liquid nitrogen temperature. The latter two systems present very interesting and unexpected results. Their absorption capacity, N_H, depends in a distinct manner on the Al concentration, x. The variation of N_H with x is well described by a phenomenological model recently proposed by us. This model demonstrates the importance of nearest neighbour effects and suggests that a cluster model approach should be used in determining the hydrogen absorption properties. The relative stability of the hydrides of these pseudobinary compounds was determined by measuring the desorption isotherms at room temperature.

In view of these and other experimental results the influence of Al on the hydrogen sorption properties of intermetallic compounds is discussed.

INTRODUCTION

Hydrogen-metal systems for various practical purposes should be cheap, and possess a large hydrogen weight percentage. One of the desirable elements of such systems is Al. Indeed, attempts are being made to study the influence of Al on the hydrogen sorption properties in different intermetallic compounds. Takeshita and Wallace [1] studied the hydrogen absorption in $Th(Ni, Al)_5$ ternaries. They obtained the hydrides $ThNi_4AlH_{2.5}$ and $ThNi_3Al_2H_{2.7}$ and no hydrides of $ThNi_5$ and $ThNi_2Al_3$. Mendelsohn et al. [2] reported sharp reduction of the plateau pressures of the system $LaNi_{5-x}Al_x(0 \leq x \leq 1.5)$, when increasing x, without imparing the kinetics or the hydrogen capacity of $LaNi_5$. Gualtieri and Wallace [3] found a rapid decrease of the hydrogen capacity, when Al was substituted for Fe, in $ErFe_2$.

The present work reports on the hydrogen sorption properties of some of the Laves phase compounds $Zr(Al_xB_{1-x})_2$ (B=V,Cr,Fe,Co), $Gd(Al_xCo_{1-x})_2$ and $CaAl_2$. These compounds may be divided into two main groups according to their behaviour with respect to hydrogen absorption:

1. $Zr(Al_xB_{1-x})_2$ (B=V,Cr) and $Gd(Al_xCo_{1-x})_2$. This group is characterized by large hydrogen capacities of the Al free compounds which form the hydrides $ZrV_2H_{5.2}$, $ZrCr_2H_{3.8}$ [4,5] and $GdCo_2H_{4.1}$ [6]. Partial substitution of V, Cr or Co by Al in these compounds decreases monotonically the hydrogen capacity. Only a few compounds in these systems were studied, namely, $Zr(Al_xV_{1-x})_2$ (x=0.2, 0.5), $Zr(Al_{0.3}Cr_{0.7})_2$ and $Gd(Al_xCo_{1-x})_2$ (x=0.3, 0.5). The maximum hydrogen pressure used was 20 atm at room temperature. The influence of the Al on the stability of the obtained intermetallic hydrides was investigated by measuring desorption isotherms at various temperatures.

2. $Zr(Al_xB_{1-x})_2$ (B=Fe,Co). This group is characterized by small hydrogen capacities of the Al free compounds which form the hydrides $ZrFe_2H_{0.15}$ and $ZrCo_2H_{0.35}$. Because of the very interesting results, the hydrogen absorption in these two systems was extensively investigated for different x values in the range $0 \leq x \leq 1$. The hydrogen capacities were measured at hydrogen pressures up to 70 atm at room temperature and 40 atm at liquid nitrogen temperature. A dramatic rise in the hydrogen capacity occurs for small x values similar to previous results for the systems $Zr(A_xB_{1-x})_2$ (A=V,Cr,Mn; B=Fe,Co; $0 \leq x \leq 1$) [4,5]. The maximum hydrogen content in both systems is achieved for $x \approx 1/12$ at 40 atm and 80 K. Further increase of the Al content leads, however, to a steep decrease in the hydrogen capacity. This general behaviour is well described by a phenomenological model, recently proposed by us [5]. The significant hydrogen absorption obtained for several $Zr(Al_xB_{1-x})_2$ compounds seems at first sight to be rather surprising as the ZrB_2 (B=Fe,Co) [4] and $ZrAl_2$ compounds absorb only small quantities of hydrogen. (Previously to this work, $ZrAl_2$ was reported [7] to be inert to hydrogen). We offer an explanation to this behaviour.

We report also hydrogen absorption in $CaAl_2$ and $GdAl_2$.

The importance of this work is associated not only with the observation of new intermetallic hydrides; in addition, we draw general conclusions concerning the influence of Al on the hydrogen sorption properties of different intermetallic compounds. We show that results obtained by other authors [1,2,3] confirm our conclusions. This work opens wide opportunities for predicting the sorption properties of Al containing intermetallic hydrides.

EXPERIMENTAL PROCEDURE

The different compounds investigated in this work were prepared by arc melting under argon on a water cooled copper hearth. Some of the compounds were annealed at a temperature of $1000^{\circ}C$ for 24 hours. Crystal structures and lattice constants were confirmed or determined by x-ray analysis. The hydriding facility used has been described previously [4]. The hydrogen capacity of the $Zr(Al_xB_{1-x})_2$ systems (B=Fe,Co) was measured at ~ 70 atm, 293 K, and at ~ 40 atm, 80 K. The hydrogen pressure used for the rest of the pseudobinary compounds was ~ 20 atm at room temperature. The accuracy at 293K is usually better than 0.1 H atom/formula unit, and at 80 K it is only ~ 0.2 H atoms/formula unit, as there were some experimental difficulties in determining the temperature corrections.

An activation procedure of heating for several hours the $CaAl_2$, $GdAl_2$ and $Zr(Al_xB_{1-x})$ compounds (B=Fe, Co; $0.5 \leq x \leq 0.75$) to 200 - $400^{\circ}C$ was usually needed to hydrogenate these compounds. The activation temperature used for $ZrAl_2$ was 800 - $900^{\circ}C$. At this temperature the gas diffused relatively fast through the stainless reactor. In this case the amount of hydrogen absorbed was determined by measuring the amount of hydrogen released at low pressure.

Desorption isotherms were measured down to minimum pressure of ~ 0.02 atm. X-ray analysis was carried out for some low pressure hydrides which did not react with oxygen when exposed to air.

EXPERIMENTAL RESULTS AND ANALYSIS

Hydrides of the Compounds $Zr(Al_{x-1}V_{1-x})_2$ (x=0.2,0.5), $Zr(Al_xCr_{1-x})_2$ (x=0.3), $Gd(Al_xCo_{1-x})_2$ (x=0.3, 0.5,1) and $CaAl_2$

Table 1 lists the above pseudobinary compounds, their crystal structures, lattice parameters and hydrogen capacities, N_H, at room temperature and hydrogen pressure of 20 atm. For comparison, Table 1 also exhibits the values of N_H of the corresponding Al free compounds. As the tendency of lowering the hydrogen capacity upon substitution of Al was clear enough, we did not extend the measurements to additional compounds of the systems $Zr(Al_xB_{1-x})_2$ (B=V,Cr) and $Gd(Al_xCo_{1-x})_2$. We have tried to estimate the influence of Al on the stability of the intermetallic hydrides by measuring their desorption characteristics. This was a difficult task as the hydrides did not show "plateau behaviour" upon desorption. Desorption isotherms for some of the hydrides studied are shown in Figs. 1-4. The amounts of hydrogen released by some of the intermetallic hydrides at nearly the same pressures and temperatures are listed in Table 1. It is clearly seen that in all cases the Al free intermetallic hydrides release relatively more hydrogen than the corresponding Al containing hydrides. Hence, the general tendency of the substituted Al is to increase the hydride stability.

X-ray analysis made on most of the hydrides in this group, revealed increase of the unit cell without change of the original crystal structure.

We have also observed hydrogen absorption in $CaAl_2$ and $GdAl_2$ where a heating to 300 - 400°C at $\sim$ 100 atm was necessary for absorbing approximately 2 and 1 H atoms/molecule respectively. The $CaAl_2$ hydride released a significant amount of hydrogen at 2.7 atm and 500°C. The X-ray analysis of the hydrogenated $CaAl_2$ revealed the presence of Al and an additional hexagonal phase with the following lattice constants a=6.2 Å c=8.0 Å. We could not identify the X-ray pattern of the $GdAl_2$ hydride. (About 50% of the lines of the x-ray pattern were indexed as a hexagonal phase with lattice constants a=6.323 Å c=9.315 Å).

Hydrides of the Compounds $Zr(Al_xB_{1-x})_2$ (B=Fe,Co; $0 \leqslant x \leqslant 1$)

Table 2 lists all the $Zr(Al_xB_{1-x})_2$ (B=Fe,Co) compounds studied, their crystal structures, lattice parameters, hydrogen capacities, N_H, and their dissociation pressures at room temperature, when available. The existence of some of these pseudobinary Laves phase compounds, their magnetic properties and lattice constants were reported by Muraoka et al. [8]. The variation of the hydrogen capacity, N_H, with x (at $\sim$ 70 atm, 293 K, and at $\sim$ 40 atm, 80 K) for the two systems $Zr(Al_xFe_{1-x})_2$ and $Zr(Al_xCo_{1-x})_2$ is presented in Fig. 5. Desorption isotherms at room temperature ($\sim$ 20°C) are plotted in Fig. 6.
The general behaviour of the two systems is similar and they exhibit the following features:
1) There is a sharp increase in the hydrogen capacity as x increases for small x values, the maximum N_H being reached for $x \sim 2/12$. The maximum hydrogen compositions $Zr(Al_{0.2}Fe_{0.8})_2 H_3$ and $Zr(Al_{2/12}Co_{10/12})_2 H_{2.4}$ were obtained at $\sim$ 40 atm, 80 K.
2) A sharp decrease of N_H is observed for x>0.25.
3) An increase in the hydrogen capacity, N_H, is obtained for small x values when the sample is cooled from room temperature to 80 K. The largest relative increase occurs for x=1/12 (see Table 2). We measured the absorption capacity at 80 K in an attempt to approach as close as possible the saturated hydride β phase of the intermetallic hydrides. (Cooling of the sample at nearly constant pressure is equivalent to exponential increase of the pressure at constant temperature). We believe this to be a necessary condition for meaningful comparison of the hydrogen capacities.

The hydrogen capacity of $Zr(Al_{1/12}Fe_{11/12})_2$ at room temperature and at $P_H \approx 70$ atm increased from 0.6 to 0.8 H atoms, after the sample has been cooled to liquid nitrogen temperature. Similar increase was observed also in $Zr(Al_{1/12}Co_{11/12})_2$. These differences in N_H can be attributed to irreversibilities (hysteresis) between the absorption and the desorption isotherms. The other compounds studied, showed no such hysteresis.

4) The equilibrium dissociation pressure increases with decreasing x, while the critical temperature T_c [9] is lowered. The critical temperature, T_c, of a hydrogen-metal system is the temperature above which no $\alpha \leftrightarrow \beta$ phase transition occurs. For x=1/12, T_c is obviously below room temperature.

5) The x-ray analysis of some of the hydrides revealed increase of the lattice constants without change in the crystal structure. We suppose that all the compounds studied exhibit the same behaviour. Similar features were observed for the hydrides of the systems $Zr(A_xB_{1-x})_2$ (A=V, Cr, Mn; B=Fe, Co) [5].

Phenomenological Model for the Hydrogen Absorption Capacity

The phenomenological model for the hydrogen absorption capacity in pseudobinary Laves phase compounds, was presented recently [5] and has been successfully applied to the systems $Zr(A_xB_{1-x})_2$ (A=V, Cr, Mn; B=Fe, Co; $0 \leq x \leq 1$). We apply the same model to the systems $Zr(Al_xB_{1-x})_2$ (B=Fe, Co) and here briefly sketch its main features. The model assumes a local absorption capacity, N_H, around each Zr atom. The number of nearest neighbours (Al and B atoms) to each Zr atom in the cubic C15 structure, is 12. In the hexagonal C14 phase, the nearest neighbour and the next nearest neighbour Al or B atoms to each Zr atom have almost the same Al-Zr (or B-Zr) distances, and their sum is also 12. For simplicity, we assume that the total number of nearest neighbours is the same in both structures and equals 12. The local capacity, n_H, depends on the particular nearest neighbour configuration around the Zr atom concerned and is represented by a double-step function:

$$
n_H = \begin{cases}
n_H^{(1)} & \text{for } n_{Al}^{(1)} > n_{Al} \geq 0 \\
n_H^{(2)} & \text{for } n_{Al}^{(2)} > n_{Al} \geq n_{Al}^{(1)} \\
n_H^{(3)} & \text{for } 12 \geq n_{Al} \geq n_{Al}
\end{cases}
$$

n_{Al} is the number of nearest neighbour atoms of type Al to a Zr atom. The Al and B atoms are assumed to be binomially distributed aroud a Zr atom; hence the total macroscopic absorption capacity $N_H(x)$, for a compound $Zr(Al_xB_{1-x})_2$ is given by

$$
N_H(x) = \sum_{n_{Al}=0}^{12} n_H \binom{12}{n_{Al}} x^{n_{Al}} (1-x)^{12-n_{Al}} \tag{1}
$$

The values of $n_H^{(1)}$ and $n_H^{(3)}$ are determined from the total macroscopic absorption capacity, N_H for $x=0$ and $x=1$, respectively. The values of $n_H^{(2)}$, $n_{A1}^{(1)}$ and $n_{A1}^{(2)}$ best suited to the experimental results are determined by a fitting procedure. We were able to fit our results by an almost unique set of these three parameters, allowing only a small tolerance in $n_H^{(2)}$. The solid curves in Fig. 5 yield the best fit of our experimental results to the phenomenological model, eq. (1). Table 3 gives the values obtained for $n_H^{(i)}$ (i=1,2,3) and $n_{A1}^{(i)}$ (i=1,2). The main differences between our present results and the previous ones for $Zr(A_xB_{1-x})_2$ (A=V, Cr, Mn; B=Fe, Co) are:
a) the lack of a broad range of x values, for which there is almost no change in the absorption capacity at the vicinity of $n_H^{(2)}$.
b) $n_H^{(3)} < n_H^{(2)}$.

The large jump of $n_H^{(1)}$ to $n_H^{(2)}$ occurring at $n_{A1}=1$ (see Table 3) indicates that according to our model, ordering between the Al and the Fe (Co) atoms should lead to a significant increase of the absorption capacity for small x values. Indeed, we have obtained an increase of the order of 0.1 - 0.2 H atoms/formula unit for several low x valued intermetallics after annealing at 1000°C for 24h. We have not continued our efforts in this direction, but it would be an interesting experiment to be performed.

DISCUSSION

The agreement of the present experimental results with the phenomenological model, supports the idea that the hydrogen capacity is determined by short-range neighbouring effects. The validity of the model is confirmed by additional experimental results on $Er(Fe_xCo_{1-x})_2$ taken from ref. 3. We were able to fit these experimental results to our model using the procedure explained above (see Fig. 7 and Table 3).

We also suggest that this agreement supports a cluster model approach for the hydrogen capacity in intermetallic compounds. In a recent work [10] we proposed a procedure for determining the specific interstitial sites occupied by the H atoms in intermetallic compounds. For LaNi$_5$ our predictions were verified by experimental results [11]. In the cubic or the hexagonal $Zr(A_xB_{1-x})_2$ intermetallic compounds (A=Al, V, Cr, Mn; B=Fe, Co) the tetrahedral e or k, f sites respectively, are supposed to be preferentially occupied. These interstitial sites are composed of a Zr atom and 3 A and (or) B atoms. Using our phenomenological model, we could not fit the hydrogen capacity in the above pseudobinary systems, if only a single e (k,f) interstitial site (only 3 and not the 12 nearest neighbours to a Zr atom) is considered. On the other hand, a Zr atom and its 12 nearest neighbours of type A and (or) B form a cluster including all its e (k,f) interstitial sites around the Zr atom. This situation indicates that the cluster model approach is preferable to the single interstitial site approach in determining the hydrogen absorption capacity in intermetallic compounds. Comparison [12] between the Quasi-Chemical approximation and the Lacher model can account for this behaviour: The Q.C. approximation assuming couple ordering of nearest hydrogen atoms, fits better experimental results, than the Lacher model which assumes a random distribution of single hydrogen atoms.

The experimental results show lowering of T_c upon substitution of Fe or Co instead of A atoms (A=V, Cr, Mn, Al) in $Zr(A_xFe_{1-x})_2$ and $Zr(A_xCo_{1-x})_2$ (see also [4]). According to the Q.C. approximation, T_c is given by [12]

$$T_c = w/2k \ln \frac{z-2}{z} \qquad (2)$$

where w is the attractive energy between nearest hydrogen atoms, z is the number of nearest hydrogen interstitials to a certain hydrogen interstitial and k is the Boltzman constant. Thus, the Fe and Co atoms act to reduce the intensity of the attraction between nearest hydrogen atoms in those intermetallics. This behaviour suggests that further increase of the hydrogen pressure at low ambient temperatures should lead to formation of hydride phases also in $ZrFe_2$ and $ZrCo_2$. This assumption is going to be checked experimentally.

Hydrogenation of the Al-rich $Zr(Al_xFe_{1-x})_2$ and $Zr(Al_xCo_{1-x})_2$ compounds required an activation procedure of heating in hydrogen atmosphere. This suggests a rather high activation energy for the hydriding process. The activation energy E_a may be defined through the diffusion constant D (see, for example [13])

$$D=D_0 \exp (-E_a/kT).$$

The Al atoms are supposed to be the main reason for this high activation energy. Confirmation of this assumption may be found in $CaAl_2$ and $GdAl_2$, where a heating to 350 - 400°C at hydrogen pressure of ~ 100 atm is necessary for absorbing apprxoimately 2 and 1 H atoms. molecule^{-1} respectively. A high activation energy may be the reason for the hydrogen inertness of $ThNi_2Al_3$ [1], $Er(Al_{0.4}Fe_{0.6})_2$ and $Er(Al_{0.5}Fe_{0.5})_2$ [3] at room temperature. It seems that the hydrogen capacity in the $Zr(Al_xFe_{1-x})_2$ and $Zr(Al_xCo_{1-x})_2$ systems is determined by a competition between the mutual hydrogen attraction and the activation energy for the hydriding process. In the Fe (Co) rich side, there seems to be only a weak attractivity between nearest hydrogen atoms (Eq. 2) but a rather low activation energy. This low activation energy accounts for the very significant increase in the capacity of $Zr(Al_{1/12}B_{11/12})_2$ (B=Fe, Co) upon cooling to 80 K. The opposite situation probably occurs in the Al-rich side.

The relative stabilities of the previously investigated hydrides of the systems $Zr(A_xB_{1-x})_2$ (A=V, Cr, Mn; B=Fe, Co) [4] are in qualitative agreement with the rule of reverse stability of Van-Mal et al. [14]. It states that among a series of isostructural inter-metallic compounds, the most stable one forms the least stable hydride. Table 4 gives the estimated heats of formation of $ZrFe_2$, $ZrCo_2$ and $ZrAl_2$ as calculated from Miedema's theory [15,16] for binary alloys. According to the rule of reverse stability the systems $Zr(Al_xB_{1-x})_2$ (B=Fe, Co) should not form hydrides at all, while this work shows the existence of hydrides in the $Zr(Al_xB_{1-x})_2$ system and their stability increases towards the Al-rich side. The above rule also does not predict the actual behaviour of the systems $Zr(Al_xB_{1-x})_2$ (B=V, Cr) and $Gd(Al_xCo_{1-x})_2$.

CONCLUSIONS

We have studied the formation and the properties of the hydrides of the Laves phase intermetallic compounds $CaAl_2$ $Zr(Al_xV_{1-x})_2$ (x=0.2, 0.5), $Zr(Al_xCr_{1-x})_2$ (x=0.3), $Gd(Al_xCo_{1-x})$ (x=0.5, 0.7, 1) and $Zr(Al_xB_{1-x})_2$ (B=Fe, Co; $0 \leq x \leq 1$). Some of the observed hydrides may be used for practical purposes; of special interest is the increase by a factor of 3 of the absorption capacity in $Zr(Al_{1/12}Fe_{11/12})_2$ when it is cooled from room temperature to 80 K.

The agreement of the experimental results for the systems $Zr(Al_xB_{1-x})_2$ (B=Fe, Co; $0 \leq x \leq 1$) with a previously presented phenomenological model, emphasizes the importance of nearest-neighbour effects and suggests a cluster model approach to be used when hydrogen absorption in intermetallic compounds is treated.

In light of our experimental results and their analysis, we draw the following conclusions concerning the influence of Al in intermetallic compounds:

1.   Partial substitution of 3d transition metals by Al in intermetallic compounds increases the stability of the corresponding hydrides; i.e. in materials where a plateau pressure exists, the substituted Al decreases the plateau pressure, and in some cases the decrease may be several orders of magnitude.

2.   The substituted Al gradually increases the activation energy for the hydriding process.

This is the general behaviour of all the compounds studied in this work and is most clearly demonstrated throughout the $Zr(Al_xB_{1-x})_2$ (B=Fe, Co) series. This behaviour does not depend on the influence of the Al on the stability of the intermetallic compounds themselves.

We find a confirmation for our conclusions in the results of several authors. The $Th(Ni,Al)_5$ ternaries [1] closely resemble the $Zr(Al_xB_{1-x})_2$ (B=Fe, Co) systems. As predicted by us, the $ThNi_4AlH_{2.5}$ is less stable than $ThNi_3Al_2H_{2.7}$. Similar to the $Zr(Al_{1/12}B_{11/12})$ (B=Fe, Co) hydrides, the $ThNi_4AlH_{2.5}$ demonstrates a low critical temperature. Hence we suppose (see discussion above) that $ThNi_5$ should form a hydride at high hydrogen pressures. We also suppose that the hydrogen capacity of $ThNi_4Al$ and may be of $ThNi_3Al_2$ should be increased upon lowering the temperature from 293 K to 80 K. This has to be checked experimentally. The lowering of the hydrogen decomposition pressures in $LaNi_{5-x}Al_x$ with increasing of x [2] confirms further our inferences.

According to our predictions a high activation energy is the reason for the hydrogen inertness of $ThNi_2Al_3$ [1] at room temperature. Indeed, we succeeded to obtain $ThNi_2Al_3H_{0.3}$ by heating the original intermetallic compound $ThNi_2Al_3$ to 300°C for several hours under hydrogen pressure of approximately 100 atm. We argue this to be the reason for the hydrogen inertness of $Er(Al_{0.4}Fe_{0.6})$ and $Er(Al_{0.5}Fe_{0.5})_2$ [3] at room temperature.

The above two features of the substituted Al may serve as an important guide in search for Al containing hydrides throughout the various $A(Al_xB_{1-x})_n$ compounds. We believe that this work opens wide opportunities for discovering of new intermetallic hydrides, many of which may be of great practical interest.

* Supported by a grant from the National Council for Research and Development, Israel, and the KFA, Julich, Germany.

Acknowledgments
The authors wish to thank R. Cohen-Arazi for the x-ray analysis, I. Miloslavski for the preparation of many of the compounds studied and J. Shinar for interesting discussions.

REFERENCES

1. T. Takeshita and W.E. Wallace, J. Less-Common Metals 55 (1977) 61.
2. M.H. Mendelsohn, M.D. Gruen and A.E. Dwight, Nature 269 (1977) 45.
3. D.M. Gualtieri and W.E. Wallace, J. Less-Common Metals 55 (1977) 53.
4. D. Shaltiel, I. Jacob and D. Davidov, J. Less-Common Metals 53 (1977) 117.
5. I. Jacob, D. Shaltiel, D. Davidov and I. Miloslavski, Solid State Commun. 23 (1977) 669.
6. H.H. Van-Mal, K.H.J. Buschow and A.R. Miedema, J. Less-Common Metals 49 (1976) 473.
7. R.L. Beck, Investigation of Hydriding Characteristics of Intermetallic Compounds, University of Denver, Denver Research Institute, Denver, Colorado, Rept. LAR-55 (1961), and Rept. DRI - 2059 (1962).
8. Y. Muraoka, M. Shiga and Y. Nakamura, Phys. Stat. Sol. A42 (1977) 369.
9. H.H. Van-Mal, Ph.D. Thesis, Delft (1976).
10. J. Shinar, I. Jacob, D. Davidov and D. Shaltiel, Int. Symp. on Hydrides for Energy Storage, Geilo, Norway (1977).
11. A. Furrer, P. Fischer and W. Halg, Int. Symp. on Hydrides for Energy Storage, Geilo, Norway (1977); P. Fischer, A. Furrer, G. Busch and L. Schlapbach, Helvetia Physica Acta 50 (1977) 421.
12. M.H. Mintz, Ph.D. Thesis, Tel-Aviv (1975).
13. C. Kittel, Introduction to Solid State Physics, p. 646, John Wiley and Sons, N.Y. (1971).
14. H.H. Van-Mal, K.H.J. Buschow and A.R. Miedema, J. Less-Common Metals 35 (1974) 65.
15. A.R. Miedema, R. Boom and F.R. De Baer, J. Less-Common Metals 41 (1975) 283.
16. A.R. Miedema, J. Less-Common Metals 46 (1976) 67.

Table 1

Crystal Structures (C 14 or C 15 type), Lattice Constants, Hydrogen Capacity, N_H, at 20 atm, 293 K, and Amount of Hydrogen Released (at the pressures and temperatures indicated) of Some of the Compounds $Zr(Al_x B_{1-x})_2$ (B=V,Cr) and $Gd(Al_x Co_{1-x})_2$

Compound	$a(\overset{o}{A})$	$c(\overset{o}{A})$	N_H (H atoms molecule^{-1})	Numbers of H atoms molecule^{-1} released at the pressures and temperatures indicated in the brackets	Percentage of the hydrogen released relatively to the hydrogen capacity at 20atm, 293K
ZrV_2	7.442	-	5.2	2.8(0.05atm,286°C)*	54%
$Zr(Al_{0.2}V_{0.8})_2$	5.281	8.635	3.6	1.2(0.05atm,296°C)	33%
$Zr(Al_{0.5}V_{0.5})_2$	5.311	8.674	2.2		
$Zr\,Cr_2$	7.204		3.8	3.7(0.05atm,292°C)*	97%
$Zr(Al_{0.3}Cr_{0.7})_2$	5.173	8.463	2.6	2(0.05atm,320°C)	77%
$GdCo_2$	7.255		4.1	2.1(0.008atm,294°C)	51%
$Gd(Al_{0.5}Co_{0.5})_2$	5.340	8.577	3.1		
$Gd(Al_{0.7}Co_{0.3})_2$	5.487	8.678	2.1	0.9(0.005atm,300°C)	41%

* Derived from the results of A. Pebler and E.A. Gulbransen,TMS-AIME 239 (1967) 1593; Electrochem. Technol. 4 (1966) 211.

Table 2

Crystal Structure (C14 or C15 Type), Lattice Constants
Hydrogen Capacity, N_H, (at the pressures and temperatures indicated),
and Equilibrium Dissociation Pressure at Room Temperature of the
Compounds $Zr(Al_x B_{1-x})_2$ (B=Fe, Co; 0 x 1).

x	$a(\overset{o}{A})$	$c(\overset{o}{A})$	N_H(H atoms molecule^{-1}) 70 atm(293K)[b]	40 atm(80K)	Equilibrium dissociation pressure at room temperature[d] (Atm)
$Zr(Al_x Fe_{x-1})_2$					
0	7.064		0.15	0.15	
1/12	7.091		0.8	2.4	25
2/12	7.124		2.2	2.9	7
2a/12	7.125		2.2	3	
0.2	7.126		2.2	3	3
0.25	7.129		1.95		0.1
0.25	5.063	8.252	2	2.7	
0.4	5.110	8.322	1.55	1.8	
0.5c	5.150	8.372	1.25	1.25	<< 0.1
0.75c	5.232	8.516	0.53	0.53	<< 0.1
1c	5.271	8.758	0.5	0.5	<< 0.1
$Zr(Al_x Co_{1-x})_2$					
0	6.953		0.35	0.35	
1/12	6.971		1.1	1.7	30
1a/12	6.988		1.4	1.8	
2/12	7.025		1.9	2.4	6
2a/12	7.039		2	2.5	2
0.25	5.025	8.109	1.9	2.2	0.4
0.4	5.084	8.184	1.35	1.45	
0.5c	5.125	8.256	1.05	1.05	<< 0.1
0.75c	5.218	8.536	0.6	0.6	<< 0.1

[a] Samples annealed for 24h at about 1000°C.

[b] N_H was measured after the samples have been cooled to 80 K.

[c] Activation procedure of heating in hydrogen atmosphere was needed for these compounds.

[d] Obtained approximately at the middle of the desorption isotherm, which did not always exhibit a plateau.

Table 3

The Parameters $n_H^{(1)}$ (i=1,2,3) and $n_A^{(i)}$ (i=1,2) Extracted
From Our Experimental Data and from Ref. 3 or by the
Fitting Procedure as Explained in the Text

System	$n_H^{(1)}$	$n_H^{(2)}$	$n_H^{(3)}$	$n_A^{(1)}$	$n_A^{(2)}$
$Zr(Al_xFe_{1-x})_2$	0.5	3.4	0.15	1	4
$Zr(Al_xCo_{1-x})_2$	0.5	2.7	0.35	1	4
$Er(Fe_xCo_{1-x})_2$	3.92	4.5	3.65	4	7

Table 4

The Heat of Formation ΔH Calculated Using the Theory of
Miedema et al. [15,16] for the Various
$Zr\ M_2$ (M=Fe, Co, Al) Compounds

Compound	ΔH(kcal/g-at. alloy)
$ZrFe_2$	-8.0
$ZrCo_2$	-13.7
$ZrAl_2$	-17.9

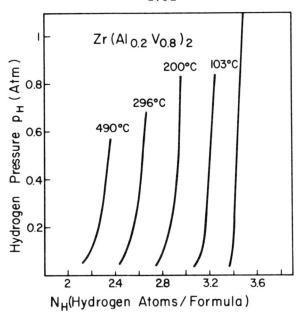

Fig. 1. Hydrogen dissociation isotherms of $Zr(Al_{0.2}V_{0.8})_2$ hydride. The hydrogen dissociation pressure p_H is expressed in atmospheres. The temperatures at which the isotherms were measured are indicated in the figure.

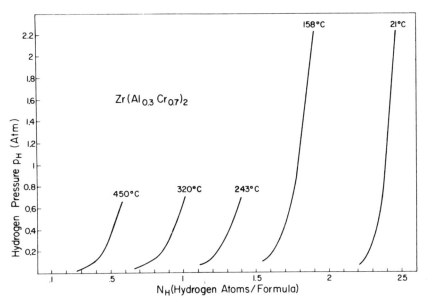

Fig. 2. Hydrogen dissociation isotherms of $Zr(Al_{0.3}Cr_{0.7})_2$ hydride. The hydrogen dissociation pressure p_H is expressed in atmospheres. The temperatures at which the isotherms were measured are indicated in the figures.

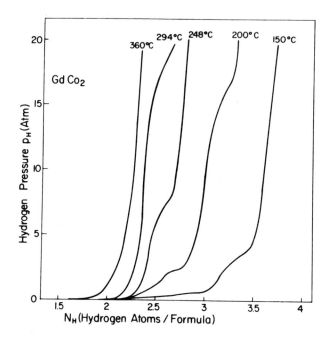

Fig. 3. Hydrogen dissociation isotherms of GdCo$_2$ hydride.
The hydrogen dissociation pressure p_H is expressed in
atmospheres. The temperature at which the isotherms
were measured are indicated in the figure.

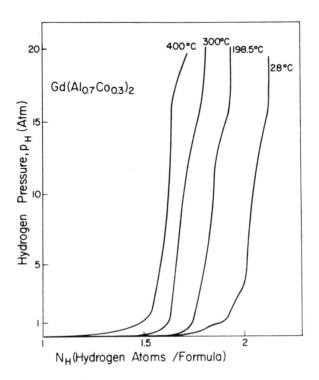

Fig. 4. Hydrogen dissociation isotherms of Gd(Al$_{0.7}$Co$_{0.3}$)$_2$ hydride. The hydrogen dissociation pressure p$_H$ is expressed in atmospheres. The temperature at which the isotherms were measured are indicated in the figure.

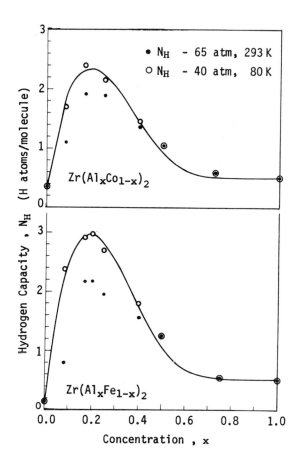

Fig. 5. The hydrogen absorption capacity of $Zr(Al_xFe_{1-x})_2$ and
$Zr(Al_xCo_{1-x})_2$ as a function of x at room temperature and
70 atm (full marks), and at liquid nitrogen temperature and
40 atm (open marks). The solid lines are the theoretical
results obtained from our model (Eq. 1) using the parameters
given in Table 3.

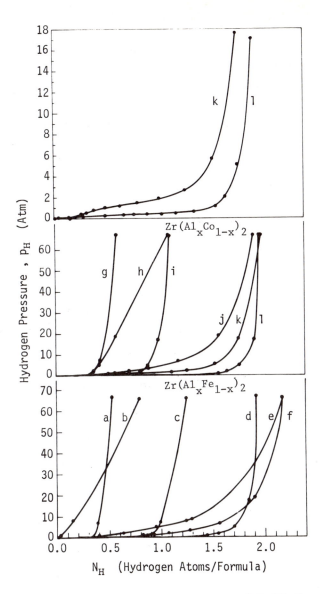

Fig. 6. Desorption isotherms at room temperature of $Zr(Al_xFe_{1-x})_2$ and $Zr(Al_xCo_{1-x})_2$ pseudobinary compounds for the following x values:
1/12 - b; 2/12 - **e**; 0.2 - f; 0.25 - d; 0.5 - c; 0.75 - a; 1/12 - h; 2/12 - j; 2/12 annealed - k; 0.25 - 1; 0.5 - i; 0.75 - g.

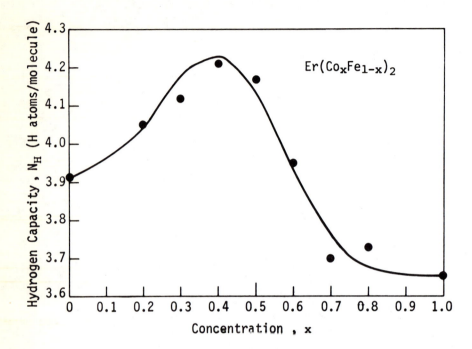

Fig. 7. The hydrogen capacity of $Er(Co_xFe_{1-x})_2$ as a function of x. The full marks are the experimental results of Gualtieri and Wallace [3]. The solid line is the theoretical fit obtained from our model (Eq. 1) using the parameters given in Table 3.